# Advances in Mechanics and Mathematics

## Volume 34



More information about this series at http://www.springer.com/series/5613

Grzegorz Łukaszewicz • Piotr Kalita

# Navier–Stokes Equations

## An Introduction with Applications

Springer

Grzegorz Łukaszewicz
Faculty of Mathematics,
Informatics, and Mechanics
University of Warsaw
Warszawa, Poland

Piotr Kalita
Faculty of Mathematics and Computer
Science
Jagiellonian University in Krakow
Krakow, Poland

ISSN 1571-8689 ISSN 1876-9896 (electronic)
Advances in Mechanics and Mathematics
ISBN 978-3-319-27758-5 ISBN 978-3-319-27760-8 (eBook)
DOI 10.1007/978-3-319-27760-8

Library of Congress Control Number: 2015960205



Printed on acid-free paper

This Springer imprint is published by Springer Nature
The registered company is Springer International Publishing AG Switzerland

*To Renata, Agata, and Jacek, with love*
*(Grzegorz)*

*To my beloved wife Kasia*
*(Piotr)*

# Preface

*Admittedly, as useful a matter as the motion of fluid and related sciences has always been an object of thought. Yet until this day neither our knowledge of pure mathematics nor our command of the mathematical principles of nature have a successful treatment.*

–Daniel Bernoulli

Incompressible Navier–Stokes equations describe the dynamic motion (flow) of incompressible fluid, the unknowns being the velocity and pressure as functions of location (space) and time variables. To solve those equations would mean to predict the behavior of the fluid under knowledge of its initial and boundary states. These equations are one of the most important models of mathematical physics. Although they have been a subject of vivid research for more than 150 years, there are still many open problems due to the nature of nonlinearity present in the equations. The nonlinear convective term present in the equations leads to phenomena such as eddy flows and turbulence. In particular the question of solution regularity for three-dimensional problem was appointed by Clay Mathematics Institute as one of the Millennium Problems, that is, the key problems in modern mathematics. This is, on one hand, due to the fact that the problem remains challenging and fascinating for mathematicians and, on the other hand, that the applications of the Navier–Stokes equations range from aerodynamics (drag and lift forces), through design of watercrafts and hydroelectric power plants, to the medical applications of the models of flow of blood in vessels.

This book is aimed at a broad audience of people interested in the Navier–Stokes equations, from students to engineers and mathematicians involved in the research on the subject of these equations.

It originated in part from a series of lectures of the first author given over the past 15 years at the Faculty of Mathematics, Informatics and Mechanics of the University of Warsaw; at summer schools at UNICAMP, Campinas, Brasil; and at Université Jean Monnet, Saint-Etienne, France. The lectures were based on the leading books on the then young theory of infinite dimensional dynamical systems, focused on mathematical physics, in particular, on Temam [220]; Chepyzhov and Vishik [61]; Doering and Gibbon [88]; Foiaş, Manley, Rosa, and Temam [99]; and Robinson [197].

The lectures at the Mathematics Faculty of the University of Warsaw were also attended by students and PhD students from the Faculty of Physics and Faculty of Geophysics, and it became clear that a routine mathematical lecture had to be extended to include additional aspects of hydrodynamics. Some students asked for “more physics and motivation” and “more real applications”; others were mainly interested in the mathematics of the Navier–Stokes equations, and yet others would like to see the Navier–Stokes equations in a more general context of evolution equations and to learn the theory of infinite dimensional dynamical systems on the research level. These several aspects of hydrodynamics well suited the tastes and interests of the lecturer, and also the second author was welcomed to join the project of the book at a later stage.

In consequence, the audience of the book is threesome:

Group I: Mathematicians, physicists, and engineers who want to learn about the Navier–Stokes equations and mathematical modeling of fluids

Group II: University teachers who may teach a graduate or PhD course on fluid mechanics basing on this book or higher-level students who start research on the Navier–Stokes equations

Group III: Researchers interested in the exchange of current knowledge on dynamical systems approach to the Navier–Stokes equations

Although, in principle, all these three groups can find interest in all chapters of the book, Chaps. 2–7 are primarily targeted at Group I, Chaps. 3, 4, 7, 8, 11, and 12 aimed mainly at Group II, and Chaps. 7–16 for Group III.

For a reader with reasonable background on calculus, functional analysis, and theory of weak solutions for PDEs, the whole book should be understandable.

The book was planned to be a monograph which could also be used as a textbook to teach a course on fluid mechanics or the Navier–Stokes equations. Typical courses could be “Navier–Stokes equations”, “partial differential equations”, “fluid mechanics”, “infinite dimensional dynamics systems.” To this end many chapters of this book include exercises. Moreover, we did not restrain ourselves to include a number of figures to liven the text and make it more intuitive and less formal. We believe that the figures will be helpful. Special care was undertaken to keep the individual chapters self-contained as far as possible to allow the reader to read the book linearly (in linear portions). That demanded several small repetitions here and there.

To understand the first chapters of this book, just the basic knowledge on calculus, that can be learned from any calculus textbook, should be enough.

The book is planned to be self-contained, but, to understand its last chapters, some knowledge from a textbook like “Partial Differential Equations” by L.C. Evans (which contains all necessary knowledge on functional analysis and PDEs) would be helpful. Each chapter contains an introduction that explains in simple words the nature of presented results and a section on bibliographical notes that will place it in the context of past and current research.

Several people greatly contributed, knowingly or not knowingly, to the creation of the book. Our thanks go to our colleagues and collaborators: Guy Bayada, Mahdi Boukrouche, Thomas Caraballo, José Langa, Pepe Real, James Robinson etc.

The first author is grateful to Guy Bayada who introduced him to the problems of lubrication theory and flows in narrow films during his visits at INSA, Lyon, and to Mahdi Boukrouche with whom he collaborated for several years on this subject. Thomas Caraballo, José Langa, and Pepe Real introduced him to the subject of pullback attractors during his visit at the University of Seville. Thanks for the opportunity to give the summer courses in Campinas and Saint-Etienne go to Marco Rojas-Medar and Mahdi Boukrouche, respectively. Many thanks go also to Chunyou Sun, Meihua Yang, and Yongqin Xie for their kind invitation of the first author to give several lectures at the University of Lanzhou, then at Huazhong University of Science and Technology in Wuhan and at The University in Changsha, China, in June 2013. Inspirational discussions and exchange of ideas with these Chinese friends, including also Qingfeng Ma and Yuejuan Wang, contributed to the form of the last chapters of the book.

The research group at Jagiellonian University with their leader, Stanisław Migórski, greatly motivated the authors as regards contact problems. The second author owes a lot to his colleagues and teachers from Jagiellonian University; he would like to express his thanks especially to Zdzisław Denkowski and Stanisław Migórski. He would also like to thank Robert Schaefer who first introduced to him the topics of fluid mechanics. He is also grateful for inspiring discussions in the field of contact mechanics to his collaborators from the University of Perpignan, Mircea Sofonea and Mikaël Barboteu.

We would like to thank Wojciech Pociecha for his help with the preparation of the figures.

The work was in part supported by the National Science Center of Poland under the Maestro Advanced Project no. DEC-2012/06/A/ST1/00262.

Finally, we express our gratitude to the AMMA Series editor, David Y. Gao, and to Marc Strauss and the editors of Springer Publishing House for their care and encouragement during the preparation of the book.

Warszawa, Poland
Kraków, Poland
October 2015

Grzegorz Łukaszewicz
Piotr Kalita

# Contents

# 1 Introduction and Summary

*When you put together the science of movements of water, remember to put beneath each proposition its applications, so that such science may not be without uses.*

– Leonardo da Vinci

This chapter provides, for the convenience of the reader, an overview of the whole book, first of its structure and then of the content of the individual chapters.

The outline of the book structure is as follows.

Chapter 2 shows the derivation of the Navier–Stokes equations from the principles of physics and discusses their physical and mathematical properties and examples of the solutions for some particular cases, without going into complicated mathematics. This part is aimed to fill in a gap between an engineer and mathematician and should be understood by anybody with basic knowledge of calculus no more complicated than the Stokes theorem.

In Chap. 3, a necessary mathematical background including these parts of functional analysis and theory of Sobolev spaces which are needed to understand modern research on the Navier–Stokes equations is presented. Chapters 4–6 comprise three examples of stationary problems.

Then we smoothly move to the research level part of the book (Chaps. 7–16) which presents the analysis from the point of view of global attractors of the asymptotic (in time) behavior of the velocities being the solutions of the Navier–Stokes system. Roughly speaking we endeavor to show how the modern theory of global attractors can be used to construct the mathematical objects that enclose the seemingly chaotic and unordered eddy and turbulent flows. We tame these flows by showing their fine properties like the finite dimensionality of global attractors, which means that the description of unrestful and turbulent states can be done by finite number of parameters or existence of invariant measures which means that the flow becomes, in statistical sense, stationary.

We deal with non-autonomous problems using the recent and elegant theory of so-called pullback attractors that allows to cope with flows with changing in time sources, sinks, or boundary data. We also solve problems with multivalued boundary conditions that allow to model various contact conditions between the fluid and enclosing object, such as the stick/slip frictional boundary behavior.



G. Łukaszewicz, P. Kalita, *Navier–Stokes Equations*, Advances in Mechanics and Mathematics 34, DOI 10.1007/978-3-319-27760-8_1

The analysis is primarily done for the two-dimensional Navier–Stokes equations, where, as a model, the problem from lubrication theory, that can be reduced to two dimensions is used. The study with various types of boundary conditions, including multivalued ones, is presented.

Some of the results presented in this part are based on the previous and already published work of the authors, but some results, that are the subjects of current research, are yet unpublished elsewhere.

Below we present the content of the book in some more detail.

In Chap. 2 we give an overview of the equations of classical hydrodynamics. We provide their derivation, discuss the associated physical quantities, comment on the constitutive laws, stress tensor, and thermodynamics, finally we present some elementary properties of the derived system and also some cases where it is possible to calculate the exact solutions of the following system of the incompressible Navier–Stokes equations,

$$\begin{aligned} \frac{\partial u}{\partial t} - \nu \Delta u + (u \cdot \nabla)u + \frac{1}{\rho}\nabla p &= f, \\ \operatorname{div} u &= 0, \end{aligned}$$

which are the main subject of the book.

In Chap. 3 we introduce the basic preliminary mathematical tools to study the Navier–Stokes equations, including results from linear and nonlinear functional analysis (e.g., the Lax–Milgram lemma, fixed point theorems) as well as the theory of function spaces (e.g., compactness theorems). We present, in particular, some of the most frequently used in the sequel embedding theorems and inequalities. We discuss the versions of the Gronwall lemma used in the sequel, and provide some necessary background in the theory of Clarke subdifferential and differential inclusions.

In Chaps. 4–6 we consider stationary problems. Chapter 4 is devoted to the stationary Navier–Stokes equations in a bounded three-dimensional domain $Q$,

$$\begin{aligned} -\nu \Delta u + (u \cdot \nabla)u + \nabla p = f \quad &\text{in} \quad Q, \\ \operatorname{div} u = 0 \quad &\text{in} \quad Q, \end{aligned}$$

with one of the boundary conditions:

1. $Q = [0, L]^3$ in $\mathbb{R}^3$ and we assume periodic boundary conditions, or
2. $Q$ is a bounded domain in $\mathbb{R}^3$, with smooth boundary, and we assume the homogeneous boundary condition $u = 0$ on $\partial Q$.

This basic problem serves as an introduction to the mathematical theory of the Navier–Stokes equations. We introduce the suitable function spaces in which we (usually) seek solutions of the stationary problem, then we present the weak formulation. It allows us to use the theories of linear and nonlinear functional analysis (Lax–Milgram lemma and fixed point theorems, respectively) to prove the existence of solutions.

To show typical methods used when dealing with nonlinear problems, we present a number of proofs based on various linearizations and fixed point theorems in standard function spaces. The solutions, due to the nonlinearity of the Navier–Stokes equations are not in general unique, however, under some restriction on the mass force and viscosity coefficient (quite intuitive from the physical point of view), one can prove their uniqueness.

In Chap. 5 we consider the stationary Navier–Stokes equations with friction in the three-dimensional bounded domain $\Omega$. The domain boundary $\partial\Omega$ is divided into two parts, namely the boundary $\Gamma_D$ on which we assume the homogeneous Dirichlet boundary condition and the contact boundary $\Gamma_C$ on which we decompose the velocity into the normal and tangent directions. In the normal direction we assume $u \cdot n = 0$, i.e., there is no leak through the boundary, while in the tangent direction we set $-T_\tau \in h(x, u_\tau)$, the multivalued relation between the tangent stress and tangent velocity. This relation is the general form of the friction law on the contact boundary. The existence of solution is shown by the Kakutani–Fan–Glicksberg fixed point theorem and some cut-off techniques.

In Chap. 6 we consider a typical problem for the hydrodynamic equations coming from the lubrication theory. In this theory the domain of the flow is usually very thin and the engineers are interested in the distribution of the pressure therein. Because of the thinness of the domain, in practice one assumes that the pressure would depend only on two and not three independent variables. The pressure distribution is governed then by the Reynolds equation, which depends on the original boundary data. Our aim is to obtain the Reynolds equation starting from the stationary Stokes equations, considered in the three-dimensional domains $\Omega^\varepsilon$, $\varepsilon > 0$,

$$-\nu\Delta u + \nabla p = f \quad \text{in} \quad \Omega^\varepsilon,$$
$$\operatorname{div} u = 0 \quad \text{in} \quad \Omega^\varepsilon,$$

with a Fourier boundary condition on the top $\Gamma_F^\varepsilon$ and Tresca boundary conditions on the bottom part of the boundary $\Gamma_C$, respectively. We show how to pass, in a precise mathematical way, as $\varepsilon \to 0$, from the three-dimensional Stokes equations to a two-dimensional Reynolds equation for the pressure distribution (see Fig. 1.1).

The passage from the three-dimensional problem to a two-dimensional one depends on several factors and additional scaling assumptions. The existence of solutions of the limit equations follows from the existence of solutions of the original three-dimensional problem. Finally, we show the uniqueness of the limit solution.

In Chaps. 7–10 we consider the nonstationary autonomous Navier–Stokes equations

$$u_t - \nu\Delta u + (u \cdot \nabla)u + \nabla p = f, \tag{1.1}$$
$$\operatorname{div} u = 0, \tag{1.2}$$

in two-dimensional domains. In these chapters the solutions are global in time and unique.

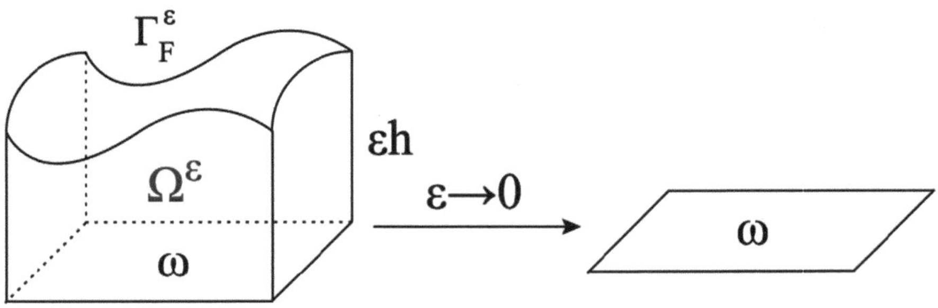


**Fig. 1.1** Schematic view of the problem considered in Chap. 6. Incompressible static Stokes equation is solved on the three-dimensional domain $\Omega^\varepsilon$. The domain thickness is given by $\varepsilon h$, where $h$ is a function of the point in the two-dimensional domain $\omega$. As $\varepsilon \to 0$, Reynolds equation on $\omega$ is recovered

Chapter 7 is a general introduction to evolutionary two-dimensional Navier–Stokes equations. We prove some basic properties of solutions assuming that the external forces do not depend on time and that the domain of the flow is bounded. The boundary conditions are either periodic or homogeneous Dirichlet ones. In this chapter we introduce the notion of the global attractor, one of the main objects to study also in Chaps. 8–10.

In Chap. 8 we prove the existence of invariant measures associated with two-dimensional autonomous Navier–Stokes equations. The invariant measures are supported on the global attractor. Then we introduce the notion of a stationary statistical solution and prove that every invariant measure is also such a solution. Existence of the invariant measures (stationary statistical solutions) supported on the global attractor reveals the statistical properties of the potentially chaotic fluid flow after a long time of evolution when the external forces do not depend on time. The non-autonomous case is considered in Chap. 12.

In Chaps. 9–10 we consider system (1.1) and (1.2) in the domain $\Omega$ depicted in Fig. 1.2, with homogeneous condition $u = 0$ on $\Gamma_D$, periodic condition $u(0, x_2) = u(L, x_2)$ on $\Gamma_L$, and several contact boundary conditions on $\Gamma_C$. The motivation for such problem setup comes again from problems in contact mechanics, the theory of lubrication and shear flows in narrow films.

One may look at the domain $\Omega$ as a rectification of the ring-like domain considered in the theory of lubrication, where it represents a cross section of an infinite journal bearing. The problem reduces to describing a flow between two cylinders. The outer cylinder is at rest and the inner cylinder rotates providing a driving force to the fluid (lubricant). Since the cylinders are infinitely long it can be assumed in the first approximation that the flow is two-dimensional. Described domain geometry is schematically presented in Fig. 1.3.

The boundary conditions on $\Gamma_C$ include the following ones.

In Chap. 9 we pose

$$u = Ue_1, \quad U \in \mathbb{R}, \quad U > 0 \quad \text{on} \quad \Gamma_C,$$

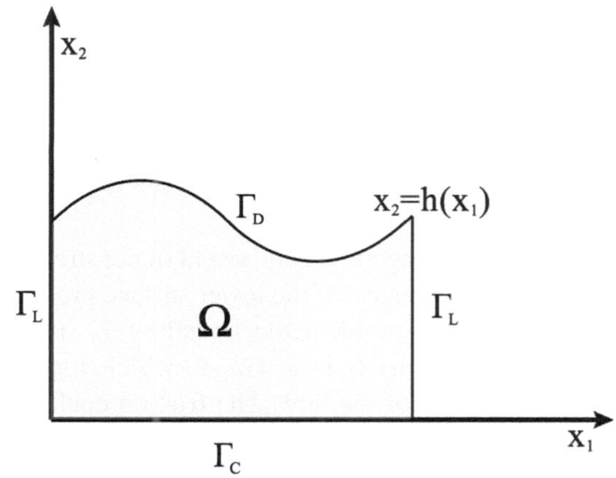


**Fig. 1.2** Schematic view of the flow domain and its boundaries in Chaps. 9 and 10

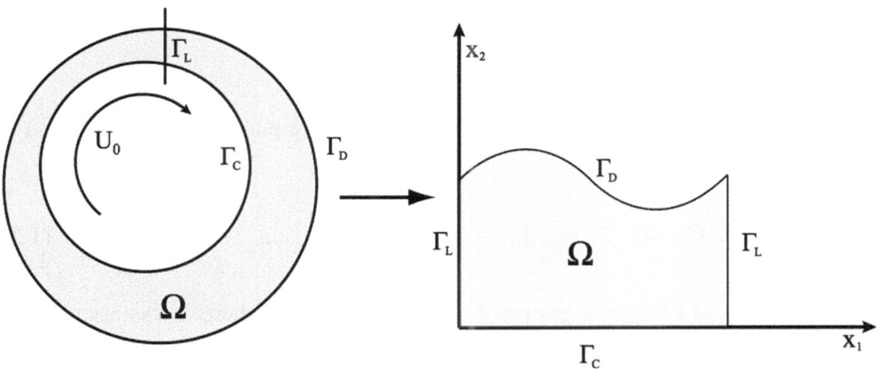


**Fig. 1.3** Three-dimensional infinite ring-like domain and its rectification considered in Chaps. 9 and 10

by which we mean that the boundary $\Gamma_C$ is moving with a constant velocity $U_0 e_1 = (U_0, 0)$ and the velocity of the fluid at the boundary equals the velocity of the boundary.

We prove the existence of a global attractor and estimate from above its fractal dimension in terms of given data and geometry of the domain of the flow.

In Chap. 10 we consider two problems. We assume that there is no flux across $\Gamma_C$ so that the normal component of the velocity on $\Gamma_C$ satisfies

$$u \cdot n = 0 \quad \text{on} \quad \Gamma_C,$$

and that the tangential component of the velocity $u_\tau$ on $\Gamma_C$ is unknown and satisfies the Tresca friction law with a constant and positive maximal friction coefficient $k$. This means that

$$\left.\begin{array}{l} |T_\tau(u,p)| \le k \\ \\ |T_\tau(u,p)| < k \Rightarrow u_\tau = U_0 e_1 \\ \\ |T_\tau(u,p)| = k \Rightarrow \exists \lambda \ge 0 \text{ such that } u_\tau = U_0 e_1 - \lambda T_\tau(u,p) \end{array}\right\} \quad \text{on} \quad \Gamma_C, \tag{1.3}$$

where $T_\tau$ is the tangential component of the stress tensor on $\Gamma_C$ and $U_0 e_1 = (U_0, 0)$, $U_0 \in \mathbb{R}$, is the velocity of the lower surface producing the driving force of the flow.

In the second problem the boundary $\Gamma_C$ is also assumed to be moving with the constant velocity $U_0 e_1 = (U_0, 0)$ which, together with the mass force, produces the driving force of the flow. The friction coefficient $k$ is assumed to be related to the slip rate through the relation $k = k(|u_\tau - U_0|)$, where $k : \mathbb{R}^+ \to \mathbb{R}^+$. If there is no slip between the fluid and the boundary then the friction is bounded by the threshold $k(0)$

$$u_\tau = U_0 \Rightarrow |T_\tau| \le k(0) \quad \text{on} \quad \Gamma_C, \tag{1.4}$$

while if there is a slip, then the friction force density (equal to tangential stress) is given by the expression

$$u_\tau \ne U_0 \Rightarrow -T_\tau = k(|u_\tau - U_0|)\frac{u_\tau - U_0}{|u_\tau - U_0|} \quad \text{on} \quad \Gamma_C. \tag{1.5}$$

Note that (1.4) and (1.5) generalize the Tresca law (1.3) where $k$ was assumed to be a constant. Here $k$ depends of the slip rate, this dependence represents the fact that the kinetic friction is less than the static one, which holds if $k$ is a decreasing function.

We prove that for both problems above there exist exponential attractors, in particular the global attractors of finite fractal dimension.

In Chaps. 11–13 we consider the time asymptotics of solutions of the two-dimensional Navier–Stokes equations. First, in Chap. 11 we prove two properties of the equations in a bounded domain, concerning the existence of determining modes and nodes. Then we study the equations in an unbounded domain, in the framework of the theory of infinite dimensional non-autonomous dynamical systems, introducing the notion of the pullback attractor.

Chapter 12 presents a construction of invariant measures and statistical solutions for the non-autonomous Navier–Stokes equations in bounded and some unbounded domains in $\mathbb{R}^2$. More precisely, we construct the family of probability measures $\{\mu_t\}_{t\in\mathbb{R}}$ and prove the relations $\mu_t(E) = \mu_\tau(U(t,\tau)^{-1}E)$ for $t, \tau \in \mathbb{R}$, $t \ge \tau$ and Borel sets $E$ in $H$. The support of each measure $\mu_t$ is included in the section $A(t)$ of the pullback attractor. We prove also the Liouville and energy equations. Finally, we consider statistical solutions of the Navier–Stokes equations supported on the pullback attractor.

In Chap. 13 we consider the problem of existence and finite dimensionality of the pullback attractor for a class of two-dimensional turbulent boundary driven flows. We generalize here the results from Chap. 9 to the non-autonomous problem. The new element in our study with respect to that in Chap. 9 is the allowance of the velocity of $\Gamma_C$ to depend on time, i.e.,

$$u = U(t)e_1, \quad U(t) \in \mathbb{R} \quad \text{on} \quad \Gamma_C.$$

Our aim is to study the time asymptotics of solutions in the frame of the dynamical systems theory. We prove the existence of the pullback attractor and estimate its fractal dimension. We shall apply the results from Chap. 11, reformulated here in the language of evolutionary processes.

Chapters 14–16 are devoted to global in time solutions of the Navier–Stokes equations which are not necessary unique. We introduce theories of global attractors for multivalued semiflows and multivalued processes to include this situation. We study further examples of contact problems in both autonomous and non-autonomous cases.

In Chap. 14 we consider two-dimensional nonstationary Navier–Stokes shear flows in the domain $\Omega$ as in Fig. 1.2, with nonmonotone and multivalued boundary conditions on $\Gamma_C$. Namely, we assume the following subdifferential boundary condition

$$\tilde{p}(x,t) \in \partial j(u_n(x,t)) \quad \text{on} \quad \Gamma_C,$$

where $\tilde{p} = p + \frac{1}{2}|u|^2$ is the Bernoulli (total) pressure, $u_n = u \cdot n$, $j : \mathbb{R} \to \mathbb{R}$ is a given locally Lipschitz superpotential, and $\partial j$ is a Clarke subdifferential of $j$.

Our considerations are motivated here by feedback control problems for fluid flows in domains with semipermeable walls and membranes and by the theory of lubrication. We prove the existence of global in time solutions of the considered problem which is governed by a partial differential inclusion, and then we prove the existence of a trajectory attractor and a weak global attractor for the associated multivalued semiflow.

In Chap. 15 we study the three-dimensional problem in a bounded domain $\Omega$. The problem domain is the three-dimensional counterpart of the one presented in Fig. 1.2. The boundary of $\Omega$ is divided into three parts: the lateral one $\Gamma_L$ on which we assume the periodic boundary conditions, the homogeneous Dirichlet one and, finally, the contact one $\Gamma_C$ on which we consider a general form of multivalued frictional type boundary conditions $-T_\tau \in g(u_\tau)$. We prove the existence of the Leray–Hopf weak solutions and, using the framework of evolutionary systems, existence of the weak global attractor.

Finally, in Chap. 16 we consider further non-autonomous and multivalued evolution problems, this time in the frame of the theory of pullback attractors for multivalued processes. First we prove an abstract theorem on the existence of pullback $\mathscr{D}$-attractor and then apply it to study a two-dimensional incompressible

Navier–Stokes flow with a general form of multivalued frictional contact conditions on $\Gamma_C$. We assume that there is no flux across $\Gamma_C$ and hence we have

$$u_n(t) = 0 \quad \text{on} \quad \Gamma_C,$$

and that the tangential component of the velocity $u_\tau$ on $\Gamma_C$ is in the following relation with the tangential stresses $T_\tau$,

$$-T_\tau(t) \in \partial j(x, t, u_\tau(t)) \quad \text{on} \quad \Gamma_C.$$

In above formula $j : \Gamma_C \times \mathbb{R} \times \mathbb{R} \to \mathbb{R}$ is a potential which is locally Lipschitz and not necessarily convex with respect to the last variable, and $\partial$ is the subdifferential in the sense of Clarke taken with respect to the last variable $u_\tau$.

The tangent conditions on $\Gamma_C$ in Chaps. 15 and 16 represent the frictional contact between the fluid and the wall, where the friction force depends in a nonmonotone and even discontinuous way on the slip rate, and are a generalization of the conditions considered in Chap. 10. For this case we prove the existence of the attractor.

Most chapters are devoted to two-dimensional problems. Three-dimensional problems are considered only in Chaps. 2, 4, 5, 6, and 15. One reason for that is associated with the character of the Navier–Stokes equations, namely the fact that in the two-dimensional problems it is relatively easy to prove the uniqueness of the solutions which allows us to use the well-developed theory of infinite dimensional dynamical systems for semigroup and processes, while the uniqueness of the three-dimensional Navier–Stokes equations is in general an open question. We also consider the two- and three-dimensional problems without assuming the solution uniqueness in the framework of (more recent) theories of trajectory attractors, multivalued semiflows, evolutionary systems, and multivalued processes.

The other reason to focus on the two-dimensional flows concern the simplicity. Our aim was to test first the more elementary two-dimensional models of some real engineering problems. The word "test" here means not only checking the well posedness of a particular problem. In Chap. 9 we estimate the attractor dimension and show how it depends on the shape of the domain (cf. [24, 26], where the upper bounds of the attractor dimension depend also on the geometry of $\Gamma_D$). Assume that the answer to the question on the dependence of the attractor dimension on the geometry of the boundary is such that in the two-dimensional case the estimate from above of the attractor dimension is independent of geometry (for example, on the roughness of the boundary represented by the oscillations of the function $h = h(x_1)$ describing $\Gamma_D$). Such a result would be contradictory to our intuition, provided the intuitive hypotheses

*attractor dimension* $\sim$ *level of chaos in the flow* $\sim$ *geometry of the flow domain*

where "$\sim$" means "is related to," are justified.

Such a contradiction with our intuition could be resolved in the following ways:

1. there is no such contradiction in the "real" three-dimensional case, it appears only in the two-dimensional case (but where lies the difference?),
2. the attractor dimension does not represent the level of chaos in the fluid flow described by the (good) model of the Navier–Stokes equations,
3. the Navier–Stokes equations model is not good enough to give the right answer to the problems of chaotic movement of the classical fluids.

The close correspondence between the level of chaos in the fluid flow and the geometry of the domain is evident as a physical phenomenon (recall observing a flow of water in the river).

To confirm the agreement of the results provided by the modeling with our physical intuition or else to confront the above potential possibilities motivated us to study the problems of the existence and properties of the attractor. There are still many interesting and important problems close to these considered in the book and we were able to touch only a few ones. One example is to further study the relations between the (type of) boundary conditions and the attractor dimension.

Finally, we remark that this book is devoted to incompressible flows, for the mathematical treatise of compressible ones see, e.g., [157, 187].

# 2
# Equations of Classical Hydrodynamics

*The neglected borderline between two branches of knowledge is often that which best repays cultivation, or, to use a metaphor of Maxwell's, the greatest benefits may be derived from a cross-fertilization of the sciences.*

– John William Strutt, 3rd Baron Rayleigh

In this chapter we give an overview of the equations of classical hydrodynamics. We provide their derivation, comment on the stress tensor, and thermodynamics, finally we present some elementary properties and also some exact solutions of the Navier–Stokes equations.

## 2.1 Derivation of the Equations of Motion

Fluid flow may be represented mathematically as a *continuous transformation* of three-dimensional Euclidean space into itself. The transformation is parametrized by a real parameter $t$ representing time.

Let us introduce a fixed rectangular coordinate system $(x_1, x_2, x_3)$. We refer to the coordinate triple $(x_1, x_2, x_3)$ as the *position* and denote it by $x$. Now consider a particle $P$ moving with the fluid, and suppose that at time $t = 0$ it occupies a position $X = (X_1, X_2, X_3)$ and that at some other time $t$, $-\infty < t < +\infty$, it has moved to a position $x = (x_1, x_2, x_3)$. Then $x$ is determined as a function of $X$ and $t$

$$x = x(X, t) \quad \text{or} \quad x_i = x_i(X, t)\,. \tag{2.1}$$

If $X$ is fixed and $t$ varies, Eq. (2.1) specifies the *path* of the particle initially at $X$. On the other hand, for fixed $t$, (2.1) determines a transformation of a region initially occupied by the fluid into its position at time $t$.



G. Łukaszewicz, P. Kalita, *Navier–Stokes Equations*, Advances in Mechanics and Mathematics 34, DOI 10.1007/978-3-319-27760-8_2

We assume that the transformation (2.1) is *continuous* and *invertible*, that is, there exists its inverse

$$X = X(x,t), \quad (\text{or} \quad X_i = X_i(x,t)) .$$

Also, to be able to differentiate, we assume that the functions $x_i$ and $X_i$ are sufficiently smooth.

From the condition that the transformation (2.1) possess a differentiable inverse it follows that its Jacobian

$$J = J(X,t) = \det\left(\frac{\partial x_i}{\partial X_j}\right)$$

satisfies

$$0 < J < \infty . \tag{2.2}$$

The initial coordinates $X$ of the particle will be referred to as the *material coordinates* of the particle. The *spatial coordinates* $x$ may be referred to as its *position*, or *place*. The representation of fluid motion as a *point transformation* violates the concept of the *kinetic theory* of fluids, as in this theory the particles are molecules, and they are in random motion. In the theory of *continuum mechanics* the state of motion at a given point $x$ and at a given time $t$ is described by a number of functions such as $\rho = \rho(x,t)$, $u = u(x,t)$, $\theta = \theta(x,t)$ representing density, velocity, temperature, and other *hydrodynamical variables*.

Due to the transformation (2.1), each such variable $f$ can also be expressed in terms of material coordinates

$$f(x,t) = f(x(X,t),t) = F(X,t) . \tag{2.3}$$

The *velocity* $u$ at time $t$ of a particle initially at $X$ is given, by definition, as

$$u(x,t) = U(X,t) = \frac{d}{dt}x(X,t) , \quad (x = x(X,t)) . \tag{2.4}$$

Above, $X$ is treated as a parameter representing a given fixed particle, and this is the reason that we use the ordinary derivative in (2.4).

Having the velocity field $u(x,t)$, we can (in principle) determine the transformation (2.1), solving the ordinary differential equation

$$\frac{d}{dt}x(X,t) = u(x(X,t),t),$$

with $x(X,0) = X$, where $X$ is a parameter.

We shall always write

$$\frac{d}{dt}F(X,t) \quad \text{and} \quad \frac{\partial}{\partial t}f(x,t) ,$$

where $F$ and $f$ are related by (2.3). We have thus

$$\frac{d}{dt}F(X,t) = \frac{d}{dt}f(x(X,t),t) = \frac{\partial f}{\partial x_i}(x(X,t),t)\frac{dx_i}{dt} + \frac{\partial f}{\partial t}(x(X,t),t)\,,$$

so that by (2.4) we obtain a general formula

$$\frac{d}{dt}F(X,t) = \frac{D}{Dt}f(x,t)\,, \tag{2.5}$$

where $\frac{D}{Dt}f(x,t) \equiv \frac{\partial f}{\partial t}(x,t) + u(x,t)\cdot\nabla f(x,t)$ is called the *material derivative* of $f$.

**Transport Theorem** Let $\Omega(t)$ denote an arbitrary volume that is moving with the fluid and let $f(x,t)$ be a scalar or vector function of position and time. The *transport theorem* states that

$$\frac{d}{dt}\int_{\Omega(t)} f(x,t)\,dx \tag{2.6}$$

$$= \int_{\Omega(t)} \left\{\frac{\partial f}{\partial t}(x,t) + u(x,t)\cdot\nabla f(x,t) + f(x,t)\,\mathrm{div}\,u(x,t)\right\}\,dx\,.$$

For the proof consider the transformation

$$x : \Omega(0) \to \Omega(t), \quad x = x(X,t)\,,$$

as in (2.1). Then

$$\int_{\Omega(t)} f(x,t)\,dx$$

$$= \int_{\Omega(0)} f(x(X,t),t)J(X,t)\,dX = \int_{\Omega(0)} F(X,t)J(X,t)\,dX\,,$$

so that

$$\frac{d}{dt}\int_{\Omega(t)} f(x,t)dx = \frac{d}{dt}\int_{\Omega(0)} F(X,t)J(X,t)\,dX \tag{2.7}$$

$$= \int_{\Omega(0)} \left\{\frac{d}{dt}F(X,t)J(X,t) + F(X,t)\frac{d}{dt}J(X,t)\right\}\,dX\,.$$

By (2.5) we have

$$\int_{\Omega(0)} \frac{d}{dt} F(X,t)J(X,t)\,dX$$

$$= \int_{\Omega(0)} \left\{ \frac{\partial f}{\partial t}(x(X,t),t) + u(x(X,t),t) \cdot \nabla f(x(X,t),t) \right\} J(X,t)\,dX$$

$$= \int_{\Omega(t)} \left\{ \frac{\partial f}{\partial t}(x,t) + u(x,t) \cdot \nabla f(x,t) \right\} dx\,.$$

To prove (2.6) it remains to prove the *Euler formula*

$$\frac{d}{dt} J(X,t) = \operatorname{div} u(x(X,t),t) J(X,t)\,, \tag{2.8}$$

the proof of which we leave to the reader as an exercise.

The fluid is called *incompressible* if for any domain $\Omega(0)$ and any $t$,

$$\text{volume } (\Omega(t)) = \text{volume } (\Omega(0))\,.$$

From (2.7) with $f(x,t) \equiv 1$ we have

$$\frac{d}{dt}\text{volume } (\Omega(t)) = \frac{d}{dt}\int_{\Omega(t)} dx = \int_{\Omega(0)} \frac{d}{dt} J(X,t)dX\,,$$

hence by (2.8), (2.2), and the arbitrariness of choice of the domain $\Omega(t)$ via $\Omega(0)$ a necessary and sufficient condition for the fluid to be incompressible is

$$\operatorname{div} u(x,t) = 0\,.$$

**Exercise 2.1.** Prove that the transport theorem can be written in the form

$$\frac{d}{dt}\int_{\Omega(t)} f(x,t)\,dx = \int_{\Omega(t)} \frac{\partial f}{\partial t}(x,t)\,dx + \int_{\partial\Omega(t)} f(x,t)u(x,t)\cdot n(x,t)\,dS\,,$$

where $n(x,t)$ is the outward unit normal to $\partial\Omega(t)$ at $x \in \partial\Omega(t)$.

**Equation of Continuity** Let $\rho = \rho(x,t)$ be the mass per unit volume of a fluid at point $x$ and time $t$. Then the mass of any finite volume $\Omega$ is

$$m = \int_{\Omega} \rho(x,t)\,dx\,.$$

The *principle of conservation of mass* says that the mass of a fluid in a material volume $\Omega$ does not change as $\Omega$ moves with the fluid; that is,

$$\frac{d}{dt}\int_{\Omega(t)} \rho(x,t)\,dx = 0\,.$$

From the transport theorem (2.6) it follows that

$$\int_{\Omega(t)} \left\{\frac{\partial \rho}{\partial t} + \operatorname{div}(\rho u)\right\} dx = 0\,,$$

whence

$$\frac{\partial \rho}{\partial t} + \operatorname{div}(\rho u) = 0\,. \tag{2.9}$$

Sometimes the principle of conservation of mass is expressed as follows. Let $\Omega$ be a fixed volume. Then

$$\frac{d}{dt}\int_{\Omega} \rho(x,t)\,dx = -\int_{\partial\Omega} \rho u\cdot n\,dS\,, \tag{2.10}$$

that is, the rate of change of mass in a fixed volume $\Omega$ is equal to the mass flux through its surface.

We notice also the general formula

$$\frac{d}{dt}\int_{\Omega(t)} \rho f\,dx = \int_{\Omega(t)} \rho\frac{D}{Dt}f\,dx\,. \tag{2.11}$$

**Exercise 2.2.** Derive (2.9) from (2.10).

**Exercise 2.3 (Cf. [212]).** Show that in material coordinates the equation of continuity is

$$\frac{d}{dt}\{\rho(X,t)J(X,t)\} = 0\,,$$

or

$$\rho(X,t)J(X,t) = \rho(X,0)\,.$$

**Exercise 2.4 (Cf. [5]).** Show that if $\rho_0(X)$ is the distribution of density of the fluid at time $t = 0$ and $\nabla(\operatorname{div} u) = 0$, then

$$\rho(x,t) = \rho_0(X(x,t))\exp\left\{-\int_0^t \operatorname{div} u(x,t)\,dt\right\}\,.$$

**Exercise 2.5.** Find $\rho(x,t)$ for the motion

$$u_i = \frac{x_i}{1 + a_i t} \quad (a_1 = 2\,, a_2 = 1\,, a_3 = 0),$$

if $\rho_0(X)$ is the distribution of density of the fluid at time $t = 0$.

**Exercise 2.6.** Prove (2.11).

**Principle of Conservation of Linear Momentum** We assume that the forces acting on an element of a continuous medium are of two kinds. *External*, or *body, forces*, such as gravitation or electromagnetic forces, can be regarded as reaching into the medium and acting throughout the volume. If $f$ represents such a force *per unit mass*, then it acts on an element $\Omega$ as

$$\int_\Omega \rho f \, dx \,.$$

The *internal*, or *contact, forces* are to be regarded as acting on an element of volume $\Omega$ through its bounding surface. Let $n$ be the unit outward normal at a point of the surface $\partial\Omega$, and $t_n$ the force *per unit area* exerted there by the material volume outside $\partial\Omega$. Then the surface force exerted on the volume $\Omega$ can be expressed by the integral

$$\int_{\partial\Omega} t_n \, dS \,.$$

The *Cauchy principle* says that $t_n$ depends at any given time only on the position and the orientation of the surface element $dS$; in other words,

$$t_n = t_n(x, t, n) \,.$$

The *principle of conservation of linear momentum* says that the rate of change of linear momentum of a material volume equals the resultant force on the volume

$$\frac{d}{dt} \int_{\Omega(t)} \rho u dx = \int_{\Omega(t)} \rho f \, dx + \int_{\partial\Omega(t)} t_n \, dS \,, \tag{2.12}$$

where $f$ is assumed to be known.

By (2.11), (2.12) yields

$$\int_{\Omega(t)} \rho \frac{Du}{Dt} dx = \int_{\Omega(t)} \rho f \, dx + \int_{\partial\Omega(t)} t_n \, dS \,. \tag{2.13}$$

From this equation we derive a very important fact, namely, that the vector $t_n$ (called *normal stress*) can be expressed as a linear function of $n$, in the form

$$t_n(x, t, n) = n(x, t) T(x, t) \,, \tag{2.14}$$

where $T = \{T_{ij}\}$ is a matrix called the *stress tensor*. This will allow us to pass from the integral form (2.13) of the equation of conservation of linear momentum to a differential one.

Let $l^3$ be the volume of $\Omega = \Omega(t)$. Dividing both sides of (2.13) by $l^2$ and letting the volume tend to zero we obtain

$$\lim_{|\Omega|\to 0} l^{-2} \int_{\partial\Omega} t_n \, dS = 0\,, \tag{2.15}$$

that is, the stress forces are in local equilibrium.

Let $\Omega$ be a domain containing a fluid, and consider a regular tetrahedron with vertex at an arbitrary point $x \in \partial\Omega$, and with three of its faces parallel to the coordinate planes. Let the slanted face have normal $n = (n_1, n_2, n_3)$ and area $\Sigma$. The normals to the other faces are $-e_1$, $-e_2$, and $-e_3$, and their areas are $n_1\Sigma$, $n_2\Sigma$, and $n_3\Sigma$. Applying (2.15) to the family of tetrahedrons obtained by letting $\Sigma \to 0$, we obtain

$$t(n) + n_1 t(-e_1) + n_2 t(-e_2) + n_3 t(-e_3) = 0\,, \tag{2.16}$$

where $t(n) = t_n = t_n(x, t, n)$, $t(-h) = t_{-h}$ for $h \in \{e_1, e_2, e_3\}$, and $n_i > 0$. By a continuity argument, (2.16) holds for all $n_i \geq 0$, and then we prove easily that $t(e_i) = -t(-e_i)$, $i = 1, 2, 3$, and that it holds for all $n$. This means that $t(n)$ may be expressed as a linear function of $n$; that is, we can write it in the form (2.14). Thus, by (2.13) and by the Green theorem we obtain

$$\int_{\Omega(t)} \rho \frac{Du}{Dt} \, dx = \int_{\Omega(t)} (\rho f + \operatorname{div} T) \, dx\,,$$

whence, by the arbitrariness of the domain of integration,

$$\rho \frac{Du}{Dt} = \rho f + \operatorname{div} T\,, \tag{2.17}$$

or

$$\rho \left( \frac{\partial}{\partial t} u_i + \sum_{j=1}^{3} u_j \frac{\partial}{\partial x_j} u_i \right) = \rho f_i + T_{ji,j}, \quad i = 1, 2, 3\,.$$

This is the general *Cauchy equation of motion* in differential form.

**Exercise 2.7.** Give a physical interpretation of the components of the stress tensor.

Notice that we have not specified $T$ yet, that is, we have not made any assumptions concerning the nature of forces acting on surface elements. These forces depend on the kind of fluid, or, more generally, on the kind of medium under consideration.

In the simplest model the contact forces act perpendicularly to the surface elements. We have then

$$t(n) = -p(x)n\,,$$

and call $p$ the *pressure*. The minus sign is chosen so that when $p > 0$, the contact forces on a closed surface tend to compress the fluid inside; $p$ represents the pressure exerted from outside on a surface of the element of the fluid.

In particular, all fluids at rest exhibit this stress behavior, namely that an element of area always experiences a stress normal to itself, and this stress is independent of the orientation. Such stress is called *hydrostatic*.

We call this idealized model a *perfect fluid*. The equation of motion for perfect fluids is

$$\rho\left(\frac{\partial u}{\partial t} + (u\cdot\nabla)u\right) = \rho f - \nabla p\,,$$

where

$$(u\cdot\nabla)u_i = \sum_{j=1}^{3} u_j\frac{\partial}{\partial x_j}u_i\,, \qquad i = 1,2,3\,.$$

All real fluids when in motion can exert tangential stresses across surface elements, in which case the tensor $T$ is not diagonal.

The stress tensor may always be written in the form

$$T_{ij} = -p\delta_{ij} + P_{ij}\,.$$

In this case $P_{ij}$ is called the *viscous stress tensor*.

In classical fluid dynamics it is assumed that the stress tensor is *symmetric*, that is,

$$T_{ij} = T_{ji}\,.$$

This assumption has very important consequences. It may be also considered as a theorem if we assume a specific form of the equation of conservation of angular momentum. We shall discuss this in Sect. 2.2.

**Exercise 2.8 (Cf. [5]).** Show that the Cauchy equation of motion can be written as

$$\frac{\partial}{\partial t}(\rho u_i) = \rho f_i + (T_{ji} - \rho u_j u_i)_{,j}\,,$$

and interpret it physically.

**Exercise 2.9 (Cf. [5]).** Show that if $F$ is any function of position and time, then

$$\int_{\partial\Omega} FT_{ji}n_j\,dS = \int_{\Omega}\left[T_{ji}F_{,j} + \rho F\left(\frac{Du_i}{Dt} - f_i\right)\right] dx$$

(theorem of stress means).

**Equation of Energy** *The first law of thermodynamics* in classical hydrodynamics states that the increase of total energy (we shall consider here only kinetic and internal energies) in a material volume is the sum of the heat transferred and the work done on the volume. We denote by $q$ the *heat flux* (then $-q \cdot n$ is the heat flux into the volume) and by $E$ the *specific internal energy*. Then the balance expressed by the first law of thermodynamics is, cf. [5],

$$\frac{d}{dt}\int_{\Omega(t)} \rho\left(\frac{1}{2}|u|^2 + E\right) dx \tag{2.18}$$

$$= \int_{\Omega(t)} \rho f \cdot u\,dx + \int_{\partial\Omega(t)} t_n \cdot u\,dS - \int_{\partial\Omega(t)} q \cdot n\,dS\,.$$

The first integral on the right-hand side is the rate at which the body force does work, the second integral represents the work done by the stress, and the third integral is the total heat flux into the volume.

We shall write this equation in another form. From the theorem of stress means (Exercise 2.9) we have, with $F = u_i$,

$$\int_{\partial\Omega(t)} u_i T_{ji} n_j\,dS = \int_{\Omega(t)}\left(T_{ji}u_{i,j} + \rho u_i\frac{Du_i}{Dt} - \rho f_i u_i\right) dx\,.$$

Rearranging the terms and using the transport theorem, we obtain

$$\frac{d}{dt}\int_{\Omega(t)} \rho\frac{1}{2}|u|^2\,dx = \int_{\Omega(t)} \rho\frac{1}{2}\frac{D}{Dt}|u|^2\,dx \tag{2.19}$$

$$= \int_{\Omega(t)} \rho f_i u_i\,dx - \int_{\Omega(t)} T_{ji}u_{i,j}\,dx + \int_{\partial\Omega(t)} u_i(t_n)_i\,dS\,.$$

Thus the rate of change of kinetic energy of a material volume is the sum of three parts: the rate at which the body forces do work, the rate at which the internal stresses do work, and the rate at which the surface stresses do work.

From (2.18), (2.19), the transport theorem, and the Green theorem we obtain

$$\int_{\Omega(t)}\left(\rho\frac{DE}{Dt} + \nabla \cdot q - T : (\nabla u)\right) dx = 0\,,$$

where $T : (\nabla u)$ is the dyadic notation for $T_{ji}u_{i,j}$, the scalar product of $T$ and $\nabla u$.

Thus

$$\rho\frac{DE}{Dt} = -\nabla \cdot q + T : (\nabla u)\,.$$

**Conservation Laws of Classical Hydrodynamics** Above we obtained the following system of conservation laws of classical hydrodynamics

$$\frac{D\rho}{Dt} = -\rho\nabla \cdot u\,, \tag{2.20}$$

$$\rho\frac{Du}{Dt} = \nabla \cdot T + \rho f\,, \tag{2.21}$$

$$\rho\frac{DE}{Dt} = -\nabla \cdot q + T : (\nabla u)\,. \tag{2.22}$$

They are laws of conservation of mass, momentum, and energy, respectively.

If we assume the *Fourier law* for the conduction of heat,

$$q = -k\nabla\theta \quad (k \geq 0)\,, \tag{2.23}$$

where $k$ is the *thermal conductivity* of the fluid then the energy equation takes the form

$$\rho\frac{DE}{Dt} = \nabla \cdot (k\nabla\theta) + T : (\nabla u)\,.$$

## 2.2 The Stress Tensor

In the classical hydrodynamics the stress tensor $T$ is defined by

$$T_{ij} = (-p + \lambda u_{k,k})\delta_{ij} + \mu(u_{i,j} + u_{j,i})\,. \tag{2.24}$$

If we define the *deformation tensor*

$$D_{i,j} = \frac{1}{2}(u_{i,j} + u_{j,i}), \tag{2.25}$$

then the above formula takes the form

$$T_{ij} = (-p + \lambda u_{k,k})\delta_{ij} + 2\mu D_{ij}\,. \tag{2.26}$$

*Remark 2.1.* Formula (2.26) is a consequence of a number of postulates, coming originally from G. Stokes, about the fundamental properties of fluids. These postulates can be formulated as follows (cf. [5, 212]):

(a) The stress tensor $T$ is a continuous function of the deformation tensor $D$ and the local thermodynamic state, but independent of other kinematic quantities.

(b) The fluid is homogeneous; that is, $T$ does not depend explicitly on $x$.
(c) The fluid is isotropic; that is, there is no preferred direction.
(d) When there is no deformation ($D = 0$), and the fluid is incompressible ($u_{k,k} = 0$), the stress is hydrostatic ($T = -pI$, $I$ is the unit matrix).

Fluids that satisfy these postulates are called *Stokesian*. It can be proved (cf. [5, 212]) that the most general form of the stress tensor in this case is

$$T = (-p + \alpha)I + \beta D + \gamma D^2 ,$$

where $p, \alpha, \beta, \gamma$ are some functions that depend on the thermodynamic state, $\alpha, \beta, \gamma$ being dependent as well on the invariants of the tensor $D$.

Moreover, when the dependence of the components of $T$ on the components of $D$ is postulated to be *linear*, the stress tensor can be written as

$$T = (-p + \lambda \ \operatorname{div} u)I + 2\mu D ,$$

which coincides with (2.24). Such linear Stokesian fluids are called *Newtonian*. Fluids that are not Newtonian are called *non-Newtonian*. One important example of the latter are the micropolar fluids [92, 159].

**The Stress Tensor and the Law of Conservation of Angular Momentum** Looking at the form of the equation of conservation of linear momentum

$$\frac{d}{dt}\int_{\Omega(t)} \rho u \, dx = \int_{\Omega(t)} \rho f \, dx + \int_{\partial\Omega(t)} t_n \, dS ,$$

and recalling the definition of angular momentum in mechanics of mass points or rigid particles, it seems natural to *assume* the following form of the *law of conservation of angular momentum*:

$$\frac{d}{dt}\int_{\Omega(t)} \rho(x \times u) \, dx = \int_{\Omega(t)} \rho(x \times f) \, dx + \int_{\partial\Omega(t)} x \times t_n \, dS . \tag{2.27}$$

In fact, this form of the law of conservation of angular momentum holds if we assume that all torques arise from macroscopic forces. This is the case in most common fluids, but a fluid with a strongly polar character, e.g., a polyatomic fluid, is capable of transmitting stress torques and being subjected to body torques. We call such fluids *polar*.

**Theorem 2.1.** *For an arbitrary continuous medium satisfying the continuity equation (2.9) and the dynamical equation (2.17) the following statements are equivalent:*

(i) *the stress tensor is symmetric,*
(ii) *equation (2.27) holds.*

*Remark 2.2.* In classical hydrodynamics the stress tensor is symmetric, and the law of conservation of angular momentum is defined by Eq. (2.27). In consequence, in classical hydrodynamics the law of conservation of angular momentum can be derived from the law of conservation of mass and the law of conservation of linear momentum, and as such adds nothing to the description of the fluid.

*Proof.* Let us assume (ii), and we shall deduce (i). Applying formula (2.11), we have from (2.27)

$$\begin{aligned}\frac{d}{dt}\int_{\Omega(t)} \rho(x\times u)\,dx \qquad\qquad (2.28)\\ = \int_{\Omega(t)} \rho\frac{D}{Dt}(x\times u)\,dx = \int_{\Omega(t)} \rho\left(x\times\frac{Du}{Dt}\right)dx\\ = \int_{\Omega(t)} \rho(x\times f)\,dx + \int_{\partial\Omega(t)} x\times t_n\,dS\,.\end{aligned}$$

By the Green theorem,

$$\int_{\partial\Omega(t)} x\times t_n\,dS = \int_{\Omega(t)} (x\times(\nabla\cdot T) + T_x)\,dx\,, \tag{2.29}$$

where $\nabla\cdot T$ is another notation for $\operatorname{div} T$, and $T_x$ is the vector $\epsilon_{ijk}T_{jk}$ ($\epsilon_{ijk}$ is the alternating tensor of Levi-Civita), so that by (2.28)

$$\int_{\Omega(t)} x\times\left(\rho\frac{Du}{Dt} - \rho f - \nabla\cdot T\right)dx = \int_{\Omega(t)} T_x\,dx\,.$$

The left-hand side vanishes identically by the Cauchy equation; hence the right-hand side vanishes for an arbitrary volume, and so $T_x = 0$. However, the components of $T_x$ are $T_{23} - T_{32}$, $T_{31} - T_{13}$, $T_{12} - T_{21}$, and the vanishing of these implies $T_{ij} = T_{ji}$, so that $T$ is symmetric.

We leave to the reader the proof that (i) implies (ii). □

## 2.3 Field Equations

Substituting the stress tensor (2.24) into the system (2.20)–(2.22) we obtain the system of *field equations* of classical hydrodynamics

$$\frac{D\rho}{Dt} = -\rho\nabla\cdot u\,, \tag{2.30}$$

$$\rho\frac{Du}{Dt} = -\nabla p + (\lambda+\mu)\nabla\operatorname{div}u + \mu\Delta u + \rho f\,, \tag{2.31}$$

$$\rho \frac{DE}{Dt} = -p \operatorname{div} u + \rho \Phi - \nabla \cdot q \,, \tag{2.32}$$

where

$$\rho \Phi = \lambda (\operatorname{div} u)^2 + 2\mu D : D \tag{2.33}$$

is the *dissipation function* of mechanical energy per mass unit.

Let us assume that the fluid is *viscous* and *incompressible*, namely, that $\mu > 0$ and

$$\operatorname{div} u = 0 \,, \tag{2.34}$$

that the specific internal energy of the fluid is proportional to its temperature,

$$E = c_r \theta \,, \quad \text{where} \quad c_r = \text{const} > 0 \,, \tag{2.35}$$

and that Fourier's law (2.23) (with $k = const \geq 0$) holds. With (2.34), (2.35), (2.23), and (2.33), system (2.30)–(2.32) becomes

$$\frac{\partial \rho}{\partial t} + u \cdot \nabla \rho = 0 \,, \qquad \operatorname{div} u = 0 \,, \tag{2.36}$$

$$\rho \left( \frac{\partial u}{\partial t} + (u \cdot \nabla) u \right) = -\nabla p + \mu \Delta u + \rho f \,, \tag{2.37}$$

$$\rho c_r \left( \frac{\partial \theta}{\partial t} + u \cdot \nabla \theta \right) = 2\mu D : D + k \Delta \theta . \tag{2.38}$$

## 2.4 Navier–Stokes Equations

Assuming that the density $\rho$ of the fluid is uniform and denoting $\nu = \frac{\mu}{\rho}$, $\kappa = \frac{k}{\rho}$ ($\nu$ is called the *kinematic viscosity* coefficient), Eqs. (2.36)–(2.38) reduce to

$$\frac{\partial u}{\partial t} + (u \cdot \nabla) u = -\frac{1}{\rho} \nabla p + \nu \Delta u + f \,, \tag{2.39}$$

$$\operatorname{div} u = 0 \,, \tag{2.40}$$

$$c_r \left( \frac{\partial \theta}{\partial t} + u \cdot \nabla \theta \right) = 2\nu D : D + \kappa \Delta \theta . \tag{2.41}$$

When the body forces $f$ do not depend on temperature, the first two equations of the above system,

$$\frac{\partial u}{\partial t} + (u \cdot \nabla)u = -\frac{1}{\rho}\nabla p + \nu \Delta u + f\,, \tag{2.42}$$

$$\operatorname{div} u = 0 \tag{2.43}$$

constitute a closed system of equations with respect to variables $u, p$, and are called *Navier–Stokes equations* of viscous incompressible fluids with uniform density (we shall call them just the Navier–Stokes equations). The mechanical energy of the flow governed by (2.42) and (2.43) due to viscous dissipation is lost and appears as heat. This can be seen from Eq. (2.41) in which the term $2\nu D : D$ is positive, provided the flow is not uniform. In real fluids, however, density depends on temperature, so that our system (2.39)–(2.41) may be physically impossible. In fact, due to viscosity and high velocity gradients the temperature rises in view of (2.41), and this produces density fluctuations, contrary to our assumption that density is uniform in the flow domain. Thus, reduced problems often play the role of more or less justified approximations. For more considerations of this kind cf. [109, Chap. 1].

When the body forces depend on temperature, $f = f(\theta)$, we have to take into account the whole system (2.39)–(2.41). One of the considered in the literature system of equations of heat conducting viscous and incompressible fluid are the so-called *Boussinesq equations*,

$$\frac{\partial u}{\partial t} + (u \cdot \nabla)u = -\frac{1}{\rho_0}\nabla p + \nu \Delta u + \frac{1}{\rho_0} g\alpha(\theta - \theta_0)\,, \tag{2.44}$$

$$\operatorname{div} u = 0\,, \tag{2.45}$$

$$\frac{\partial \theta}{\partial t} + u \cdot \nabla \theta = \frac{\kappa}{c_r}\Delta \theta\,, \tag{2.46}$$

where $g$ represents the vertical *gravity acceleration*, $\alpha$ is the *thermal expansion coefficient*, and $\frac{\kappa}{c_r}$ is the *thermal diffusion coefficient*. Moreover, $\rho_0$ and $\theta_0$ are some reference density and temperature, respectively. In the velocity equation the vertical buoyancy force $\frac{1}{\rho_0} g\alpha(\theta - \theta_0)$ results from changes of density associated with temperature variations $\rho - \rho_0 = -\alpha(\theta - \theta_0)$. This is the only term in the system where changes of density were taken into account. We have also abandoned the viscous dissipation term in the temperature equation.

## 2.5 Vorticity Dynamics

Taking the *curl* of the equation of motion

$$\frac{\partial u}{\partial t} + (u \cdot \nabla)u = -\frac{1}{\rho}\nabla p + \nu \Delta u,$$

we obtain

$$\frac{\partial \omega}{\partial t} + (u \cdot \nabla)\omega = (\omega \cdot \nabla)u + \nu \Delta \omega, \tag{2.47}$$

where the vector field $\omega = \nabla \times u$ is called *vorticity* of the fluid. It has a simple physical interpretation. In the case of two-dimensional motion with

$$u = (u_1(x,y), u_2(x,y), 0),$$

the vorticity reduces to

$$\omega = (0, 0, \omega_3(x_1, x_2)) = \left(0, 0, \frac{\partial u_2(x_1, x_2)}{\partial x_1} - \frac{\partial u_1(x_1, x_2)}{\partial x_2}\right),$$

where the third component represents twice the angular velocity of a small (infinitesimal) fluid element at point $(x_1, x_2)$. The vorticity field is, by definition, divergence free,

$$\operatorname{div} \omega = 0.$$

In the case of two-dimensional motions the Eq. (2.47) reduces to

$$\frac{\partial \omega}{\partial t} + (u \cdot \nabla)\omega = \nu \Delta \omega,$$

and we can see that the vorticity in the fluid is transported by two agents: *convection* and *diffusion*, just as the temperature in the system (2.44)–(2.46).

For inviscid fluids ($\nu = 0$) the vorticity field has very important properties that allow us to imagine behavior of complicated turbulent flows [83]. In this case, vorticity is a *local variable* which means that we can isolate a patch of vorticity and observe how it is transported along the velocity field trajectories with a finite speed. For two-dimensional flows this is evident as then the vorticity equation reduces to

$$\frac{\partial \omega}{\partial t} + (u \cdot \nabla)\omega = 0.$$

For more information, cf. [166].

**Exercise 2.10.** Vorticity has nothing in common with rotation of the fluid as a whole. Calculate the vorticity of the flows: (a) $u(x_1, x_2, x_3) = (u_1(x_2), 0, 0)$ and (b) $u(r, \phi, z) = (0, k/r, 0)$ for $r > 0$.

## 2.6 Thermodynamics

**Equations of State** From the point of view of thermodynamics the state of a homogeneous fluid can be described by some definite relations among a number of certain *state variables*, the most important being the volume $V$ ($V = 1/\rho$), the entropy $S$, the internal energy $E$, the pressure $p$, and the absolute temperature $\theta$, cf. [212].

In such a description one may start with a relation of the form (cf. [212])

$$E = E(S, V) \quad \text{(Gibbs relation)} \tag{2.48}$$

and define $p$ and $\theta$ by

$$p = -\frac{\partial E}{\partial V}, \quad \theta = \frac{\partial E}{\partial S}, \tag{2.49}$$

with $p, \theta > 0$ by assumption. In this case, taking the total differential in (2.48) and using (2.49), we obtain

$$dE = \theta\, dS - p\, dV \quad \text{or} \quad dE = \theta\, dS - p\frac{1}{\rho^2}\, d\rho\,. \tag{2.50}$$

A simple phase system is said to undergo a *differentiable process* if its state variables are differentiable functions of time: $V = V(t)$, $S = S(t)$, etc. Assuming such a dependence one usually assumes, together with (2.50), that

$$\frac{DE}{Dt} = \theta\frac{DS}{Dt} - p\frac{DV}{Dt}$$

or

$$\frac{DE}{Dt} = \theta\frac{DS}{Dt} - p\frac{1}{\rho^2}\frac{D\rho}{Dt}\,. \tag{2.51}$$

Relation (2.51) makes it possible to write a definite form of the balance of entropy when we know the laws of conservation of mass and internal energy. We shall use this relation in the sequel.

**Second Law of Thermodynamics and Constraints on Viscosity Coefficients** Consider the law of conservation of energy (2.32)

$$\rho\frac{DE}{Dt} = \nabla\cdot(k\nabla\theta) - p\,\operatorname{div} u + \rho\Phi\,, \tag{2.52}$$

where Fourier's law is assumed, and $\rho\Phi$ is given by (2.33). We see that the internal energy increases with the influx of heat transfer, compression, and the viscous dissipation.

From the law of conservation of mass

$$\frac{D\rho}{Dt} + \rho \operatorname{div} u = 0,$$

we have

$$\operatorname{div} u = -\frac{1}{\rho}\frac{D\rho}{Dt}. \tag{2.53}$$

Substituting (2.53) into (2.52) we can write

$$\rho\frac{DE}{Dt} - \frac{p}{\rho}\frac{D\rho}{Dt} = \nabla\cdot(k\nabla\theta) + \rho\Phi. \tag{2.54}$$

Now we shall transform the left-hand side of (2.54). From (2.51) we have

$$\rho\theta\frac{DS}{Dt} = \rho\frac{DE}{Dt} + \frac{p}{\rho}\frac{D\rho}{Dt}. \tag{2.55}$$

Thus

$$\rho\frac{DS}{Dt} = \frac{1}{\theta}\nabla\cdot(k\nabla\theta) + \frac{1}{\theta}\rho\Phi, \tag{2.56}$$

so that

$$\begin{aligned}\int_{\Omega(t)}\rho\frac{DS}{Dt}\,dx = \frac{d}{dt}\int_{\Omega(t)}\rho S\,dx \qquad\qquad (2.57)\\ = \int_{\Omega(t)}\left(\frac{1}{\theta}\nabla\cdot(k\nabla\theta) + \frac{1}{\theta}\rho\Phi\right)dx\\ = \int_{\Omega(t)}\left(\nabla\cdot\left(\frac{k}{\theta}\nabla\theta\right) + \frac{k}{\theta^2}|\nabla\theta|^2 + \frac{1}{\theta}\rho\Phi\right)dx.\end{aligned}$$

In the end we obtain the following integral form of the *balance of entropy*

$$\int_{\Omega(t)}\rho\frac{DS}{Dt}dx = \int_{\partial\Omega(t)}\frac{k}{\theta}\frac{\partial\theta}{\partial n}dS + \int_{\Omega(t)}\left(\frac{k}{\theta^2}|\nabla\theta|^2 + \frac{1}{\theta}\rho\Phi\right)dx. \tag{2.58}$$

Equation (2.58) with $\Phi$ as in (2.33) and with $\mu \geq 0$, $3\lambda + 2\mu \geq 0$ (cf. [212]) is consistent with the *second law of thermodynamics*, which says that the rate of increase of entropy is not less than the heat transfer into the material volume. The first integral on the right-hand side of (2.58) is just the heat transferred into the material volume divided by the temperature

$$\int_{\partial\Omega(t)} \frac{k}{\theta}\frac{\partial\theta}{\partial n}\, dS = \int_{\partial\Omega(t)} \frac{-q\cdot n}{\theta}\, dS\,,$$

and the second integral on the right-hand side of (2.58) is nonnegative.

*Remark 2.3.* For a perfect fluid, with no heat conductivity and viscosity, Eq. (2.56) becomes

$$\frac{DS}{Dt} = 0\,,$$

so that the entropy is conserved.

*Remark 2.4.* If we assume that the Fourier law holds, the second law of thermodynamics demands

$$\frac{k}{\theta^2}|\nabla\theta|^2 + \frac{1}{\theta}\rho\Phi \geq 0\,, \tag{2.59}$$

where $\rho\Phi$ is given by (2.33). The inequality (2.59) gives us the restrictions on the coefficients in the constitutive equation (2.24). Without assuming the Fourier law, the second law of thermodynamics states that

$$-\frac{1}{\theta}q\cdot\nabla\theta + \rho\Phi \geq 0\,.$$

This law will be satisfied if $q\cdot\nabla\theta \leq 0$ (i.e., the heat flux is not directed against the temperature gradient) and $\Phi \geq 0$ (i.e., deformation always absorbs energy converting it to heat).

**Exercise 2.11.** Prove that inequality (2.59), with $\Phi$ given in (2.33), yields the following constraints on the coefficients

$$k \geq 0, \quad 3\lambda + 2\mu \geq 0, \quad \mu \geq 0\,.$$

*Hint.* The left-hand side of inequality (2.59) must be nonnegative for an arbitrary choice of $\theta_{,i}$ and $D_{ij}$. It is clear that this implies $k \geq 0$. The nonnegativity of the terms containing $D_{ij}$ is equivalent to the condition

$$3\lambda + 2\mu \geq 0 \quad \text{and} \quad \mu \geq 0\,.$$

## 2.7 Similarity of Flows and Nondimensional Variables

Let us consider two initial boundary value problems

$$\rho\frac{Du}{Dt} = -\nabla p + \mu\Delta u, \qquad \operatorname{div} u = 0 \tag{2.60}$$

in $\Omega = [0, L]^n$, $u = u(x,t)$, $p = p(x,t)$, with initial condition $u(x, 0) = u_0(x)$, and

$$\rho' \frac{Du'}{Dt'} = -\nabla p' + \mu' \Delta u', \qquad \operatorname{div} u' = 0 \tag{2.61}$$

in $\Omega' = [0, L']^n$, $u' = u'(x', t')$, $p' = p'(x', t')$, with initial condition $u'(x', 0) = u'_0(x')$, respectively, and with periodic boundary conditions, $n = 2$ or 3.

Changing dependent and independent variables in (2.60) as follows:

$$u^\star(x^\star, t^\star) = \frac{u(x,t)}{U}, \quad p^\star(x^\star, t^\star) = \frac{p(x,t) - p_0}{\rho U^2} \tag{2.62}$$

and

$$x^\star = \frac{x}{L}, \quad t^\star = \frac{U}{L} t, \tag{2.63}$$

we obtain the system

$$\frac{Du^\star}{Dt^\star} = -\nabla p^\star + \frac{1}{Re} \Delta u^\star, \qquad \operatorname{div} u^\star = 0 \tag{2.64}$$

in $\Omega^\star = [0, 1]^n$, $u^\star = u^\star(x^\star, t^\star)$, $p^\star = p^\star(x^\star, t^\star)$, with initial condition $u^\star(x^\star, 0) = u_0(x^\star) = u_0(Lx^\star)/U$, with periodic boundary condition, and with

$$Re = \frac{\rho L U}{\mu}.$$

In the case of problem (2.60) the symbols $\rho, L, U, \mu$ denote *dimensional* characteristic quantities of the flow, describing the density of the fluid ($\rho$), characteristic linear dimension of the domain of the flow ($L$), some characteristic velocity ($U$), and viscosity ($\mu$). For $U$ we can take, for example, the square root of the average initial kinetic energy in $\Omega$, see [88] for more details. According to (2.62) and (2.63), in (2.64) all dependent and independent variables are *nondimensional*, together with the *Reynolds number Re*. The characteristic quantities of the flow in (2.64) are equal to one, with the exception of viscosity which equals $1/Re$.

In the same way we can reduce system (2.61) to system (2.64), now with the Reynolds number

$$Re' = \frac{\rho' L' U'}{\mu'}.$$

Assume now that $Re = Re'$. Then both systems (2.60) and (2.61) (including the boundary conditions) are represented, in *nondimensional* variables, by the same system (2.64). We call systems (2.60) and (2.61) *dynamically similar*. Solving system (2.64) we generate solutions to an infinite three-parameter family of systems having the same Reynolds number.

The physical interpretation of the Reynolds number is that it describes the relation between the magnitudes of the inertial term $(u \cdot \nabla)u$ and the viscous term $\nu \Delta u$ $(\nu = \mu/\rho)$ in the equation of motion, according to

$$\frac{\textit{inertial term}}{\textit{viscous term}} = O\left(\frac{|(u \cdot \nabla)u|}{|\nu \Delta u|}\right) = O\left(\frac{U(U/L)}{\nu(U/L^2)}\right) = O(Re),$$

or, saying it differently, the Reynolds number can be interpreted as a ratio of the strength of inertial forces to viscous forces.

**Exercise 2.12.** Let $u^\star, p^\star$ solve system (2.64). Show that

$$u(x,t) = Uu^\star\left(\frac{x}{L}, \frac{U}{L}t\right), \quad p(x,t) = \rho U^2 p^\star\left(\frac{x}{L}, \frac{U}{L}t\right) + p_0$$

solve system (2.60).

**Exercise 2.13.** Let $\rho = \rho'$ and $\mu = \mu'$, and let $u', p'$ solve system (2.61). Show that

$$u(x,t) = \frac{U}{U'}u'\left(\frac{L'}{L}x, \frac{L'U}{LU'}t\right),$$
$$p(x,t) = \frac{U^2}{U'^2}\left(p'\left(\frac{L'}{L}x, \frac{L'U}{LU'}t\right) - p'_0\right) + p_0$$

solve system (2.60). Here arbitrary constants $p_0, p'_0$ may represent typical values of pressures in the considered problems for systems (2.60) and (2.61), respectively, cf. (2.62).

**Exercise 2.14.** Let $u', p'$ solve the system of equations

$$\rho'\frac{Du'}{Dt} = -\nabla p' + \mu \Delta u', \quad \operatorname{div} u' = 0 \tag{2.65}$$

in the whole space. Show that

$$u(x,t) = lu'(lx, l^2t), \quad p(x,t) = l^2(p'(lx, l^2t) - p'_0) + p_0,$$

where $l > 0$, also solve this system of equations. Thus having one (nontrivial) solution of system (2.65) we can generate in principle an infinite number of different solutions of the same system.

Consider the transformations,

$$(x,t,u,p) \to \left(lx, l^2t, \frac{1}{l}u, \frac{1}{l^2}p\right), \quad l > 0, \tag{2.66}$$

or

$$\begin{aligned} x' &= lx\,, \\ t' &= l^2 t\,, \\ u' &= \frac{1}{l}u\,, \\ p' &= \frac{1}{l^2}p. \end{aligned}$$

Show that transformations (2.66) form a group. This group is called a one parameter *symmetry group* of *scalings* of system (2.65).

A solution $(u', p')$ is called *self-similar* if $u(x,t) = u'(x',t'),\ p(x,t) = p'(x',t')$, cf. [17].

**Exercise 2.15.** Let $u', p'$ solve the system of equations

$$\rho' \frac{Du'}{Dt} = -\nabla p' + \mu \Delta u', \qquad \operatorname{div} u' = 0$$

in the whole space $\mathbb{R}^n$. Show that

$$\begin{aligned} u(x,t) &= u'(x-ct,t) + c, \\ p(x,t) &= p'(x-ct,t), \end{aligned}$$

where $c$ is any vector in $\mathbb{R}^n$, also solve this system of equations (*Galilean invariance*), and

$$\begin{aligned} u(x,t) &= Q^t u'(Qx,t), \\ p(x,t) &= p'(Qx,t), \end{aligned}$$

where $Q$ is any rotation matrix ($Q^t = Q^{-1}$) also solve this system of equations (*rotation symmetry*).

## 2.8 Examples of Simple Exact Solutions

**The Flow Due to an Impulsively Moved Plain Boundary** Let the domain of the flow be half-space

$$\{(x,y,z) \in \mathbb{R}^3 \ :\ \ y > 0\},$$

and assume that at times $t < 0$ the flow is at rest, and that at time $t = 0$ the plane $y = 0$ is suddenly jerked into motion in the $x$-direction, with a constant velocity $U$.

Due to the geometrical symmetry of the problem and under additional "intuitively apparent" preconceptions, we assume that the flow is of the form $u = (u(y, t), 0, 0)$. Substituting $u$ of the above form into the Navier–Stokes equations (2.42) we obtain

$$\frac{\partial u}{\partial t} = -\frac{1}{\rho}\frac{\partial p}{\partial x} + \nu \frac{\partial^2 u}{\partial y^2}, \tag{2.67}$$

$$\frac{\partial p}{\partial y} = \frac{\partial p}{\partial z} = 0. \tag{2.68}$$

As $u$ depends only on $y$, from (2.67) together with (2.68) it follows that $\frac{\partial p}{\partial x}$ depends only on $t$. Assume additionally that $\frac{\partial p}{\partial x} = 0$ and that the flow is at rest for $y = \infty$. Then the above system together with the initial and boundary conditions reads

$$\frac{\partial u}{\partial t} = \nu \frac{\partial^2 u}{\partial y^2}, \tag{2.69}$$

with

$$u(y, 0) = 0 \quad \text{for} \quad y > 0\,, \tag{2.70}$$

$$u(0, t) = U \quad \text{for} \quad t > 0\,, \tag{2.71}$$

$$u(\infty, t) = 0 \quad \text{for} \quad t > 0. \tag{2.72}$$

We say that a solution $u = u(y, t)$ of the initial and boundary value problem (2.69)–(2.72) is *invariant* or *self-similar* with respect to the one-parameter symmetry group of scalings

$$y' = \alpha y, \quad t' = \alpha^2 t, \quad u' = u, \quad \alpha > 0, \tag{2.73}$$

if the function $u' = u'(y', t') = u(y, t)$ is a solution of the problem

$$\frac{\partial u'}{\partial t'} = \nu \frac{\partial^2 u'}{\partial y'^2},$$

with

$$u'(y', 0) = 0 \quad \text{for} \quad y' > 0\,,$$

$$u'(0, t') = U \quad \text{for} \quad t' > 0\,,$$

$$u(\infty, t') = 0 \quad \text{for} \quad t' > 0.$$

Observe that the transformation of the independent variables does not change the space-time domain.

The relation $u'(y',t') = u(y,t)$ satisfied by the invariant solution implies that the number of its independent variables can be reduced by one [17]. Together with the *dimensional homogeneity principle* [33] stating that physical phenomena must be described by laws that do not depend on the unit of measure applied to the dimensions of the variables that describe the phenomena, we arrive at the following form of the solution

$$u(y,t) = f(\eta), \quad \eta = \frac{y}{\sqrt{\nu t}}. \tag{2.74}$$

Notice that the new variable $\eta$, called *similarity variable* is nondimensional and is invariant with respect to the symmetry group of scalings $y' = \alpha y$, $t' = \alpha^2 t$.

**Exercise 2.16.** Assume that problem (2.69)–(2.72) is invariant with respect to a *dilation group*

$$y' = e^a y, \quad t' = e^b t, \quad u' = e^c u,$$

for some $a, b, c$, that is if $u(x,t)$ is a solution of the problem then $u'(x',t')$ is also a solution of the same problem. Then we get (2.73).

Observe, that if a problem is uniquely solvable and invariant with respect to a symmetry group of scalings $x \to x', t \to t', u \to u'$ then its unique solution is also invariant.

Substituting function of the form (2.74) to problem (2.69)–(2.72) we obtain

$$f'' + \frac{1}{2}\eta f' = 0,$$

with

$$f(0) = U, \quad f(\infty) = 0.$$

Integrating this equations we obtain

$$u(y,t) = f(\eta) = U\left(1 - \frac{1}{\sqrt{\pi}}\int_0^\eta e^{\frac{-s^2}{4}}\, ds\right), \tag{2.75}$$

or, in the nondimensional form

$$\frac{u(y,t)}{U} = 1 - \frac{1}{\sqrt{\pi}}\int_0^\eta e^{\frac{-s^2}{4}}\, ds.$$

The example of the velocity profile given by (2.75) is depicted in Fig. 2.1.

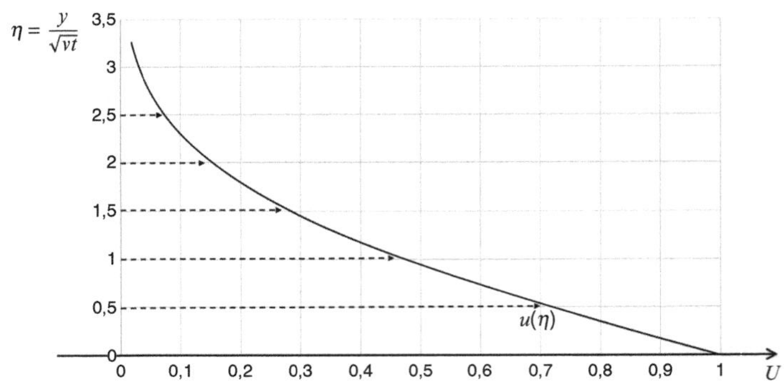


**Fig. 2.1** Velocity profile for the flow in the half-space due to the impulsively moved boundary. In the plot we have $U = 1$. The variable on the horizontal axis is the velocity $u$, and on the vertical axis $\eta = \frac{y}{\sqrt{\nu t}}$

**Exercise 2.17.** Check that the pair $(u, p) = (u((y, t), 0, 0), C)$ where $C$ is a constant and $u$ is the function given in (2.75) is a solution of the Navier–Stokes equations (2.42) and (2.43) with initial and boundary conditions (2.70)–(2.72). Observe that we have found only one of, perhaps, many other solutions of the problem. We cannot claim that the solution is *unique* in $u$.

From (2.75) it follows that the velocity of the flow is a nonlocal variable, that is, it spreads information through the region of the flow with infinite speed. Although at negative times the flow was at rest, at any, however small, positive time $t$ and any, however large, $y > 0$, we have $u(y, t) > 0$. The same observation concerns the vorticity vector (the third and only nonzero component of which is $\omega = -\frac{\partial u}{\partial y}$) as both functions $u$ and $\omega$ satisfy the diffusion equation.

**The Poiseuille Flow** Consider a *stationary* flow $u = (u(y), 0, 0)$ in the domain

$$\{(x, y, z) \in \mathbb{R}^3 \; : \; 0 < y < h\},$$

and with homogeneous boundary conditions $u = 0$ at $y = 0$ and $y = h$. Substituting $u$ of the above form to the Navier–Stokes equations we obtain the equations of motion

$$\nu \frac{\partial^2 u}{\partial y^2} = \frac{1}{\rho} \frac{\partial p}{\partial x},$$

$$\frac{\partial p}{\partial y} = \frac{\partial p}{\partial z} = 0,$$

from which it follows that

$$\nu \frac{\partial^2 u}{\partial y^2} = \frac{1}{\rho} \frac{\partial p}{\partial x} = C \tag{2.76}$$

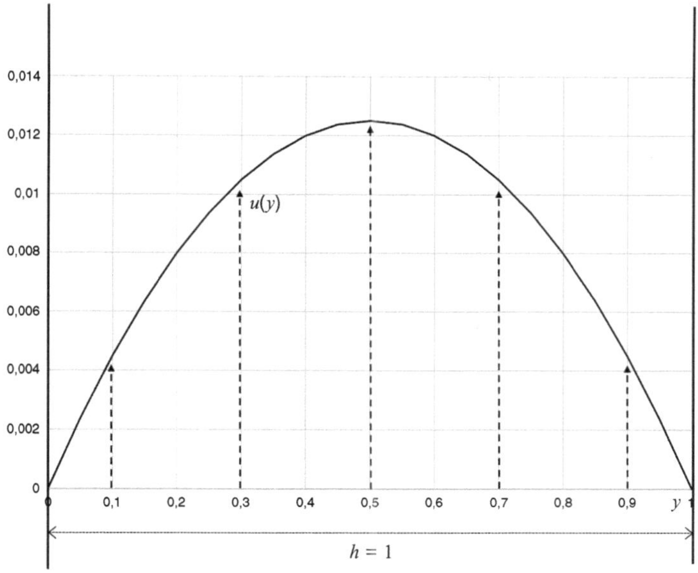


**Fig. 2.2** Parabolic velocity profile for the Poiseuille flow. The constants were chosen as $h = 1$, $\nu = 10$, and $C = -1$. Velocity $u(y)$ is on the vertical axis, and the space variable $y$ is on the horizontal axis

for some constant $C$. Taking into account the boundary conditions we obtain

$$u(y) = \frac{C}{2\nu} y(y-h)\,,$$

where

$$C = \frac{1}{\rho}\frac{\partial p}{\partial x}.$$

From the last equation we have $p(x) = C\rho x + B$, for an arbitrary constant $B$. Let us take $B = 0$. We can see that there is an *infinite number of solutions* $(u, p)$ of the form

$$u(y) = \frac{C}{2\nu} y(y-h), \quad p(x) = C\rho x. \tag{2.77}$$

We call them the *Poiseuille flows*. For $C = 0$ the flow is just at rest, both velocity and pressure equal zero. For $C < 0$ the fluid flows towards the increasing $x$-direction, $u > 0$, driven by the force due to the pressure gradient (as $C < 0$ the pressure decreases with increasing $x$). The example of the velocity profile given by (2.77) for negative value of $C$ is depicted in Fig. 2.2.

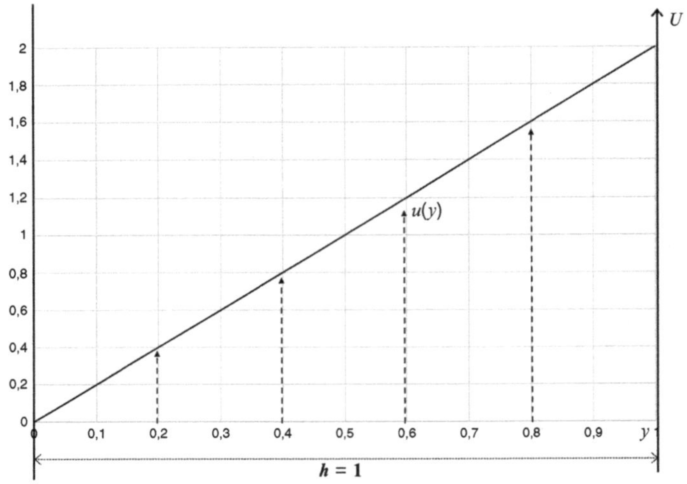


**Fig. 2.3** Linear velocity profile for the Couette flow. The constants were chosen as $h = 1$ and $C = 2$. Velocity $u(y)$ is on the vertical axis, and the space variable $y$ is on the horizontal axis

**The Couette Flow** Let us consider a flow $u = (u(y), 0, 0)$ in the same region as in the preceding example, this time, however, with different boundary conditions. We pose $u = 0$ at $y = 0$ and $u = U$ at $y = h$. As the driving force now comes from the motion of the upper boundary $y = h$, we pose $C = 0$ which corresponds to $p = const$ in the whole domain. Solving Eq. (2.76) in $u$ with $C = 0$ and with the new boundary conditions we obtain the solution

$$u(y) = \frac{U}{h}y, \quad p = const. \tag{2.78}$$

of the problem, with a uniform pressure distribution. We call this flow the *Couette flow*. The velocity profile given by (2.78) for negative value of $C$ is depicted in Fig. 2.3.

## 2.9 Comments and Bibliographical Notes

**Remarks on Modeling** Exact solutions serve, among other things, to test a given theory of hydrodynamics. By *theory* we mean here a set of governing equations (conservation laws) together with additional conditions (boundary or initial and boundary conditions), that is, a boundary or initial and boundary value problem. In the following chapters we shall prove several results about such theories of classical hydrodynamics.

A particular theory of hydrodynamics is *well-set*, that is, that it *possesses a unique solution* depending *continuously* upon the data (boundary data, external forces, etc.).

As was observed by Birkhoff in [16], "until it has been shown that the boundary value problem is well-set, one cannot conclude that its equations are erroneous." The latter means "false" according to the following definition, cf. [16]: A theory of rational hydrodynamics will be called *incomplete* if its conditions do not uniquely determine the flow, *overdetermined* if its conditions are mathematically incompatible, *false* if it is well-set but gives grossly incorrect predictions.

Since from a particular theory in its whole generality, it is in most cases quite impossible to predict qualitative properties of the flow (it is sometimes easier, however, to compare various flows among themselves), we are forced to "simplify," whatever that means. In this connection, there are very often exact solutions of special problems that serve to test whether a given theory gives correct predictions and consequently is not a false theory.

We enumerate a number of such simplifications (which we have already used in Sect. 2.8):

1. linearization of governing equations,
2. setting a special form of boundary conditions,
3. assumption of some symmetry in space (which reduces the number of spatial dimensions),
4. assumption that the problem is time-independent,
5. assumption that acting forces are potential or absent.

After suitable simplifications have been done, the original problem reduces to the one that can be (luckily) solved explicitly by solving, e.g., a linear system of ordinary differential equations with constant coefficients. Having the exact solutions, we try to compare them with experiments.

When simplifying, however, one has to be very careful in order not to oversimplify and obtain paradoxes, for example. It is well known (cf. [16]) that the various plausible intuitive assumptions we make *implicitly* when simplifying may be fallacious. Following [16], we state the assumptions that are especially suggestive:

1. intuition suffices for determining which physical variables require consideration,
2. small causes produce small effects, and infinitesimal causes produce infinitesimal effects,
3. symmetric causes produce effects with the same symmetry,
4. the flow topology can be guessed by intuition,
5. the processes of analysis can be used freely: the functions of rational hydrodynamics can be freely integrated, differentiated, expanded in series, integrals, etc.,
6. mathematical problems suggested by intuitive physical ideas are well-set.

In Chaps. 14–16 we consider problems whose solutions might not be unique.

For more information about classical hydrodynamics we refer the reader to [1, 14, 83, 149].

For a history of hydrodynamics and the derivation and development of the Navier–Stokes equations, in particular, see [81].

# 3
# Mathematical Preliminaries

In this chapter we introduce the basic preliminary mathematical tools to study the Navier–Stokes equations, including results from linear and nonlinear functional analysis as well as the theory of function spaces. We present, in particular, some of the most frequently used in the sequel embedding theorems and differential inequalities.

## 3.1 Theorems from Functional Analysis

**Theorems from Linear Functional Analysis** The core of this subsection is the Lax–Milgram lemma (which we state for separable Hilbert spaces) as well as its proof using the Galerkin method. This method will be very useful in our subsequent considerations, in the proofs of the existence of solutions of both linear and nonlinear problems.

**Theorem 3.1.** *In a reflexive Banach space every closed ball is weakly compact.*

*Proof.* A proof can be found, e.g., in [106], for separable spaces in [118]. □

In particular, let $\{x_n\}$ be a sequence in a reflexive Banach space $B$ such that $\|x_n\|_B \le M$ for some $M > 0$. Then there exists a subsequence $\{x_\mu\}$ of the sequence $\{x_n\}$ and an element $x$ in $B$ with $\|x\|_B \le M$ such that $\{x_\mu\}$ converges weakly to $x$ in $B$; that is, for every linear and continuous functional $L$ on $B$,

$$\lim_{\mu\to\infty} L(x_\mu) = L(x)\,.$$


G. Łukaszewicz, P. Kalita, *Navier–Stokes Equations*, Advances in Mechanics and Mathematics 34, DOI 10.1007/978-3-319-27760-8_3

**Theorem 3.2 (Lax–Milgram Lemma for Separable Spaces).** *Let $H$ be a separable Hilbert space with a norm $\|\cdot\|$, $L \in H'$ a linear functional on $H$, $a(u, v)$ a bilinear and continuous form on $H \times H$, coercive, that is, such that for some $\alpha > 0$ and for all $u \in H$*

$$a(u, u) \geq \alpha \|u\|^2 .$$

*Then there exists a unique element $u \in H$ such that*

$$a(u, v) = L(v) \quad \text{for all} \quad v \in H . \tag{3.1}$$

*Proof.* We use the *Galerkin method*, to which we shall refer in the following chapters. It consists of proving the existence of elements $u_m \in H_m$ such that

$$a(u_m, v) = L(v) \quad \text{for all} \quad v \in H_m , \tag{3.2}$$

where $H_m$ are *finite dimensional* subspaces of $H$ such that $H_1 \subset H_2 \subset \cdots \subset H_m \subset \cdots$ and $\bigcup_{m=1}^{\infty} H_n$ is a dense subset of $H$, and then by *passing to the limit* with $n$ to obtain (3.1).

Let $\omega_1$, $\omega_2$, $\omega_3$, … be a basis of $H$ and $H_m = \operatorname{span}\{\omega_1, \ldots, \omega_m\}$, $m = 1, 2, 3, \ldots$.

1. We shall show that for each positive integer $m$ there exists $u_m \in H_m$ such that (3.2) holds. Let

$$u_m = \sum_{k=1}^{m} \xi_k \omega_k .$$

Then (3.2) is equivalent to the system of linear equations

$$\sum_{k=1}^{m} \xi_k a(\omega_k, \omega_l) = L(\omega_l) , \quad l = 1, 2, \ldots, m .$$

This system has a unique solution $(\xi_1, \ldots, \xi_m)$ for every right-hand side if and only if the matrix $\{a(\omega_k, \omega_l)\}_{k,l \leq m}$ is nonsingular. We shall show that the homogeneous system

$$\sum_{k=1}^{m} \xi_k a(\omega_k, \omega_l) = 0 , \quad l = 1, 2, \ldots, m ,$$

has a unique solution $\xi = (0, 0, \ldots, 0)$. We multiply the $l$th equation by $\xi_k$ and add the equations to get

$$\sum_{l=1}^{m} \sum_{k=1}^{m} \xi_l \xi_k a(\omega_k, \omega_l) = a\left( \sum_{k=1}^{m} \xi_k \omega_k, \sum_{l=1}^{m} \xi_l \omega_l \right) = a(u_m, u_m) = 0 .$$

From the *coerciveness* of the form $a(\cdot,\cdot)$ we obtain $u_m = 0$. As the vectors $\omega_1, \ldots, \omega_m$ are linearly independent, we conclude that $\xi_1 = \xi_2 = \cdots = \xi_m = 0$. Hence the matrix $\{a(\omega_k, \omega_l)\}_{k,l \le m}$ is nonsingular, and for each positive integer $m$ there exists an *approximate solution* $u_m \in H_m$.

2. Convergence of the sequence $\{u_m\}$. We have

$$\alpha \|u_m\|^2 \le a(u_m, u_m) = L(u_m) \le \|L\|_{H'} \|u_m\| ,$$

whence

$$\|u_m\| \le \frac{1}{\alpha}\|L\|_{H'}$$

for each positive integer $m$. From Theorem 3.1 it follows that there exists a subsequence $\{u_\mu\}$ of the sequence of approximate solutions $\{u_m\}$ and an element $u \in H$ such that $u_\mu \to u$ weakly in $H$. For $\mu \ge j$ we have

$$a(u_\mu, v) = L(v) \quad \text{for all} \quad v \in H_j \subset H_\mu ,$$

so that

$$\lim_{\mu\to\infty} a(u_\mu, v) = a(u, v) = L(v) \quad \text{for} \quad v \in \bigcup_{j=1}^{\infty} H_j .$$

As $\bigcup_{j=1}^{\infty} H_j$ is a dense subset of $H$, we conclude from the continuity of $a(\cdot,\cdot)$ and $L(\cdot)$ that (3.1) holds for every $v \in H$.

3. Uniqueness of $u$. Let us suppose that we have two different elements $u_1$ and $u_2$ such that $a(u_1, v) = L(v)$ and $a(u_2, v) = L(v)$ for all $v \in H$. Thus $a(u_1 - u_2, v) = 0$, and taking $v = u_1 - u_2$ we obtain

$$0 = a(u_1 - u_2, u_1 - u_2) \ge \alpha \|u_1 - u_2\|^2 ,$$

whence $u_1 = u_2$. We have come to a contradiction, which proves the uniqueness. □

*Remark 3.1.* If $L$ is a linear and continuous functional on the Banach space $B$, we will write $L \in B'$, the dual space of $B$, and we will use the notation $L(v) = \langle L, v\rangle$ for $v \in B$.

**Exercise 3.1.** Prove that the whole sequence of approximate solutions converges weakly to $u$.

As a corollary we obtain the following

**Theorem 3.3 (Riesz–Fréchet Theorem for Separable Spaces).** *Let $H$ be a separable Hilbert space with scalar product $(\cdot,\cdot)$ and norm $\|\cdot\|$ and let $L \in H'$ be a linear and continuous functional on $H$. Then there exists a unique element $u \in H$ such that*

$$(u, v) = L(v) \quad \text{for each} \quad v \in H$$

*and $\|u\| = \|L\|_{H'}$.*

*Remark 3.2.* In the above two theorems the separability of the space $H$ is not essential; for the proofs see [110].

**Fixed Point Theorems** Fixed point theorems are the basic tool that we shall use. We begin with the Banach contraction principle, the only theorem in this subsection that guarantees the uniqueness of the fixed point. The existence of a unique fixed point is a strong condition, however, and in most cases we shall make use of the Schauder or the Leray–Schauder theorems. They both follow from the Brouwer fixed point theorem.

**Theorem 3.4 (Banach Contraction Principle).** *Let $T$ be an operator defined on the Banach space $X$ with values in $X$. Assume that $T$ is a contraction, i.e., that a number $\alpha$, $0 \leq \alpha < 1$, exists such that for all pairs of elements $u, v \in X$ we have*

$$\|Tu - Tv\|_X \leq \alpha \|u - v\|_X .$$

*Then there exists a unique element $u \in X$ such that $Tu = u$.*

*Proof.* Let $u_0$ be an arbitrary element of the space $X$. Then the sequence $u_0, u_1, u_2, \ldots$, where for $n \geq 1$, $u_n$ are defined recursively by $u_n = Tu_{n-1}$, converges in $X$ to the fixed point of the operator $T$. □

**Exercise 3.2.** Provide the details of the proof and show that

$$\|u_n - u\|_X \leq \frac{\alpha^{n-1}}{1-\alpha} \|u_1 - u_0\|_X ,$$

for $n = 1, 2, \ldots$.

**Theorem 3.5 (Brouwer).** *Let $K$ be a nonempty, convex, and compact set in $\mathbb{R}^n$. If $T : K \to K$ is a continuous mapping, then it has at least one fixed point, that is, there exists $u_0 \in K$ such that $T(u_0) = u_0$.*

*Proof.* For a proof see, e.g., [118, 158]. □

**Theorem 3.6 (Schauder).** *Let $K$ be a closed, nonempty, bounded, and convex set in a Banach space $X$. Let $T$ be a completely continuous (that is, compact and continuous) operator defined on $K$ such that*

$$T(K) \subset K .$$

*Then there exists at least one element $u_0 \in K$ such that $T(u_0) = u_0$.*

*Remark 3.3.* If $X = \mathbb{R}^n$, then the Schauder theorem reduces to the Brouwer theorem.

*Proof.* The closure of the set $T(K)$ is compact. For each $n \in \mathbb{N}$ there exists a sequence $x_1, x_2, \ldots, x_{r(n)}$ in $T(K)$ such that

$$\min_{1 \leq i \leq r(n)} \|x - x_i\|_X < \varepsilon = \frac{1}{n} \quad \text{for each} \quad x \in T(K) .$$

Let $X_n$ be the linear hull of the set $\{x_1, \ldots, x_{r(n)}\}$, that is, the set of all points of the form $\lambda_1 x_1 + \ldots + \lambda_{r(n)} x_{r(n)}$, where $\lambda_1, \ldots, \lambda_{r(n)}$ are arbitrary real numbers. We will write $X_n = \operatorname{span}\{x_1, \ldots, x_{r(n)}\}$. Set $K_n = K \cap X_n$. The set $K_n$ is convex, bounded, closed, nonempty, and lies in a finite dimensional subspace of $X$. We define the map $T_n : K \to K_n$ by

$$T_n(x) = F_{\frac{1}{n}} T(x) ,$$

where

$$F_\varepsilon(x) = \frac{\sum_{i=1}^{r(n)} m_i(x) x_i}{\sum_{i=1}^{r(n)} m_i(x)} ,$$

$$m_i(x) = \begin{cases} \varepsilon - \|x - x_i\|_X & \text{when} \quad \|x - x_i\|_X < \varepsilon , \\ 0 & \text{when} \quad \|x - x_i\|_X \geq \varepsilon . \end{cases}$$

We have $F_\varepsilon(K) \subset K_n$. The map $T_n$ is continuous, as the functions $m_i$ are continuous. For an arbitrary $x \in K$ we have

$$\|Tx - T_n x\|_X = \left\| \frac{\sum m_i(Tx) Tx - \sum m_i(Tx) x_i}{\sum m_i(Tx)} \right\|_X$$

$$\leq \frac{\sum m_i(Tx) \|Tx - x_i\|_X}{\sum m_i(Tx)} < \frac{1}{n} . \tag{3.3}$$

Thus, we have approximated in a uniform way the compact operator $T$ by a finite dimensional operator $T_n$.

Since $T_n : K_n \to K_n$ satisfies the hypothesis of the Brouwer fixed point theorem, there exist $\bar{x}_n$ such that $T_n \bar{x}_n = \bar{x}_n$. As $T$ is compact, the sequence $\{\bar{x}_n\}$ has a convergent subsequence $\bar{x}_{n_k} \to \bar{x}$. From the continuity of $T$ it follows that $T\bar{x}_{n_k} \to T\bar{x}$. From (3.3) we conclude that $T\bar{x}_{n_k} - T_{n_k}\bar{x}_{n_k} \to 0$. Thus $T\bar{x} = \bar{x}$, and $T$ has a fixed point in $K$. □

**Exercise 3.3.** Prove, by giving simple examples, that all assumptions in the Schauder theorem are necessary.

**Theorem 3.7 (Leray–Schauder).** *Let $T$ be a completely continuous mapping of a Banach space $X$ into itself, and suppose that there exists a constant $M$ such that*

$$\|x\|_X < M \tag{3.4}$$

*for all $x \in X$ and $\sigma \in [0, 1]$ satisfying $x = \sigma Tx$. Then $T$ has a fixed point.*

*Proof (cf. [110]).* We may assume without loss of generality that $M = 1$. Let us define the map $T^*$ by

$$T^*x = \begin{cases} Tx & \text{if} \quad \|Tx\|_X \le 1\,, \\ \dfrac{Tx}{\|Tx\|_X} & \text{if} \quad \|Tx\|_X > 1\,. \end{cases}$$

Then, $T^*$ is a continuous mapping of the closed unit ball $B(1) = \{x \in X : \|x\|_X \le 1\}$ into itself. Since the set $TB(1)$ is precompact, the same is true for $T^*B(1)$.

From the Schauder fixed point theorem it follows that $T^*$ has a fixed point $x$. We shall show that $x$ is a fixed point of $T$. In fact, if $\|Tx\|_X > 1$, then $x = T^*x = \sigma Tx$ with $\sigma = \frac{1}{\|Tx\|_X}$ and $\|x\|_X = \|T^*x\|_X = 1$, which contradicts (3.4). □

In Chaps. 5 and 15 we will use the following Kakutani–Fan–Glicksberg fixed point theorem (see [3], Corollary 17.55) to prove the existence of weak solutions.

**Theorem 3.8 (Kakutani–Fan–Glicksberg).** *Let $S \subset X$ be nonempty, compact, and convex set, where $X$ is a locally convex Hausdorff topological vector space and let the multifunction $\varphi : S \to 2^S$ have nonempty convex values and closed graph. Then the set of fixed point of $\varphi$ (i.e., $\{x \in S : x \in \varphi(x)\}$) is nonempty and compact.*

## 3.2 Sobolev Spaces and Distributions

We assume that $\Omega$ is a nonempty open subset of $\mathbb{R}^n$, $n$ a positive integer.

**Definition 3.1.** By $L^p(\Omega)$, $1 \le p < \infty$, we denote a linear space of real-valued functions $f$ defined on $\Omega$ such that $|f|^p$ is integrable on $\Omega$ with respect to Lebesgue measure. The functional

$$f \mapsto \|f\|_{L^p(\Omega)} = \left( \int_\Omega |f|^p \, dx \right)^{1/p}$$

is a norm in $L^p(\Omega)$ (provided that we identify functions that are equal to each other almost everywhere in $\Omega$).

The space $L^p(\Omega)$ is a Banach space. For $p = 2$ it is a Hilbert space with the scalar product

$$(f, g) = \int_\Omega fg \, dx.$$

We have

**Theorem 3.9.** *Let $\{f_n\}$ be a Cauchy sequence in $L^p(\Omega)$, that is,*

$$\lim_{m,n\to\infty} \|f_m - f_n\|_{L^p(\Omega)} = 0.$$

*Then there exists $f \in L^p(\Omega)$ and a subsequence $\{f_\nu\}$ of the sequence $\{f_n\}$ such that*

$$\lim_{\nu\to\infty} \|f_\nu - f\|_{L^p(\Omega)} = 0$$

*(completeness of $L^p(\Omega)$), and $f_\nu(x) \to f(x)$ for almost all $x \in \Omega$. Moreover there exists $h \in L^p(\Omega)$ with $h(x) \geq 0$ a.e. in $\Omega$ such that $|f_\nu(x)| \leq h(x)$ for a.e. $x \in \Omega$.*

**Definition 3.2.** By $L^\infty(\Omega)$ we denote the set of measurable real functions $f$ on $\Omega$ for which

$$\operatorname{ess\,sup}\{|f(x)| \,:\, x \in \Omega\} \equiv \inf\{k > 0 \,:\, \mu(\{x \in \Omega \,:\, |f(x)| > k\}) = 0\} < \infty.$$

The linear space $L^\infty(\Omega)$ with norm

$$\|f\|_{L^\infty(\Omega)} = \operatorname{ess\,sup}\{|f(x)| \,:\, x \in \Omega\}$$

is a Banach space.

We say that a real-valued function $f$ on $\Omega$ is *locally integrable* on $\Omega$, and write $f \in L^1_{loc}(\Omega)$, if for every compact $K \subset \Omega$, $f$ is integrable on $K$.

Similarly, we define spaces $L^p_{loc}(\Omega)$, for $1 < p < \infty$.

For $f$ defined on $\Omega$ we define the *support* of $f$ as the closure of the set $\{x \in \Omega \,:\, f(x) \neq 0\}$, and denote it by $\operatorname{supp} f$.

Let $\alpha = (\alpha_1, \ldots, \alpha_n)$ be a multi-index the coordinates of which are nonnegative integers. We write $|\alpha| = \alpha_1 + \cdots + \alpha_n$, and

$$D^\alpha = \frac{\partial^{|\alpha|}}{\partial^{\alpha_1} x_1 \cdots \partial^{\alpha_n} x_n}.$$

We say that $f \in C^k(\Omega)$ (resp. $f \in C^k(\overline{\Omega})$) if there exist partial derivatives $D^\alpha f$ for all $\alpha$ with $|\alpha| \leq k$ that are continuous in $\Omega$ (resp. $\overline{\Omega}$).

Moreover,

$$C^\infty(\Omega) = \bigcap_{k=1}^{\infty} C^k(\Omega).$$

By $C^\infty_0(\Omega)$ we denote the set of $f \in C^\infty(\Omega)$ for which $\operatorname{supp} f \subset \Omega$. If $\Omega$ is bounded, $\operatorname{supp} f$ is a compact subset of $\Omega$.

**Lemma 3.1 (cf. [2, 142]).** *The set $C^\infty_0(\Omega)$ is dense in $L^p(\Omega)$ for $1 \leq p < \infty$.*

**Lemma 3.2 (Du Bois–Reymond, cf. [2, 182]).** *Let $f \in L^1_{loc}(\Omega)$, and let*

$$\int_\Omega f\varphi \, dx = 0$$

*for each $\varphi \in C^\infty_0(\Omega)$. Then $f = 0$ almost everywhere in $\Omega$.*

**Definition 3.3 (Generalized Derivative).** Let $\alpha = (\alpha_1, \dots, \alpha_n)$ be a multi-index the coordinates of which are nonnegative integers. We call a function $f^\alpha \in L^1_{loc}(\Omega)$ the $\alpha$th generalized (or weak) derivative in $\Omega$ of $f \in L^1_{loc}(\Omega)$ if for each $\varphi \in C_0^\infty(\Omega)$

$$\int_\Omega f D^\alpha \varphi \, dx = (-1)^{|\alpha|} \int_\Omega f^\alpha \varphi \, dx.$$

**Exercise 3.4.** Using Lemma 3.2 show that if $f \in C^{|\alpha|}(\Omega)$, then $f^\alpha = D^\alpha f$; that is, there exists the generalized derivative $f^\alpha$, and it equals the usual classical partial derivative $D^\alpha f$.

**Exercise 3.5.** Prove that generalized partial derivatives are uniquely determined (provided that they exist).

**Exercise 3.6.** Let $\Omega$ be the interval $(0, 1)$ on the real line, and let $f(x) = |x|$. Prove that $f$ has the generalized first derivative $f'$, and $f'(x) = \operatorname{sgn} x$ almost everywhere in $\Omega$.

In the sequel we shall write $D^\alpha f$ instead of $f^\alpha$ for generalized derivatives.

**Definition 3.4 (Sobolev Space $W^{m,p}(\Omega)$).** Let $m$ be a positive integer, $1 \le p < \infty$. By $W^{m,p}(\Omega)$ we denote a linear subspace of elements $f$ in $L^p(\Omega)$ for which generalized partial derivatives $D^\alpha f$ exist for all $|\alpha| \le m$ and belong to $L^p(\Omega)$, with norm

$$\|f\|_{W^{m,p}(\Omega)} = \left( \sum_{|\alpha| \le m} \|D^\alpha f\|^p_{L^p(\Omega)} \right)^{1/p}. \tag{3.5}$$

**Lemma 3.3 (cf. [2, 142]).** *The space $W^{m,p}(\Omega)$ is a Banach space. For $1 < p < \infty$ $W^{m,p}(\Omega)$ is reflexive.*

**Exercise 3.7.** Prove that the space $W^{m,p}(\Omega)$ is complete.

*Hint.* Use Theorem 3.9, Lemma 3.2, and Definition 3.4.

**Lemma 3.4.** *For $1 < p < \infty$ every bounded sequence in $W^{m,p}(\Omega)$ contains a weakly convergent subsequence.*

*Proof.* The space $W^{m,p}(\Omega)$ is reflexive. □

**Definition 3.5.** By $W_0^{m,p}(\Omega)$ we denote the closure of the set $C_0^\infty(\Omega)$ in the norm of $W^{m,p}(\Omega)$.

The spaces $W^{m,p}(\Omega)$ are also denoted by $W_p^m(\Omega)$. The spaces $W^{m,2}(\Omega)$ and $W_0^{m,2}(\Omega)$, denoted also by $H^m(\Omega)$ and $H_0^m(\Omega)$, respectively, are Hilbert spaces with scalar product

$$(f, g) = \sum_{|\alpha| \le m} \int_\Omega D^\alpha f D^\alpha g \, dx.$$

**Definition 3.6.** Let $m$ be a positive integer, $1 < p < \infty$, and $1/p + 1/q = 1$. By $W^{-m,q}(\Omega)$ we denote the space of linear continuous functionals on the space $W_0^{m,p}(\Omega)$.

The spaces $W^{-m,2}(\Omega)$ are also denoted by $H^{-m}(\Omega)$.

Now we shall define distributions and distributional derivatives.

Let us introduce in the set $C_0^\infty(\Omega)$ the following notion of convergence.

**Definition 3.7.** We say that a sequence $\{\varphi_n\} \subset C_0^\infty(\Omega)$ converges to zero if there exists a compact $K \subset \Omega$ such that

(i) $\operatorname{supp} \varphi_n \subset K$ for each $\varphi_n$,
(ii) $\lim_{n\to\infty} D^\alpha \varphi_n = 0$ for each multi-index $\alpha$, uniformly on $\Omega$.

We denote by $\mathscr{D}(\Omega)$ the set $C_0^\infty(\Omega)$ together with the convergence introduced above.

**Definition 3.8.** We call the map $T : \mathscr{D}(\Omega) \to \mathbb{R}$ a distribution on $\Omega$ if it is linear, that is,

$$T(\alpha\varphi + \beta\psi) = \alpha T(\varphi) + \beta T(\psi)$$

for all $\alpha, \beta \in \mathbb{R}$ and $\varphi, \psi \in \mathscr{D}(\Omega)$, and if

$$\lim_{n\to\infty} T(\varphi_n) = 0$$

for each sequence $\{\varphi_n\} \subset \mathscr{D}(\Omega)$ that converges to zero in $\mathscr{D}(\Omega)$.

We denote by $\mathscr{D}'(\Omega)$ the set of all distributions on $\Omega$ and write $\langle T, \varphi\rangle$ instead $T(\varphi)$. For $f \in L^1_{loc}(\Omega)$, the map on $\mathscr{D}(\Omega)$ defined by

$$\langle T_f, \varphi\rangle = \int_\Omega f\varphi\, dx$$

is a distribution. By Lemma 3.2 we can identify it with $f$, and we write

$$\langle f, \varphi\rangle = \int_\Omega f\varphi\, dx\,.$$

Distributions that can be represented by locally integrable functions we call *regular*.

An example of a distribution that is not regular is the map

$$\delta(x_0) : \mathscr{D}(\Omega) \ni \varphi \to \delta_{x_0}(\varphi) = \varphi(x_0) \in \mathbb{R},$$

for some $x_0 \in \Omega$, called the *Dirac delta*.

**Definition 3.9 (Distributional Derivative).** Let $T \in \mathscr{D}'(\Omega)$. Then the map

$$\frac{\partial T}{\partial x_i} : \mathscr{D}(\Omega) \to \mathbb{R},$$

$i \in \{1, \dots, n\}$, defined by

$$\left\langle \frac{\partial T}{\partial x_i}, \varphi \right\rangle = -\left\langle T, \frac{\partial \varphi}{\partial x_i} \right\rangle$$

for each $\varphi \in \mathscr{D}(\Omega)$, is a distribution. Similarly, for an arbitrary multi-index $\alpha$ we define $D^\alpha T \in \mathscr{D}'(\Omega)$ by

$$\langle D^\alpha T, \varphi \rangle = (-1)^{|\alpha|} \langle T, D^\alpha \varphi \rangle$$

for each $\varphi \in \mathscr{D}(\Omega)$.

For $T \in \mathscr{D}'(\Omega)$, we call $D^\alpha T$ the $\alpha$th distributional derivative of $T$.

**Exercise 3.8.** Prove that each distribution has all distributional derivatives of any order.

**Exercise 3.9.** Show that $f''(x) = 2\delta(x)$ for $f$ from Exercise 3.6 above. One can show that there is no locally integrable function $g$ on $\Omega$ such that $f'' = g$.

**Definition 3.10.** We say that an open and bounded subset $\Omega \subset \mathbb{R}^n$ belongs to the class $C^{k,\alpha}$, $0 \leq \alpha \leq 1$, $k$ a nonnegative integer, if for every point $x_0 \in \partial\Omega$ there exists a ball $B$ centered at $x_0$ and a one-to-one mapping $\psi$ of $B$ on $D \subset \mathbb{R}^n$ such that

(i) $\psi(B \cap \Omega) \subset \mathbb{R}^n_+ = \{x = (x_1, \dots, x_n) \in \mathbb{R}^n : x_n > 0\}$,
(ii) $\psi(B \cap \partial\Omega) \subset \partial\mathbb{R}^n_+$,
(iii) $\psi \in C^{k,\alpha}(B)$, $\psi^{-1} \in C^{k,\alpha}(D)$.

If the boundary of $\Omega$ is smooth enough, say $\Omega \in C^{0,1}$, then the Sobolev spaces $W^{m,p}(\Omega)$ defined as in Definition 3.4 above can be described alternatively as the closure of the set $C^m(\overline{\Omega})$ in the norm (3.5), cf. [142]. From this new definition of $W^{m,p}(\Omega)$ it is clear that the set $C^m(\overline{\Omega})$ is dense in $W^{m,p}(\Omega)$ (provided that the boundary of $\Omega$ is smooth enough).

Still another description, equivalent to Definition 3.4, of Sobolev spaces can be given in terms of distributions. We define the space $W^{m,p}(\Omega)$ as the linear subspace of those $f \in L^p(\Omega)$ for which all distributional derivatives $D^\alpha f$, $|\alpha| \leq m$, belong to $L^p(\Omega)$, with norm (3.5).

## 3.3 Some Embedding Theorems and Inequalities

We begin with three general results.

**Theorem 3.10 ([110, Theorem 7.26], [185]).** *Let $\Omega$ be an open and bounded subset of $\mathbb{R}^n$ with Lipschitz boundary, that is, $\Omega \in C^{0,1}$. Then*

(i) *if* $kp < n$, *then the space* $W^{k,p}(\Omega)$ *is continuously embedded in the space* $L^{p^*}(\Omega)$, $p^* = np/(n-kp)$, *and compactly embedded in* $L^q(\Omega)$ *for* $q < p^*$;
(ii) *if* $0 \le m < k - n/p < m+1$, *then the space* $W^{k,p}(\Omega)$ *is continuously embedded in* $C^{m,\alpha}(\overline{\Omega})$, $\alpha = k - n/p - m$, *and compactly embedded in* $C^{m,\beta}(\overline{\Omega})$, *for* $\beta < \alpha$.

**Lemma 3.5 (A Version of the Rellich Theorem, cf. [182]).** *Let* $\Omega$ *be an open and bounded subset of* $\mathbb{R}^n$. *Then the embedding of* $H_0^1(\Omega)$ *in* $L^2(\Omega)$ *is compact.*

**Theorem 3.11 ([105, Theorem 10.1], [235, p. 1034]).** *Let* $\Omega$ *be a bounded domain in* $\mathbb{R}^n$ *with Lipschitz boundary, and let* $u$ *be any function in* $W^{m,r}(\Omega) \cap L^q(\Omega)$, $1 \le r, q \le \infty$. *For any multi-index* $j$, $0 \le |j| < m$, *and for any number* $\theta$ *from the interval* $|j|/m \le \theta \le 1$, *set*

$$\frac{1}{p} = \frac{|j|}{m} + \theta\left(\frac{1}{r} - \frac{m}{n}\right) + (1-\theta)\frac{1}{q}.$$

*If* $m - |j| - n/r$ *is not a nonnegative integer, then*

$$\|D^j u\|_{L^p(\Omega)} \le C\|u\|^{\theta}_{W^{m,r}(\Omega)}\|u\|^{1-\theta}_{L^q(\Omega)}. \tag{3.6}$$

*If* $m - |j| - n/r$ *is a nonnegative integer, then inequality (3.6) holds for* $\theta = |j|/m$. *The constant* $C$ *depends only on* $\Omega$, $r$, $q$, $m$, $j$, $\theta$.

Now we shall state some special cases of Theorems 3.10 and 3.11, useful in the following chapters.

Let $\Omega \subset \mathbb{R}^3$ be as in Theorem 3.10. Then

1. $H^1(\Omega)$ is continuously embedded in $L^6(\Omega)$,

$$\|u\|_{L^6(\Omega)} \le C\|u\|_{H^1(\Omega)}$$

   (cf. Lemma 3.9 below).
2. $H^1(\Omega)$ is compactly embedded in $L^4(\Omega)$.
3. $H^2(\Omega)$ is continuously embedded in $C^{0,1/2}(\overline{\Omega})$, and $C^{0,1/2}(\overline{\Omega})$ is compactly embedded in $C(\overline{\Omega})$; in particular, we have

$$\operatorname*{ess\,sup}_{x\in\Omega} |u(x)| \le C\|u\|_{H^2(\Omega)}.$$

**Exercise 3.10.** Show that $W^{2,3/2}(\Omega)$ is continuously embedded in $W^{1,3}(\Omega)$ for $\Omega \subset \mathbb{R}^3$ as in Theorem 3.10.

Let $\Omega$ be a bounded domain in $\mathbb{R}^3$ with Lipschitz boundary. Then

$$\|u\|_{L^4(\Omega)} \le C\|u\|^{3/4}_{H^1(\Omega)}\|u\|^{1/4}_{L^2(\Omega)}$$

[cf. inequality (3.10)].

**Lemma 3.6 (Poincaré Inequality).** *Let $\Omega$ be an open and bounded subset of $\mathbb{R}^n$, $n$ a positive integer, $d = \operatorname{diam}\Omega = \sup\{|x-y| : x, y \in \Omega\}$. Then, for each $u \in H_0^1(\Omega)$ and $i \in \{1, \dots, n\}$,*

$$\|u\|_{L^2(\Omega)} \leq \frac{d}{\sqrt{2}} \left\| \frac{\partial u}{\partial x_i} \right\|_{L^2(\Omega)} . \tag{3.7}$$

*Proof.* Let $u \in C_0^\infty(\Omega)$, and fix $i = 1$. We have

$$u(x_1, \dots, x_n) = \int_{-\infty}^{x_1} \frac{\partial u}{\partial t}(t, x_2, \dots, x_n) dt.$$

By the Schwarz inequality,

$$|u(x_1, \dots, x_n)| \leq \int_{x_1^*}^{x_1} \left| \frac{\partial u}{\partial t}(t, x_2, \dots, x_n) \right|^2 dt\, (x_1 - x_1^*)$$

$$\leq \int_{-\infty}^{+\infty} \left| \frac{\partial u}{\partial x_1}(x) \right|^2 dx_1\, (x_1 - x_1^*) ,$$

where $x_1^* = \inf\{t : u(t, x_2, \dots, x_n) = 0,\ (t, x_2, \dots, x_n) \in \Omega\}$. Integrating with respect to $x_1$ we obtain

$$\int_{-\infty}^{+\infty} |u(x)|^2 dx_1 \leq \frac{d^2}{2} \int_{-\infty}^{+\infty} \left| \frac{\partial u}{\partial x_1}(x) \right|^2 dx_1 .$$

Now, integrating with respect to $x_2, \dots, x_n$ we obtain inequality (3.7) for $i = 1$ and $u \in C_0^\infty(\Omega)$. Let $u \in H_0^1(\Omega)$. There exists a sequence $\{u_n\} \subset C_0^\infty(\Omega)$ converging to $u$ in $H_0^1(\Omega)$. We prove (3.7) for $u$, passing to the limit in

$$\|u_n\|_{L^2(\Omega)} \leq \frac{d}{\sqrt{2}} \left\| \frac{\partial u_n}{\partial x_i} \right\|_{L^2(\Omega)} ,$$

for $i \in \{1, \dots, n\}$. □

**Corollary 3.1.** *Let $\Omega \subset \mathbb{R}^n$ be an open and bounded set. Then the following two norms are equivalent in $H_0^1(\Omega)$:*

$$\|u\| = \left( \|u\|_{L^2(\Omega)}^2 + \sum_{|\alpha|=1} \|D^\alpha u\|_{L^2(\Omega)}^2 \right)^{1/2} ,$$

*and*

$$|||u||| = \left( \sum_{|\alpha|=1} \|D^\alpha u\|_{L^2(\Omega)}^2 \right)^{1/2} .$$

**Lemma 3.7 (Ladyzhenskaya Inequality in Two Dimensions).** *Let $\Omega \subset \mathbb{R}^2$ be an open and bounded set. Then for all $u \in H_0^1(\Omega)$*

$$\|u\|_{L^4(\Omega)} \leq 2^{1/4}\|u\|_{L^2(\Omega)}^{1/2}\|Du\|_{L^2(\Omega)}^{1/2}. \tag{3.8}$$

*Proof (cf. [146]).* We shall prove inequality (3.8) for smooth functions with compact support in $\Omega$, and the general case will follow immediately from the density of these functions in $H_0^1(\Omega)$. Let $u \in C_0^\infty(\Omega)$. For convenience we consider $u$ as defined on the whole space $\mathbb{R}^2$ and equal zero outside of $\Omega$.

We have

$$u^2(x_1, x_2) = 2\int_{-\infty}^{x_k} uu_{x_k}\, dx_k \quad \text{for} \quad k = 1, 2,$$

so that

$$\max_{x_k} u^2(x_1, x_2) \leq 2\int_{-\infty}^{\infty} |uu_{x_k}|\, dx_k \quad \text{for} \quad k = 1, 2, \tag{3.9}$$

and by the Schwarz inequality,

$$\begin{aligned}\int_{\mathbb{R}^2} u^4 dx &\leq \int_{-\infty}^{+\infty} \max_{x_2} u^2 dx_1 \int_{-\infty}^{+\infty} \max_{x_1} u^2 dx_2\\ &\leq 4\int_{\mathbb{R}^2} |uu_{x_2}| dx \int_{\mathbb{R}^2} |uu_{x_1}| dx\\ &\leq 2\int_{\mathbb{R}^2} |u|^2 dx \int_{\mathbb{R}^2} |Du|^2 dx,\end{aligned}$$

which proves the lemma for smooth functions with compact support in $\Omega$. □

**Lemma 3.8 (Ladyzhenskaya Inequality in Three Dimensions).** *Let $\Omega \subset \mathbb{R}^3$ be an open and bounded set. Then for all $u \in H_0^1(\Omega)$*

$$\|u\|_{L^4(\Omega)} \leq \sqrt{2}\|u\|_{L^2(\Omega)}^{1/4}\|Du\|_{L^2(\Omega)}^{3/4}. \tag{3.10}$$

*Proof (cf. [146]).* As in the above lemma, it suffices to prove inequality (3.10) for smooth functions with compact support. Let $u \in C_0^\infty(\Omega)$. We have, by (3.8) and (3.9),

$$\begin{aligned}\int_{\mathbb{R}^3} u^4 dx &\leq 2\int_{-\infty}^{+\infty} \left(\int_{\mathbb{R}^2} u^2 dx_1 dx_2 \int_{\mathbb{R}^2} (u_{x_1}^2 + u_{x_2}^2) dx_1 dx_2\right) dx_3\\ &\leq 2\max_{x_3} \int_{\mathbb{R}^2} u^2 dx_1 dx_2 \int_{\mathbb{R}^3} |Du|^2 dx\end{aligned}$$

$$\leq 4 \int_{\mathbb{R}^3} |uu_{x_3}| dx \int_{\mathbb{R}^3} |Du|^2 dx$$

$$\leq 4 \left( \int_{\mathbb{R}^3} |u|^2 dx \right)^{1/2} \left( \int_{\mathbb{R}^3} |Du|^2 dx \right)^{3/2},$$

which gives inequality (3.10). □

**Lemma 3.9 (cf. [146]).** *Let $\Omega \subset \mathbb{R}^3$ be an open and bounded set. Then for all $u \in H_0^1(\Omega)$,*

$$\|u\|_{L^6(\Omega)} \leq 48^{1/6} \|Du\|_{L^2(\Omega)}. \tag{3.11}$$

*Proof.* Without loss of generality we can assume that $u \in C_0^\infty(\Omega)$ and $u \geq 0$. We have

$$I \equiv \int_{\mathbb{R}^3} u^6 dx = \int_{-\infty}^{+\infty} \left( \int_{\mathbb{R}^2} u^3 u^3 dx_2 dx_3 \right) dx_1$$

$$\leq \int_{-\infty}^{+\infty} \left( \int_{-\infty}^{+\infty} \max_{x_2} u^3 dx_3 \int_{-\infty}^{+\infty} \max_{x_3} u^3 dx_2 \right) dx_1$$

$$\leq 9 \int_{-\infty}^{+\infty} \left( \int_{\mathbb{R}^2} |u^2 u_{x_2}| dx_2 dx_3 \int_{\mathbb{R}^2} |u^2 u_{x_3}| dx_2 dx_3 \right) dx_1$$

$$\leq 9 \int_{\mathbb{R}^1} \left\{ \int_{\mathbb{R}^2} u^4 dx_2 dx_3 \left( \int_{\mathbb{R}^2} u_{x_2}^2 dx_2 dx_3 \right)^{1/2} \left( \int_{\mathbb{R}^2} u_{x_3}^2 dx_2 dx_3 \right)^{1/2} \right\} dx_1.$$

Now we use the Schwarz inequality and proceed as follows:

$$I \leq 9 \max_{x_1} \int_{\mathbb{R}^2} u^4 dx_2 dx_3 \left( \int_{\mathbb{R}^3} u_{x_2}^2 dx \right)^{1/2} \left( \int_{\mathbb{R}^3} u_{x_3}^2 dx \right)^{1/2}$$

$$\leq 36 \int_{\mathbb{R}^3} |u^3 u_{x_1}| dx \left( \int_{\mathbb{R}^3} |u_{x_2}^2| dx \right)^{1/2} \left( \int_{\mathbb{R}^3} |u_{x_3}^2| dx \right)^{1/2}$$

$$\leq 36 \sqrt{I} \prod_{i=1}^{3} \left( \int_{\mathbb{R}^3} |u_{x_i}^2| dx \right)^{1/2}.$$

In the end,

$$\sqrt{I} \leq 36 \prod_{i=1}^{3} \|u_{x_i}\|_{L^2(\Omega)} = 36 \left\{ \prod_{i=1}^{3} \|u_{x_i}\|_{L^2(\Omega)}^2 \right\}^{1/2}$$

$$\leq 36 \left\{ 3^{-3} \left( \sum_{i=1}^{3} \|u_{x_i}\|_{L^2(\Omega)}^2 \right)^3 \right\}^{1/2}$$

(as $(a_1 a_2 a_3)^{1/3} \leq (a_1 + a_2 + a_3)/3$ for $a_i \geq 0$), which gives inequality (3.11). □

**Theorem 3.12 (Hardy Inequality, cf. [112, 142]).** *Let $a, b \in \mathbb{R}$, $a < b$, $u \in L^p(a, b)$ with $p > 1$. Then*

$$\int_a^b \left( \frac{1}{x-a} \int_a^x |u(s)| ds \right)^p dx \leq \left( \frac{p}{p-1} \right)^p \int_a^b |u(x)|^p dx, \tag{3.12}$$

*and*

$$\int_a^b \left( \frac{1}{b-x} \int_x^b |u(s)| ds \right)^p dx \leq \left( \frac{p}{p-1} \right)^p \int_a^b |u(x)|^p dx. \tag{3.13}$$

*Proof.* Let $u_n(x) = 0$ for $x \in \left(a, a + \frac{1}{n}\right)$ and $u_n(x) = u(x)$ for $x \in \left(a + \frac{1}{n}, b\right)$. We prove the first inequality. Integration by parts and Hölder inequality give

$$\int_a^b \left( \frac{1}{x-a} \int_a^x |u_n(s)| ds \right)^p dx = \frac{1}{(b-a)^{p-1}(1-p)} \left( \int_a^b |u_n(s)| ds \right)^p$$
$$+ \frac{p}{p-1} \int_a^b \frac{1}{(x-a)^{p-1}} \left( \int_a^x |u_n(s)| ds \right)^{(p-1)} |u_n(x)| dx$$
$$\leq \frac{p}{p-1} \int_a^b \frac{1}{(x-a)^{p-1}} \left( \int_a^x |u(s)| ds \right)^{p-1} |u_n(x)| dx$$
$$\leq \frac{p}{p-1} \left( \int_a^b \frac{1}{(x-a)^p} \left( \int_a^x |u_n(s)| ds \right)^p dx \right)^{(p-1)/p} \left( \int_a^b |u_n(x)|^p dx \right)^{1/p}.$$

Thus,

$$\int_a^b \left( \frac{1}{x-a} \int_a^x |u_n(s)| ds \right)^p dx \leq \left( \frac{p}{p-1} \right)^p \int_a^b |u(x)|^p dx \tag{3.14}$$

for $n \in \mathbb{N}$. Applying the Fatou lemma we obtain (3.12). In a similar way we obtain the second inequality.

**Exercise 3.11.** Deduce inequality (3.12) from inequality (3.14) and prove inequality (3.13).

**Exercise 3.12.** Prove that for $v \in W^{1,p}(a,b)$ with $v(a) = 0, p > 1$, the following inequality holds:

$$\int_a^b \left|\frac{v(x)}{x-a}\right|^p dx \leq \left(\frac{p}{p-1}\right)^p \int_a^b |v'(x)|^p dx.$$

## 3.4 Sobolev Spaces of Periodic Functions

We consider Sobolev spaces $H_p^s(Q)$ of $L$-periodic functions on the $m$-dimensional domain $Q = [0, L]^m$. Let $C_p^\infty(Q)$ be the space of restrictions to $Q$ of infinitely differentiable functions which are $L$-periodic in each direction, that is, $u(x + Le_j) = u(x)$, $j = 1, \dots, m$ (by $e_j$ we denote the vectors of the canonical basis, $e_j = (0, \dots, 1, \dots, 0)$, where 1 is on the $j$th coordinate).

**Definition 3.11.** For an arbitrary nonnegative integer $s$ we define the Sobolev space $H_p^s(Q)$ as the completion of $C_p^\infty(Q)$ in the norm

$$\|u\|_{H_p^s(Q)} = \left\{ \sum_{0 \leq |\alpha| \leq s} \|D^\alpha u\|_{L^2(Q)}^2 \right\}^{1/2}.$$

We denote, $\alpha = (\alpha_1, \dots, \alpha_m)$ for nonnegative integers $\alpha_1, \dots, \alpha_m$, $|\alpha| = \alpha_1 + \dots + \alpha_m$, $\alpha! = \alpha_1!\alpha_2! \cdots \alpha_m!$. For $x = (x_1, \dots, x_m)$, $x^\alpha = x_1^{\alpha_1} x_2^{\alpha_2} \cdots x_m^{\alpha_m}$, $D^\alpha = \frac{\partial^{|\alpha|}}{\partial x^\alpha}$.

In this notation the Taylor series in $\mathbb{R}^m$ can be written just as

$$f(x) = \sum_{|\alpha| \geq 0} \frac{1}{\alpha!} D^\alpha f(x_0)(x - x_0)^\alpha.$$

**Characterization of Sobolev Spaces $H_p^s(Q)$ by Using the Fourier Series** Each function $u \in C_p^\infty(Q)$ has a representation

$$u(x) = \sum_{k \in \mathbb{Z}^m} c_k e^{2\pi i k \frac{x}{L}},$$

where $c_k$ are complex numbers. We consider real functions, for which $c_{-k} = \overline{c_k}$. It is known that for functions from $C_p^\infty(Q)$ the above series converges uniformly.

We have

$$D^\alpha u(x) = \left(\frac{2\pi i}{L}\right)^{|\alpha|} \sum_{k \in \mathbb{Z}^m} c_k k^\alpha e^{2\pi i k \frac{x}{L}}.$$

From the Parseval identity

$$\|D^\alpha u\|_{L^2(Q)}^2 = L^m \left(\frac{2\pi}{L}\right)^{2|\alpha|} \sum_{k\in\mathbb{Z}^m} |c_k|^2 k^{2\alpha}.$$

We introduce a new norm in $H_p^s(Q)$ by

$$\|u\|_{H_f^s(Q)} = \left\{ \sum_{k\in\mathbb{Z}^m} \left(1 + |k|^{2s}\right) |c_k|^2 \right\}^{1/2}.$$

**Lemma 3.10.** *The norms $\|\cdot\|_{H_p^s}$ and $\|\cdot\|_{H_f^s(Q)}$ are equivalent, that is*

$$c_1 \|u\|_{H_f^s(Q)} \le \|u\|_{H_p^s(Q)} \le c_2 \|u\|_{H_f^s(Q)}$$

*for all $u \in H_p^s(Q)$ and some positive constants $c_1, c_2$.*

*Proof.* For every integer $s$ there exist positive constants $C_1$, $C_2$ such that for all $k \in \mathbb{R}^m$

$$C_1 \le \frac{\sum_{0\le|\alpha|\le s} k^{2\alpha}}{1 + |k|^{2s}} \le C_2. \tag{3.15}$$

Now,

$$\|u\|_{H_p^s(Q)}^2 = \sum_{0\le|\alpha|\le s} \|D^\alpha u\|_{L^2(Q)}^2 = L^m \sum_{0\le|\alpha|\le s} \left(\frac{2\pi}{L}\right)^{2|\alpha|} \left(\sum_{k\in\mathbb{Z}^m} |c_k|^2 k^{2\alpha}\right)$$

$$\le c \sum_{k\in\mathbb{Z}^m} \left(\sum_{0\le|\alpha|\le s} |k|^{2|\alpha|}\right) |c_k|^2 \le c C_2 \sum_{k\in\mathbb{Z}^m} \left(1 + |k|^{2s}\right) |c_k|^2 = c C_2 \|u\|_{H_f^s(Q)}^2.$$

Setting $c_2 = cC_2$ we have the second inequality of the lemma. Using once again (3.15) we obtain the first inequality of the lemma. □

We have thus

**Lemma 3.11.** *The space $H_p^s(Q)$ coincides with*

$$\left\{ u \in L^2(Q) \ : \ u(x) = \sum_{k\in\mathbb{Z}^m} c_k e^{2\pi i k \frac{x}{L}}, \quad \overline{c_k} = c_{-k}, \sum_{k\in\mathbb{Z}^m} |k|^{2s} |c_k|^2 < \infty \right\}.$$

**Lemma 3.12.** *In the space*

$$\dot{H}_p^s(Q) = \left\{ u \in H_p^s(Q) \ : \ \int_Q u(x)dx = 0 \right\}$$

*(of these $u \in H_p^s(Q)$ for which $c_0 = 0$ in their series representation) the expression*

$$\|u\|_{\dot{H}_p^s(Q)} = \left( \sum_{k \in \mathbb{Z}^m} |k|^{2s} |c_k|^2 \right)^{1/2}$$

*is a norm equivalent to the norm $\| \cdot \|_{H_p^s(Q)}$.*

*Proof.* The proof follows from the *Poincaré inequality*

$$\|u\|_{L^2(Q)} \leq \frac{L}{2\pi} \|Du\|_{L^2(Q)}, \tag{3.16}$$

which holds for all $u$ in $\dot{H}_p^1(Q)$. In fact, we have

$$\begin{aligned} \|Du\|_{L^2(Q)}^2 &= \sum_{|\alpha|=1} \|D^\alpha u\|_{L^2(Q)}^2 = \sum_{|\alpha|=1} L^m \left(\frac{2\pi}{L}\right)^2 \sum_{k \in \mathbb{Z}^m \setminus \{0\}} |c_k|^2 |k|^2 \\ &\geq \left(\frac{2\pi}{L}\right)^2 L^m \sum_{k \in \mathbb{Z}^m \setminus \{0\}} |c_k|^2 = \left(\frac{2\pi}{L}\right)^2 \|u\|_{L^2(Q)}, \end{aligned}$$

which ends the proof. □

In the space $\| \cdot \|_{\dot{H}_p^s(Q)}$ we introduce the scalar product

$$(u, v)_{\dot{H}_p^s(Q)} = \sum_{k \in \mathbb{Z}^m} |k|^{2s} c_k \overline{d_k},$$

for $u(x) = \sum_{k \in \mathbb{Z}^m} c_k e^{2\pi i k \frac{x}{L}}$ and $v(x) = \sum_{k \in \mathbb{Z}^m} d_k e^{2\pi i k \frac{x}{L}}$. We have also

$$c_k = \hat{u}(k) = \frac{1}{L^m} \int_Q u(x) e^{-2\pi i k \frac{x}{L}} dx.$$

The *Fourier coefficients* $\hat{u}$ of $u$ are *complex amplitudes* of harmonics $e^{2\pi i k \frac{x}{L}}$ associated to *wave numbers* $\frac{2\pi k}{L}$. The length scale related to the wave number $\frac{2\pi k}{L}$ equals $\frac{L}{|k|}$. Notice that

$$\frac{L}{\sqrt{m} k_{max}} \leq \frac{L}{|k|} = \frac{L}{\sqrt{k_1^2 + \cdots + k_m^2}} \leq \frac{L}{k_{min}}.$$

**Theorem 3.13 (Sobolev Embedding Theorem).** *Let $u \in H_p^s(Q)$, $s > \frac{m}{2}$. Then $u \in C_p^0(Q)$ and*

$$\sup_{x \in Q} |u(x)| = \|u\|_{L^\infty(Q)} \leq C(s) \|u\|_{H_p^s(Q)}.$$

*Proof.* Let $u(x) = \sum_{k\in\mathbb{Z}^m} c_k e^{2\pi i k \frac{x}{L}}$. Then, for $s > \frac{m}{2}$ we have

$$\|u\|_{L^\infty(Q)} \le \sum_{k\in\mathbb{Z}^m} |c_k| \le \sum_{k\in\mathbb{Z}^m} \frac{1}{(1+|k|^{2s})^{1/2}} \left(1+|k|^{2s}\right)^{1/2} |c_k|$$

$$\le \left\{\sum_{k\in\mathbb{Z}^m} \frac{1}{1+|k|^{2s}}\right\}^{1/2} \left\{\sum_{k\in\mathbb{Z}^m} \left(1+|k|^{2s}\right)|c_k|^2\right\}^{1/2} = C(s)\|u\|_{H^s_p(Q)}.$$

Moreover, the absolute convergence of the series of coefficients $\sum_{k\in\mathbb{Z}^m} |c_k|$ implies the uniform convergence of the Fourier series of $u$ and, in consequence, the continuity of $u$. □

**Exercise 3.13.** Prove that for $s > \frac{m}{2}$ the series $\sum_{k\in\mathbb{Z}^m} \frac{1}{1+|k|^{2s}}$ is convergent.

**Exercise 3.14.** Prove that if $s > \frac{m}{2} = j$ and $u \in H^s_p(Q)$ then $u$ is $j$ times continuously differentiable function in $Q$ (we write $u \in C^j(Q)$) and

$$\|u\|_{C^j(Q)} = \sum_{|\alpha|\le j} \sup_{x\in Q} |D^\alpha u(x)| \le C(s)\|u\|_{H^s_p(Q)}.$$

We shall need also the Rellich–Kondrachov theorem on compact embedding.

**Theorem 3.14 (Rellich–Kondrachov).** *The space $H^1_p(Q)$ is compactly embedded in the space $L^2(Q)$ which means that for every bounded in $H^1_p(Q)$ sequence of functions there exists its subsequence that is convergent in $L^2(Q)$.*

*Proof.* Let $\{u_n\}$, $u_n(x) = \sum_{k\in\mathbb{Z}^m} c_{n,k} e^{2\pi i k \frac{x}{L}}$, $n \in \mathbb{N}$ be a bounded sequence in $H^1_p(Q)$, that is,

$$\sum_{k\in\mathbb{Z}^m} \left(1+|k|^2\right)|c_{n,k}|^2 \le M$$

for all $n$ and some positive constant $M$. We have to prove that there exists a subsequence $u_j$ which converges to some $u^\star = \sum_{k\in\mathbb{Z}^m} c^\star_k e^{2\pi i k \frac{x}{L}} \in H^1_p(Q)$ strongly in $L^2(Q)$. It is known that from every bounded sequence in a Hilbert space one can choose a weakly convergent subsequence. Assume that $u_j$ converges weakly to $u^\star$ in $H^1_p(Q)$. Then $u^\star$ satisfies

$$\sum_{k\in\mathbb{Z}^m} \left(1+|k|^2\right)|c^\star_k|^2 \le M.$$

**Exercise 3.15.** Let the sequence $u_j$ in a Hilbert space $H$ converge weakly to $u^\star \in H$, that is, for every $v \in H$, $(u_j - u^\star, v)_H \to 0$ as $j \to \infty$, where $(\cdot,\cdot)_H$ denotes the scalar product in $H$. Prove that the norm of $u^\star$ in $H$ satisfies

$$\|u^\star\|_H \le \liminf_{j\to\infty} \|u_j\|_H.$$

**Exercise 3.16.** Let $u_j$ converge weakly to $u^\star$ in $H_p^1(Q)$. Prove that it is equivalent to the convergence of all Fourier coefficients, namely, $c_{j,k} \to c_k^\star$ as $j \to \infty$, for every $k \in \mathbb{Z}^m$.

We shall prove that the subsequence $u_j$ converges to $u^\star$ in $L^2(Q)$. Observe that

$$\sum_{k\in\mathbb{Z}^m} \left(1+|k|^2\right) |c_{j,k} - c_k^\star|^2 \leq 4M,$$

whence

$$\begin{aligned}\|u_j - u^\star\|^2_{L^2(Q)} = \sum_{k\in\mathbb{Z}^m} |c_{j,k} - c_k^\star|^2 &\leq \sum_{|k|\leq K} |c_{j,k} - c_k^\star|^2 + \frac{1}{K^2} \sum_{|k|>K} |c_{j,k} - c_k^\star|^2 |k|^2 \\ &\leq \sum_{|k|\leq K} |c_{j,k} - c_k^\star|^2 + \frac{4M}{K^2}.\end{aligned}$$

For every $\varepsilon > 0$ we can take $K$ large enough so that $\frac{4M}{K^2} < \frac{\varepsilon}{2}$. Then, in view of Exercise 3.16 we can take $j$ large enough so that the first term on the right-hand side is smaller than $\frac{\varepsilon}{2}$. This proves the convergence $u_j \to u^\star$ in $L^2(Q)$. □

**Exercise 3.17.** Prove that $H^{s+1}(Q)$ is compactly embedded in $H^s(Q)$.

**Laplace Equation in Sobolev Spaces of Periodic Functions** Let $f \in L^2(Q)$. Assume that we look for $u \in H_p^s(Q)$ satisfying equation

$$-\Delta u = f. \tag{3.17}$$

If $u \in H_p^s(Q)$ is a solution of (3.17) then for an arbitrary constant $c$, $u + c$ is also a solution in the same space. To guarantee uniqueness we restrict our attention to solutions satisfying the additional property

$$\int_Q u(x)dx = 0, \tag{3.18}$$

which is equivalent to $u \in \dot{H}_p^s(Q)$. But then we have also

$$\int_Q \Delta u(x)dx = 0. \tag{3.19}$$

**Exercise 3.18.** Prove that if $u \in H_p^s(Q)$ satisfies (3.18) then also (3.19) holds.

It then follows that

$$\int_Q f(x)dx = 0,$$

whence $f$ is in $\dot{L}^2(Q) = \dot{H}_p^0(Q)$. For any $f \in L^2(Q)$ there exists a constant $c$ such that $f + c \in \dot{L}^2(Q)$. In consequence, to have the solution uniqueness, we shall assume that $f \in \dot{L}^2(Q)$. Let

$$f(x) = \sum_{k \in \mathbb{Z}^m} f_k e^{2\pi i k \frac{x}{L}}, \quad f_0 = 0$$

and

$$u(x) = \sum_{k \in \mathbb{Z}^m} c_k e^{2\pi i k \frac{x}{L}}, \quad u_0 = 0. \tag{3.20}$$

We compute

$$-\Delta u(x) = \left(\frac{2\pi}{L}\right)^2 \sum_{k \in \mathbb{Z}^m} |k|^2 u_k e^{2\pi i k \frac{x}{L}} = \sum_{k \in \mathbb{Z}^m} f_k e^{2\pi i k \frac{x}{L}} = f,$$

and comparing the coefficients we obtain

$$u_k = \frac{L^2}{4\pi^2} \frac{f_k}{|k|^2} \quad \text{for} \quad k \neq 0. \tag{3.21}$$

We can prove now the following:

**Lemma 3.13.** *If $f \in \dot{L}^2(Q)$ and $u \in \dot{H}_p^1(Q)$ are a generalized solution of the Laplace equation, that is, $u$ satisfies the integral identity*

$$\int_Q \nabla u : \nabla v \, dx = \int_Q f v \, dx \tag{3.22}$$

*for all $v$ in $\dot{C}_p^1(Q)$, then $u \in \dot{H}_p^2(Q)$ and*

$$\|u\|_{\dot{H}_p^2(Q)} \leq C \|f\|_{L^2(Q)}. \tag{3.23}$$

*Proof.* First, observe that in the above integral identity we can take the test functions from the space $\dot{H}_p^1(Q)$. Then the existence of a unique generalized solution in $\dot{H}_p^1(Q)$ follows easily from the Riesz–Frèchet theorem. This solution is given by (3.20) where the coefficients are given in (3.21). Moreover, the solution belongs to $\dot{H}_p^2(Q)$, as

$$\begin{aligned} \|u\|^2_{\dot{H}_p^2(Q)} &= \sum_{k \in \mathbb{Z}^m} |k|^4 |u_k|^2 = \sum_{k \in \mathbb{Z}^m} |k|^4 \left(\frac{L^2}{4\pi^2} \frac{f_k}{|k|^2}\right)^2 \\ &= \frac{L^4}{16\pi^4} \sum_{k \in \mathbb{Z}^m} |f_k|^2 = \frac{L^4}{16\pi^4} \|f\|_{L^2(Q)}. \end{aligned}$$

This ends the proof. □

Looking on the equation $-\Delta u = f$ as on the abstract operator equation $Au = f$ in the space $\dot{L}^2(Q)$ with operator $A : \dot{L}^2(Q) \supset D(A) \to \dot{L}^2(Q)$ we see that

$$D(A) = \{u \in \dot{H}^1_p(Q) \ : \ Au \in \dot{L}^2(Q)\} = \dot{H}^2_p(Q).$$

In fact, from (3.23) it follows that $D(A) \subset \dot{H}^2_p(Q)$, on the other hand, for every $u$ in $\dot{H}^2_p(Q)$, $Au \in \dot{L}^2(Q)$, whence $\dot{H}^2_p(Q) \subset D(A)$.

**Exercise 3.19.** Prove that if for all $v \in \dot{H}^1_p(Q)$ functions $f \in \dot{H}^s_p(Q)$ and $u \in \dot{H}^1_p(Q)$ satisfy (3.22) then $u \in H^{s+2}_p(Q)$ and (3.23) holds.

**Lemma 3.14.** *The operator $A$ is invertible on $\dot{L}^2(Q)$ and $A^{-1} : \dot{L}^2(Q) \to \dot{L}^2(Q)$ is compact, that is, it maps bounded sets in $\dot{L}^2(Q)$ to precompact sets in $\dot{L}^2(Q)$.*

*Proof.* From the above considerations we know that the operator $A^{-1}$ maps $\dot{L}^2(Q)$ onto $\dot{H}^2_p(Q)$ and that the correspondence is one to one. From Theorem 3.14 it follows that $\dot{H}^2_p(Q)$ is compactly embedded in $\dot{L}^2(Q)$, whence $A^{-1}$ is compact. □

**Lemma 3.15.** *In the considered case of periodic boundary conditions the eigenfunctions of the Laplacian operator in $\dot{L}^2(Q)$ belong to $\dot{C}^\infty_p(Q)$.*

*Proof.* Let $Aw = \lambda w \in \dot{L}^2(Q)$. From Lemma 3.13 it follows that $w \in \dot{H}^2_p(Q)$. Thus $Aw = \lambda w \in \dot{H}^2_p(Q)$. Making use of Exercise 3.19 we conclude that $w \in \dot{H}^4_p(Q)$. By induction we obtain $w \in \dot{H}^s_p(Q)$ for each integer $s$. Now, it suffices to use the Sobolev embedding theorem to conclude that $w$ belongs to $\dot{C}^\infty_p(Q)$. □

**Exercise 3.20.** Prove that if $f \in \dot{C}^\infty_p(Q)$, $u \in \dot{H}^1_p(Q)$, and $Au = f$ then $u$ is infinitely smooth, namely, $u \in \dot{C}^\infty_p(Q)$.

**Theorem 3.15 (cf. [197, Corollary 3.26]).** *Let $H$ be a Hilbert space and let $A$ be a symmetric linear operator in $H$ whose range is all of $H$, and suppose further that its inverse is defined and compact. Then $A$ has an infinite set of real eigenvalues $\lambda_n$ with corresponding eigenfunctions $\omega_n$: $A\omega_n = \lambda_n\omega_n$. If the eigenvalues are ordered so that $|\lambda_{n+1}| \geq |\lambda_n|$ one has $\lim_{n\to\infty} |\lambda_n| = \infty$. Furthermore, the $\omega_n$ can be chosen so that they form an orthonormal basis for $H$, and in terms of this basis the operator $A$ can be represented by*

$$Au = \sum_{j=1}^{\infty} \lambda_j (u, \omega_j)_H \omega_j$$

*on its domain*

$$D(A) = \left\{ u \ : \ u = \sum_{j=1}^{\infty} c_j\omega_j, \ \sum_{j=1}^{\infty} |c_j|^2\lambda_j^2 < \infty \right\}.$$

*$D(A)$ is a Hilbert space with the inner product $(u, v)_{D(A)} = (Au, Av)_H$ and corresponding norm $\|u\|_{D(A)} = \|Au\|_H$.*

For the operator $Au = -\Delta u$ in $\dot{L}^2(Q)$, whence we have

$$u(x) = \sum_{k\in\mathbb{Z}^m\setminus\{0\}} c_k e^{2\pi i k\frac{x}{L}} = \sum_{k\in\mathbb{Z}^m\setminus\{0\}} c_k\left(\cos\left(2\pi k\frac{x}{L}\right) + i\sin\left(2\pi k\frac{x}{L}\right)\right)$$

$$= \sqrt{\frac{L^m}{2}} \sum_{k\in(\mathbb{Z}^m\setminus\{0\})/2} \left(Re(c_k)\omega_k^{(c)} - Im(c_k)\omega_k^{(s)}\right),$$

where

$$\omega_k^{(c)}(x) = \sqrt{\frac{2}{L^m}}\cos\left(2\pi k\frac{x}{L}\right), \quad \omega_k^{(s)}(x) = \sqrt{\frac{2}{L^m}}\sin\left(2\pi k\frac{x}{L}\right),$$

and $A\omega_k^{(\cdot)} = \frac{4\pi^2|k|^2}{L^2}\omega_k^{(\cdot)}$, $\|\omega_k^{(\cdot)}\|_{L^2(Q)} = 1$.

For example, for $m = 2$, the smallest eigenvalue is $\frac{4\pi^2}{L^2}|k|^2$ with $|k| = \sqrt{k_1^2 + k_2^2} = 1$. It repeats four times, $\lambda_1 = \lambda_2 = \lambda_3 = \lambda_4$ in the sequence

$$0 < \lambda_1 \le \lambda_2 \le \lambda_3 \le \ldots \le \lambda_n \le \ldots.$$

The corresponding eigenfunctions are given as $\omega_1 = \omega_{(1,0)}^{(c)}$, $\omega_2 = \omega_{(1,0)}^{(s)}$, $\omega_3 = \omega_{(0,1)}^{(c)}$, $\omega_4 = \omega_{(0,1)}^{(s)}$.

## 3.5 Evolution Spaces and Their Useful Properties

The solutions of evolution problems are the functions which depend on both space and time variables. Typically, the space variable belongs to a certain domain $\Omega \subset \mathbb{R}^d$ and the time variable belongs to the time interval $I = [t_0, t_1]$. A typical technique is to deal with these variables differently. Namely, assume that $u$ is a function of $(x,t) \in \Omega \times I$ and freeze $t \in I$. Then $u(\cdot,t)$ is only a function of the space variable. Usually we assume that this function belongs to a certain Banach space of functions defined on $\Omega$, let us denote this space by $V$. We write $u(t) \in V$. If we "unfreeze" the time variable now, we can write $u : I \to V$, so the function of $(x,t) \in \Omega \times I$ is treated as the Banach space valued function of a time variable only. This chapter is devoted to the recollection of the definition and basic properties of spaces of such functions.

**Definition 3.12.** Let $V$ be a separable Banach space. The function $u : I \to V$ is strongly measurable if there exists a sequence of simple functions $u_n : I \to V$ such that

$$\lim_{n\to\infty} u_n(t) = u(t) \quad \text{for a.e.} \quad t \in I.$$

A function $v : I \to V$ is simple if $v(t) = \sum_{k=1}^m \chi_{A_k}(t) w_k$ for all $t \in I$, where $A_k$ are Lebesgue measurable subsets of $I$, and $w_k \in V$.

**Definition 3.13.** Let $V$ be a separable Banach space. A strongly measurable function $v : I \to V$ is Bochner integrable if

$$\int_I \|v(t)\|_V \, dt < \infty.$$

The space of Bochner integrable functions is denoted by $L^1(I; V)$. It is possible to use the equivalent definition of this space, which gives rise to the so-called Bochner integral.

**Theorem 3.16 (cf. [85, Theorem 2, p. 45]).** *Let $v : I \to V$ be a strongly measurable function. Then $v \in L^1(I; V)$ is Bochner integrable if and only if there exists a sequence of simple functions $v_n : I \to V$ such that*

$$\lim_{n\to\infty} \int_I \|v_n(t) - v(t)\|_V \, dt = 0. \tag{3.24}$$

**Definition 3.14.** If the function $v : I \to V$ is simple then its Bochner integral over the Lebesgue measurable set $E \subset I$ can be defined as

$$\int_E v(t) \, dt = \sum_{i=1}^{m} w_k m(A_k \cap E).$$

If, in turn, $v \in L^1(I; V)$, then the Bochner integral is defined as

$$\int_E v(t) \, dt = \lim_{n\to\infty} \int_E v_n(t) \, dt,$$

where $v_n$ is the sequence of simple functions from Theorem 3.16.

It is easy to prove that the value of the Bochner integral does not depend on the choice of the sequence of simple functions which satisfy (3.24). More information on Bochner integral and its properties can be found in [85, 107, 234]. The definitions of Bochner integral and the space $L^1(I; V)$ do not require the separability of $V$. In such a case, however, the notion of measurability of a Banach space valued function should be defined more carefully. For the sake of simplicity, however, in this book we always use this notions for separable spaces $V$.

As a natural extension of $L^1(I; V)$ it is possible to define the spaces $L^p(I; V)$ for $p \in [1, \infty]$ as the spaces of the measurable functions, such that

$$\|f\|_{L^p(I;V)} = \begin{cases} \left(\int_I \|f(t)\|_V^p \, dt\right)^{\frac{1}{p}} < \infty & \text{for} \quad p \in [1, \infty), \\ \operatorname{ess\,sup}_{t\in I} \|f(t)\|_V < \infty & \text{for} \quad p = \infty. \end{cases}$$

Once we identify any two functions which differ only of a subset of $I$ of Lebesgue measure zero, the space $L^p(I; V)$ endowed with the norm $\| \cdot \|_{L^p(I;V)}$ becomes a

Banach space. Moreover, through a natural isometric isomorphism we can make the following identification $L^p(I; L^p(\Omega)^m) = L^p(\Omega \times I)^m$ valid for $p \neq \infty$ (note that $L^\infty(\Omega)^m$ is not a separable space, see Example 1.42 in [204] and Example 23.4 in [234]).

We have the following useful results related to the dual space to $L^p(I; V)$.

**Proposition 3.1 (cf. [234, Proposition 23.7]).** *Let $V$ be a separable and reflexive Banach space and let $1 < p < \infty$, and $\frac{1}{p} + \frac{1}{q} = 1$. Then $L^p(I; V)$ is a reflexive and separable Banach space. Moreover, the dual space of $L^p(I; V)$ is isometrically isomorphic with $L^q(I; V')$.*

**Proposition 3.2 (cf. [234, Proposition 23.9]).** *Let $V$ be a separable and reflexive Banach space and let $1 < p < \infty$, and $\frac{1}{p} + \frac{1}{q} = 1$. If $u \in L^p(I; V)$ and $v \in V'$, then*

$$\left\langle v, \int_I u(t)\, dt \right\rangle_{V' \times V} = \int_I \langle v, u(t) \rangle_{V' \times V}\, dt.$$

*Moreover from*

$$\begin{aligned} u_n &\to u \quad \text{strongly in} \quad L^p(I; V), \\ v_n &\to v \quad \text{weakly in} \quad L^q(I; V'), \end{aligned}$$

*it follows that*

$$\int_I \langle v_n(t), u_n(t) \rangle_{V' \times V}\, dt \to \int_I \langle v(t), u(t) \rangle_{V' \times V}\, dt.$$

A function $v \in L^1(I; V)$ is called a distributional time derivative of a function $u \in L^1(I; V)$, provided the following equality

$$\int_I u(t)\phi'(t)\, dt = -\int_I v(t)\phi(t)\, dt \quad \text{for all} \quad \phi \in C_0^\infty(I), \tag{3.25}$$

holds in $V$. In such a case we write $v = u' = u_t = \frac{du}{dt}$. Sometimes, the distributional time derivative of a function belonging to $L^1(I; V)$ may not exist as a function belonging to the same space, but it may have values in a larger Banach space $Z$, such that we have the embedding $V \subset Z$. Then, the function $v \in L^1(I; Z)$ is a distributional time derivative of $u \in L^1(I; V)$, provided (3.25) holds in $V$. Note that, in such a case, the integral on the right-hand side of (3.25) belongs to $V$, although the integrand has values only in $Z$, and hence not necessarily in $V$.

As it can be easily checked, the distributional derivative, provided it exists as a function $u' \in L^1(I; Z)$, is unique with respect to equality for a.e. $t \in I$. Moreover we have the following equality in $Z$ valid for all $s, t \in \overline{I}$

$$u(t) = u(s) + \int_s^t u'(r)\, dr.$$

Note that if $u \in L^1(I;V)$ and $u' \in L^1(I;Z)$ with $V \subset Z$, then, after modification of the null set we must have $u \in C(\bar{I};Z)$ and hence the pointwise values of $u$ make sense as the elements of $Z$ in the last equality.

An important role in the study of weak solutions of evolution problems is played by the so-called evolution triples (or Gelfand triples). An evolution triple consists of the reflexive Banach space $V$ (we will always assume that $V$ is also separable), the Hilbert space $H$ such that we have the continuous and dense embedding $V \subset H$, and the dual space $V'$. As the space $H$ is always identified with its dual we can write $V \subset H \subset V'$ with all embeddings being continuous and dense. From now on we will always denote the scalar product on $H$ by $(\cdot,\cdot)$ and associated norm by $|\cdot|$. The norm on $V$ is denoted by $\|\cdot\|$ and the duality pairing between $V$ and $V'$ by $\langle\cdot,\cdot\rangle$. An important property of evolution triples is the following formula

$$\langle u, v\rangle = (u, v) \quad \text{whenever} \quad u \in H \quad \text{and} \quad v \in V. \tag{3.26}$$

The following results which hold in the framework of evolution triples will be frequently used later.

**Corollary 3.2.** *Let $V \subset H \subset V'$ be an evolution triple such that $V \subset H$ is additionally compact and $V$ is Hilbert. Let $A : V \to V'$ be a linear operator which is*

- *bounded ($\|Au\|_{V'} \leq \|A\|_{\mathscr{L}(V;V')}\|u\|$ for all $u \in V$),*
- *symmetric ($\langle Au, v\rangle = \langle Av, u\rangle$ for all $u, v \in V$),*
- *coercive ($\langle Au, u\rangle \geq \alpha\|u\|^2$ for all $u \in V$ with $\alpha > 0$).*

*Then $A$ satisfies the assumptions of Theorem 3.15 and we can construct an orthogonal in $V$ and orthonormal in $H$ basis of eigenfunctions $\omega_j$ of $A$ with the corresponding eigenvalues such that $|\lambda_j| \to \infty$ as $j \to \infty$.*

*Proof.* The Lax–Milgram lemma (cf. Lemma 3.2) implies that for any $f \in H$ there exists a unique $v \in V$ such that $Av = f$. Hence the range of $A$ (treated as an operator from $D(A) \subset H$ to $H$) is whole $H$ and $A^{-1} : H \to H$ is well defined. Moreover, by (3.26), $A$, when treated as an operator on $H$, is symmetric. Obviously $A^{-1}$ is linear. To prove compactness of $A^{-1}$ let $f_n$ be a bounded sequence in $H$. Let $u_n \in V$ be such that $Au_n = f_n$. Coercivity of $A$ implies that the sequence $u_n$ is bounded in $V$ and hence, by the compactness of the embedding $V \subset H$, for a subsequence, $u_n \to u$ strongly in $H$ for certain $u \in H$. The proof of compactness is complete. □

**Theorem 3.17 (cf. [197, Lemma 7.5]).** *Let $V \subset H \subset V'$, $A : V \to V'$, and $\omega_j$ be as in Corollary 3.2. Denote $V_n = \operatorname{span}\{\omega_1, \ldots, \omega_n\}$. If we define $P_n : V' \to V_n$ by $P_n v = \sum_{i=1}^n \langle v, \omega_i\rangle w_i$ for $v \in V'$ then we have*

$$\text{if} \quad v \in V \quad \text{then} \quad \|P_n v\| \leq \|v\| \quad \text{and} \quad \lim_{n\to\infty} \|P_n v - v\| = 0,$$

$$\text{if} \quad v \in H \quad \text{then} \quad |P_n v| \leq |v| \quad \text{and} \quad \lim_{n\to\infty} |P_n v - v| = 0,$$

$$\text{if} \quad v \in V' \quad \text{then} \quad \|P_n v\|_{V'} \leq \|v\|_{V'} \quad \text{and} \quad \lim_{n\to\infty} \|P_n v - v\|_{V'} = 0.$$

We will frequently use the following two results.

**Lemma 3.16 (cf. [234, Proposition 23.23], [219, Chap. 3, Lemma 1.2]).** *Let $V \subset H \subset V'$ be an evolution triple, let $I = (t_1, t_2)$ be a finite time interval, and let $1 < p < \infty$ and $\frac{1}{p} + \frac{1}{q} = 1$. Then*

1. *The space $\mathscr{W}(I) = \{u \in L^p(I;V) : u' \in L^q(I;V')\}$ is a Banach space equipped with the norm*

$$\|u\|_{\mathscr{W}(I)} = \|u\|_{L^p(I;V)} + \|u'\|_{L^q(I;V')}.$$

2. *The space $\mathscr{W}(I)$ embeds continuously in $C(\bar{I};H)$, i.e., every function from $\mathscr{W}(I)$ is almost everywhere equal to a continuous function with values in $H$, and we have*

$$\max_{t\in\bar{I}} |u(t)| \leq C\|u\|_{\mathscr{W}(I)} \quad \text{for all} \quad u \in \mathscr{W}(I).$$

3. *We have the following integration by parts formula valid for all $u, v \in \mathscr{W}(I)$,*

$$(u(t_1), v(t_1)) - (u(t_0), v(t_0)) = \int_I \langle v'(t), u(t)\rangle + \langle u'(t), v(t)\rangle \, dt.$$

4. *For any $u \in \mathscr{W}(I)$ the following equality holds in the sense of distributions on $I$*

$$\frac{d}{dt}|u(t)|^2 = 2\langle u'(t), u(t)\rangle.$$

**Corollary 3.3.** *Let $V, H, p, q$ be as in the previous Lemma. Let $u_n \to u$ weakly in $L^p(I;V)$ and $u_n' \to v$ weakly in $L^q(I;V')$. Then $v = u'$ and $u_n(t) \to u(t)$ weakly in $H$ for all $t \in \bar{I}$.*

*Proof.* The fact that $v = u'$ follows, for example, from Proposition 23.19 in [234]. We will prove the second assertion. We have

$$u_n(s) = u_n(t) + \int_s^t u_n'(r)dr \quad \text{for all} \quad s, t \in \bar{I}.$$

Taking the duality with any $v \in V$ and integrating with respect to $t$ over $I$ we get

$$\begin{aligned}(t_1 - t_0)(u_n(s), v) &= \int_I (u_n(t), v)\, dt + \int_I \int_s^t \langle u_n'(r), v\rangle \, dr\, dt \\ &= \int_I (u_n(t), v)\, dt + \int_I \langle u_n'(t), v\phi_s(t)\rangle \, dt,\end{aligned}$$

where $\phi_s = (t_0 - s)\chi_{(t_0,s)} + (t_1 - s)\chi_{(s,t_1)}$. The assertion follows by passing to the limit $n \to \infty$ in the last formula and using the density of the embedding $V \subset H$. $\square$

We now recall two important results on the triples of spaces $V_1 \subset V_2 \subset V_3$, where the embedding $V_1 \subset V_2$ is *compact*. The first result is known as the *Ehrling lemma*.

**Lemma 3.17 (See [204, Lemma 7.6]).** *Let $V_1 \subset V_2 \subset V_3$ be three Banach spaces with all embeddings continuous and $V_1 \subset V_2$ compact. Then for any $\varepsilon > 0$ there exists a constant $C(\varepsilon)$ such that*

$$\|v\|_{V_2} \leq \varepsilon\|v\|_{V_1} + C(\varepsilon)\|v\|_{V_3} \quad \text{for all} \quad v \in V_1.$$

The next result is known as *Aubin–Lions compactness theorem.*

**Theorem 3.18 (cf. [6, 155], [216, Corollary 4], [204, Lemma 7.7], [219, Chap. 3, Theorem 2.1]).** *Let $I$ be a bounded time interval and let $1 \leq p, q < \infty$ and let $X_1 \subset X_2 \subset X_3$ be three separable Banach spaces with all embeddings being continuous. If the embedding $X_1 \subset X_2$ is compact, then the embedding*

$$\{u \in L^p(I; X_1) \ : \ u' \in L^q(I; X_3)\} \subset L^p(I; X_2)$$

*is also compact.*

Fairly general form of the above Lemma is given in [216]. Note that the assumption of separability is not required there.

A function $u : \bar{I} \to V$, where $V$ is a Banach space is said to be *weakly continuous* provided the function

$$\bar{I} \ni t \to \langle v, u(t)\rangle \in \mathbb{R}$$

is continuous for all $v \in V'$. We write $u \in C_w(\bar{I}; V)$, or $u \in C(\bar{I}; V_{weak})$, or, sometimes $C(\bar{I}; V_w)$. Of course $C(\bar{I}; V) \subset C_w(\bar{I}; V)$ with the strict inclusion. We have the following result.

**Lemma 3.18 (cf. [219, Chap. 3, Lemma 1.4]).** *Let $X \subset Y$ be two Banach spaces with a continuous embedding and let $I$ be a finite time interval. If $u \in L^\infty(I; X)$ and $u \in C_w(\bar{I}; Y)$ then $u \in C_w(\bar{I}; X)$.*

Finally we give a parabolic embedding result.

**Lemma 3.19 (see [155, Chap. 1, Lemma 6.7]).** *Let $\Omega \subset \mathbb{R}^3$ be an open and bounded set with Lipschitz boundary and let $I$ be a bounded time interval. If*

$$u \in L^2(I; H^1(\Omega)) \cap L^\infty(I; L^2(\Omega)),$$

*then $u \in L^4(I; L^3(\Omega))$.*

*Proof.* By the Sobolev embedding theorem we have $H^1(\Omega) \subset L^6(\Omega)$ with a continuous embedding, so, using the Hölder inequality we have for a.e. $t \in I$ and for a constant $c > 0$

$$\|u(t)\|_{L^3(\Omega)} \leq \|u(t)\|_{L^2(\Omega)}^{\frac{1}{2}} \|u(t)\|_{L^6(\Omega)}^{\frac{1}{2}} \leq c\|u(t)\|_{L^2(\Omega)}^{\frac{1}{2}} \|u(t)\|_{H^1(\Omega)}^{\frac{1}{2}}.$$

Thus we have

$$\int_I \|u(t)\|_{L^3(\Omega)}^4 \, dt \le c^4 \operatorname*{ess\,sup}_{t\in I}\{\|u(t)\|_{L^2(\Omega)}^2\} \int_I \|u(t)\|_{H^1(\Omega)}^2 \, dt,$$

and the assertion is proved. □

## 3.6 Gronwall Type Inequalities

In this section we deal with Gronwall type inequalities, which are the efficient and useful tool in the study of evolution problems. These inequalities are used in the derivation of the a priori estimates that are satisfied by the solution of the analyzed problem.

The first Gronwall type inequality is useful in the study of existence and uniqueness of solutions to initial and boundary value problems for evolutionary PDEs, as well as in the derivation of dissipative a priori estimates for their solutions.

**Lemma 3.20 (Gronwall Inequality).** *Let $t_0$ be a real number. Let $x \in L^1_{loc}(t_0, \infty)$ be a function such that $x' \in L^1_{loc}(t_0, \infty))$, and let $h \in L^1_{loc}(t_0, \infty)$ and $k \in L^1_{loc}([t_0, \infty))$. If*

$$x'(t) \le x(t)k(t) + h(t) \quad \text{for a.e.} \quad t \in (t_0, \infty), \tag{3.27}$$

*then*

$$x(t) \le x(t_0)e^{\int_{t_0}^t k(s)\,ds} + \int_{t_0}^t h(s)e^{\int_s^t k(r)\,dr}\,ds \quad \text{for all} \quad t \in [t_0, \infty). \tag{3.28}$$

*If, in particular, $k(t) = C$ for all $t \ge t_0$, $C$ being a real constant, then*

$$x(t) \le x(t_0)e^{-Ct_0}e^{Ct} + e^{Ct}\int_{t_0}^t h(s)e^{-Cs}\,ds \quad \text{for all} \quad t \in [t_0, \infty). \tag{3.29}$$

*If, furthermore, $h(t) = D$ for all $t \ge t_0$, $D$ being a real constant, then*

$$x(t) \le x(t_0)e^{-Ct_0}e^{Ct} + \frac{D}{C}(e^{-Ct_0}e^{Ct} - 1) \quad \text{for all} \quad t \in [t_0, \infty). \tag{3.30}$$

*Proof.* Multiplying (3.27) by the integrating factor $e^{-\int_{t_0}^t k(s)\,ds}$ we get

$$x'(t)e^{-\int_{t_0}^t k(s)\,ds} - x(t)k(t)e^{-\int_{t_0}^t k(s)\,ds} \le h(t)e^{-\int_{t_0}^t k(s)\,ds} \quad \text{a.e. } t > t_0,$$

whence

$$\frac{d}{dt}\left(x(t)e^{-\int_{t_0}^{t} k(s)\,ds}\right) \leq h(t)e^{-\int_{t_0}^{t} k(s)\,ds} \quad \text{for a.e.} \quad t > t_0.$$

Integrating the above inequality from $t_0$ to $t$ we obtain (3.28). Assertions (3.29) and (3.30) follow from (3.28) by a straightforward calculation. □

If, in (3.29) or (3.30), we have $C < 0$, then the resultant estimate is, after appropriately large time, independent on the initial condition $x(t_0)$. Such estimates are called *dissipative* and they are particularly useful in the study of global attractors. If, however, we only have (3.27) with $k(t)$ positive on $(t_0, \infty)$, then the resultant bounds grow to infinity as $t \to \infty$. In such situation, we need the following uniform Gronwall inequality.

**Lemma 3.21 (Uniform Gronwall Inequality).** *Let $t_0 \in \mathbb{R}$. Let $x \in L^1_{loc}(t_0, \infty)$ be a function such that $x' \in L^1_{loc}(t_0, \infty)$, and let $h \in L^1_{loc}(t_0, \infty)$ and $k \in L^1_{loc}([t_0, \infty))$. Let moreover $x(t) \geq 0$, $h(t) \geq 0$, and $k(t) \geq 0$ for a.e. $t > t_0$. If*

$$x'(t) \leq x(t)k(t) + h(t) \quad \text{for a.e.} \quad t \in (t_0, \infty), \tag{3.31}$$

*and*

$$\int_t^{t+r} k(s)\,ds \leq a_1, \int_t^{t+r} h(s)\,ds \leq a_2, \int_t^{t+r} x(s)\,ds \leq a_3 \quad \text{for all} \quad t \in [t_0, \infty),$$

*where $a_1, a_2, a_3, r$ are positive constants, then*

$$x(t+r) \leq \left(\frac{a_3}{r} + a_2\right) e^{a_1} \quad \text{for all} \quad t \in [t_0, \infty).$$

*Proof.* Let $t_0 \leq t \leq t + r$. We multiply (3.31) written for $p \in (t, t+r)$ by the integrating factor $e^{-\int_t^p k(\tau)\,d\tau}$, getting

$$\begin{aligned} \frac{d}{dp}\left(x(p)e^{-\int_t^p k(\tau)\,d\tau}\right) &= x'(p)e^{-\int_t^p k(\tau)\,d\tau} - x(p)k(p)e^{-\int_t^p k(\tau)\,d\tau} \\ &\leq h(p)e^{-\int_t^p k(\tau)\,d\tau} \leq h(p) \quad \text{a.e.} \quad p \in (t, t+r). \end{aligned}$$

We fix $s \in (t, t+r)$ and integrate the last inequality over the interval $(s, t+r)$. We get

$$x(t+r)e^{-\int_t^{t+r} k(\tau)\,d\tau} - x(s)e^{-\int_t^{s} k(\tau)\,d\tau} \leq \int_s^{t+r} h(p)\,dp \leq a_2,$$

whence we have the inequality

$$x(t+r) \leq x(s)e^{\int_s^{t+r} k(\tau)\,d\tau} + a_2 e^{\int_t^{t+r} k(\tau)\,d\tau} \leq x(s)e^{a_1} + a_2 e^{a_1},$$

valid for all $s \in (t, t+r)$. We integrate the last inequality with respect to $s$ over the interval $(t, t+r)$, which gives us

$$rx(t+r) \le \int_t^{t+r} x(s)\,ds\, e^{a_1} + ra_2 e^{a_1} \le a_3 e^{a_1} + ra_2 e^{a_1},$$

which immediately yields the assertion. □

If we need to estimate the norm of the unknown function $x$ by a constant that is arbitrarily small, then the following *translated Gronwall inequality* becomes useful (see [161]).

**Lemma 3.22.** *Let $\lambda > 0$ be a constant. Let $x \in L^1_{loc}(t_0, \infty)$ be such that $x' \in L^1_{loc}(t_0, \infty)$ and $x(t) \ge 0$ for a.e. $t \ge t_0$. Let moreover $h \in L^1_{loc}(t_0, \infty)$ be such that $h(t) \ge 0$ for a.e. $t \ge t_0$. If we have*

$$x'(t) + \lambda x(t) \le h(t) \quad \text{a.e.} \quad t > t_0, \tag{3.32}$$

*then for all $t \ge t_0$ and $r > 0$ we have*

$$x(t+r) \le \frac{2}{r} e^{-\lambda \frac{r}{2}} \int_t^{t+\frac{r}{2}} x(s)\,ds + e^{-\lambda(r+t)} \int_t^{t+r} h(s) e^{\lambda s}\,ds.$$

*In particular, taking $r = 2$, we get, for all $t \ge t_0$*

$$x(t+2) \le e^{-\lambda} \int_t^{t+1} x(s)\,ds + e^{-\lambda(t+2)} \int_t^{t+2} h(s) e^{\lambda s}\,ds.$$

*Proof.* Multiplying (3.32) written for $p \ge t_0$ by the integrating factor $e^{\lambda p}$, we get

$$\frac{d}{dp}(x(p)e^{\lambda p}) = x'(p)e^{\lambda p} + x(p)\lambda e^{\lambda p} \le h(p)e^{\lambda p} \quad \text{a.e.} \quad p \ge t_0.$$

Let $t \ge t_0$ and $r > 0$. We fix $s \in \left(t, t + \frac{r}{2}\right)$ and integrate the last inequality over the interval $(s, t+r)$ leading to

$$x(t+r)e^{\lambda(t+r)} - x(s)e^{\lambda s} \le \int_s^{t+r} h(p)e^{\lambda p}\,dp,$$

whence we have

$$\begin{aligned} x(t+r) &\le x(s)e^{\lambda(s-t-r)} + e^{-\lambda(t+r)} \int_s^{t+r} h(p)e^{\lambda p}\,dp \\ &\le x(s)e^{\lambda(s-t-r)} + e^{-\lambda(t+r)} \int_t^{t+r} h(p)e^{\lambda p}\,dp. \end{aligned}$$

We integrate the last inequality with respect to $s$ over the interval $(t, t+\frac{r}{2})$, which gives

$$\frac{r}{2}x(t+r) \leq \int_t^{t+\frac{r}{2}} x(s)e^{\lambda(s-t-r)}\,ds + \frac{r}{2}e^{-\lambda(t+r)}\int_t^{t+r} h(s)e^{\lambda s}\,ds,$$

and the assertion follows immediately. □

We shall use also the following lemma.

**Lemma 3.23 (Generalized Gronwall Inequality).** *Let $\alpha = \alpha(t)$ and $\beta = \beta(t)$ be locally integrable real-valued functions on $[0,\infty)$ that satisfy the following conditions for some $T > 0$:*

$$\liminf_{t\to\infty} \frac{1}{T}\int_t^{t+T} \alpha(\tau)d\tau > 0, \tag{3.33}$$

$$\limsup_{t\to\infty} \frac{1}{T}\int_t^{t+T} \alpha^-(\tau)d\tau < \infty, \tag{3.34}$$

$$\lim_{t\to\infty} \frac{1}{T}\int_t^{t+T} \beta^+(\tau)d\tau = 0, \tag{3.35}$$

*where $\alpha^-(t) = \max\{-\alpha(t), 0\}$, $\beta^+(t) = \max\{\beta(t), 0\}$. Suppose that $\xi = \xi(t)$ is an absolutely continuous nonnegative function on $[0,\infty)$ that satisfies the following inequality almost everywhere on $(0,\infty)$,*

$$\frac{d\xi(t)}{dt} + \alpha(t)\xi(t) \leq \beta(t). \tag{3.36}$$

*Then $\xi(t) \to 0$ as $t \to \infty$.*

For the proof, see [99].

## 3.7 Clarke Subdifferential and Its Properties

We start this section from the recollection of the definition and properties of the Clarke subdifferential, a generalization of Gâteaux derivative to functionals which are only locally Lipschitz. The Clarke subdifferential at a given point assumes multiple values, so it is a *multifunction*. Then, we give an example of a multivalued partial differential equation, where the semilinear term has the form of a Clarke subdifferential. Finally, we show several examples how to compute Clarke subdifferentials and we discuss the usefulness of the Clarke subdifferential to express various boundary conditions in mechanics. Throughout this chapter $X$ is always assumed to be a Banach space.

**Definition 3.15.** A functional $j : X \to \mathbb{R}$ is locally Lipschitz if for every $x \in X$ there exists an open neighborhood $\mathcal{O} \in \mathcal{N}(x)$ and a constant $k_{\mathcal{O}}$ such that for all $z, y \in \mathcal{O}$ $|j(z) - j(y)| \leq k_{\mathcal{O}} \|z - y\|_X$.

The following two definitions were introduced by Clarke (see [71, 72]).

**Definition 3.16.** Let $j : X \to \mathbb{R}$ be locally Lipschitz and let $x, h \in X$. A generalized directional derivative in the sense of Clarke of the functional $j$ at the point $x \in X$ and in the direction $h \in X$, denoted by $j^0(x; h)$ is defined as

$$j^0(x; h) = \limsup_{y \to x, \lambda \to 0+} \frac{j(y + \lambda h) - j(y)}{\lambda}.$$

**Definition 3.17.** Let $j : X \to \mathbb{R}$ be locally Lipschitz. The generalized subdifferential in the sense of Clarke of the functional $j$ at the point $x \in X$ is the set $\partial j(x) \subset X'$ defined as

$$\partial j(x) = \{\xi \in X' : \langle \xi, h \rangle_{X' \times X} \leq j^0(x; h) \quad \text{for all} \quad h \in X\}. \tag{3.37}$$

We recall some results on the properties of the Clarke subdifferential and Clarke generalized directional derivative.

**Proposition 3.3 (cf. [72, Proposition 2.1.1]).** *If $j : X \to \mathbb{R}$ is locally Lipschitz, then $j^0 : X \times X \to \mathbb{R}$ is upper semicontinuous, i.e.,*

$$\xi_n \to x, h_n \to h \Rightarrow \limsup_{n \to \infty} j^0(x_n; h_n) \leq j^0(x; h) \quad \text{for all} \quad x, h \in X.$$

*Moreover, $j^0(x; \cdot)$ is Lipschitz on $X$ with the Lipschitz constant equal to the local Lipschitz constant $k_{\mathcal{O}}$ of $j$ at $x$ given in Definition 3.15.*

**Proposition 3.4 (cf. [72, Proposition 2.1.2]).** *If $j : X \to \mathbb{R}$ is locally Lipschitz, then for every $x \in X$, the set $\partial j(x)$ is nonempty, weakly-* compact, and convex. Moreover, for every $\xi \in \partial j(x)$ we have $\|\xi\|_X \leq k_{\mathcal{O}}$ (where $k_{\mathcal{O}}$ is the local Lipschitz constant of $j$ at $x$ given in Definition 3.15) and we have*

$$j^0(x; h) = \max_{\xi \in \partial j(x)} \langle \xi, h \rangle_{X' \times X} \quad \text{for all} \quad x, h \in X.$$

**Proposition 3.5 (cf. [72, Proposition 2.1.5]).** *If $j : X \to \mathbb{R}$ is locally Lipschitz, then the graph of a multifunction $X \ni X \to \partial j(x) \in 2^{X'}$ is a strong-weak-* sequentially closed set.*

For a functional defined on $\mathbb{R}^d$ one can calculate its Clarke subdifferential from the following characterization.

**Theorem 3.19 (see [72, Theorem 2.5.1]).** *Let $j : \mathbb{R}^d \to \mathbb{R}$ be locally Lipschitz then*

$$\partial j(s) = \overline{\mathrm{conv}}\{\lim_{n\to\infty} \nabla j(s_n) : \; s_n \to s, \; s_n \notin S \cup N_j\} \quad \text{for} \quad s \in \mathbb{R}^d, \tag{3.38}$$

*where $S$ is any Lebesgue null set, and $N_j$ is the Lebesgue null set on which $j$ is not differentiable.*

We recall the Lebourg mean value theorem (cf. [72, Theorem 2.3.7])

**Theorem 3.20.** *Let $X$ be a Banach space and $j : X \to \mathbb{R}$ be a locally Lipschitz functional. For any $x, y \in X$ there exists $\theta \in (0,1)$ and $\xi \in \partial j(\theta x + (1-\theta)y)$ such that $\langle \xi, y - x\rangle_{X'\times X} = j(y) - j(x)$.*

We are in position to prove the approximation theorem for Clarke subdifferentials.

**Theorem 3.21.** *Let $j : \mathbb{R}^d \to \mathbb{R}$ be a locally Lipschitz functional such that the growth condition $|\xi| \leq C(1 + |s|)$ holds for all $s \in \mathbb{R}^d$ and all $\xi \in \partial j(s)$ with $C > 0$. Let moreover $\rho_n : \mathbb{R}^d \to \mathbb{R}$ be a sequence of mollifier kernels given by $\rho_n(r) = n^d \rho(nr)$ for $r \in \mathbb{R}^d$, where $\rho : \mathbb{R}^d \to \mathbb{R}$ is a function such that $\rho \in C^\infty(\mathbb{R}^d)$, $\rho(s) \geq 0$ on $\mathbb{R}^d$, $\int_{\mathbb{R}^d} \rho(s)\,ds = 1$, and $\operatorname{supp}\rho \subset [-1,1]^d$. Define*

$$j_n(s) = \int_{\mathbb{R}^d} \rho_n(r) j(s-r)\,dr.$$

*Then*

*(A)* $j_n \in C^\infty(\mathbb{R}^d)$,
*(B)* $|\nabla j_n(s)| \leq D(1+|s|)$ *for all* $s \in \mathbb{R}^d$ *with* $D > 0$ *independent of* $n$,
*(C)* *if $(\Omega, \Sigma, \mu)$ is a finite and complete measure space and $v_n \in L^2(\Omega)^d$, is a sequence such that $v_n \to v$ strongly in $L^2(\Omega)^d$ and $\nabla j_n(v_n) \to \xi$ weakly in $L^2(\Omega)^d$, then $\xi(x) \in \partial j(v(x))$ $\mu$-a.e. in $\Omega$.*

*Proof.* The assertion $(A)$ follows, for example, from [93], Theorem 6, Appendix C. Let us prove $(B)$. Let $h \in \mathbb{R}^d$. We have

$$\nabla j_n(s)\cdot h = \lim_{\lambda\to 0} \frac{j_n(s+\lambda h) - j_n(s)}{\lambda} = \lim_{\lambda\to 0} \int_{\operatorname{supp}\rho_n} \rho_n(r) \frac{j(s+\lambda h - r) - j(s-r)}{\lambda}\,dr.$$

Using the Lebourg mean value theorem we get

$$\left|\frac{j(s+\lambda h - r) - j(s-r)}{\lambda}\right| \leq |z(s,\lambda h, r)||h| \leq C(1 + |s| + |r| + \theta|\lambda||h|)|h|,$$

where $z(s,\lambda h, r) \in \partial j(s - r + \theta\lambda h)$ with $\theta \in (0,1)$ possibly changing from point to point. We get

$$|\nabla j_n(s)\cdot h| \leq C \lim_{\lambda\to 0} \int_{\operatorname{supp}\rho_n} \rho_n(r)(1 + |s| + |r| + |\lambda||h|)|h|\,dr.$$

As $\operatorname{supp}\rho_n \subset \left[-\frac{1}{n}, \frac{1}{n}\right]^d$ we have

$$|\nabla j_n(s)| \leq C \lim_{\lambda\to 0} \int_{\operatorname{supp}\rho_n} \rho_n(r)\left(1 + |s| + \frac{1}{n} + |\lambda|\right) dr \leq C(2 + |s|),$$

and the assertion $(B)$ follows. We pass to the proof of $(C)$. As $v_n \to v$ in $L^2(\Omega)^d$ we also have, for a subsequence, $v_n(x) \to v(x)$ and $|v_n(x)| \leq h(x)$ for $\mu$-a.e. $x \in \Omega$ with $h \in L^2(\Omega)$. The further argument will be done for this subsequence. Take $w \in L^\infty(\Omega)^d$. From the weak convergence $\nabla j_n(v_n) \to \xi$ it follows that

$$\begin{aligned}\int_\Omega \xi(x)\cdot w(x)\, d\mu(x) &= \lim_{n\to\infty} \int_\Omega \nabla j_n(v_n(x))\cdot w(x)\, d\mu(x)\\ &\leq \int_\Omega \limsup_{n\to\infty} \nabla j_n(v_n(x))\cdot w(x)\, d\mu(x),\end{aligned}$$

where the use of the Fatou lemma is justified by the growth condition $(B)$ and the bound $|v_n(x)| \leq h(x)$. We estimate the integrand for $\mu$-a.e. $x \in \Omega$ as follows

$$\begin{aligned}\limsup_{n\to\infty} \nabla j_n(v_n(x))\cdot w(x) &= \limsup_{n\to\infty} \lim_{\lambda\to 0+} \frac{j_n(v_n(x) + \lambda w(x)) - j_n(v_n(x))}{\lambda}\\ &= \limsup_{n\to\infty} \lim_{\lambda\to 0+} \int_{\operatorname{supp}\rho_n} \rho_n(s) \frac{j(v_n(x) + \lambda w(x) - s) - j(v_n(x) - s)}{\lambda}\, ds\\ &\leq \limsup_{\substack{n\to\infty\\ \lambda\to 0+}} \sup_{r\in \operatorname{supp}\rho_n} \frac{j(v_n(x) + \lambda w(x) - r) - j(v_n(x) - r)}{\lambda}\\ &= \limsup_{\substack{n\to\infty\\ \lambda\to 0+}} \frac{j(v_n(x) + \lambda w(x) - r_{n,\lambda}) - j(v_n(x) - r_{n,\lambda})}{\lambda},\end{aligned}$$

where $r_{n,\lambda} \in \operatorname{supp}\rho_n$ is such that the supremum over $r$ of the difference quotient is achieved. Observe that as $n \to \infty$ we must have $v_n(x) - r_{n,\lambda} \to v(x)$. From the definition of the Clarke directional derivative we get

$$\limsup_{n\to\infty} \nabla j_n(v_n(x))\cdot w(x) \leq \limsup_{\substack{z\to v(x)\\ \lambda\to 0+}} \frac{j(z + \lambda w(x)) - j(z)}{\lambda} = j^0(v(x); w(x)),$$

whereas

$$\int_\Omega \xi(x)\cdot w(x)\, d\mu(x) \leq \int_\Omega j^0(v(x); w(x))\, d\mu(x). \tag{3.39}$$

We will show that $\xi(x)\cdot z \leq j^0(v(x); z)$ $\mu$-a.e. in $\Omega$ for all $z \in \mathbb{R}^d$ and in consequence we will have the assertion $\xi(x) \in \partial j(v(x))$ $\mu$-a.e. in $\Omega$. First we prove that $j^0(v(x); w(x)) - \xi(x) \cdot w(x) \geq 0$ for $\mu$-a.e. $x \in \Omega$ for all $w \in L^\infty(\Omega)^d$. Indeed, suppose it is not the case. Then for a certain $\bar{w} \in L^\infty(\Omega)^d$ and set $N \subset \Omega$ of positive measure we have the inequality $j^0(v(x); \bar{w}(x)) - \xi(x) \cdot \bar{w}(x) < 0$. Take $\hat{w} = \bar{w}$ on $N$ and $\hat{w} = 0$ on $\Omega \setminus N$. We have $\hat{w} \in L^\infty(\Omega)^d$ and

$$\int_\Omega j^0(v(x); \hat{w}(x)) - \xi(x) \cdot \hat{w}(x)\, d\mu(x) = \int_N j^0(v(x); \bar{w}(x)) - \xi(x) \cdot \bar{w}(x)\, d\mu(x) < 0,$$

a contradiction with (3.39). Let $\{w_n\}_{n=1}^\infty$ be a countable dense set in $\mathbb{R}^d$. By taking constant functions in place of $w$ we get $j^0(v(x); w_n) \geq \xi(x) \cdot w_n$ outside a certain null set $M_n$. If follows that $j^0(v(x); w_n) \geq \xi(x) \cdot w_n$ for all $n$ outside a null sets $\bigcup_{n=1}^\infty M_n$. By Proposition 3.3 and density of $\{w_n\}_{n=1}^\infty$ in $\mathbb{R}^d$ it follows that $\xi(x) \cdot z \leq j^0(v(x); z)$ for all $z \in \mathbb{R}^d$ on a set of full measure and the proof is complete. □

## 3.8 Nemytskii Operator for Multifunctions

In this section we prove a result, that will be useful in passing to the limit in the terms containing the Clarke subdifferential. Before we state the theorem, however, we need to recall a useful result on pointwise behavior of weakly convergence sequences (see Proposition 3.6 in [177]). To formulate it we will need the notion of Kuratowski–Painlevé upper limit of a sequence of sets $A_n \subset \mathbb{R}^d$, given by

$$K\text{-}\limsup_{n\to\infty} A_n = \{x \in \mathbb{R}^d : \ x = \lim_{k\to\infty} x_{n_k},\ x_{n_k} \in A_{n_k},\ 0 < n_1 < n_2 < \ldots < n_k < \ldots\}.$$

**Theorem 3.22.** *Let $(\Theta, \Sigma, \mu)$ be a $\sigma$-finite measure space, and let $d$ be a positive natural number, and $1 \leq p < \infty$. Let $u_n \to u$ weakly in $L^p(\Theta)^d$ and $u_n(\theta) \in G(\theta)$ for $\mu$-a.e. $\theta \in \Theta$ and all $n \in \mathbb{N}$, where $G(\theta) \subset \mathbb{R}^d$ is a nonempty, compact set for $\mu$-a.e. $\theta \in \Theta$. Then*

$$u(x) \in \overline{\mathrm{conv}}\left(K\text{-}\limsup_{n\to\infty} \{u_n(x)\}\right).$$

**Theorem 3.23.** *Let $(\Theta, \Sigma, \mu)$ be a $\sigma$-finite measure space and let $W = L^p(\Theta)^d$ with a positive natural number $d$ and $1 < p < \infty$. Assume that the multifunction $h : \Theta \times \mathbb{R}^d \to 2^{\mathbb{R}^d}$ satisfies the assumptions*

*$(h1)$ for every $s \in \mathbb{R}^d$ the multifunction $\theta \to h(\theta, s)$ has a measurable selection, i.e., there exists a measurable function $f : \Omega \to \mathbb{R}^d$ such that $f(\theta) \in h(\theta, s)$ for $\mu$-a.e. $\theta \in \Theta$,*

*$(h2)$ the set $h(\theta, s)$ is nonempty, compact, and convex in $\mathbb{R}^d$, for all $s \in \mathbb{R}^d$ and for $\mu$-a.e. $\theta \in \Theta$,*

*$(h3)$ graph of $s \to h(\theta, s)$ is a closed set in $\mathbb{R}^{2d}$ for $\mu$-a.e. $\theta \in \Theta$,*

*$(h4)$ we have $|\xi| \leq c_1(\theta) + c_2|s|^{p-1}$ for all $s \in \mathbb{R}^d$, $\mu$-a.e. $\theta \in \Theta$ and all $\xi \in h(\theta, s)$, with $c_2 > 0$ and $c_1 \in L^q(\Theta)$, where $\frac{1}{q} + \frac{1}{p} = 1$.*

*Then the multivalued operator* $H: W \to 2^{W'}$ *defined by*

$$H(u) = \{ \xi \in W' : \ \xi(\theta) \in h(\theta, u(\theta)) \quad \mu\text{-a.e. on} \quad \Theta \},$$

*satisfies*

$(H1)$ *H has nonempty and convex values,*
$(H2)$ *the graph of H is sequentially closed in* $(W, \text{strong}) \times (W', \text{weak})$ *topology,*
$(H3)$ $\|\xi\|_W \le \|c_1\|_{L^q(\Theta)} + c_2\|u\|_W^{p-1}$ *for all* $w \in W$ *and* $\xi \in H(u)$.

*Proof.* Let $u \in W$. First we will prove $(H3)$. Let $\xi \in H(u)$. We have

$$\|\xi\|_{W'} = \sup_{\|w\|_W=1} \int_\Theta \xi(\theta) \cdot w(\theta)\, d\mu \le \sup_{\|w\|_W=1} \int_\Theta (c_1(\theta) + c_2|u(\theta)|^{p-1})|w(\theta)|\, d\mu.$$

Hence we have

$$\|\xi\|_{W'} \le \sup_{\|w\|_W=1} (\|c_1\|_{L^q(\Theta)}\|w\|_W + c_2\|u\|_W^{p-1}\|w\|_W) = \|c_1\|_{L^q(\Theta)} + c_2\|u\|_W^{p-1}, \tag{3.40}$$

and $(H3)$ is proved. We pass to the proof of $(H1)$. Let $u \in W$. Consider a multifunction $\Theta \ni \theta \to h(\theta, u(\theta)) \subset \mathbb{R}^d$. We will show that this multifunction has a measurable selection. Let $v_n : \Theta \to \mathbb{R}^d$ be a sequence of simple (i.e., piecewise constant and measurable) functions such that $|v_n(\theta)| \le |u(\theta)|$ and $v_n(\theta) \to u(\theta)$ a.e. $\theta \in \Theta$. By $(h1)$ it follows that there exists the sequence of measurable functions $\xi_n : \Theta \to \mathbb{R}^d$ such that $\xi_n(\theta) \in h(\theta, v_n(\theta))$ a.e. $\theta \in \Theta$. By (3.40) it follows that

$$\|\xi_n\|_{W'} \le \|c_1\|_{L^q(\Theta)} + c_2\|v_n\|_W^{p-1} \le \|c_1\|_{L^q(\Theta)} + c_2\|u\|_W^{p-1},$$

and hence, for a subsequence, not renumbered, we have

$$\xi_n \to \xi \quad \text{weakly in} \quad W' \quad \text{for certain} \quad \xi \in W'. \tag{3.41}$$

Since for a.e. $\theta \in \Theta$ we have

$$|\xi_n(\theta)| \le c_1(\theta) + c_2|v_n(\theta)| \le c_1(\theta) + c_2|u(\theta)|,$$

we can define $G(\theta) = \{z \in \mathbb{R}^d : |z| \le c_1(\theta) + c_2|u(\theta)|\}$ and we can use Theorem 3.22 concluding that

$$\xi(\theta) \in \overline{\text{conv}} \left( K\text{-}\limsup_{n\to\infty} \{\xi_n(\theta)\} \right) \quad \text{a.e.} \quad \theta \in \Theta.$$

Now $z \in K\text{-}\limsup_{n\to\infty}\{\xi_n(\theta)\}$, whenever $\xi_{n_k}(\theta) \to z$. Keeping in mind that $v_{n_k}(\theta) \to u(\theta)$ and $\xi_{n_k}(\theta) \in h(\theta, v_{n_k}(\theta))$ and we can use $(h3)$ to conclude that $z \in h(\theta, u(\theta))$ for a.e. $\theta \in \Theta$. Hence we have

$$\xi(\theta) \in \overline{\text{conv}}\, h(\theta, u(\theta)) = h(\theta, u(\theta)) \quad \text{a.e.} \quad \theta \in \Theta, \tag{3.42}$$

$\xi$ being the required measurable selection. Moreover, (3.41) implies that $\xi \in W'$ and the proof of non-emptiness of $H(u)$ is complete. Convexity of $H(u)$ follows straightforwardly from the convexity of $h(\theta, s)$ in $(h2)$, and the proof of $(H1)$ is complete.

Finally, we will establish $(H2)$. Let $u_n \to u$ in $W$ and $\xi_n \to \xi$ weakly in $W'$ with $\xi_n \in H(u_n)$. Passing to a subsequence, we may assume that $u_n(\theta) \to u(\theta)$ for a.e. $\theta \in \Theta$ and $|u_n(\theta)| \leq f(\theta)$ for a.e. $\theta \in \Theta$ with $f \in L^p(\Theta)$. Since we already know that $\xi \in W$ we must prove that $\xi(\theta) \in h(\theta, u(\theta))$ for a.e. $\theta \in \Theta$. The proof of this assertion exactly follows the lines of the proof of (3.42) in $(H1)$, which completes the proof of the theorem. □

*Remark 3.4.* The assertion $(H2)$ of the above Lemma also follows directly by the application of the convergence theorem of Aubin and Cellina (see Theorem 7.2.1 in [8] or Theorem 1.4.1 in [7]).

We will use the notation

$$H(u) = S^q_{h(\cdot,u(\cdot))},$$

as $H(u)$ is the set of all selections of the class $L^q(\Theta)^d$ of the multifunction $\Theta \ni \theta \to h(\theta, u(\theta)) \subset \mathbb{R}^d$. The operator $H$ is known as the *Nemytskii operator for a multifunction h.*

Now we present a version of above theorem valid if the multifunction $h$ has the form of a Clarke subdifferential. In such a case the above theorem simplifies. Note that for a functional of more than one variable, say $\varphi(\theta, s)$, by $\partial\varphi(\theta, s)$ we will always understand the Clarke subdifferential taken with respect to the *last* variable $s$.

**Theorem 3.24.** *Let $(\Theta, \Sigma, \mu)$ be a finite and complete measure space and denote $W = L^p(\Theta)^d$ with $d \in \mathbb{N}$ and $1 < p < \infty$. Assume that the functional $\varphi : \Theta \times \mathbb{R}^d \to \mathbb{R}$ satisfies the assumptions*

($\varphi 1$) *for every $s \in \mathbb{R}^d$ the function $\Theta \ni \theta \to \varphi(\theta, s)$ is measurable,*
($\varphi 2$) *for $\mu$-almost all $\theta \in \Theta$ the function $\mathbb{R}^d \ni s \to \varphi(\theta, s)$ is locally Lipschitz,*
($\varphi 3$) *$|\xi| \leq c_1(\theta) + c_2|s|^{p-1}$ for all $s \in \mathbb{R}^d$, $\mu$-a.e. $\theta \in \Theta$ and all $\xi \in \partial\varphi(\theta, s)$, with $c_2 > 0$ and $c_1 \in L^q(\Theta)$, where $\frac{1}{q} + \frac{1}{p} = 1$.*

*Then the multivalued operator $H: W \to 2^{W'}$ defined by*

$$H(u) = S^q_{\partial\varphi(\cdot,u(\cdot))} = \{\xi \in W' : \xi(\theta) \in \partial\varphi(\theta, u(\theta)) \quad \mu\text{-a.e. on} \quad \Theta\}$$

*satisfies*

$(H1)$ *H has nonempty and convex values,*
$(H2)$ *the graph of H is sequentially closed in* $(W, \text{strong}) \times (W', \text{weak})$ *topology,*
$(H3)$ *we have $\|\xi\|_W \leq \|c_1\|_{L^q(\Theta)} + c_2\|u\|_W^{p-1}$ for all $w \in W$ and $\xi \in H(u)$.*

*Proof.* We will use Theorem 3.23. The hypothesis $(h4)$ follows from $(\varphi 3)$ while the hypotheses $(h2)$ and $(h3)$ follow from the properties of the Clarke subdifferential (Propositions 3.4 and 3.5). It remains to prove $(h1)$. To this end consider the integral functional

$$I(u) = \int_{\Theta} \varphi(\theta, u(\theta))\, d\mu \quad \text{for} \quad u \in L^p(\Theta)^d.$$

We use Theorem 5.6.39 in [84] (see also Theorem 2.7.5 in [72], where, however, $c_1$ is assumed to be a constant and not a function of $\theta$), whence it follows that $I$ is locally Lipschitz and

$$\partial I(u) \subset S^q_{\partial\varphi(\cdot,u(\cdot))}.$$

Let $s \in \mathbb{R}^d$ and let $u_s(\theta) = s$ for $\theta \in \Theta$. As the measure $\mu$ is finite, $u_s \in L^p(\Theta)^d$. Since, by Proposition 3.4, $\partial I(u_s)$ is nonempty, then there exists $\xi \in L^q(\Theta)^d$ such that $\xi(\theta) \in \partial\varphi(\theta, s)$ for $\mu$-a.e. $\theta \in \Theta$ and the proof is complete. □

## 3.9 Clarke Subdifferential: Examples

As an example of the application of the Clarke subdifferential in PDEs let us consider the following example. Let $\Omega$ be a bounded domain in $\mathbb{R}^d$ with Lipschitz boundary and let $g \in L^\infty_{loc}(\mathbb{R})$. Consider the following initial and boundary value problem

$$\begin{aligned} \frac{\partial u(x,t)}{\partial t} - \Delta u(x,t) + g(u(x,t)) &= f(x,t) \quad \text{on} \quad \Omega \times (0,T), \\ u(x,t) &= 0 \quad \text{on} \quad \partial\Omega \times (0,T), \\ u(x,0) &= u_0(x) \quad \text{on} \quad \Omega. \end{aligned}$$

The question arises how to understand the weak solution of the above problem when $g$ is *discontinuous*. It turns out that the function $g$ can be replaced with its multivalued regularization constructed in the following way. First we define the functional

$$j(s) = \int_0^s g(s)\, ds \quad \text{for} \quad s \in \mathbb{R}.$$

As $g$ is locally bounded, $j$ must be locally Lipschitz and it makes sense to define $\partial j(s)$. We replace the above initial and boundary value problem by the following one

$$\begin{aligned}\frac{\partial u(x,t)}{\partial t} - \Delta u(x,t) + \xi(x,t) = f(x,t) \quad &\text{on} \quad \Omega \times (0,T),\\ \xi(x,t) \in \partial j(u(x,t)) \quad &\text{on} \quad \Omega \times (0,T),\\ u(x,t) = 0 \quad &\text{on} \quad \partial\Omega \times (0,T),\\ u(x,0) = u_0(x) \quad &\text{on} \quad \Omega.\end{aligned}$$

This problem is called a *multivalued partial differential equation*, a *partial differential inclusion*, or a *subdifferential inclusion* and it turns out, that under appropriate assumptions of $f, u_0$ and the growth condition on $\partial j$ it has a weak solution understood in the following sense.

*Find* $u \in L^2(0,T;H_0^1(\Omega))$ *with* $u' \in L^2(0,T;H^{-1}(\Omega))$ *and* $\xi \in L^2(0,T;L^2(\Omega))$ *such that* $u(0) = u_0$, $\xi(t) \in S^2_{\partial j(u(t))}$, *and for a.e.* $t \in (0,T)$ *and all* $v \in H_0^1(\Omega)$,

$$\langle u'(t), v\rangle_{H^{-1}(\Omega)\times H_0^1(\Omega)} + (\nabla u(t), \nabla v)_{L^2(\Omega)^d} + (\xi(t) - f(t), v)_{L^2(\Omega)} = 0,$$

Recall that for $v \in L^2(\Omega)$ by $S^2_{\partial j(v)}$ we mean the set of all functions $h \in L^2(\Omega)$ such that $h(x) \in \partial j(v(x))$ a.e. in $\Omega$. If we identify $L^2(0,T;L^2(\Omega)) = L^2(\Omega \times (0,T))$, we can also write $S^2_{\partial j(u(t))} = S^2_{\partial j(u(\cdot,t))}$.

We can associate with the above problem the following problem known as *hemivariational inequality*.

*Find* $u \in L^2(0,T;H_0^1(\Omega))$ *with* $u' \in L^2(0,T;H^{-1}(\Omega))$ *such that* $u(0) = u_0$, *and for a.e.* $t \in (0,T)$ *and all* $v \in H_0^1(\Omega)$,

$$\langle u'(t), v\rangle_{H^{-1}(\Omega)\times H_0^1(\Omega)} + (\nabla u(t), \nabla v)_{L^2(\Omega)^d} + \int_\Omega j^0(u(x,t); v(x))\, dx \geq (f(t), v)_{L^2(\Omega)}.$$

It follows directly from the definition (3.37) that every solution of above partial differential inclusion is also a solution of the hemivariational inequality. The converse also holds true, but this has been shown only recently by Carl (see [49]): his proof is more involved and uses the notion of upper and lower solutions.

The Clarke subdifferential is very easy to compute in finite dimensions using the characterization (3.38), when the number of points in which the functional is not continuously differentiable is a null Lebesgue set.

*Example 1.* Let us consider the example of the functional $\varphi : \mathbb{R} \to \mathbb{R}$

$$\varphi(s) = \begin{cases} -s-2 & \text{for} \quad s < -1,\\ s & \text{for} \quad s \in [-1,1],\\ -s+2 & \text{for} \quad s > 1.\end{cases}$$

This functional is of class $C^1$ on a neighborhood of every $s \notin \{-1, 1\}$, and for the points $\{-1, 1\}$ we take convex hull of left and right derivatives

$$\partial\varphi(s) = \begin{cases} \{-1\} & \text{for} \quad s < -1, \\ [-1, 1] & \text{for} \quad s = -1, \\ \{1\} & \text{for} \quad s \in (-1, 1), \\ [-1, 1] & \text{for} \quad s = 1, \\ \{-1\} & \text{for} \quad s > 1. \end{cases}$$

The potential and its subdifferential are both depicted in Fig. 3.1.

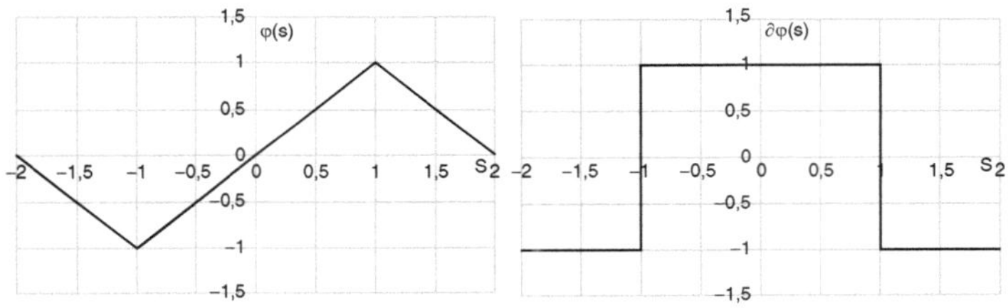


**Fig. 3.1** The example of a simple locally Lipschitz function, which is nondifferentiable in two points, and its Clarke subdifferential

*Example 2.* Let $\mu : [0, \infty) \to \mathbb{R}$ be a continuous function such that $\mu(0) > 0$. Let $s_0 \in \mathbb{R}^d$. Define the functional $\varphi : \mathbb{R}^d \to \mathbb{R}$ by the formula

$$\varphi(s) = \int_0^{|s-s_0|} \mu(r)\, dr. \tag{3.43}$$

It is easy to verify, using the mean value theorem for integrals, that $\varphi$ is locally Lipschitz. It is also easy to calculate the gradient of $\varphi$ for all $s \neq s_0$, which establishes the fact that $\varphi$ is of class $C^1$ on the neighborhood of every $s \neq s_0$. The value of the Clarke subdifferential at $s_0$ is obtained by taking the convex hull of the sphere obtained from limits of the gradients of all sequences converging to $s_0$. We get

$$\partial\varphi(s) = \begin{cases} \left\{\mu(|s-s_0|)\frac{s-s_0}{|s-s_0|}\right\} & \text{for} \quad s \neq s_0, \\ \{y \in \mathbb{R}^d : |y| \leq \mu(0)\} = \bar{B}(0, \mu(0)) & \text{for} \quad s = s_0. \end{cases}$$

Graphs of $\varphi$ and $\partial\varphi$ for $d = 1$, $s_0 = 1$, and $\mu(r) = 1 - r$ are depicted in Fig. 3.2.

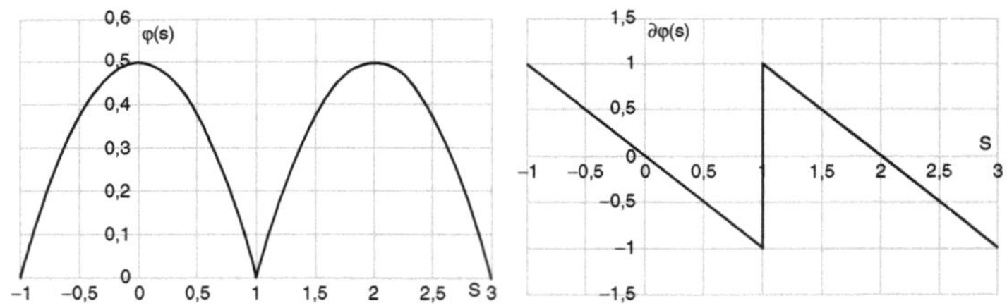


**Fig. 3.2** The second example of a nondifferentiable nonconvex potential and its Clarke subdifferential

*Example 3.* Note that it is not required in Example 2 for $\mu$ to be a continuous function. For example, we can consider the following function $\mu$

$$\mu(r) = \begin{cases} \mu_1 & \text{for} \quad r \in [0, M), \\ \mu_2 & \text{for} \quad r \in [M, \infty), \end{cases}$$

where $\mu_1, M > 0$, and $\mu_2 \in \mathbb{R}$. Then, $\varphi$ calculated by (3.43) is given by

$$\varphi(s) = \begin{cases} \mu_1 |s - s_0| & \text{for} \quad |s - s_0| \in [0, M), \\ \mu_1 M + \mu_2(|s - s_0| - M) & \text{for} \quad |s - s_0| \in [M, \infty). \end{cases}$$

For $s \neq s_0$ and $|s - s_0| \neq M$ the functional $\varphi$ is continuously differentiable. Using this fact and the characterization (3.38) we can calculate its Clarke subdifferential as

$$\partial\varphi(s) = \begin{cases} \bar{B}(0, \mu_1) & \text{for} \quad s = s_0, \\ \left\{\mu_1 \frac{s - s_0}{|s - s_0|}\right\} & \text{for} \quad |s - s_0| \in (0, M), \\ \left\{\frac{\lambda}{M}(s - s_0) : \lambda \in [\mu_1, \mu_2]\right\} & \text{for} \quad |s - s_0| = M, \\ \left\{\mu_2 \frac{s - s_0}{|s - s_0|}\right\} & \text{for} \quad |s - s_0| \in (M, \infty). \end{cases}$$

The plot of $\varphi$ and its Clarke subdifferential for $d = 1$, $s_0 = 0$, $\mu_1 = 2$, $\mu_2 = 1$, and $M = 1$ is presented in Fig. 3.3.

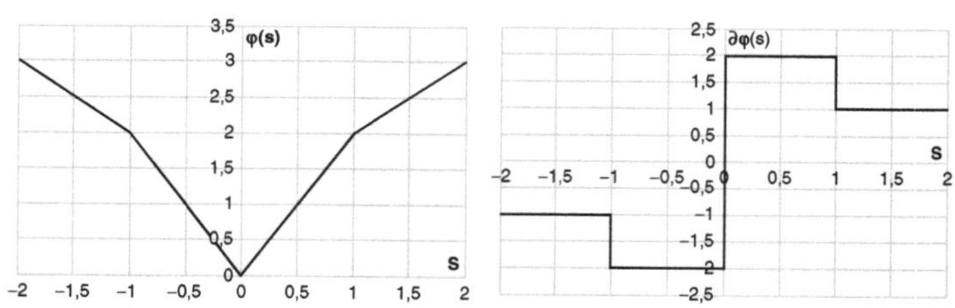


**Fig. 3.3** The third example of a locally Lipschitz potential and its Clarke subdifferential

Finally, we present a brief comparison of the Clarke subdifferential with the convex subdifferential. If the functional $\varphi : X \to \mathbb{R} \cup \{\infty\}$, where $X$ is a Banach space, is convex, then its convex subdifferential is defined as

$$\partial_{conv}\varphi(s) = \begin{cases} \emptyset & \text{if } s \notin \operatorname{dom}\varphi, \\ \{\xi \in X' : \langle \xi, v - s\rangle_{X' \times X} \leq \varphi(v) - \varphi(s) \;\; \forall\, v \in \operatorname{dom}\varphi\} & \text{otherwise,} \end{cases}$$

where $\operatorname{dom}\varphi$ is the set of all points in which $\varphi$ in not equal to $\infty$. If a functional $\varphi : X \to \mathbb{R}$ is both convex and locally Lipschitz then both subdifferentials must coincide (see, e.g., Proposition 2.2.7 in [72]). There are, however, examples where only one of the two notions makes sense. For the functionals which are locally Lipschitz and nonconvex we can speak only of the Clarke subdifferential, while for convex functionals which assume infinite values (and such functionals are useful, for example, in constrained optimization or in unilateral problems in solid mechanics) only the convex subdifferential makes sense.

## 3.10 Comments and Bibliographical Notes

The preliminary material provided in this chapter (and much more) can be found, for example, in the following monographs devoted to the Navier–Stokes equations: [13, 75, 88, 99, 108, 146, 156, 157, 218, 219, 221]. More information about differential inclusions and theory of multifunctions can be found in [7, 8]. For the theory of Clarke subdifferential and hemivariational inequalities we refer the reader to [50, 71, 72, 114, 173, 177, 184]. The applications of this theory in contact mechanics are described in [177, 217].

# 4 Stationary Solutions of the Navier–Stokes Equations

> *Now I think hydrodynamics is to be the root of all physical science, and is at present second to none in the beauty of its mathematics.*
>
> – William Thomson, 1st Baron Kelvin

In this chapter we introduce some basic notions from the theory of the Navier–Stokes equations: the function spaces $H$, $V$, and $V'$, the Stokes operator $A$ with its domain $D(A)$ in $H$, and the bilinear form $B$. We apply the Galerkin method and fixed point theorems to prove the existence of solutions of the nonlinear stationary problem, and we consider problems of uniqueness and regularity of solutions.

## 4.1 Basic Stationary Problem

Let $Q = [0, L]^3$ and define the spaces

$$H = \{u = (u_1, u_2, u_3) \in \dot{L}^2_p(Q)^3 \ : \ \operatorname{div} u = 0\},$$

and

$$V = \{u = (u_1, u_2, u_3) \in \dot{H}^1_p(Q)^3 \ : \ \operatorname{div} u = 0\},$$

with the norms

$$\|u\|_H = |u| = \sqrt{\int_Q |u(x)|^2 dx}, \quad |u|^2 = (u, u),$$

and

$$\|u\|_V = \|u\| = \sqrt{\int_Q |\nabla u(x)|^2 dx}, \quad \|u\|^2 = (\nabla u, \nabla u),$$


G. Łukaszewicz, P. Kalita, *Navier–Stokes Equations*, Advances in Mechanics and Mathematics 34, DOI 10.1007/978-3-319-27760-8_4

respectively. $V$ and $H$ are Hilbert spaces and $V \subset H$. If we identify $H$ with its dual $H'$ in view of the Riesz–Fréchet representation theorem, we have $V \subset H \subset V'$ with continuous embeddings and with each space dense in the following one.

### 4.1.1 The Stokes Operator

We define the Stokes operator $A : V \to V'$ by

$$\langle Au, v\rangle = \int_Q \nabla u \cdot \nabla v \, dx \quad \text{for all} \quad v \in V. \tag{4.1}$$

Using this definition we can write the weak formulation of the Stokes problem: Given $f \in V'$, find $u \in V$ and $p \in \dot{L}^2(Q)$ such that

$$-\nu \Delta u + \nabla p = f, \quad \operatorname{div} u = 0, \tag{4.2}$$

in the weak sense, that is,

$$\nu(\nabla u, \nabla v) - (p, \operatorname{div} v) = \langle f, v\rangle \quad \text{for all} \quad v \in \dot{H}^1_p(Q)^3, \tag{4.3}$$

in an equivalent abstract form

$$\nu\langle Au, v\rangle = \langle f, v\rangle \quad \text{for all} \quad v \in V. \tag{4.4}$$

We shall prove that for $f \in H$, the Stokes operator coincides with the minus Laplacian, $Au = -\Delta u$. To this end we use the explicit spectral representation of functions in $\dot{L}^2(Q)^3$. Assume in the beginning that $f \in \dot{L}^2(Q)^3$. Then

$$f(x) = \sum_{k\in Z^3} f_k e^{2\pi i k \frac{x}{L}}, \quad \sum_{k\in Z^3} |f_k|^2 < \infty, \quad f_{-k} = \bar{f}_k, \qquad f_0 = 0, \tag{4.5}$$

where $f_k = (f_k^1, f_k^2, f_k^3)$. Let $u \in V$ and $p \in \dot{L}^2(Q)$ satisfy (4.3). Assuming that

$$u(x) = \sum_{k\in Z^3} u_k e^{2\pi i k \frac{x}{L}}, \quad p(x) = \sum_{k\in Z^3} p_k e^{2\pi i k \frac{x}{L}} \tag{4.6}$$

with $u_0 = 0$, $k \cdot u_k = 0$, $p_0 = 0$, $u_{-k} = \bar{u}_k$, $p_{-k} = \bar{p}_k$ we obtain,

$$\nu \frac{4\pi^2 |k|^2}{L^2} u_k + \frac{2\pi i k}{L} p_k = f_k.$$

Multiplying this equation by $k$ and using the relation $k \cdot u_k = 0$ we obtain

$$p_k = \frac{L}{2\pi i} \frac{f_k \cdot k}{|k|^2} \quad \text{for} \quad k \neq 0,$$

and then we compute

$$u_k = \frac{1}{\nu}\frac{L^2}{4\pi^2|k|^2}\left(f_k - k\frac{f_k\cdot k}{|k|^2}\right) \quad \text{for} \quad k \neq 0.$$

Now, if $f \in H$, then $f_k \cdot k = 0$ and the equation for $u_k$ reduces to

$$u_k = \frac{1}{\nu}\frac{L^2}{4\pi^2|k|^2}f_k \quad \text{for} \quad k \neq 0.$$

Thus, the weak solution $u$ of the Stokes problem (4.4) (with $\nu = 1$) coincides with the weak solution of the problem $-\Delta u = f$ for $f \in H$. From this we see immediately that $u \in \dot{H}^2_p(Q)^3$. Moreover, from the above explicit formulas for the Fourier coefficients $u_k$ and $p_k$ we have the following regularity theorem.

**Theorem 4.1.** *If $f \in \dot{L}^2_p(Q)^3$, then $u \in \dot{H}^2_p(Q)^3$, $p \in \dot{H}^1_p(Q)$, and*

$$\|u\|^2_{\dot{H}^2_0(Q)^3} + \|p\|^2_{\dot{H}^1_0(Q)} \leq \frac{L^2}{4\pi^2}\left(\frac{L^2}{\pi^2\nu^2}+1\right)\|f\|^2_{\dot{L}^2(Q)^3}.$$

*Proof.* We have

$$\begin{aligned}
\|u\|^2_{\dot{H}^2_0(Q)^3} &= \sum_{k\in Z^3}|k|^4|u_k|^2 = \sum_{k\in Z^3}|k|^4\frac{1}{\nu^2}\frac{L^4}{16\pi^4|k|^4}\left|f_k - k\frac{f_k\cdot k}{|k|^2}\right|^2 \\
&\leq 2\sum_{k\in Z^3}|k|^4\frac{1}{\nu^2}\frac{L^4}{16\pi^4|k|^4}|f_k|^2 + 2\sum_{k\in Z^3}|k|^4\frac{1}{\nu^2}\frac{L^4}{16\pi^4|k|^4}|f_k|^2 \\
&= \frac{L^4}{4\pi^4\nu^2}\|f\|^2_{\dot{L}^2(Q)^3},
\end{aligned}$$

and

$$\begin{aligned}
\|p\|^2_{\dot{H}^1_p(Q)} = \sum_{k\in Z^3}|k|^2|p_k|^2 &= \sum_{k\in Z^3}|k|^2\frac{L^2}{4\pi^2}\left|\frac{f_k\cdot k}{|k|^2}\right|^2 \\
&\leq \sum_{k\in Z^3}\frac{L^2}{4\pi^2}|f_k|^2 = \frac{L^2}{4\pi^2}\|f\|^2_{\dot{L}^2(Q)^3},
\end{aligned}$$

which ends the proof. □

*Remark 4.1.* Observe that setting $f$, $u$, and $p$ of the form (4.5) and (4.6), respectively, directly to Eqs. (4.2) we would get the same spectral representations of the solution $u, p$.

Thus, in particular,

$$D(A) = \{u \in V \;:\; Au \in H\} = V \cap \dot{H}^2_p(Q)^3.$$

### 4.1.2 The Nonlinear Problem

Let us consider the stationary problem:

$$\begin{aligned} -\nu\Delta u + (u\cdot\nabla)u + \nabla p = f \quad &\text{in} \quad Q\,, \\ \operatorname{div} u = 0 \quad &\text{in} \quad Q\,, \end{aligned}$$

with one of the boundary conditions:

1. $Q = [0, L]^3$ in $\mathbb{R}^3$ and we assume periodic boundary conditions, or
2. $Q$ is a bounded domain in $\mathbb{R}^3$, with smooth boundary, and we assume the homogeneous Dirichlet boundary condition $u = 0$ on $\partial Q$.

In the second case we define

$$H = \{u = (u_1, u_2, u_3) \in L^2(Q)^3 \;:\; \operatorname{div} u = 0,\; u_n = u\cdot n|_{\partial Q} = 0\},$$

and

$$V = \{u = (u_1, u_2, u_3) \in H_0^1(Q)^3 \;:\; \operatorname{div} u = 0\}\,,$$

with the norms

$$\|u\|_H = |u| = \sqrt{\int_Q |u(x)|^2\,dx}, \quad |u|^2 = (u, u),$$

and

$$\|u\|_V = \|u\| = \sqrt{\int_Q |\nabla u(x)|^2\,dx}, \quad \|u\|^2 = (\nabla u, \nabla u).$$

The Stokes operator $A$ is defined in the same way as in the periodic case (cf. (4.1)) and we have

$$D(A) = \{u \in V \;:\; Au \in H\} = V \cap H^2(Q)^3.$$

For both cases, let us denote $b(u, v, w) = ((u\cdot\nabla v), w)$ for $u, v, w \in V$ and define a nonlinear operator $B : V \to V'$ by $(Bu, v) = b(u, u, v)$ for all $v \in V$. We shall also write $((u, v))$ instead of $(\nabla u, \nabla v)$.

**Exercise 4.1.** Prove that for all $u, v, w \in V$ it is

$$b(u, v, w) = -b(u, w, v),$$

and thus $b(u, v, v) = 0$.

**Exercise 4.2.** Prove that for all $u, v, w \in V$,

$$|b(u, v, w)| \leq C(Q)\|u\| \|v\| \|w\|.$$

*Hint.* Use the Hölder inequality and the continuous embedding of $L^4(Q)$ in $V$.

Let $H$ and $V$ be the corresponding function spaces. Then the weak formulation of the above nonlinear stationary problem is as follows.

**Problem 4.1.** For $f \in H$ (or $f \in V'$) find $u \in V$ such that

$$\nu((u, v)) + b(u, u, v) = \langle f, v\rangle \quad \text{for all} \quad v \in V$$

or, equivalently,

$$\nu Au + Bu = f \quad \text{in} \quad H \quad \text{or} \quad V'.$$

*Remark 4.2.* Concerning regularity of solutions one can prove that if $f \in H$ then all solutions $u$ belong to $D(A)$.

We shall prove the following existence theorem.

**Theorem 4.2.** *Let $Q = [0, L]^3$ (for the periodic problem) or let $Q$ be a bounded domain in $\mathbb{R}^3$ with smooth boundary (for the Dirichlet problem). Then*

*(i) For every $f \in V'$ and $\nu > 0$ there exists at least one solution of Problem 4.1.*
*(ii) If $\nu^2 \geq c_1(Q)\|f\|_{V'}$, then the solution of Problem 4.1 is unique.*

*Proof.* ad ($i$). To prove existence of solutions we use the Galerkin method. For every $m \in \mathbb{N}$ let

$$u_m(x) = \sum_{j=1}^{m} \xi_{j,m}\omega_j(x), \quad \xi_{j,m} \in \mathbb{R},$$

be the approximate solution, where $\omega_j$, $j \in \mathbb{N}$ are the eigenfunctions of the Stokes operator corresponding to the considered problem. This means that

$$\nu((u_m, v)) + b(u_m, u_m, v) = \langle f, v\rangle \quad \text{for all} \quad v \in V_m\,, \tag{4.7}$$

where $V_m = \text{span}\{\omega_1, \ldots, \omega_m\}$. Such solution exists in view of the Brouwer fixed point theorem.

**Exercise 4.3.** Prove that the operator $A : H \supset D(A) \to H$ is a self-adjoint positive operator in $H$ and is an isomorphism from $D(A)$ onto $H$, and its inverse $A^{-1} : H \to H$ is continuous, compact, and self-adjoint so there exists a sequence $\lambda_j, j \in \mathbb{N}, 0 < \lambda_1 \leq \lambda_2 \leq \lambda_3 \leq \ldots, \lambda_j \to \infty$, and a family of elements $\omega_j \in D(A)$ orthonormal in $H$ such that $A\omega_j = \lambda_j\omega_j$ for $j \in \mathbb{N}$, cf. Theorem 3.15.

**Exercise 4.4.** Prove existence of $u_m$ using the Brouwer fixed point theorem.

*Solution to Exercise 4.4.* Consider the ball $B = \{u \in V_m : \|u\| \le \frac{1}{\nu}\|f\|_{V'}\}$ in $V_m$ and the map $\Phi : \hat{u} \to u$ where $\hat{u} \in B$ and $u$ is the unique solution of the linear problem

$$\nu((u, v)) + b(\hat{u}, u, v) = \langle f, v\rangle \quad \text{for all} \quad v \in V_m.$$

Setting $v = u$ in this identity we obtain that $u \in B$. We shall show that $\Phi$ is continuous in $V_m$. To this end, let

$$\nu((w, v)) + b(\hat{w}, u, v) = \langle f, v\rangle \quad \text{for all} \quad v \in V_m.$$

Subtracting the equation for $w$ from that for $u$, setting $v = u - w$, and using the estimate for the solution $v$, we obtain

$$\|u - w\| \le \frac{C(Q)}{\nu^2}\|f\|_{V'}\|\hat{u} - \hat{w}\|,$$

which proves the continuity of $\Phi$. As $V_m$ is a finite dimensional space and $\Phi$ is a continuous map from a convex and compact set $B$ to $B$, we conclude existence of $u_m$, solution of (4.7) in view of the Brouwer fixed point theorem.
From (4.7) we conclude

$$\|u_m\| \le \frac{\|f\|_{V'}}{\nu}.$$

Thus, for a subsequence,

$$u_m \to u \quad \text{weakly in} \quad V,$$

and

$$u_m \to u \quad \text{strongly in} \quad H,$$

as $V$ is compactly embedded in $H$. Passing with $m$ to the limit in Eq. (4.7) we obtain equation

$$\nu((u, v)) + b(u, u, v) = \langle f, v\rangle$$

for all $v$ being a finite linear combination of elements of the basis $\omega_k$, $k \in \mathbb{N}$. Since such elements $v$ are dense in $V$, the existence part of the theorem has been proved.

Proof of *(ii)*. To prove uniqueness of solutions for large viscosity coefficients with respect to mass forces, $\nu^2 > c_1(Q)\|f\|_{V'}$, let us assume that there are two distinct solutions $u_1$ and $u_2$, that is,

$$\nu((u_1, v)) + b(u_1, u_1, v) = \langle f, v\rangle,$$

and

$$\nu((u_2, v)) + b(u_2, u_2, v) = \langle f, v\rangle$$

for all $v \in V$. Setting $v = u_1 - u_2$ and subtracting the second equation from the first one we obtain

$$\begin{aligned}\nu\|u_1 - u_2\|^2 = -b(u_1 - u_2, u_2, u_1 - u_2) &\leq c_1(Q)\|u_1 - u_2\|^2\|u_2\| \\ &\leq \frac{c_1(Q)}{\nu}\|f\|_{V'}\|u_1 - u_2\|^2,\end{aligned}$$

as $\|u_2\| \leq \frac{1}{\nu}\|f\|_{V'}$, whence

$$\left(\nu - \frac{c_1(Q)}{\nu}\|f\|_{V'}\right)\|u_1 - u_2\|^2 \leq 0\,,$$

which gives a contradiction. Thus $u_1 = u_2$. □

**Exercise 4.5.** Prove that Theorem 4.2 is valid also in the two-dimensional case, when $Q \subset \mathbb{R}^2$ is a square (in the periodic case) or a bounded domain with smooth boundary (in the case of the Dirichlet boundary conditions).

In the end we shall show how one can proceed differently to prove existence of the Galerkin approximate solutions $u_m$ satisfying (4.7).

**Theorem 4.3.** *Let $X$ be a finite dimensional Hilbert space, with a scalar product $[\cdot,\cdot]$ and the associated norm $[\cdot]$, and let $P : X \to X$ be a continuous map such that*

$$[P(\xi), \xi] > 0 \quad \text{for} \quad [\xi] = k > 0$$

*for some $k$. Then there exists $\xi \in X$ with $[\xi] \leq k$ for which $P(\xi) = 0$.*

*Proof.* Let $B = B(0, k)$ be a closed ball in $X$, centered at zero and with radius $k$. Assume that $P$ is different from zero in this ball. Then the map

$$\xi \to S(\xi) = -\frac{kP(\xi)}{[P(\xi)]}, \quad S : B \to B \tag{4.8}$$

is continuous in $X$. From the Brouwer fixed point theorem it follows that $S$ has a fixed point in $B$, that is,

$$\xi_0 = -\frac{kP(\xi_0)}{[P(\xi_0)]} \tag{4.9}$$

for some $\xi_0 \in B$. We have $[\xi_0] = k$. Multiplying (4.9) by $\xi_0$ we get

$$[\xi_0]^2 = -\frac{k[P(\xi_0), \xi_0]}{[P(\xi_0)]}\,,$$

which contradicts $[P(\xi), \xi] > 0$ for $[\xi] = k$. Thus $P(\xi) = 0$ for some $\xi \in B$. □

Now, let $X = V_m$ with the norm induced from $V$. Let us define $P = P_m : V_m \to V_m$ by the relation,

$$[P_m(u), v] = ((P_m(u), v)) = \nu((u, v)) + b(u, u, v) - \langle f, v\rangle \tag{4.10}$$

for all $u, v \in V_m$. The map is well defined in view of the Riesz–Fréchet representation theorem. In fact, for every $u$ in $V_m$ the map $v \to \nu((u, v)) + b(u, u, v) - \langle f, v\rangle$ defines a linear and continuous functional on $V_m$. Thus, there exists a unique $P_m(u)$ in $V_m$ satisfying the above relation.

The map $P_m$ is continuous in $V_m$ and

$$\begin{aligned}[P_m(u), u] &= \nu\|u\|^2 + b(u, u, u) - \langle f, u\rangle = \nu\|u\|^2 - \langle f, u\rangle \\ &\geq \nu\|u\|^2 - \|f\|_{V'}\|u\| = \|u\|\{\nu\|u\| - \|f\|_{V'}\}.\end{aligned}$$

Whence, for $k > \frac{\|f\|_{V'}}{\nu}$ and $\|u\| = k$ we have $[P_m(u), u] > 0$. In view of our auxiliary theorem, there exists $u_m$ in $V_m$ such that $P_m(u_m) = 0$, that is, (4.7) is satisfied. □

**Exercise 4.6.** Prove that the maps $S$ defined in (4.8) and $P_m$ defined in (4.10) are continuous.

### *4.1.3 Other Topological Methods to Deal with the Nonlinearity*

In this subsection we shall use the Schauder and, alternatively, the Leray–Schauder fixed points theorems to prove directly the existence of a solution of the nonlinear problem, thus bypassing the Galerkin approximate solutions. When the viscosity is large enough with respect to external forces, we shall use the Banach contraction principle to prove the existence of the unique solution of the nonlinear problem.

**Exercise 4.7.** Let $f \in V'$, $Q \subset \mathbb{R}^3$ as above. Prove that there exists $u \in V$ such that

$$\nu((u, v)) + b(u, u, v) = \langle f, v\rangle \quad \text{for all} \quad v \in V$$

by using

*(a)* the Schauder fixed point theorem,
*(b)* the Leray–Schauder fixed point theorem,
*(c)* assuming that the viscosity is large enough with respect to external forces, use the Banach contraction principle to prove the unique solution $u$.

*Solution to Exercise 4.7. (a)* Let us define a map $T$ in $V$ by

$$\nu((Tu, v)) + b(u, Tu, v) = \langle f, v\rangle \quad \text{for all} \quad v \in V\,.$$

Using the Lax–Milgram lemma we conclude that the map $T$ is well defined. Setting $v = Tu$ we get

$$\nu \|Tu\|^2 = \langle f, Tu \rangle \leq \|f\|_{V'} \|Tu\| ,$$

whence

$$\|Tu\| \leq \frac{1}{\nu} \|f\|_{V'} \equiv r.$$

Define $K = \{v \in V : \|v\| \leq r\}$. Then $T(K) \subset K$. Let us consider the map $T : K \to K$. We shall show that $T$ is continuous and compact in the topology of $V$. Let $u_1, u_2$ be in $K$. We have,

$$\nu((Tu_1, v)) + b(u_1, Tu_1, v) = \langle f, v \rangle \quad \text{for all} \quad v \in V,$$

and

$$\nu((Tu_2, v)) + b(u_2, Tu_2, v) = \langle f, v \rangle \quad \text{for all} \quad v \in V.$$

Subtracting the second equation from the first one we get

$$\nu((Tu_1 - Tu_2, v)) + b(u_1 - u_2, Tu_2, v) + b(u_1, Tu_1 - Tu_2, v) = 0$$

for all $v$ in $V$. Set $v = Tu_1 - Tu_2$ and we get, since $b(u_1, Tu_1 - Tu_2, Tu_1 - Tu_2) = 0$,

$$\|Tu_1 - Tu_2\| \leq \frac{C}{\nu} \|Tu_2\| |u_1 - u_2|_{L^4(Q)^3} \leq \frac{C_1}{\nu} r \|u_1 - u_2\|.$$

The continuity follows immediately. To prove compactness observe that any sequence $(u_k) \subset K$ contains a subsequence which converges strongly in the $L^4$ topology. Thus, compactness follows easily from the first of the above inequalities.
*(b)* Let us define the map $u \to Tu$ in $V$ by

$$\nu((Tu, v)) + b(u, u, v) = \langle f, v \rangle \quad \text{for all} \quad v \in V . \tag{4.11}$$

Using the Riesz–Fréchet representation theorem we conclude that the map $T$ is well defined.

Assume that $\lambda Tu = u$, where $\lambda \in [0, 1]$. We shall prove that all such $u$ are in some ball. If $\lambda = 0$, then $u = 0$. Now, for $\lambda > 0$, $Tu = \frac{1}{\lambda} u$. Setting $v = u$ in (4.11) we obtain $\nu \|u\|^2 \leq \lambda \|f\|_{V'} \|u\|$, whence $\|u\| \leq \frac{1}{\nu} \|f\|_{V'} \equiv M$.

Now, we shall prove continuity and compactness of $T$ in $V$. Let $u_m, u_k$ be in $V$. Then, subtracting the equations for $Tu_m$ and $Tu_k$, and setting $v = Tu_m - Tu_k$ we obtain

$$\nu\|Tu_m - Tu_k\|^2 = b(u_m, Tu_m - Tu_k, u_m - u_k) + b(u_m - u_k, Tu_m - Tu_k, u_k)$$
$$\leq |u_m|_{L^4(Q)^3}\|Tu_m - Tu_k\||u_m - u_k|_{L^4(Q)^3} +$$
$$|u_m - u_k|_{L^4(Q)^3}\|Tu_m - Tu_k\||u_k|_{L^4(Q)^3},$$

whence,

$$\|Tu_m - Tu_k\| \leq \frac{1}{\nu}(|u_m|_{L^4(Q)^3} + |u_k|_{L^4(Q)^3})|u_m - u_k|_{L^4(Q)^3}$$
$$\leq \frac{C}{\nu}(|u_m|_{L^4(Q)^3} + |u_k|_{L^4(Q)^3})\|u_m - u_k\|.$$

To prove the continuity of $T$ in $V$ observe that if $u_m \to u_k$ in $V$ then from the above estimate, $\|Tu_m - Tu_k\| \to 0$. To prove the compactness of $T$ in $V$, let $\{u_m\}$ be a bounded sequence in $V$. Then, it contains a convergent subsequence $\{u_\mu\}$ in $L^4(Q)^3$. Thus, from the above estimate it follows that the sequence $\{Tu_\mu\}$ is convergent.

Using the Leray–Schauder fixed point theorem we conclude that there exists $u$ in $V$ such that $Tu = u$. This $u$ is of course a solution of our stationary Navier–Stokes problem.

*(c)* Let us define the map $T : V \to V$ by

$$\nu((Tu, v)) + b(u, Tu, v) = \langle f, v\rangle \quad \text{for all} \quad v \in V. \tag{4.12}$$

The map is well defined. We shall prove that it is contracting in a ball $B(0, r)$ in $V$ provided the viscosity is large enough with respect to external forces. Let

$$\nu((Tu_1, v)) + b(u_1, Tu_1, v) = \langle f, v\rangle \quad \text{for all} \quad v \in V.$$

We get

$$\nu((Tu - Tu_1, v)) + b(u, Tu - Tu_1, v) + b(u - u_1, Tu_1, v) = 0$$

for all $v \in V$, whence, with $v = Tu - Tu_1$,

$$\|Tu - Tu_1\| \leq \frac{C}{\nu}\|Tu_1\|\|u - u_1\|.$$

From (4.12) it follows that for every $u \in V$, $\|Tu\| \leq \frac{1}{\nu}\|f\|_{V'}$. Therefore,

$$\|Tu - Tu_1\| \leq \frac{C}{\nu^2}\|f\|_{V'}\|u - u_1\|.$$

Consider the closed ball $B = B(0, r)$ in $V$ with $r = \frac{1}{\nu}\|f\|_{V'}$. Assume moreover that $\frac{C}{\nu^2}\|f\|_{V'} < 1$. Then the map $T$ has the following properties: $T : B \to B$ and is contracting on $B$. By the Banach contraction principle, there exists a unique $u \in B(0, r)$ such that $Tu = u$. This $u$ is the unique solution of the stationary Navier–Stokes problem.

**Exercise 4.8.** Prove the Ladyzhenskaya inequality (3.8) for all functions $u \in \dot{H}^1_p(Q)$, $Q = [0,L]^2$, knowing that it holds for all functions $u \in H^1_0(\Omega)$, where $\Omega$ is a bounded set in $\mathbb{R}^2$.

**Exercise 4.9.** Using the methods from Exercise 4.7 prove existence of solutions of the stationary Navier–Stokes problems in the two-dimensional case, for periodic and homogeneous boundary conditions, respectively. You may find useful the Ladyzhenskaya inequality (3.8).

## 4.2 Comments and Bibliographical Notes

There are many other important and interesting problems concerning stationary solutions of the Navier–Stokes equations. We mention problems of flows with nonhomogeneous Dirichlet boundary conditions or with other types of boundary conditions, flows in nonsmooth, unbounded, or exterior domains. Still other problems concern behavior of solutions when the viscosity tends to zero, bifurcation of solutions, and the structure of the set of stationary solutions.

In Chaps. 5 and 6 we shall consider stationary flows appearing in applications in the theory of lubrication.

For thorough treatment of the basic mathematical problems of the Navier–Stokes equations we refer the reader to monographs: [75, 88, 99, 108, 146, 156, 218, 219], and, for the compressible case, to [157, 187].

# 5
# Stationary Solutions of the Navier–Stokes Equations with Friction

In this chapter we consider the three-dimensional stationary Navier–Stokes equations with multivalued friction law boundary conditions on a part of the domain boundary. We formulate two existence theorems for the formulated problem. The first one uses the Kakutani–Fan–Glicksberg fixed point theorem, and the second one, with the relaxed assumptions, is based on the cut-off argument.

## 5.1 Problem Formulation

Let $\Omega \subset \mathbb{R}^3$ be a bounded domain with Lipschitz boundary as schematically presented in Fig. 5.1. The boundary $\partial\Omega$ is divided into two disjoint parts $\partial\Omega = \overline{\Gamma_D} \cup \overline{\Gamma_C}$. Assume that the viscosity $\nu$ is a positive constant and the mass force density $f$ does not depend on time. Consider the stationary system of the Navier–Stokes equations

$$-\nu\Delta u + (u \cdot \nabla)u + \nabla p = f \text{ in } \Omega, \tag{5.1}$$

$$\operatorname{div} u = 0 \text{ in } \Omega, \tag{5.2}$$

with the following boundary conditions.

On $\Gamma_D$ we assume the Dirichlet boundary condition $u = 0$. Moreover we let

$$u_n = 0 \text{ on } \Gamma_C, \tag{5.3}$$

$$-T_\tau \in h(u_\tau) \text{ on } \Gamma_C, \tag{5.4}$$

where $u_n = u \cdot n$ is the outward normal component of the velocity vector, its tangent component is given by $u_\tau = u - u_n n$ on $\Gamma_C$, and $T_\tau = Tn - (Tn \cdot n)n$ is the tangential part of the stress vector $Tn$



G. Łukaszewicz, P. Kalita, *Navier–Stokes Equations*, Advances in Mechanics and Mathematics 34, DOI 10.1007/978-3-319-27760-8_5

$$T = -pI + 2\nu D(u), \tag{5.5}$$

[cf. (2.24)–(2.25)], and $h : \Gamma_C \times \mathbb{R}^3 \to 2^{\mathbb{R}^3}$ is a *friction multifunction*.

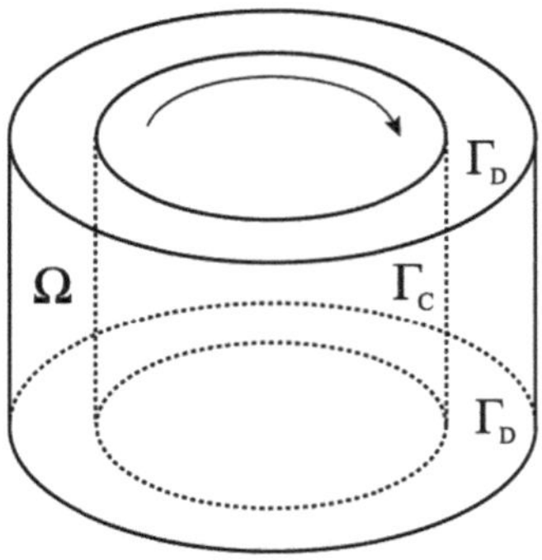


**Fig. 5.1** The domain of the studied problem. The set $\Omega$ is a space between two cylinders. The Dirichlet boundary is the wall of the outer cylinder, and two rings bounding $\Omega$ at *top* and *bottom*. The contact boundary $\Gamma_C$ is the wall of the inner, rotating, cylinder. On $\Gamma_D$ we impose the homogeneous Dirichlet boundary condition, and on $\Gamma_C$ we impose the frictional contact condition which allows for the occurrence of either stick, or slip between the liquid and the rotating cylinder

**Exercise 5.1.** Prove that $T_\tau = 2\nu D(u)_\tau$ on $\Gamma_C$, where $D(u)_\tau = D(u)n - D(u)n \cdot n$.

## 5.2 Friction Operator and Its Properties

We will prove the existence of a weak solution for two cases. In the first case, in Sect. 5.4 we will assume the *linear growth condition* on the friction multifunction $h$ (see ($h4$) below) with the restriction on the growth constant (see ($h5$) below) and we will use the Kakutani–Fan–Glicksberg theorem to obtain the solution. Next, in Sect. 5.5 we will relax the conditions on $h$, namely instead of the linear growth condition we will assume the weaker growth condition with exponent $p-1$ where $2 \le p < 4$ (see ($h6$) below) This possibility comes from the fact that in three space dimensions the trace operator is compact from the space $H^1(\Omega)$ to $L^p(\partial\Omega)$ with $1 \le p < 4$. Moreover we will need no restriction on the constant in the growth condition. This restriction is clearly nonphysical, i.e., it gives a limitation on the magnitude of the boundary friction force no matter if this force is "dissipative" (i.e., directed opposite to the velocity) or "excitatory" (i.e., directed the same as the velocity). We will replace ($h5$) with its one sided counterpart denoted by ($h7$) below. This condition enforces the limitation on the friction force only if it has the same direction as the velocity. The proof of existence for the relaxed assumptions will use the cut-off argument.

We assume the following properties of the friction multifunction $h$:

$(h1)$ for every $s \in \mathbb{R}^3$ the multifunction $x \to h(x,s)$ has a measurable selection, i.e., there exists $g_s : \Gamma_C \to \mathbb{R}^3$, a measurable function, such that $g_s(x) \in h(x,s)$ for all $x \in \Gamma_C$,

$(h2)$ the set $h(x,s)$ is nonempty, compact, and convex in $\mathbb{R}^3$, for all $s \in \mathbb{R}^3$ and for a.e. $x \in \Gamma_C$,

$(h3)$ the graph of $s \to h(x,s)$ is closed in $\mathbb{R}^6$ for a.e. $x \in \Gamma_C$,

$(h4)$ for all $s \in \mathbb{R}^3$, a.e. $x \in \Gamma_C$ and all $\xi \in h(x,s)$, we have $|\xi| \le c_1(x) + c_2|s|$, with some $c_2 \ge 0$ and $c_1 \in L^2(\Gamma_C)$.

Denote $W = L^2(\Gamma_C)^3$ and define a multivalued operator $H\colon W \to 2^W$ by

$$H(u) = \{\, \xi \in W : \ \xi(x) \in h(x,u(x)) \quad \text{a.e. on} \quad \Gamma_C \,\}. \tag{5.6}$$

The following Lemma is a straightforward consequence of Theorem 3.23.

**Lemma 5.1.** *Assume the multifunction h satisfies* $(h1)$–$(h4)$. *Then the operator H given by (5.6) satisfies*

$(H1)$ *H has nonempty and convex values,*

$(H2)$ *the graph of H is sequentially closed in* $(W, \text{strong}) \times (W, \text{weak})$ *topology,*

$(H3)$ *for all* $w \in W$ *and* $\xi \in H(u)$, $\|\xi\|_W \le \|c_1\|_{L^2(\Gamma_C)} + c_2\, \|u\|_W$.

For the first existence result we will need the additional bound on the constant $c_2$ in the growth condition $(h4)$, dependent on the viscosity coefficient and the norm of the trace operator $\|\gamma\|$,

$(h5)$ $c_2\|\gamma\|^2 < 2\nu$.

For the second existence result, the assumptions $(h4)$ and $(h5)$ will be replaced by the following, weaker ones:

$(h6)$ there exists $2 \le p < 4$ such that for all $s \in \mathbb{R}^3$ we have

$$\max_{\xi \in h(x,s)} |\xi| \le c_3(x) + c_4|s|^{p-1} \quad \text{for a.e.} \ \ x \in \Gamma_C,$$

with some $c_3 \in L^2(\Gamma_C)$ and $c_4 \ge 0$,

and

$(h7)$ For all $s \in \mathbb{R}^d$, some $d_1 \in \left[0, \frac{2\nu}{\|\gamma\|^2}\right)$ and $d_2 \in L^2(\Gamma_C)$ we have

$$\min_{\xi \in h(x,s)} \xi \cdot s \ge -d_1|s|^2 - d_2(x) \quad \text{for a.e.} \ \ x \in \Gamma_C.$$

When $(h4)$ is replaced by $(h6)$, the multivalued friction operator is defined on $L^p(\Gamma_C)^3$ in place of $L^2(\Gamma_C)^3$. We will denote $U = L^p(\Gamma_C)^3$. Then $U' = L^q(\Gamma_C)^3$ with $\frac{1}{p} + \frac{1}{q} = 1$ and $q \in \left(\frac{4}{3}, 2\right]$. Define the multifunction $G : U \to 2^{U'}$ by

$$G(u) = \{\xi \in U' \ : \ \xi(x) \in h(x,u(x)) \text{ for a.e. } x \in \Gamma_C\}. \tag{5.7}$$

Analogously to Lemma 5.1 we have the following result which is a straightforward consequence of Theorem 3.23.

**Lemma 5.2.** *Assume the multifunction h satisfies* $(h1)$–$(h4)$. *Then the operator G given by (5.7) satisfies*

(G1) *G has nonempty and convex values,*
(G2) *the graph of G is sequentially closed in* $(U, \text{strong}) \times (U', \text{weak})$ *topology,*
(G3) *for all* $w \in W$ *and* $\xi \in G(u)$, $\|\xi\|_{U'} \leq \|c_3\|_{L^q(\Gamma_C)} + c_4 \|u\|_U$.

## 5.3 Weak Formulation

Let

$$\tilde{V} = \{u \in C^\infty(\overline{\Omega})^3 : \ \operatorname{div} u = 0, \ \ u = 0 \text{ on } \Gamma_D, \ \ u_n = 0 \text{ on } \Gamma_C\}$$

and

$$V = \text{ closure of } \tilde{V} \text{ in } H^1(\Omega)^3 \text{ norm}, \quad W = L^2(\Gamma_C)^3.$$

If $v \in V$ then we have the following Korn inequality (see, for instance, [194] for a proof of this inequality in a general setting)

$$\|v\|^2_{H^1(\Omega)^3} \leq c_K \int_\Omega |D(u)|^2 \, dx. \tag{5.8}$$

Note that the Korn inequality implies that $\|\cdot\|$, where

$$\|u\| = \sqrt{\int_\Omega |D(u)|^2 \, dx} \tag{5.9}$$

is a norm on the space $V$ equivalent to the standard $H^1(\Omega)^3$ norm.
We denote by $\gamma : V \to W$ the trace operator and by $\|\gamma\|$ its norm $\|\gamma\|_{\mathscr{L}(V;W)}$.
We can pass to the derivation of the weak formulation of the studied problem. To this end we assume that $u$, $p$, and $T$ are smooth enough and satisfy (5.1)–(5.4). We multiply (5.1) by $v \in \tilde{V}$ and integrate over $\Omega$, which gives

$$-\nu \int_\Omega u_{i,jj} v_i \, dx + \int_\Omega u_j u_{i,j} v_i \, dx + \int_\Omega p_{,i} v_i \, dx = \int_\Omega f_i v_i \, dx. \tag{5.10}$$

Integrating by parts the term with pressure we obtain

$$\int_\Omega p_{,i} v_i \, dx = -\int_\Omega p v_{i,i} \, dx + \int_{\partial\Omega} p v_n \, dS.$$

Using the fact that div $v = 0$ and $v_n = 0$ on the whole $\partial\Omega$, it follows that

$$\int_\Omega p_{,i} v_i \, dx = 0. \tag{5.11}$$

As div $u = 0$, we have, for smooth $u$, that $u_{j,ij} = 0$ and hence

$$u_{i,jj} v_i = (u_{i,jj} + u_{j,ij}) v_i = (u_{i,j} + u_{j,i})_{,j} v_i \, .$$

From the following Green formula valid for a smooth matrix valued function $F$ ($F : \Omega \to M^{3\times3}$) and $v \in V$

$$\int_\Omega F_{ij,j} v_i + F_{ij} v_{i,j} \, dx = \int_{\partial\Omega} F_{ij} v_i n_j \, dS, \tag{5.12}$$

we have

$$\begin{aligned}
-\nu \int_\Omega u_{i,jj} v_i \, dx &= -2\nu \int_\Omega D(u)_{ij,j} v_i \, dx \\
&= 2\nu \int_\Omega D(u)_{ij} v_{i,j} \, dx - 2\nu \int_{\partial\Omega} D(u)_n v_n + D(u)_\tau \cdot v_\tau \, dS \\
&= 2\nu \int_\Omega D(u)_{ij} v_{i,j} \, dx - \int_{\Gamma_C} T_\tau \cdot v_\tau \, dS.
\end{aligned}$$

Finally, by the symmetry of $D(u)$

$$\begin{aligned}
\int_\Omega D(u)_{ij} v_{i,j} \, dx &= \frac{1}{2} \int_\Omega D(u)_{ij} v_{i,j} \, dx + \frac{1}{2} \int_\Omega D(u)_{ji} v_{j,i} \, dx \\
&= \frac{1}{2} \int_\Omega D(u)_{ij} v_{i,j} \, dx + \frac{1}{2} \int_\Omega D(u)_{ij} v_{j,i} \, dx \\
&= \int_\Omega D(u)_{ij} D(v)_{ij} \, dx.
\end{aligned} \tag{5.13}$$

In this way we have proved that

$$-\nu \int_\Omega u_{i,jj} v_i \, dx = 2\nu \int_\Omega D(u)_{ij} D(v)_{ij} \, dx - \int_{\Gamma_C} T_\tau \cdot v_\tau \, dS. \tag{5.14}$$

**Exercise 5.2.** Prove the Green formula (5.12).

Using (5.11) and (5.14) in (5.10) we can justify the following weak formulation of the considered problem.

**Problem 5.1.** Let $f \in L^2(\Omega)^3$. Find $u \in V$ and $\xi \in H(u_\tau)$ such that for every $v \in V$

$$2\nu \int_\Omega D(u)_{ij} D(v)_{ij}\, dx + \int_{\Gamma_C} \xi \cdot v_\tau\, dS + \int_\Omega (u \cdot \nabla) u \cdot v\, dx = \int_\Omega f \cdot v\, dx. \tag{5.15}$$

*Remark 5.1.* If $\Gamma_C$ is a piecewise flat surface (i.e., $\overline{\Gamma_C} = \bigcup_{k=1}^K \overline{\Gamma_C^k}$, where $n = \text{const}$ on each $\Gamma_C^k$), then for $u, v \in V$

$$2 \int_\Omega D(u)_{ij} D(v)_{ij}\, dx = \int_\Omega \nabla u \cdot \nabla v\, dx, \tag{5.16}$$

and (5.15) could be replaced by

$$\nu \int_\Omega \nabla u \cdot \nabla v\, dx + \int_{\Gamma_C} \xi \cdot v_\tau\, d\Gamma + \int_\Omega (u \cdot \nabla) u \cdot v\, dx = \int_\Omega f \cdot v\, dx.$$

Let us prove (5.16). To this end first assume that $u \in V$ and $v \in \tilde{V}$. On each $\Gamma_C^k$ we can construct an orthonormal basis $n, \tau_1, \tau_2$, where both vectors $\tau_1, \tau_2$ lie in the surface $\Gamma_C^k$. Take $v \in \tilde{V}$. On $\Gamma_C^k$ we have $\nabla v_n = \frac{\partial v_n}{\partial n} n + \frac{\partial v_n}{\partial \tau_1} \tau_1 + \frac{\partial v_n}{\partial \tau_1} \tau_1 = \frac{\partial v_n}{\partial n} n$, where the last equality follows from the fact that $v_n = 0$ on $\Gamma_C^k$. Now, we have

$$\int_\Omega u_{j,i} v_{i,j}\, dx = -\int_\Omega u_j v_{i,ij}\, dx + \int_{\partial\Omega} u_j v_{i,j} n_i\, dS.$$

As $\operatorname{div} v = 0$, and $u = 0$ on $\Gamma_D$, we have

$$\int_\Omega u_{j,i} v_{i,j}\, dx = \int_{\Gamma_C} u_j v_{i,j} n_i\, dS = \int_{\Gamma_C} u \cdot \frac{\partial v_n}{\partial n} n\, dS = \int_{\Gamma_C} (u_n n + u_\tau) \cdot \frac{\partial v_n}{\partial n} n\, dS = 0.$$

The last equality follows from the fact that $u_n = 0$ and $u_\tau$ is orthogonal to $n$ on $\Gamma_C$. Using (5.13) we have

$$\begin{aligned}\int_\Omega D(u)_{ij} D(v)_{ij}\, dx &= \int_\Omega D(u)_{ij} v_{i,j}\, dx = \frac{1}{2}\left(\int_\Omega u_{i,j} v_{i,j}\, dx + \int_\Omega u_{j,i} v_{i,j}\, dx\right) \\ &= \frac{1}{2}\int_\Omega u_{i,j} v_{i,j}\, dx,\end{aligned}$$

and the assertion follows by the density of $\tilde{V}$ in $V$.

We introduce the linear operator $A : V \to V'$ and the bilinear operator $B : V \times V \to V'$ given by

$$\langle Au, v\rangle = 2\int_\Omega D(u)_{ij} D(v)_{ij}\, dx \quad \text{for} \quad u, v \in V, \tag{5.17}$$

and

$$\langle B(u,w), v\rangle = \int_\Omega (u\cdot\nabla)w\cdot v\,dx \quad \text{for} \quad u,w,v\in V,$$

respectively, and the functional $F \in V'$ given by

$$\langle F, v\rangle = \int_\Omega f\cdot v\,dx \quad \text{for} \quad v\in V.$$

Note that, for $u, v, w \in V$ the integral in the definition of $B$ is well defined, as from the Sobolev embedding theorem it follows that $H^1(\Omega) \subset L^4(\Omega)$. Using this notation, Problem 5.1 can be equivalently reformulated as follows.

**Problem 5.2.** Find $u \in V$ and $\xi \in H(u_\tau)$ such that for every $v \in V$ we have

$$\langle \nu Au + B(u,u), v\rangle + (\xi, v_\tau)_W = \langle F, v\rangle.$$

We end this section by proving several auxiliary properties of the operators $A$ and $B$.

**Lemma 5.3.** *The operator $A : V \to V'$ defined by (5.17) is bilinear, continuous, and coercive. Moreover, we have*

$$\langle Au, u\rangle = 2\|u\|^2 \quad \textit{for} \quad u\in V.$$

*Proof.* The assertion follows in a straightforward way from the definition of $A$ and the definition of the norm in $V$, cf. (5.9). □

**Lemma 5.4.** *For $u, v, w \in V$ we have*

$$\langle B(u,w), v\rangle \le c_B\|u\|\,\|w\|\,\|v\|, \tag{5.18}$$

$$\langle B(u,v), v\rangle = 0, \tag{5.19}$$

$$\langle B(u,w), v\rangle = -\langle B(u,v), w\rangle. \tag{5.20}$$

*Proof.* We first prove (5.18). By the Hölder inequality

$$\langle B(u,w), v\rangle \le \|u\|_{L^4(\Omega)^3}\|\nabla w\|_{L^2(\Omega)^{3\times3}}\|v\|_{L^4(\Omega)^3}.$$

From the Sobolev embedding theorem it follows that $\|z\|_{L^4(\Omega)^3} \le C\|z\|_{H^1(\Omega)^3}$ for all $z \in H^1(\Omega)^3$ with a constant $C > 0$. Hence, we have

$$\langle B(u,w), v\rangle \le C^2\|u\|_{H^1(\Omega)^3}\|w\|_{H^1(\Omega)^3}\|v\|_{H^1(\Omega)^3},$$

and (5.18) follows by the Korn inequality (5.8).

For the proof of (5.19), let $u, v \in \tilde{V}$. We have

$$\begin{aligned}\langle B(u,v),v\rangle &= \int_\Omega u_j v_{i,j} v_i \, dx = \frac{1}{2}\int_\Omega u_j \left(\sum_{j=1}^{3} v_i^2\right)_{,j} dx \\ &= -\frac{1}{2}\int_\Omega u_{j,j}\left(\sum_{i=1}^{3} v_i^2\right) dx + \frac{1}{2}\int_{\partial\Omega} u_j n_j \sum_{i=1}^{3} v_i^2 \, dS = 0,\end{aligned}$$

where the last equality follows from the fact that $\operatorname{div} u = 0$ in $\Omega$ and $u_n = u \cdot n = 0$ on $\partial\Omega$. Now as $\tilde{V}$ is dense in $V$, for $u, v \in V$ we can construct sequences $u_m, v_m \in \tilde{V}$ such that $\|v - v_m\| \to 0$ and $\|u - u_m\| \to 0$ as $m \to \infty$. Hence

$$\begin{aligned}\langle B(u,v),v\rangle &= \langle B(u - u_m + u_m, v - v_m + v_m), v - v_m + v_m\rangle \\ &= \langle B(u-u_m, v), v\rangle + \langle B(u_m, v - v_m), v\rangle \\ &\quad + \langle B(u_m, v_m), v - v_m\rangle + \langle B(u_m, v_m), v_m\rangle.\end{aligned}$$

We use the fact that $\langle B(u_m, v_m), v_m\rangle = 0$ and (5.18) to get

$$|\langle B(u,v),v\rangle| \le c_B(\|u - u_m\|\,\|v\|^2 + \|v - v_m\|\,\|u_m\|(\|v\| + \|v_m\|)).$$

The sequences $u_m, v_m$ are bounded in $V$ as they converge in this space, and the sequences $\|u - u_m\|$ and $\|v - v_m\|$ converge to zero, whence the assertion is proved. Finally, (5.20) is a straightforward consequence of (5.19). For the proof it suffices to substitute $v + w$ in place of $v$ in (5.19). □

**Lemma 5.5.** *The mapping $V^2 \ni (w,u) \to B(w,u) \in V'$ is sequentially weakly continuous, i.e.*

$$\begin{aligned}&w_k \to w \quad \text{and} \quad u_k \to u \quad \text{weakly in} \quad V \qquad (5.21)\\ &\qquad\Rightarrow \langle B(w_k,u_k),v\rangle \to \langle B(w,u),v\rangle \quad \text{for all} \quad v \in V.\end{aligned}$$

*Proof.* First, take $v \in \tilde{V}$ and let $u_k \to u$, $w_k \to w$ weakly in $V$. We will prove that

$$\langle B(w_k, v), u_k\rangle \to \langle B(w,v),u\rangle. \qquad (5.22)$$

As the embedding $V \subset L^2(\Omega)^3$ is compact we have $u_k \to u$ and $w_k \to w$ strongly in $L^2(\Omega)^3$. Moreover,

$$\langle B(w_k,v),u_k\rangle - \langle B(w,v),u\rangle = \langle B(w_k,v),u_k - u\rangle + \langle B(w_k - w, v), u\rangle,$$

whence

$$\begin{aligned}&|\langle B(w_k,v),u_k\rangle - \langle B(w,v),u\rangle| \le \\ &\qquad \|v\|_{C^1(\overline{\Omega})^3}(\|w_k\|_{L^2(\Omega)^3}\|u_k - u\|_{L^2(\Omega)^3} + \|w_k - w\|_{L^2(\Omega)^3}\|u\|_{L^2(\Omega)^3}),\end{aligned}$$

whereas (5.22) is proved, and by (5.20)

$$\langle B(w_k, u_k), v\rangle \to \langle B(w, u), v\rangle. \tag{5.23}$$

Now, let $v \in V$ and let $v_m \in \tilde{V}$ be a sequence such that $\|v_m - v\| \to 0$ as $m \to \infty$. We have

$$\langle B(w_k, u_k), v\rangle = \langle B(w_k, u_k), v_m\rangle + \langle B(w_k, u_k), v - v_m\rangle.$$

From the fact that the sequences $w_k, u_k$ are bounded in $V$ it follows that

$$\lim_{m\to\infty} \limsup_{k\to\infty} |\langle B(w_k, u_k), v - v_m\rangle| \leq \lim_{m\to\infty} \limsup_{k\to\infty} c_B \|w_k\| \|u_k\| \|v - v_m\| = 0,$$

which yields (5.21) and the proof is complete. □

## 5.4 Existence of Weak Solutions for the Case of Linear Growth Condition

Define

$$S = \left\{ (\xi, w) \in W \times V \ : \|\xi\|_W \leq \frac{2\nu \|c_1\|_{L^2(\Gamma_C)} + c_2 \|\gamma\| \|F\|_{V'}}{2\nu - c_2 \|\gamma\|^2}, \right.$$
$$\left. \|w\| \leq \frac{\|\gamma\| \|c_1\|_{L^2(\Gamma_C)} + \|F\|_{V'}}{2\nu - c_2 \|\gamma\|^2} \right\}. \tag{5.24}$$

We formulate the main theorem of this section.

**Theorem 5.1.** *Assume (h1)–(h5). Then Problem 5.1 has a solution $(\xi, u) \in S$. Moreover, any solution of Problem 5.1 must belong to S.*

The proof of the above theorem will be done through several lemmas. First, we formulate the following auxiliary problem.

**Problem 5.3.** Fix $\xi \in W$ and $w \in V$. Find $u \in V$ such that for every $v \in V$ we have

$$\langle \nu A u + B(w, u), v\rangle + (\xi, v_\tau)_W = \langle F, v\rangle. \tag{5.25}$$

**Lemma 5.6.** *For a given $(\xi, w) \in W \times V$ the auxiliary Problem 5.3 has exactly one solution $u \in V$.*

*Proof.* Define the mapping $M : V \to V'$ by $M(u) = \nu A u + B(w, u)$. Obviously this mapping is linear. By Lemma 5.3 and (5.18), $M$ is bounded, and hence also continuous. Moreover, by Lemma 5.3 and (5.13)

$$\langle Mu, u\rangle = \nu\langle Au, u\rangle = \nu 2\|u\|^2,$$

whence $M$ is also coercive. Consider the functional $G : V \to \mathbb{R}$ defined as $G(v) = \langle F, v\rangle - (\xi, v_\tau)_W$. Then $G$ is linear and

$$|G(v)| \le \|F\|_{V'}\|v\| + \|\xi\|_W\|v_\tau\|_W \le (\|F\|_{V'} + \|\xi\|_W\|\gamma\|)\|v\|,$$

whence $G$ is bounded and, in consequence, it belongs to $V'$. By the Lax–Milgram lemma, it follows that Problem 5.3 has exactly one solution $u \in V$. □

In view of the unique solvability of Problem 5.3, we can define the mapping $\Lambda_1 : W \times V \to V$ which assigns to the pair $(\xi, w) \in W \times V$ the unique solution $u \in V$ of (5.25). We can write

$$\langle \nu A\Lambda_1(\xi, w) + B(w, \Lambda_1(\xi, w)), v\rangle + (\xi, v_\tau)_W = \langle F, v\rangle. \tag{5.26}$$

**Lemma 5.7.** *For every $(\xi, w) \in W \times V$ we have*

$$\|\Lambda_1(\xi, w)\| \le \frac{\|\gamma\|}{2\nu}\|\xi\|_W + \frac{\|F\|_{V'}}{2\nu}.$$

*Proof.* Taking the test function $v = \Lambda_1(\xi, w)$ in (5.26) we get

$$2\nu\|\Lambda_1(\xi, w)\|^2 + (\xi, \Lambda_1(\xi, w)_\tau)_W = \langle F, \Lambda_1(\xi, w)\rangle.$$

It follows that

$$2\nu\|\Lambda_1(\xi, w)\|^2 \le \|\xi\|_W\|\gamma\|\|\Lambda_1(\xi, w)\| + \|F\|_{V'}\|\Lambda_1(\xi, w)\|,$$

whence we have the assertion. □

**Lemma 5.8.** *If $\xi_k \to \xi$ weakly in W, $w_k \to w$ weakly in $V$, and $\Lambda_1(\xi_k, w_k) \to \zeta$ weakly in $V$, then $\zeta = \Lambda_1(\xi, w)$, i.e., the graph of $\Lambda_1$ is a sequentially closed set in* $(W, \text{weak}) \times (V, \text{weak}) \times (V, \text{weak})$ *topology.*

*Proof.* Let $\xi_k \to \xi$ weakly in W, $w_k \to w$ weakly in $V$, and $\Lambda_1(\xi_k, w_k) \to \zeta$ weakly in $V$. We have

$$\langle \nu A\Lambda_1(\xi_k, w_k) + B(w_k, \Lambda_1(\xi_k, w_k)), v\rangle + (\xi_k, v_\tau)_W = \langle F, v\rangle,$$

for $v \in V$. We pass to the limit in all terms in the last equation. As $A$ is linear and continuous, it is also weakly sequentially continuous, and we have

$$\langle \nu A\Lambda_1(\xi_k, w_k), v\rangle \to \langle \nu A\zeta, v\rangle.$$

Lemma 5.5 implies that

$$\langle B(w_k, \Lambda_1(\xi_k, w_k)), v\rangle \to \langle B(w, \zeta), v\rangle,$$

and weak convergence $\xi_k \to \xi$ in $W$ implies that

$$(\xi_k, v_\tau)_W \to (\xi, v_\tau)_W.$$

Summarizing, we get

$$\langle \nu A\zeta + B(w, \zeta), v \rangle + (\xi, v_\tau)_W = \langle F, v \rangle,$$

whereas $\zeta = \Lambda_1(\xi, w)$ and the proof is complete. □

We define the multifunction $\Lambda : W \times V \to 2^{W \times V}$ by the formula

$$\Lambda(\xi, w) = H(w_\tau) \times \{\Lambda_1(\xi, w)\}.$$

We will show that $\Lambda$ satisfies all assumptions of the Kakutani–Fan–Glicksberg fixed point theorem, cf. Theorem 3.8, and, in consequence, $\Lambda$ has a fixed point $(\xi, w)$. We observe that every fixed point of $\Lambda$ must be a solution of Problem 5.1. Indeed, assume that $(\xi, w)$ is a fixed point of $\Lambda$. Then $\xi \in H(w_\tau)$ and $w = \Lambda_1(\xi, w)$. From the definition of $\Lambda_1$ it follows that

$$\langle \nu Aw + B(w, w), v \rangle + (\xi, v_\tau)_W = \langle F, v \rangle \text{ for all } v \in V.$$

Clearly, $(\xi, w)$ solves Problem 5.1. Hence, be showing the existence of a fixed point of $\Lambda$ we prove the existence of a solution for Problem 5.1.

**Lemma 5.9.** *Assume (h1)–(h5). Then $\Lambda(S) \subset S$ and if $x \in \Lambda(x)$, then $x \in S$.*

*Proof.* First we prove the assertion that $\Lambda(S) \subset S$. Let $(\eta, u) \in \Lambda(\xi, w)$, where $(\xi, w) \in S$. Then $\eta \in H(w_\tau)$ and $u = \Lambda_1(\xi, w)$. From Lemma 5.7 it follows that

$$\|u\| \le \frac{\|\gamma\|}{2\nu}\|\xi\|_W + \frac{\|F\|_{V'}}{2\nu}. \tag{5.27}$$

By Lemma 5.1 it is possible to use the growth condition ($H3$). We get

$$\|\eta\|_W \le \|c_1\|_{L^2(\Gamma_C)} + c_2\|\gamma\|\|w\|. \tag{5.28}$$

As $(\xi, w) \in S$, we can use the bounds given in the definition (5.24). The estimate (5.27) implies that

$$\|u\| \le \frac{\|\gamma\|}{2\nu}\frac{2\nu\|c_1\|_{L^2(\Gamma_C)} + c_2\|\gamma\|\|F\|_{V'}}{2\nu - c_2\|\gamma\|^2} + \frac{\|F\|_{V'}}{2\nu} = \frac{\|\gamma\|\|c_1\|_{L^2(\Gamma_C)} + \|F\|_{V'}}{2\nu - c_2\|\gamma\|^2},$$

and the estimate (5.28) implies that

$$\|\eta\|_W \le \|c_1\|_{L^2(\Gamma_C)} + c_2\|\gamma\|\frac{\|\gamma\|\|c_1\|_{L^2(\Gamma_C)} + \|F\|_{V'}}{2\nu - c_2\|\gamma\|^2} = \frac{2\nu\|c_1\|_{L^2(\Gamma_C)} + c_2\|\gamma\|\|F\|_{V'}}{2\nu - c_2\|\gamma\|^2}.$$

We have proved that $u$ and $\eta$ satisfy the estimates given in definition (5.24) of the set $S$, whence $(\eta, u) \in S$. The proof of the first assertion is complete.

For the proof of the second assertion assume that $(\eta, u) \in \Lambda(\eta, u)$. We must show that $(\eta, u) \in S$, i.e., $(\eta, u)$ satisfy the bounds given in definition (5.24). By Lemma 5.7 and the growth condition $(H3)$ in Lemma 5.1 we have

$$\|u\| \leq \frac{\|\gamma\|}{2\nu}\|\eta\|_W + \frac{\|F\|_{V'}}{2\nu}, \quad \|\eta\|_W \leq \|c_1\|_{L^2(\Gamma_C)} + c_2\|\gamma\|\,\|u\|.$$

Combining the above two inequalities in two different ways we obtain

$$\|u\| \leq \frac{\|\gamma\|}{2\nu}(\|c_1\|_{L^2(\Gamma_C)} + c_2\|\gamma\|\|u\|) + \frac{\|F\|_{V'}}{2\nu},$$

$$\|\eta\|_W \leq \|c_1\|_{L^2(\Gamma_C)} + c_2\|\gamma\|\left(\frac{\|\gamma\|}{2\nu}\|\eta\|_W + \frac{\|F\|_{V'}}{2\nu}\right).$$

A straightforward calculation proves that $(\eta, u)$ satisfy the bounds given in the definition (5.24), whereas indeed $(\eta, u) \in S$. The proof is complete. □

The space $W \times V$ endowed with the $(W, \text{weak}) \times (V, \text{weak})$ topology is a locally convex Hausdorff topological vector space, and the set $S$ is nonempty, compact, and convex in this space. Assuming $(h1)$–$(h4)$, the values of $\Lambda$ are nonempty and convex (see $(H1)$ of Lemma 5.1).

**Lemma 5.10.** *Assume $(h1)$–$(h5)$. Then the graph of $\Lambda$ is sequentially closed in $(W, \text{weak}) \times (V, \text{weak}) \times (W, \text{weak}) \times (V, \text{weak})$.*

*Proof.* Let $\xi_k \to \xi$ weakly in $W$ and $w_k \to w$ weakly in $V$. Let moreover $(\eta_k, u_k)$ be such that $\eta_k \to \eta$ weakly in $W$, $u_k \to u$ weakly in $V$ and $(\eta_k, u_k) \in \Lambda(\xi_k, w_k)$. The last assertion means that $\eta_k \in H(w_{k\tau})$ and $u_k \in \Lambda_1(\xi_k, w_k)$. Lemma 5.8 implies that $u \in \Lambda_1(\xi, w)$. Since the trace operator $\gamma : V \to W$ is compact, we have $w_{k\tau} \to w_\tau$ strongly in $W$. It remains to use $(H2)$ of Lemma 5.1 to conclude that $\eta \in H(w_\tau)$, whence $(\eta, u) \in \Lambda(\xi, w)$. The proof is complete. □

**Lemma 5.11.** *Assume $(h1)$–$(h5)$. Then the graph of $\Lambda|_S$ is a closed set in topology $(W, \text{weak}) \times (V, \text{weak}) \times (W, \text{weak}) \times (V, \text{weak})$.*

*Proof.* Lemma 5.10 implies that the graph of $\Lambda|_S$ is sequentially closed in topology $(W, \text{weak}) \times (V, \text{weak}) \times (W, \text{weak}) \times (V, \text{weak})$. The graph of $\Lambda|_S$ is contained in $S \times S$, a set which is compact, and, by the Eberlein–Šmulyan theorem also sequentially compact in $(W, \text{weak}) \times (V, \text{weak}) \times (W, \text{weak}) \times (V, \text{weak})$ topology. Thus the graph of $\Lambda|_S$ must be sequentially compact, and by the Eberlein–Šmulyan theorem, it is also compact and hence closed. □

We have all the ingredients to prove that main theorem of this section.

*Proof (of Theorem 5.1).* The assertion of the theorem is a straightforward application of the Kakutani–Fan–Glicksberg fixed point theorem and Lemmata 5.9 and 5.11. □

## 5.5 Existence of Weak Solutions for the Case of Power Growth Condition

As in this section we no longer impose the linear growth condition $(h4)$, but we replace it with the more general power growth condition $(h6)$, we cannot interpret anymore the term $\int_{\Gamma_C} \xi \cdot v_\tau \, dS$ in (5.15) as the scalar product in $W$. Instead, we represent this term using the duality between $U$ and $U'$, and the multivalued operator $G$ given by (5.7). We are in position to formulate the following problem.

**Problem 5.4.** Find $u \in V$ and $\xi \in G(u_\tau)$ such that for every $v \in V$ we have

$$\langle \nu A u + B(u,u), v\rangle + \langle \xi, v_\tau\rangle_{U' \times U} = \langle F, v\rangle. \tag{5.29}$$

The main result of this section is the following existence theorem for the above problem.

**Theorem 5.2.** *Assume* $(h1)$–$(h3)$ *and* $(h6)$, $(h7)$. *Then Problem 5.4 has a solution* $(\xi, u) \in W \times V$.

*Proof.* The proof consists of two steps. In the first step we will define the *cut-off friction multifunction* $h_m$ and the associated cut-off Problem 5.5. In the second step we pass to infinity with the cut-off constant and we recover the solution of the original problem.

**Step 1.** Let us first choose a natural number $m$ and define

$$h_m(x,s) = \begin{cases} h(x,s) & \text{if } \ |s| \le m, \\ h\left(x, \frac{ms}{|s|}\right) & \text{if } \ |s| > m. \end{cases}$$

We will show that $h_m$ satisfies $(h1)$–$(h5)$. Naturally, $h_m$ satisfies $(h1)$ and $(h2)$. We will prove that $h_m$ satisfies $(h3)$, i.e., the graph of $s \to h_m(x,s)$ is closed in $\mathbb{R}^6$ for a.e. $x \in \Gamma_C$. To this end, let $s_k \to s$ and $\xi_k \to \xi$ with $\xi_k \in h_m(x, s_k)$. Define

$$r_k = \begin{cases} s_k & \text{if } \ |s_k| \le m, \\ \frac{ms_k}{|s_k|} & \text{if } \ |s_k| > m. \end{cases}$$

From the fact that $\xi_k \in h_m(x, s_k)$ and the definition of $h_m$ it follows that $\xi_k \in h(x, r_k)$. We will consider two cases. If $|s| \le m$, then we have $r_k \to s$ and we use $(h3)$ for $h$ whence it follows that $\xi \in h(x,s) = h_m(x,s)$ for a.e. $x \in \Gamma_C$. In the second case we have $|s| > m$. Then we have $r_k \to \frac{ms}{|s|}$ and we can use $(h3)$ for $h$ to deduce that $\xi \in h\left(x, \frac{ms}{|s|}\right) = h_m(x,s)$ for a.e. $x \in \Gamma_C$. The proof that $h_m$ satisfies $(h3)$ is complete. We will now demonstrate that $h_m$ satisfies $(h4)$ and $(h5)$. Indeed, let $s \in \mathbb{R}^3$ and $\xi \in h_m(x,s)$. By the definition of $h_m$ and $(h6)$ for a.e. $x \in \Gamma_C$ we have $|\xi| \le c_3(x) + c_4 m^{p-1}$, whence $(h4)$ holds with $c_1 = c_3(\cdot) + c_4 m^{p-1} \in L^2(\Gamma_C)$ and $c_2 = 0$, which implies that $h_m$ satisfies also $(h5)$.

We have proved that $h_m$ satisfies $(h1)$–$(h5)$. We can define the multivalued operator $H_m : W \to 2^W$ by the formula

$$H_m(u) = \{\xi \in W \ : \ \xi(x) \in h_m(x, u(x)) \quad \text{a.e. on} \quad \Gamma_C\}.$$

We formulate the following cut-off version of Problem 5.4, where the cut-off parameter is denoted by $m$.

**Problem 5.5.** Find $u_m \in V$ and $\xi_m \in H_m(u_{m\tau})$ such that for every $v \in V$ we have

$$\langle \nu A u_m + B(u_m, u_m), v\rangle + (\xi_m, v_\tau)_W = \langle F, v\rangle. \tag{5.30}$$

As $h_m$ satisfies $(h1)$–$(h5)$ we can use Theorem 5.1 to deduce that for each $m$ there exists $(\xi_m, u_m) \in W \times V$, a solution of Problem 5.5.

**Step 2.** We pass with the cut-off constant $m$ to infinity and we recover the solution of Problem 5.4. Taking $v = u_m$ in (5.30) we get

$$2\nu \|u_m\|^2 + (\xi_m, u_{m\tau})_W \leq \|F\|_{V'} \|u_m\|. \tag{5.31}$$

Now we use $(h7)$ to obtain

$$\begin{aligned}
(\xi_m, u_{m\tau})_W &= \int_{\Gamma_C} \xi_m \cdot u_{m\tau}\, dS \\
&= \int_{\{|u_{m\tau}|\leq m\}} \xi_m \cdot u_{m\tau}\, dS + \int_{\{|u_{m\tau}|>m\}} \frac{|u_{m\tau}|}{m}\xi_m \cdot \frac{m u_{m\tau}}{|u_{m\tau}|}\, dS \\
&\geq -\int_{\{|u_{m\tau}|\leq m\}} d_1|u_{m\tau}|^2 + d_2\, dS - \int_{\{|u_{m\tau}|>m\}} \frac{|u_{m\tau}|}{m}(d_1 m^2 + d_2)\, dS \\
&\geq -d_1\|u_{m\tau}\|_W^2 - \|d_2\|_{L^1(\Gamma_C)} - \frac{1}{m}\|d_2\|_{L^2(\Gamma_C)}\|u_{m\tau}\|_W.
\end{aligned}$$

Using the last inequality in (5.31) we get, by the trace theorem

$$(2\nu - d_1\|\gamma\|^2)\|u_m\|^2 \leq \|F\|_{V'}\|u_m\| + \|d_2\|_{L^1(\Gamma_C)} + \frac{1}{m}\|\gamma\|\|d_2\|_{L^2(\Gamma_C)}\|u_m\|.$$

It follows that the sequence $u_m$ is bounded in $V$ and thus, for a subsequence, denoted by the same index, $u_m \to u$ weakly in $V$. From the fact that $p \in [2,4)$ it follows that the trace operator $\gamma_p : V \to U$ is compact (see, for example, Theorem 2.79 in [50]) and $u_{m\tau} \to u_\tau$ strongly in $U$.

Using $(h6)$ we estimate

$$\begin{aligned}
\|\xi_m\|_{U'}^q &\leq \int_{\{|u_{m\tau}|\leq m\}} (c_3 + |u_{m\tau}|^{p-1})^q\, dS + \int_{\{|u_{m\tau}|>m\}} \left(c_3 + \left|\frac{m u_{m\tau}}{|u_{m\tau}|}\right|^{p-1}\right)^q dS \\
&\leq \int_{\Gamma_C} 2^{q-1}(c_3^q + |u_{m\tau}|^p)\, dS.
\end{aligned}$$

Since $c_3 \in L^2(\Gamma_C) \subset L^q(\Gamma_C)$ and the sequence $u_{m\tau}$ which converges in $U$ strongly is norm bounded in this space, it follows that $\xi_m$ is bounded in $U'$ and hence, for a subsequence, not renumbered, $\xi_m \to \xi$ weakly in $U'$. We can rewrite (5.30) as

$$\langle \nu A u_m + B(u_m, u_m), v\rangle + \langle \xi_m, v_\tau\rangle_{U'\times U} = \langle F, v\rangle,$$

and we can pass to the limit to obtain (5.29). We only need to show that $\xi \in G(u_\tau)$. Since $u_{m\tau} \to u_\tau$ strongly in $U$, then, for a subsequence, denoted by the same index, $u_{m\tau}(x) \to u_\tau(x)$ with $|u_{m\tau}(x)| \le g(x)$ for a.e. $x \in \Gamma_C$, where $g \in L^p(\Gamma_C)$. Fix $\varepsilon > 0$. By the Luzin theorem there exists a closed set $M_\varepsilon \subset \Gamma_C$, with $m_2(\Gamma_C \setminus M_\varepsilon) < \varepsilon$ and $g$ continuous on $M_\varepsilon$. Define $m_\varepsilon = \max_{x\in M_\varepsilon} g(x)$. For all natural $m \ge m_\varepsilon$ we have $|u_m(x)| \le g(x) \le m_\varepsilon \le m$ a.e. on $M_\varepsilon$ and hence $\xi_m(x) \in h(x, u_{m\tau}(x))$ a.e. on $M_\varepsilon$. We can use $(H2)$ of Theorem 3.23 for the space $Z = L^p(M_\varepsilon)^3$ and the map

$$Z \ni u \to \{\xi \in Z' \; : \; \xi(x) \in h(x, u(x)) \quad \text{a.e. on} \quad M_\varepsilon\} \in 2^{Z'}$$

(indeed $u_{m\tau} \to u_\tau$ strongly in $Z$ and $\xi_m \to \xi$ weakly in $Z'$), whereas a.e. on $M_\varepsilon$ we have $\xi(x) \in h(x, u_\tau(x))$. Since $\varepsilon > 0$ was arbitrary, it follows that $\xi(x) \in h(x, u_\tau(x))$ a.e. on $\Gamma_C$ and the assertion is proved. This completes the proof of Theorem 5.2. □

## 5.6 Comments and Bibliographical Notes

In this chapter the multivalued term has the form of an arbitrary multifunction. In most typical applications this multifunction is a Clarke subdifferential of a certain locally Lipschitz functional, usually nondifferentiable in a discrete or Lebesgue null set of points. In such a case, the multivalued PDE is equivalent to a certain *hemivariational inequality*, as it has been proved by Carl [49] (the proof in [49] is done for an evolution problem but it works well for the static case, too). There are many techniques to prove existence of weak solutions for the multivalued elliptic partial differential equations or hemivariational inequalities. The idea based on the use of surjectivity results for pseudomonotone operators has been extensively exploited in [184]. The application of this technique to stationary incompressible Navier–Stokes problem with the multivalued feedback control boundary condition in the normal direction involving the Bernoulli pressure was presented in [175]. The approach based on the Galerkin method and the finite element method was studied in [114], and the techniques using upper and lower solutions were thoroughly analyzed in [50]. Yet another interesting approach to study the existence of weak solutions using the Kuratowski–Knaster–Mazurkiewicz (KKM) principle was proposed in [78]. The approach based on the Kakutani–Fan–Glicksberg fixed point theorem is, to our knowledge, new.

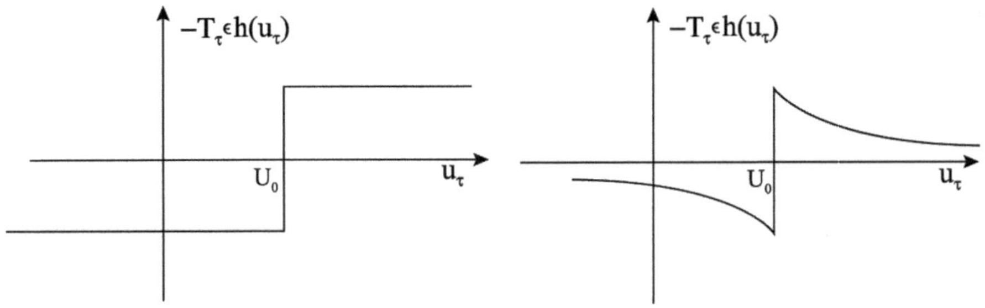


**Fig. 5.2** Two examples of a frictional contact condition. The *vertical line* in both plots represents the stick effect, where the tangential velocity of the fluid $u_\tau$ is equal to the given velocity $U_0$. In such a case the friction condition is actually a (nonhomogeneous) Dirichlet condition. If the stick does not occur, i.e., $u_\tau \neq U_0$ we have a slip effect. In such a case, the tangential stress $T_\tau$ is given as a function of $u_\tau$. The *right plot* depicts the possibility of taking into account the dependence of the friction coefficient of the slip rate. It is a priori unknown which part of the contact boundary is a stick zone, and which part is the slip zone. Of course, in the case of evolution problems, both zones can change with time

The book [177] is devoted to the applications of multivalued PDEs in the continuum mechanics and presents many examples of contact conditions in the solid mechanics leading to multivalued PDEs and hemivariational inequalities. An example of the law that can be used in the model presented in this chapter is the *Tresca* friction law

$$h(x, u_\tau) = \begin{cases} \left\{k_\tau \frac{u_\tau - U_0}{|u_\tau - U_0|}\right\} & \text{for } u_\tau \neq U_0, \\ \overline{B}(0, k_\tau) & \text{otherwise.} \end{cases}$$

The interpretation of this law is the following: on the contact boundary the foundation is moving with the given velocity $U_0 \in \mathbb{R}^3$. If there is no slip, i.e., $u_\tau = U_0$ then the magnitude of the friction force cannot exceed the friction bound $k_\tau$. When the friction force is large enough, its absolute value equal to $k_\tau$, then the slip appears and we have $u_\tau \neq U_0$. The direction of the friction force is always opposite to the vector $u_\tau - U_0$. The Tresca law is presented in the left plot in Fig. 5.2.

Obviously, this example satisfies all the assumptions $(h1)$–$(h5)$, as we have $(h4)$ with $c_2 = 0$. It is possible to generalize the Tresca law, making the friction bound $k_\tau$ dependent on $|u_\tau - U_0|$. In such a case we speak of *slip rate dependent friction*. In particular the situation when $k_\tau$ is decreasing with increasing $|u_\tau - U_0|$ reflects the fact that the kinetic friction is less than the static friction. This, generalized, version of the Tresca law is presented in the right plot in Fig. 5.2.

The dependence of the $k_\tau$ on $|u_\tau - U_0|$ can also be discontinuous, to account for flattening of asperities on the surface. Similar laws were used, for example, in [12, 213] for static problems in elasticity, see Fig. 4.6 in [12] and Fig. 2a in [213] for examples of particular nonmonotone friction laws.

# 6
# Stationary Flows in Narrow Films and the Reynolds Equation

In this chapter we study a typical problem from the theory of lubrication, namely, the Stokes flow in a thin three-dimensional domain $\Omega^\varepsilon$, $\varepsilon > 0$. We assume the Fourier boundary condition (only the friction part) at the top surface and a nonlinear Tresca interface condition at the bottom one.

When $\varepsilon \to 0$ the domains $\Omega^\varepsilon$ become in the limit a two-dimensional domain $\omega$, the mutual bottom of the domains $\Omega^\varepsilon$. Let $(u^\varepsilon, p^\varepsilon)$ be the solution of the considered Stokes problem in $\Omega^\varepsilon$. An important problem in the theory of lubrication, which deals with flows in thin films, is to describe the behavior of solutions $(u^\varepsilon, p^\varepsilon)$ as $\varepsilon \to 0$ in terms of some limit functions $(u^\star, p^\star)$. It happens that the pressure $p^\star$ depends only on two variables which define the surface $\omega$ and satisfies the Reynolds equation, a fundamental equation in this theory. We can say that the Reynolds equation describes a distribution of the pressure "in an infinitely thin domain." Moreover, the boundary conditions imposed on the solutions of the Stokes equation in $\Omega^\varepsilon$ tend in an appropriate sense to some boundary conditions for the Reynolds equation. Our aim is to provide a strict mathematical analysis of this asymptotic problem.

## 6.1 Classical Formulation of the Problem

We consider a boundary value problem for the Stokes equations in $\Omega^\varepsilon$,

$$-\nu\Delta u + \nabla p = 0 \quad \text{in} \quad \Omega^\varepsilon, \tag{6.1}$$

$$\operatorname{div} u = 0 \quad \text{in} \quad \Omega^\varepsilon, \tag{6.2}$$


G. Łukaszewicz, P. Kalita, *Navier–Stokes Equations*, Advances in Mechanics and Mathematics 34, DOI 10.1007/978-3-319-27760-8_6

The domain $\Omega^\varepsilon$ is as follows. Let $\omega \subset \mathbb{R}^2$ be an open bounded set with sufficiently smooth boundary. Then

$$\Omega^\varepsilon = \{(x_1,x_2,x_3) \in \mathbb{R}^3 : \quad (x_1,x_2) \in \omega, \quad 0 < x_3 \leq \varepsilon h(x_1,x_2)\}.$$

The boundary of $\Omega^\varepsilon$ consists of three parts: the bottom $\omega$, the lateral part $\Gamma_L^\varepsilon$, and the top surface $\Gamma_F^\varepsilon$, $\partial\Omega^\varepsilon = \overline{\omega} \cup \overline{\Gamma_L^\varepsilon} \cup \overline{\Gamma_F^\varepsilon}$. Our boundary conditions are:

- At the bottom $\omega$ we assume

$$u_n = u \cdot n = 0 \quad \text{on} \quad \omega. \tag{6.3}$$

The tangential velocity on $\omega$ is unknown and satisfies the *Tresca boundary condition* with the *friction coefficient* $k^\varepsilon$,

$$\begin{aligned} |T_\tau| = k^\varepsilon &\Longrightarrow \exists\lambda \geq 0 \quad u_\tau = s - \lambda T_\tau, \\ |T_\tau| < k^\varepsilon &\Longrightarrow \qquad u_\tau = s, \end{aligned} \tag{6.4}$$

where $|.|$ denotes the Euclidean norm in $\mathbb{R}^2$, and $s$ is the velocity of the lower surface $\omega$, $n = (n_1,n_2,n_3)$ is the outward unit normal to $\Omega^\varepsilon$, and

$$u_n = u \cdot n, \quad u_\tau = u - u_n n.$$

Boundary condition (6.3)–(6.4) is a form of the well-known Coulomb solid–solid interface law [91] applied here for the fluid–solid interface. Another variant, taking into account also the normal stresses, was considered in [21].
- On the lateral boundary $\Gamma_L^\varepsilon$ we assume

$$u = G^\varepsilon \quad \text{on} \quad \Gamma_L^\varepsilon. \tag{6.5}$$

- At the top surface $\Gamma_F^\varepsilon$ we assume the *Fourier boundary condition*

$$T_\tau + l^\varepsilon u = 0 \quad \text{on} \quad \Gamma_F^\varepsilon, \tag{6.6}$$

and

$$u \cdot n = 0 \quad \text{on} \quad \Gamma_F^\varepsilon, \quad \Gamma_F^\varepsilon \quad \text{given by} \quad x_3 = h^\varepsilon(x_1,x_2) = \varepsilon h(x_1,x_2). \tag{6.7}$$

Two-dimensional cross section of the domain of the studied problem is presented in Fig. 6.1.

We have to specify the data $G^\varepsilon$ in (6.5), $l^\varepsilon$ in (6.6), and $h^\varepsilon$ in (6.7), depending on $\varepsilon$. We will specify this dependence in Sect. 6.3.

We assume that $G^\varepsilon \in (H^1(\Omega^\varepsilon))^3$ with $\operatorname{div} G^\varepsilon = 0$ in $\Omega^\varepsilon$, $G^\varepsilon \cdot n = 0$ on $\Gamma_F^\varepsilon \cup \omega$. We assume also that $h$ satisfies

$$0 < h(x) \le h_M, \quad h \in C^2(\overline{\omega}), \tag{6.8}$$

and that $l^\varepsilon > 0$. In (6.4) and (6.6) $T_\tau$ is the tangential part of the stress vector $Tn$ on the boundary $\partial\Omega^\varepsilon$, where the stress tensor $T$ is given by

$$T_{ij} = -p\,\delta_{ij} + 2\nu D_{ij}(u) \quad \text{where} \quad D_{ij}(u) = \frac{1}{2}\left(\frac{\partial u_i}{\partial x_j} + \frac{\partial u_j}{\partial x_i}\right) \quad \text{for} \quad 1 \le i,j \le 3,$$

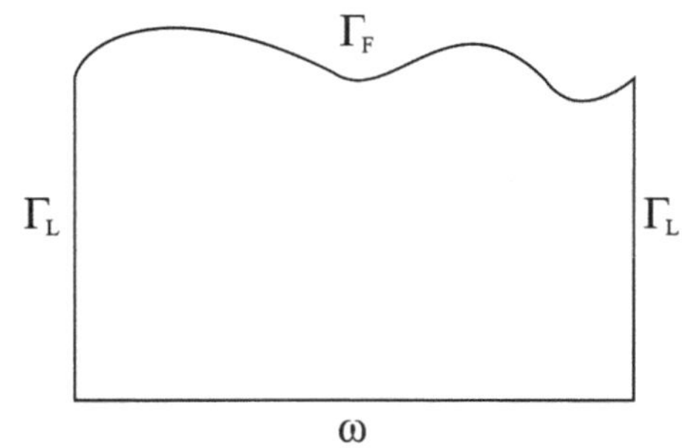


**Fig. 6.1** Two-dimensional cross section of the domain of the studied problem. For $\varepsilon = 1$ we denote $\Gamma_F^\varepsilon = \Gamma_F$ and $\Gamma_L^\varepsilon = \Gamma_L$. On $\Gamma_F$ we assume the Fourier condition, on $\Gamma_L$, the (nonhomogeneous) Dirichlet condition, and on $\omega$, the Tresca condition

namely

$$T_\tau = Tn - T_n n, \tag{6.9}$$

where $T_n = (Tn) \cdot n$ is the normal part of the stress tensor on the boundary $\partial\Omega^\varepsilon$.

*Remark 6.1.* In [15, 21] a similar problem was considered, it differed from the above in a boundary condition at the top surface; namely, in these papers it was assumed that

$$u = 0 \quad \text{at} \quad \Gamma_F^\varepsilon. \tag{6.10}$$

Condition (6.10) helps much in obtaining the solution. With (6.6) and (6.7) in place of (6.10) the problem is more difficult due to the lack of the usual forms of the Poincaré and Korn inequalities. What is important is to obtain other forms of these inequalities, in which the constants do not explode as $\varepsilon$ tends to zero.

The plan of the rest of the chapter is as follows. In Sect. 6.2 we provide a weak formulation of the problem, prove suitable forms of the Korn and Poincaré inequalities, and obtain key estimates. In Sect. 6.3, taking into account a small parameter, we introduce a scaling, and using estimates from Sect. 6.2 prove a convergence theorem for the rescaled functions. In Sect. 6.4 we establish the limit variational inequality, give precise characterization of the sets to which its solution and admissible test functions belong, and prove the strong convergence of the velocity fields. In Sect. 6.5 the Reynolds equation and the limit boundary conditions are obtained. In the last section we prove the uniqueness of solutions of the limit problem.

## 6.2 Weak Formulation and Main Estimates

We start from remarks on the notation in this chapter. We will denote by $x = (x_1, x_2)$ variables belonging to the set $\omega$. Surface element in integrals will be denoted by $dS$, or, in case of integrals over $\omega$ by $dx$ or $dx_1\, dx_2$. The real variable in the vertical direction, orthogonal to $\omega$ will be denoted by $y$ or $x_3$ and the corresponding length element in integrals by $dy$ or $dx_3$. Volume element in volume integrals will be denoted by $dx\, dy$ or $dx\, dx_3$. Let us define

$$K = \{\varphi \in (H^1(\Omega^\varepsilon))^3 : \quad \varphi - G^\varepsilon = 0 \quad \text{on} \quad \Gamma_L^\varepsilon, \quad \varphi \cdot n = 0 \quad \text{on} \quad \Gamma_F^\varepsilon \cup \omega\}, \tag{6.11}$$

where

$$G^\varepsilon \in (H^1(\Omega^\varepsilon))^3 \quad \text{with} \quad \operatorname{div} G^\varepsilon = 0 \quad \text{in} \quad \Omega^\varepsilon, \quad G^\varepsilon \cdot n = 0 \quad \text{on} \quad \Gamma_1^\varepsilon \cup \omega, \tag{6.12}$$

$$K_d = \{v \in K : \quad \operatorname{div} v = 0 \quad \text{in} \quad \Omega^\varepsilon\},$$

and

$$L_0^2(\Omega^\varepsilon) = \{p \in L^2(\Omega^\varepsilon) : \quad \int_{\Omega^\varepsilon} p\, dx\, dx_3 = 0\}.$$

Our aim is to justify the following weak formulation of problem (6.1)–(6.7).

**Definition 6.1.** *The weak solution of problem (6.1)–(6.7) is a pair $(u, p)$ such that*

$$u \in K_d, \tag{6.13}$$

$$p \in L_0^2(\Omega^\varepsilon),$$

$$a(u, \varphi - u) - \int_{\Omega^\varepsilon} p \operatorname{div} \varphi\, dx\, dx_3 + \int_{\Gamma_F^\varepsilon} l^\varepsilon u(\varphi - u)\, dS$$
$$+ \int_\omega k^\varepsilon (|\varphi - s| - |u - s|)\, dx \geq 0 \quad \text{for all} \quad \varphi \in K, \tag{6.14}$$

where

$$a(u, v) = 2\nu \int_{\Omega^\varepsilon} D_{ij}(u) D_{ij}(v)\, dx\, dx_3. \tag{6.15}$$

*Remark 6.2.* In (6.14), $k^\varepsilon$ comes from the Tresca condition. It is a positive constant and its dependence on $\varepsilon$ shall be defined in Sect. 6.3.

Observe that from condition (6.13) it follows that:

$$u = G^\varepsilon \quad \text{on} \quad \Gamma_L^\varepsilon \quad (\text{cf. } (6.5)),$$

$$u \cdot n = 0 \quad \text{on} \quad \Gamma_F^\varepsilon \cup \omega \quad (\text{cf. } (6.3) \quad \text{and} \quad (6.7)).$$

Tresca and slip conditions are included in (6.14). To obtain (6.14) we observe that

$$-\nu \Delta u_i + p_{,i} = \left[-2\nu(u_{i,j} + u_{j,i}) + p\delta_{ij}\right]_{,j} \equiv -T_{ij,j},$$

as $u_{j,ij} = (u_{j,j})_{,i} = 0$ from $\operatorname{div} u = 0$.
We multiply the equations of motion

$$-T_{ij,j} = 0 \quad (i = 1, 2, 3)$$

by $\varphi_i - u_i$, integrate by parts in $\Omega^\varepsilon$ and add for $i = 1, 2, 3$, to get

$$\int_{\Omega^\varepsilon} T_{ij} \frac{\partial(\varphi_i - u_i)}{\partial x_j}\, dx\, dx_3 - \int_{\partial\Omega^\varepsilon} Tn \cdot (\varphi - u)\, dS = 0. \tag{6.16}$$

Now

$$\int_{\Omega^\varepsilon} T_{ij} \frac{\partial(\varphi_i - u_i)}{\partial x_j}\, dx\, dx_3 = \int_{\Omega^\varepsilon} T_{ij} D_{ij}(\varphi_i - u_i)\, dx\, dx_3,$$

since

$$T_{ij} D_{ij}(v) = \frac{1}{2} T_{ij}(v_{i,j} + v_{j,i}) = \frac{1}{2} T_{ij} v_{i,j} + \frac{1}{2} T_{ji} v_{j,i} = T_{ij} v_{i,j}$$

by symmetry.
From (6.16) and (6.15),

$$\begin{aligned}
\int_{\Omega^\varepsilon} T_{ij} \frac{\partial(\varphi_i - u_i)}{\partial x_j}\, dx\, dx_3 &= \int_{\Omega^\varepsilon} (2\nu D_{ij}(u) - p\, \delta_{ij}) D_{ij}(\varphi - u)\, dx\, dx_3 \\
&= a(u, \varphi - u) - \int_{\Omega^\varepsilon} p \operatorname{div}(\varphi - u)\, dx\, dx_3 \\
&= a(u, \varphi - u) - \int_{\Omega^\varepsilon} p \operatorname{div} \varphi\, dx\, dx_3,
\end{aligned} \tag{6.17}$$

as $\operatorname{div} u = 0$ in $\Omega^\varepsilon$.

The second term in (6.16) is divided into three parts, as $\partial\Omega^\varepsilon = \overline{\Gamma_F^\varepsilon} \cup \overline{\omega} \cup \overline{\Gamma_L^\varepsilon}$. We have, by (6.9) and (6.6),

$$
\begin{aligned}
-\int_{\Gamma_F^\varepsilon} T_{ij}n_j(\varphi_i - u_i)\, dS &= -\int_{\Gamma_F^\varepsilon} [T_\tau + T_n n]\cdot((\varphi-u)_\tau + (\varphi-u)_n n)\, dS \\
&= -\int_{\Gamma_F^\varepsilon} (-l^\varepsilon u)\cdot(\varphi-u)_\tau\, dS \\
&= l^\varepsilon \int_{\Gamma_F^\varepsilon} u\cdot(\varphi-u)\, dS,
\end{aligned}
$$

as $(\varphi-u)_n = n\cdot(\varphi-u) = 0$ on $\Gamma_F^\varepsilon$ by (6.11) and (6.12).

Integration over $\omega$ and the Tresca condition yield

$$
\int_\omega k^\varepsilon(|\varphi - s| - |u-s|)\, dx \geq -\int_\omega Tn\cdot(\varphi-u)\, dx, \tag{6.18}
$$

where

$$
s = (s_1, s_2, 0) = G^\varepsilon|_\omega. \tag{6.19}
$$

Observe that $G^\varepsilon \cdot n = 0$ on $\omega$ by (6.12) so that $G_3^\varepsilon = 0$.

The third integral is

$$
-\int_{\Gamma_L^\varepsilon} Tn\cdot(\varphi-u)\, dS = 0, 1
$$

as $\varphi = u = G^\varepsilon$ on $\Gamma_L^\varepsilon$ by (6.11) and (6.13).

In this way, from (6.16)–(6.18) we get

$$
\begin{aligned}
a(u, \varphi-u) - \int_{\Omega^\varepsilon} p \operatorname{div}\varphi\, dx\, dx_3 + l^\varepsilon \int_{\Gamma_F^\varepsilon} u(\varphi-u)\, dS& \\
+ k^\varepsilon \int_\omega (|\varphi-s| - |u-s|)\, dx \geq 0.&
\end{aligned} \tag{6.20}
$$

*Remark 6.3.* Observe that in $a(.,.)$ there is the Newtonian viscosity $\nu$. In the sequel we set $\nu = 1$.

We use (6.20) to obtain the basic estimate of the velocity given in Lemma 6.4 below.

**Lemma 6.1.** *We have*

$$
\frac{1}{2}a(u,u) + \frac{l^\varepsilon}{2}\int_{\Gamma_F^\varepsilon} |u|^2\, dS + k^\varepsilon \int_\omega |u-s|\, dx \leq \frac{l^\varepsilon}{2}\int_{\Gamma_F^\varepsilon} |G^\varepsilon|^2\, dS + 2\int_{\Omega^\varepsilon} |\nabla G^\varepsilon|^2\, dx\, dx_3. \tag{6.21}
$$

*Proof.* We set $\varphi = G^\varepsilon$ in (6.20) to obtain,

$$a(u,u) + l^\varepsilon \int_{\Gamma_F^\varepsilon} |u|^2\, dS + k^\varepsilon \int_\omega |u - s|\, dx \le a(u, G^\varepsilon) + l^\varepsilon \int_{\Gamma_F^\varepsilon} uG^\varepsilon\, dS, \qquad (6.22)$$

as $\operatorname{div} G^\varepsilon = 0$, and $G^\varepsilon = s$ on $\omega$ by (6.19). Moreover, as $\nu = 1$,

$$\begin{aligned}
a(u, G^\varepsilon) &= 2\int_{\Omega^\varepsilon} D_{ij}(u)D_{ij}(G^\varepsilon)\, dx\, dx_3 \\
&\le 2\int_{\Omega^\varepsilon} (D_{ij}(u)D_{ij}(u))^{\frac{1}{2}}(4D_{ij}(G^\varepsilon)D_{ij}(G^\varepsilon))^{\frac{1}{2}}\, dx\, dx_3 \\
&\le \int_{\Omega^\varepsilon} D_{ij}(u)D_{ij}(u)\, dx\, dx_3 + 4\int_{\Omega^\varepsilon} D_{ij}(G^\varepsilon)D_{ij}(G^\varepsilon))\, dx\, dx_3 \\
&= \frac{1}{2}a(u,u) + 2a(G^\varepsilon, G^\varepsilon) \le \frac{1}{2}a(u,u) + 2\int_{\Omega^\varepsilon} |\nabla G^\varepsilon|^2\, dx\, dx_3. \qquad (6.23)
\end{aligned}$$

From (6.22) and (6.23) we obtain immediately (6.21). □

The desired estimate of the velocity will follow from (6.21) and a form of the Korn inequality, suitable for our problem.

**Lemma 6.2 (A Form of the Korn Inequality).** *We have*

$$\int_{\Omega^\varepsilon} |\nabla(u - G^\varepsilon)|^2\, dx\, dx_3 \le a(u - G^\varepsilon, u - G^\varepsilon) + C(\Gamma_F^\varepsilon)\int_{\Gamma_F^\varepsilon} |u - G^\varepsilon|^2\, dS, \qquad (6.24)$$

*where*

$$C(\Gamma_F^\varepsilon) = 2\|D_2 h^\varepsilon\|_{C(\overline{\omega})}\left(1 + \|D_1 h^\varepsilon\|^2_{C(\overline{\omega})}\right). \qquad (6.25)$$

Observe that we use here $h \in C^2(\overline{\omega})$, cf. (6.8), and that $C(\Gamma_F^\varepsilon)$ is of order $\varepsilon$ if $h^\varepsilon = \varepsilon h$ as in (6.7).

*Proof.* Assume that $v = u - G^\varepsilon$ is a smooth function. The result will follow from density of smooth functions in $K_d$. We have

$$\begin{aligned}
a(v,v) &= \frac{1}{2}\int_{\Omega^\varepsilon}\left(\frac{\partial v_i}{\partial x_k} + \frac{\partial v_k}{\partial x_i}\right)^2 dx\, dx_3 \\
&= \int_{\Omega^\varepsilon}\left(\frac{\partial v_i}{\partial x_k}\frac{\partial v_i}{\partial x_k} + \frac{\partial v_i}{\partial x_k}\frac{\partial v_k}{\partial x_i}\right) dx\, dx_3 \\
&= \int_{\Omega^\varepsilon} |\nabla v|^2\, dx\, dx_3 + \int_{\Omega^\varepsilon} \frac{\partial v_i}{\partial x_k}\frac{\partial v_k}{\partial x_i}\, dx\, dx_3
\end{aligned}$$

$$
\begin{aligned}
&= \int_{\Omega^\varepsilon} |\nabla v|^2\,dx\,dx_3 - \int_{\Omega^\varepsilon} \frac{\partial^2 v_i}{\partial x_k \partial x_i} v_k\,dx\,dx_3 + \int_{\partial\Omega^\varepsilon} \frac{\partial v_i}{\partial x_k} v_k n_i\,dS \\
&= \int_{\Omega^\varepsilon} |\nabla v|^2\,dx\,dx_3 + \int_{\Omega^\varepsilon} \frac{\partial v_i}{\partial x_i}\frac{\partial v_k}{\partial x_k}\,dx\,dx_3 \\
&\quad - \int_{\partial\Omega^\varepsilon} \frac{\partial v_i}{\partial x_i} v_k n_k\,dS + \int_{\partial\Omega^\varepsilon} \frac{\partial v_i}{\partial x_k} v_k n_i\,dS.
\end{aligned}
$$

If div $v = 0$, then we have

$$
\int_{\Omega^\varepsilon} |\nabla v|^2\,dx\,dx_3 = a(v,v) + \int_{\partial\Omega^\varepsilon} \frac{\partial v_i}{\partial x_i} v_k n_k\,dS - \int_{\partial\Omega^\varepsilon} \frac{\partial v_i}{\partial x_k} v_k n_i\,dS. \tag{6.26}
$$

For $v = u - G^\varepsilon$ we have $(u-G^\varepsilon)\cdot n = 0$ on $\partial\Omega^\varepsilon$ as $u \in K_d$ and $G^\varepsilon$ is as in (6.12) thus the first surface integral on the right-hand side equals zero, and the second surface integral reduces to that over $\Gamma_F^\varepsilon \cup \omega$, as $v = u - G^\varepsilon = 0$ on $\Gamma_L^\varepsilon$. We have to estimate the integral

$$
\int_{\Gamma_F^\varepsilon \cup \omega} \frac{\partial(u_i - G_i^\varepsilon)}{\partial x_k}(u_k - G_k^\varepsilon) n_i\,dS = \int_{\Gamma_F^\varepsilon \cup \omega} (u_k - G_k^\varepsilon)\frac{\partial(u_i - G_i^\varepsilon)}{\partial x_k} n_i\,dS. \tag{6.27}
$$

Since $(u - G^\varepsilon)\cdot n = 0$ on $\Gamma_F^\varepsilon \cup \omega$, then $\frac{\partial(u-G^\varepsilon)\cdot n}{\partial x_k} = 0$, i.e.,

$$
\sum_{i=1}^{3} \frac{\partial(u_i - G_i^\varepsilon)}{\partial x_k} n_i = -\sum_{i=1}^{3} (u_i - G_i^\varepsilon) n_{i,k}. \tag{6.28}
$$

But $n_{i,k} = 0$ on $\omega$, so that from (6.27) and (6.28) we have

$$
\left| \int_{\Gamma_F^\varepsilon} (u_k - G_k^\varepsilon)\frac{\partial(u_i - G_i^\varepsilon)}{\partial x_k} n_i\,dS \right| \le \tilde{C}(\Gamma_F^\varepsilon) \int_{\Gamma_F^\varepsilon} |u - G^\varepsilon|^2\,dS, \tag{6.29}
$$

where

$$
\tilde{C}(\Gamma_F^\varepsilon) \le 2 \max_{q \in \Gamma_F^\varepsilon} |n_{i,k}(q)|.
$$

As $\Gamma_F^\varepsilon$ is given by $x_3 = h^\varepsilon(x_1, x_2)$,

$$
n(q) = \frac{\left(-\frac{\partial h^\varepsilon}{\partial x_1}(x_1,x_2), -\frac{\partial h^\varepsilon}{\partial x_2}(x_1,x_2), 1\right)}{\sqrt{1 + |\nabla h^\varepsilon(x)|^2}} = n(x_1, x_2). \tag{6.30}
$$

We compute

$$n_{3,i}(x_1,x_2) = \frac{\partial n_3}{\partial x_i}(x_1,x_2) = -(1+|\nabla h^\varepsilon|^2)^{-\frac{3}{2}}\left(\frac{\partial h^\varepsilon}{\partial x_1}\frac{\partial^2 h^\varepsilon}{\partial x_i \partial x_1} + \frac{\partial h^\varepsilon}{\partial x_2}\frac{\partial^2 h^\varepsilon}{\partial x_i \partial x_2}\right),$$

whence

$$\left|\frac{\partial n_3}{\partial x_i}\right| \leq 2|D_1 h^\varepsilon||D_2 h^\varepsilon| \leq |D_2 h^\varepsilon|(1+|D_1 h^\varepsilon|^2),$$

and similarly

$$|n_{j,i}(x_1,x_2)| = \left|\frac{\partial n_j}{\partial x_i}\right| \leq |D_2 h^\varepsilon|(1+|D_1 h^\varepsilon|^2).$$

Thus $\tilde{C}(\Gamma_F^\varepsilon) \leq C(\Gamma_F^\varepsilon)$, where $C(\Gamma_F^\varepsilon)$ is defined in (6.25). In the end, from (6.26) and (6.29) we obtain (6.24). □

Now, having inequality (6.24), we can estimate $\int_{\Omega^\varepsilon} |\nabla u|^2 \,dx\,dx_3$. We have

$$C(\Gamma_F^\varepsilon)\int_{\Gamma_F^\varepsilon} |u-G^\varepsilon|^2 \,dS \leq 2C(\Gamma_F^\varepsilon)\left(\int_{\Gamma_F^\varepsilon} |u|^2 \,dS + \int_{\Gamma_1^\varepsilon} |G^\varepsilon|^2 \,dS\right), \tag{6.31}$$

and, by (6.23),

$$\begin{aligned} a(u-G^\varepsilon, u-G^\varepsilon) &= a(u,u) + a(G^\varepsilon,G^\varepsilon) - 2a(u,G^\varepsilon) \\ &\leq 2a(u,u) + 5\int_{\Omega^\varepsilon} |\nabla G^\varepsilon|^2 \,dx\,dx_3. \end{aligned} \tag{6.32}$$

Moreover

$$\int_{\Omega^\varepsilon} |\nabla u|^2 \,dx\,dx_3 \leq 2\int_{\Omega^\varepsilon} |\nabla(u-G^\varepsilon)|^2 \,dx\,dx_3 + 2\int_{\Omega^\varepsilon} |\nabla G^\varepsilon|^2 \,dx\,dx_3. \tag{6.33}$$

In the end, from (6.24), along with (6.31)–(6.33) we have

**Lemma 6.3.**

$$\begin{aligned} \int_{\Omega^\varepsilon} |\nabla u|^2 \,dx\,dx_3 \leq 4\,a(u,u) &+ 12\int_{\Omega^\varepsilon} |\nabla G^\varepsilon|^2 \,dx\,dx_3 \\ &+ 4C(\Gamma_F^\varepsilon)\left(\int_{\Gamma_F^\varepsilon} |u|^2 \,dS + \int_{\Gamma_F^\varepsilon} |G^\varepsilon|^2 \,dS\right). \end{aligned} \tag{6.34}$$

The first term on the right-hand side as well as the surface integral $\int_{\Gamma_F^\varepsilon} |u|^2 \,dS$ is estimated in (6.21). From (6.21) and (6.34) we have

**Lemma 6.4.**

$$\int_{\Omega^\varepsilon} |\nabla u|^2 \, dx\, dx_3 \leq \left(28 + \frac{16C(\Gamma_F^\varepsilon)}{l^\varepsilon}\right) \int_{\Omega^\varepsilon} |\nabla G^\varepsilon|^2 \, dx\, dx_3 \qquad (6.35)$$
$$+ 4\left(l^\varepsilon + 2C(\Gamma_F^\varepsilon)\right) \int_{\Gamma_F^\varepsilon} |G^\varepsilon|^2 \, dS.$$

*Remark 6.4.* If we set $\varepsilon l^\varepsilon = \hat{l} \in (0, \infty)$, then $l^\varepsilon \to \infty$ as $\varepsilon \to 0$, and

$$28 + \frac{16C(\Gamma_F^\varepsilon)}{l^\varepsilon} = O(1) \quad \text{as} \quad \varepsilon \to 0,$$
$$4(l^\varepsilon + 2C(\Gamma_F^\varepsilon)) = O\left(\frac{1}{\varepsilon}\right) \quad \text{as} \quad \varepsilon \to 0.$$

Later on we shall show that $\varepsilon \int_{\Omega^\varepsilon} |\nabla u|^2 \, dx\, dx_3 = O(1)$ as $\varepsilon \to 0$. This will come from estimates of the integrals on the right-hand side of (6.35), by means of the $H^1(\Omega)^3$ norm of a given function $\hat{G}$ in $H^1(\Omega)^3$, to which the family of functions $G^\varepsilon$ is suitably related (here and in the sequel we put $\Omega = \Omega^1$).

To be able to estimate $\int_\Omega |u|^2 \, dx\, dy$ we need a suitable form of the Poincaré inequality.

**Lemma 6.5 (A Form of the Poincaré Inequality).** *Set* $h_M^\varepsilon = \max_{\overline{\omega}} |h^\varepsilon(x_1, x_2)|$. *Then*

$$\int_{\Omega^\varepsilon} |u|^2 \, dx\, dx_3 \leq 2h_M^\varepsilon \int_{\Gamma_F^\varepsilon} |u|^2 \, dS + 2(h_M^\varepsilon)^2 \int_{\Omega^\varepsilon} \left|\frac{\partial u}{\partial x_3}\right|^2 dx\, dx_3. \qquad (6.36)$$

*Proof.* Let $(x_1, x_2, x_3) = (x, t)$, $x = (x_1, x_2)$ for short. Then

$$u(x, t) = u(x, h^\varepsilon(x)) - \int_t^{h^\varepsilon(x)} \frac{\partial u}{\partial z}(x, z)\, dz,$$

so that

$$|u(x, t)| \leq 2|u(x, h^\varepsilon(x))|^2 + 2h^\varepsilon(x) \int_0^{h^\varepsilon(x)} \left|\frac{\partial u}{\partial z}(x, z)\right|^2 dz.$$

We integrate over $t \in [0, h^\varepsilon(x)]$ to obtain

$$\int_0^{h^\varepsilon(x)} |u(x, x_3)|^2 dx_3 \leq 2h^\varepsilon(x)|u(x, h^\varepsilon(x))|^2 + 2(h^\varepsilon(x))^2 \int_0^{h^\varepsilon(x)} \left|\frac{\partial u}{\partial z}(x, x_3)\right|^2 dx_3,$$

and, after integration over $\omega$, we obtain (6.36). □

We have now all basic estimates. In Sect. 6.3 we shall state them once again but in new variables: $\hat{u}$ instead of $u^\varepsilon$, etc., defined in $\Omega^1$.

We have the following existence theorem in $\Omega^\varepsilon$.

**Theorem 6.1.** *For the data as in Sects. 6.1 and 6.2, there exists a unique weak solution* $(u,p) = (u^\varepsilon, p^\varepsilon)$ *of problem (6.1)–(6.7),*

$$u^\varepsilon \in K_d \quad \text{and} \quad p^\varepsilon \in L_0^2(\Omega^\varepsilon).$$

*Proof.* For the proof, which requires some more advanced variational methods to be presented in this place, we refer the reader to [15] where the functional $j$ is here given by

$$j(\varphi) = \int_\omega k^\varepsilon |\varphi(x,0) - s|\, dx + l^\varepsilon \int_{\Gamma_F^\varepsilon} u\varphi \, dS. \qquad \square$$

## 6.3 Scaling and Uniform Estimates

Let $\Omega = \Omega^1$ and define, for a velocity $u$ in $\Omega^\varepsilon$,

$$y = \frac{x_3}{\varepsilon}, \quad \hat{u}_i(x,y) = u_i(x,x_3), \quad i = 1,2 \quad \text{and} \quad \hat{u}_3(x,y) = \frac{1}{\varepsilon} u_3(x,x_3),$$

$\hat{u}(x,y) = (\hat{u}_1(x,y), \hat{u}_2(x,y), \hat{u}_3(x,y))$ being a new velocity in $\Omega$.

For the pressure $p$ in $\Omega^\varepsilon$ we define $\hat{p}(x,y) = \varepsilon^2 p(x,x_3)$. Let

$$\hat{l} = \varepsilon l^\varepsilon \quad \text{and} \quad \hat{k} = \varepsilon k^\varepsilon,$$

where $\hat{l}$ and $\hat{k}$ are some positive numbers, $l^\varepsilon$ stands in (6.6), and $k^\varepsilon$ appears in the Tresca condition (6.4).

Now, for a given function $\hat{G}(x,y) = (\hat{G}_1(x,y), \hat{G}_2(x,y), \hat{G}_3(x,y))$ in $H^1(\Omega)^3$ such that $\hat{G}\cdot n = 0$ on $\Gamma_F \cup \omega \quad (\Gamma_F = \Gamma_F^1)$ and

$$\frac{\partial \hat{G}_1}{\partial x_1} + \frac{\partial \hat{G}_2}{\partial x_2} + \frac{\partial \hat{G}_3}{\partial y} = 0 \quad (\operatorname{div} \hat{G} = 0),$$

we define a family of functions $G^\varepsilon$ in $\Omega^\varepsilon$, by

$$\hat{G}_i(x,y) = G_i^\varepsilon(x,x_3), \quad i = 1,2, \quad \hat{G}_3(x,y) = \frac{1}{\varepsilon} G_3^\varepsilon(x,x_3).$$

By $\hat{K}$ we define the set

$$\hat{K} = \left\{ \varphi \in (H^1(\Omega))^3 : \quad \varphi - \hat{G} = 0 \quad \text{on} \quad \Gamma_L, \quad \varphi \cdot n = 0 \quad \text{on} \quad \Gamma_F \cup \omega \right\},$$

where $\Gamma_L = \Gamma_L^1$. Observe that $\operatorname{div} \hat{G} = 0$ in $\Omega$ implies $\operatorname{div} G^\varepsilon = 0$ in $\Omega^\varepsilon$.

It is clear that $\hat{u} = \hat{G}$ on $\Gamma_L$ implies $u = G^\varepsilon$ on $\Gamma_L^\varepsilon$, and $\hat{u} \cdot n = 0$ on $\omega$ implies $u \cdot n = 0$ on $\omega$. Moreover, from (6.30) it follows that also $\hat{u} \cdot n = 0$ on $\Gamma_F$ implies $u \cdot n = 0$ on $\Gamma_F^\varepsilon$.

We have proved in fact that

$$u \in K_d \quad \text{is equivalent to} \quad \hat{u} \in \hat{K}, \ \operatorname{div} \hat{u} = 0.$$

To compare solutions $(u^\varepsilon, p^\varepsilon)$ for various $\varepsilon$ (these solutions exist by Theorem 6.1) we transform them as above to the same domain $\Omega$. The transformed solutions $(\hat{u}^\varepsilon, \hat{p}^\varepsilon)$ are solutions of a suitably transformed problem from Definition 6.1. We shall write the problem precisely later on.
Now, we obtain estimates for $\hat{u} = \hat{u}^\varepsilon$ directly from those obtained for $u = u^\varepsilon$ in Sect. 6.2.

We shall write estimates (6.35) and (6.36) in new variables.

**Lemma 6.6.** *We have*

$$\int_{\Omega^\varepsilon} |\nabla G^\varepsilon|^2 \, dx \, dx_3 \leq \frac{1}{\varepsilon} \int_\Omega |\nabla \hat{G}|^2 \, dx \, dy \quad \text{for all} \quad \varepsilon \in (0, 1]. \tag{6.37}$$

**Exercise 6.1.** Prove (6.37).

**Lemma 6.7.** *We have*

$$\int_{\Gamma_F^\varepsilon} |G^\varepsilon|^2 \, dS \leq C_0(\Omega) \int_\Omega \left( |\hat{G}|^2 + |\nabla \hat{G}|^2 \right) dx \, dy \quad \text{for all} \quad \varepsilon \in (0, 1], \tag{6.38}$$

*where $C_0(\Omega)$ is independent of $\varepsilon$.*

*Proof.* For $i = 1, 2$ we have

$$\begin{aligned} \int_{\Gamma_F^\varepsilon} |G_i^\varepsilon|^2 \, dS &= \int_\omega |G_i(x, h^\varepsilon(x))|^2 \sqrt{1 + |\nabla h^\varepsilon(x)|^2} dx \\ &\leq \int_\omega |\hat{G}_i(x, h(x))|^2 \sqrt{1 + |\nabla h^\varepsilon(x)|^2} dx = \int_{\Gamma_F} |\hat{G}_i|^2 \, dS, \end{aligned} \tag{6.39}$$

(by definition $G_i(x, h^\varepsilon(x)) = \hat{G}_i(x, h(x))$), and

$$\begin{aligned}\int_{\Gamma_F^\varepsilon} |G_3^\varepsilon|^2\, dS &= \int_\omega |G_3^\varepsilon(x, h^\varepsilon(x))|^2 \sqrt{1 + |\nabla h^\varepsilon(x)|^2} dx \\ &\le \varepsilon^2 \int_\omega |\hat{G}_3(x, h(x))|^2 \sqrt{1 + |\nabla h(x)|^2} \frac{\sqrt{1 + |\nabla h^\varepsilon(x)|^2}}{\sqrt{1 + |\nabla h(x)|^2}} dx \\ &\le \varepsilon^2 \max_{x \in \omega} \sqrt{1 + |\nabla h^\varepsilon(x)|^2} \int_{\Gamma_F} |\hat{G}_3|^2\, dS. \qquad (6.40)\end{aligned}$$

From (6.39) and (6.40),

$$\int_{\Gamma_F^\varepsilon} |G^\varepsilon|^2\, dS \le \int_{\Gamma_F} |\hat{G}|^2\, dS. \qquad (6.41)$$

As

$$\int_{\Gamma_F} |\hat{G}|^2\, dS \le C_0(\Omega) \int_\Omega \left(|\hat{G}|^2 + |\nabla \hat{G}|^2\right) dx\, dy, \qquad (6.42)$$

where $C_0(\Omega)$ depends only on $\Omega$, (6.41) and (6.42) give (6.38). □

Using inequalities (6.37) and (6.38) in (6.35) and taking into account Remark 6.4 we see that

$$\varepsilon \int_{\Omega^\varepsilon} |\nabla u|^2\, dx\, dx_3 \le C(\Omega, h, \hat{l}) \int_\Omega \left(|\hat{G}|^2 + |\nabla \hat{G}|^2\right) dx\, dy \equiv \tilde{C}_0 \quad \text{for all} \quad \varepsilon \in (0, 1]. \qquad (6.43)$$

Writing (6.43) in terms of $\hat{u}$, we obtain

$$\varepsilon^2 \sum_{i,j=1}^2 \left\| \frac{\partial \hat{u}_i}{\partial x_j} \right\|^2_{L^2(\Omega)} + \sum_{i=1}^2 \left\| \frac{\partial \hat{u}_i}{\partial y} \right\|^2_{L^2(\Omega)} + \varepsilon^2 \left\| \frac{\partial \hat{u}_3}{\partial y} \right\|^2_{L^2(\Omega)} + \varepsilon^4 \sum_{i=1}^2 \left\| \frac{\partial \hat{u}_3}{\partial x_i} \right\|^2_{L^2(\Omega)} \le \tilde{C}_0, \qquad (6.44)$$

and also, as in [15, 21],

$$\left\| \frac{\partial \hat{p}}{\partial x_i} \right\|_{H^{-1}(\Omega)} \le C_1 \quad \text{and} \quad \left\| \frac{\partial \hat{p}}{\partial y} \right\|_{H^{-1}(\Omega)} \le \varepsilon C_2. \qquad (6.45)$$

**Exercise 6.2.** Prove estimates (6.45). First, for $\psi \in H_0^1(\Omega^\varepsilon)$ set $\varphi = u \pm \psi$ in (6.14), where $u = u^\varepsilon$, $p = p^\varepsilon$, to get

$$a(u^\varepsilon, \psi) - \int_{\Omega^\varepsilon} p^\varepsilon \operatorname{div} \psi\, dx\, dx_3 = 0.$$

Further estimates follow from the Poincaré inequality (6.36) written in new variables,

$$\int_{\Omega} |\hat{u}_i|^2 \, dx\, dy \le 2h_M \int_{\Gamma_F} |\hat{u}_i|^2 \, dS + 2h_M^2 \int_{\Omega} \left| \frac{\partial \hat{u}_i}{\partial y} \right|^2 \, dx\, dy \quad \text{for} \quad i = 1,2,3. \tag{6.46}$$

Observe that changing the variables, we obtain the same form of the Poincaré inequality.

**Exercise 6.3.** Prove (6.46).

To estimate the integrals on the left-hand side, we have to estimate the surface integrals on the right-hand side. We have,

$$\begin{aligned}\int_{\Gamma_F} |\hat{u}_i|^2 dS &= \int_{\omega} |\hat{u}_i(x,h(x))|^2 \sqrt{1+|\nabla h(x)|^2} dx \\ &= \int_{\omega} |\hat{u}_i(x,h^{\varepsilon}(x))|^2 \sqrt{1+|\nabla h^{\varepsilon}(x)|^2} \frac{\sqrt{1+|\nabla h(x)|^2}}{\sqrt{1+|\nabla h^{\varepsilon}(x)|^2}} dx \\ &\le \max_{x\in\omega} \sqrt{1+|\nabla h(x)|^2} \int_{\Gamma_F^{\varepsilon}} |u_i|^2 \, dS \quad \text{for} \quad i = 1,2, \end{aligned} \tag{6.47}$$

and

$$\begin{aligned}\int_{\Gamma_F} |\hat{u}_3|^2 \, dS &= \int_{\omega} |\hat{u}_3(x,h(x))|^2 \sqrt{1+|\nabla h(x)|^2} dx \\ &\le \frac{1}{\varepsilon^2} \max_{x\in\omega} \sqrt{1+|\nabla h(x)|^2} \int_{\Gamma_F^{\varepsilon}} |u_3|^2 \, dS. \end{aligned} \tag{6.48}$$

From (6.21) using (6.37) and (6.38) we have

$$\begin{aligned}\int_{\Gamma_F^{\varepsilon}} |u|^2 \, dS &\le \int_{\Gamma_F^{\varepsilon}} |G^{\varepsilon}|^2 \, dS + \frac{4}{l^3} \int_{\Omega^{\varepsilon}} |\nabla G^{\varepsilon}|^2 \, dx\, dx_3 \\ &\le C_0(\Omega) \int_{\Omega} (|\hat{G}|^2 + |\nabla \hat{G}|^2) \, dx\, dy + \frac{4}{\hat{l}} \int_{\Omega} |\nabla \hat{G}|^2 \, dx\, dy \\ &\equiv C(\Omega, \hat{l}, \hat{G}). \end{aligned} \tag{6.49}$$

Thus, from the Poincaré inequality (6.46) with (6.44) and (6.47)–(6.49) we obtain in the end

$$\|\hat{u}_i\|_{L^2(\Omega)}^2 \le C_3, \tag{6.50}$$

$$\varepsilon^2 \|\hat{u}_3\|_{L^2(\Omega)}^2 \le C_4. \tag{6.51}$$

Let

$$V_y = \left\{ v \in L^2(\Omega) \ : \ \frac{\partial v}{\partial y} \in L^2(\Omega) \right\}$$

be the Banach space with norm

$$\|v\|_{V_y} = \left( \|v\|^2_{L^2(\Omega)} + \left\| \frac{\partial v}{\partial y} \right\|^2_{L^2(\Omega)} \right)^{\frac{1}{2}}.$$

**Theorem 6.2.** *There exist $u_i^\star \in V_y$ for $i = 1,2$, $p^\star \in L_0^2(\Omega)$, and a subsequence $\varepsilon \to 0$ such that*

$$\hat{u}_i \to u_i^\star \quad \text{weakly in} \quad V_y \quad \text{for} \quad i = 1,2, \tag{6.52}$$

$$\varepsilon \frac{\partial \hat{u}_i}{\partial x_j} \to 0 \quad \text{weakly in} \quad L^2(\Omega) \quad \text{for} \quad i,j = 1,2, \tag{6.53}$$

$$\varepsilon \frac{\partial \hat{u}_3}{\partial y} \to 0 \quad \text{weakly in} \quad L^2(\Omega), \tag{6.54}$$

$$\varepsilon^2 \frac{\partial \hat{u}_3}{\partial x_i} \to 0 \quad \text{weakly in} \quad L^2(\Omega) \quad \text{for} \quad i = 1,2, \tag{6.55}$$

$$\hat{p} \to p^\star \quad \text{weakly in} \quad L^2(\Omega), \qquad p^\star \text{ depends only on } x, \tag{6.56}$$

$$\varepsilon \hat{u}_3 \to 0 \quad \text{weakly in} \quad L^2(\Omega). \tag{6.57}$$

*Proof.* From (6.44) and (6.50) we have (6.52), and from (6.44) and (6.52) we get (6.53). By (6.53) and $\sum_{j=1}^2 \frac{\partial \hat{u}_j}{\partial x_j} = -\frac{\partial \hat{u}_3}{\partial y}$, (6.54) holds. From (6.51) and (6.44) we deduce (6.55). From (6.45) we obtain (6.56), and from (6.51), $\operatorname{div} \hat{u} = 0$, and a suitable choice of the test function we obtain also (6.57). □

## 6.4 Limit Variational Inequality, Strong Convergence, and the Limit Equation

To obtain the variational inequality for limit functions $u^\star, p^\star$ in $\Omega$ we change variables in the inequality defining weak solutions in $\Omega^\varepsilon$ and then using Theorem 6.2, pass to zero with $\varepsilon$.

First, we write (6.20) in new variables $\hat{u}, \hat{p}$ in $\Omega$. In the place of the test function $\varphi$ in $\Omega^\varepsilon$ we put the new test function $\hat{\varphi}$ in $\Omega$, which is connected with $\varphi$ in the same way as $\hat{u}$ is connected with $u$. For $\varphi \in K$ we have $\hat{\varphi} \in \hat{K}$, so that summing over $K$ with $\varphi$ we sum over $\hat{K}$ with $\hat{\varphi}$. We have, after multiplying (6.20) by $\varepsilon$,

$$\begin{aligned}
&\frac{1}{2}\sum_{i,j=1}^{2} \int_{\Omega}\left(\varepsilon\frac{\partial \hat{u}_i}{\partial x_j}+\varepsilon\frac{\partial \hat{u}_j}{\partial x_i}\right)\left(\varepsilon\frac{\partial}{\partial x_j}(\hat{\varphi}_i-\hat{u}_i)+\varepsilon\frac{\partial}{\partial x_i}(\hat{\varphi}_j-\hat{u}_j)\right)dxdy\\
&+\frac{1}{2}\sum_{i=1}^{2}\int_{\Omega}\left(\frac{\partial \hat{u}_i}{\partial y}+\varepsilon^2\frac{\partial \hat{u}_3}{\partial x_i}\right)\left(\frac{\partial}{\partial y}(\hat{\varphi}_i-\hat{u}_i)+\varepsilon^2\frac{\partial}{\partial x_i}(\hat{\varphi}_3-\hat{u}_3)\right)dxdy\\
&+2\int_{\Omega}\varepsilon\frac{\partial \hat{u}_3}{\partial y}\varepsilon\frac{\partial}{\partial y}(\hat{\varphi}_3-\hat{u}_3)dxdy-\int_{\Omega}\hat{p}\left(\frac{\partial\hat{\varphi}_1}{\partial x_1}+\frac{\partial\hat{\varphi}_2}{\partial x_2}+\frac{\partial\hat{\varphi}_3}{\partial y}\right)dxdy\\
&+\sum_{i=1}^{2}\int_{\omega}\hat{l}\hat{u}_i(x,h(x))[\hat{\varphi}_i(x,h(x))-\hat{u}_i(x,h(x))]\sqrt{1+|\nabla h^{\varepsilon}(x)|^2}dx\\
&+\int_{\omega}\hat{k}(|\hat{\varphi}-s|-|\hat{u}-s|)dx\geq 0. \qquad (6.58)
\end{aligned}$$

If we take all nonlinear terms in (6.58) to the right-hand side and take $\liminf_{\varepsilon\to 0}$ on both sides (in fact on the left-hand side the simple $\lim_{\varepsilon\to 0}$ exists), we obtain the limit variational inequality,

$$\begin{aligned}
&\sum_{i=1}^{2}\tfrac{1}{2}\int_{\Omega}\frac{\partial u_i^\star}{\partial y}\frac{\partial(\hat{\varphi}_i-u_i^\star)}{\partial y}dxdy-\int_{\Omega}p^\star(x)\left(\frac{\partial\hat{\varphi}_1}{\partial x_1}+\frac{\partial\hat{\varphi}_2}{\partial x_2}\right)dxdy\\
&-\int_{\omega}p^\star(x)\left(\hat{\varphi}_1(x,h(x))\frac{\partial h}{\partial x_1}(x)+\hat{\varphi}_2(x,h(x))\frac{\partial h}{\partial x_2}(x)\right)dx\\
&+\sum_{i=1}^{2}\hat{l}\int_{\omega}u_i^\star(x,h(x))\left(\hat{\varphi}_i(x,h(x))-u_i^\star(x,h(x))\right)dx\\
&+\int_{\omega}\hat{k}\left(|\hat{\varphi}-s|-|u^\star-s|\right)dx\geq 0. \qquad (6.59)
\end{aligned}$$

We shall confine ourselves to present the main points of the proof.

1. In the above inequality (6.59) the third components of $u^\star$ and $\hat{\varphi}$ do not occur, so that we can write $\bar{\varphi}=(\hat{\varphi}_1,\hat{\varphi}_2)$, and (6.59) holds for all $\bar{\varphi}$ that are projections of $\hat{\varphi}\in\hat{K}$ on the first two components. Let

$$\Pi(\hat{K})=\{\bar{\varphi}=(\hat{\varphi}_1,\hat{\varphi}_2)\in(H^1(\Omega))^2:\exists\hat{\varphi}_3\in H^1(\Omega),\ \hat{\varphi}=(\hat{\varphi}_1,\hat{\varphi}_2,\hat{\varphi}_3)\in\hat{K}\}.$$

The second line in (6.59) is written in terms of $\bar{\varphi}$, and equals

$$\int_\Omega p^\star(x)\frac{\partial\hat{\varphi}_3}{\partial y}dydx = \int_\omega\int_0^{h(x)} p^\star(x)\frac{\partial\hat{\varphi}_3}{\partial y}dydx = \int_\omega p^\star(x)\hat{\varphi}_3(x,h(x))dx,$$

as $\hat{\varphi}_3(x,0) = 0$ and $\hat{\varphi}_3 = 0$, on $\Gamma_L$. Since $\hat{\varphi}_1 n_1 + \hat{\varphi}_2 n_2 + \hat{\varphi}_3 n_3 = 0$, and

$$\hat{\varphi}_3 = \frac{-1}{n_3}(\hat{\varphi}_1 n_1 + \hat{\varphi}_2 n_2) = \sqrt{1+|\nabla h(x)|^2}\,(\hat{\varphi}_1 n_1 + \hat{\varphi}_2 n_2) = \hat{\varphi}_1\frac{\partial h}{\partial x_1} + \hat{\varphi}_2\frac{\partial h}{\partial x_2},$$

(cf. (6.30)), we have

$$\int_\Omega p^\star(x)\frac{\partial\hat{\varphi}_3}{\partial y}dydx = \int_\omega p^\star(x)\left(\sum_{i=1}^2 \hat{\varphi}_i(x,h(x))\frac{\partial h}{\partial x_i}(x)\right)dx.$$

2. Integrals over $\Gamma_F$. We know that, cf. (6.21), (6.37), and (6.38),

$$l^\varepsilon\int_{\Gamma_F^\varepsilon} u^2\,dS = O\left(\frac{1}{\varepsilon}\right),$$

and from (6.47) and (6.48) we get

$$\int_{\Gamma_F} |\hat{u}_i|^2 dS \le C(h)\int_{\Gamma_F^\varepsilon} |u_i|^2 dS \quad \text{for} \quad i = 1,2,$$

$$\int_{\Gamma_F} |\hat{u}_3|^2 dS \le \frac{1}{\varepsilon^2}C(h)\int_{\Gamma_F^\varepsilon} |u_3|^2 dS,$$

whence from (6.49),

$$\int_{\Gamma_F} |\hat{u}_i|^2 dS \le C(h)C(\Omega,\hat{l},\hat{G}) \quad \text{and} \quad \int_{\Gamma_F} |\hat{u}_3|^2 dS \le \frac{1}{\varepsilon^2}C(h)C(\Omega,\hat{l},\hat{G}). \tag{6.60}$$

From (6.60) we conclude that $\hat{u}_i \to v_i^\star$ weakly in $L^2(\Gamma_F)$, and, since

$$\varepsilon l^\varepsilon\int_{\Gamma_F^\varepsilon} u_i(\varphi_i - u_i)dS$$
$$= \hat{l}\int_\omega \hat{u}_i(x,h(x))\,[\hat{\varphi}_i(x,h(x)) - \hat{u}_i(x,h(x))]\,\sqrt{1+|\nabla h^\varepsilon(x)|^2}dx,$$

we get the terms

$$\hat{l}\int_\omega u_i^\star(x,h(x))\left[\hat{\varphi}_i(x,h(x)) - u_i^\star(x,h(x))\right]dx$$

on the left-hand side of (6.59), as $\sqrt{1+|\nabla h^\varepsilon(x)|^2} \to 1$ strongly in $L^2(\omega)$ with $\varepsilon \to 0$.

The third component (for $i = 3$) is nonpositive in the limit and we can omit it. In fact,

$$\begin{aligned}
\varepsilon l^\varepsilon \int_{\Gamma_F^\varepsilon} u_3(\varphi_3 - u_3) dS &\leq \hat{l} \int_{\Gamma_F^\varepsilon} u_3 \varphi_3 dS \\
&= \hat{l} \int_\omega u_3(x, h^\varepsilon(x)) \varphi_3(x, h^\varepsilon(x)) \sqrt{1 + |\nabla h^\varepsilon(x)|^2} dx \\
&= \hat{l} \int_\omega \varepsilon \hat{u}_3(x, h(x)) \varepsilon \hat{\varphi}_3(x, h(x)) \sqrt{1 + |\nabla h^\varepsilon(x)|^2} dx,
\end{aligned}$$

and from (6.60) $\{\varepsilon \hat{u}_3\}$ is bounded in $L^2(\Gamma_F)$ and

$$\varepsilon \hat{\varphi}_3(x, h(x)) \sqrt{1 + |\nabla h^\varepsilon(x)|^2} \to 0 \quad \text{in} \quad L^2(\omega),$$

so that

$$\lim_{\varepsilon \to 0} \hat{l} \int_{\Gamma_F^\varepsilon} u_3(\varphi_3 - u_3)\, dS \leq 0.$$

## 6.5 Remarks on Function Spaces

In this subsection we give precise characterization (e.g., in terms of density theorems) of the admissible test functions in the limit inequality (6.59), crucial for further considerations (strong convergence of the velocity field, limit equations, uniqueness).

Consider the limit variational inequality (6.59). The function space to which the test functions belong is

$$\begin{aligned}
\Pi(\hat{K}) := \{\bar{\varphi} = (\hat{\varphi}_1, \hat{\varphi}_2) \in (H^1(\Omega))^2 : \exists \hat{\varphi}_3 \in H^1(\Omega),\ \hat{\varphi} = (\hat{\varphi}_1, \hat{\varphi}_2, \hat{\varphi}_3) \in \hat{K}, \\
\text{that is, } \hat{\varphi} \in (H^1(\Omega))^3,\ \hat{\varphi} - \hat{G} = 0 \text{ on } \Gamma_L,\ \hat{\varphi} \cdot n = 0 \text{ on } \Gamma_F \cup \omega\}.
\end{aligned}$$

Here, $\hat{G} \in \hat{K}$, $\operatorname{div} \hat{G} = \frac{\partial \hat{G}_1}{\partial x_1} + \frac{\partial \hat{G}_2}{\partial x_2} + \frac{\partial \hat{G}_3}{\partial y} = 0$ in $\Omega$. Denote $\bar{G} = (\hat{G}_1, \hat{G}_2)$ for $\hat{G} = (\hat{G}_1, \hat{G}_2, \hat{G}_3) \in \hat{K}$. We would like to characterize the space $\Pi(\hat{K})$ in terms of $\hat{\varphi}_1, \hat{\varphi}_2$ only. Instead of working directly with $\Pi(\hat{K})$ it will be easier to consider the space $\Pi_0(\hat{K}) := \{\bar{\varphi} \in (H^1(\Omega))^2 : \bar{\varphi} + \bar{G} \in \Pi(\hat{K})\}$. Thus,

$$\begin{aligned}
\Pi_0(\hat{K}) = \{\ \bar{\varphi} = (\hat{\varphi}_1, \hat{\varphi}_2) \in (H^1(\Omega))^2 : \quad \text{there exists} \quad \hat{\varphi}_3 \in H^1(\Omega) \quad \text{such that} \\
\hat{\varphi} = (\hat{\varphi}_1, \hat{\varphi}_2, \hat{\varphi}_3) = 0 \quad \text{on} \quad \Gamma_L, \quad \hat{\varphi} \cdot n = 0 \quad \text{on} \quad \Gamma_F \cup \omega\}. \qquad (6.61)
\end{aligned}$$

We shall characterize $\Pi_0(\hat{K})$ in terms of $(\hat{\varphi}_1, \hat{\varphi}_2)$. Let $\bar{\varphi} = (\hat{\varphi}_1, \hat{\varphi}_2)$ be an arbitrary element of $(H^1(\Omega))^2$ such that $\hat{\varphi}_1 = \hat{\varphi}_2 = 0$ on $\Gamma_L$. We shall prove that one can always find $\hat{\varphi}_3 \in H^1(\Omega)$ such that $(\hat{\varphi}_1, \hat{\varphi}_2, \hat{\varphi}_3) = \hat{\varphi}$ satisfies condition in (6.61), that is, we shall prove that $\Pi_0(\hat{K}) = \Lambda_0(\hat{K})$, where

$$\Lambda_0(\hat{K}) = \{\bar{\varphi} \in (H^1(\Omega))^2 : \quad \hat{\varphi}_1 = \hat{\varphi}_2 = 0 \quad \text{on} \quad \Gamma_L\}.$$

We can see that if $\bar{\varphi} \in \Pi_0(\hat{K})$ then $\bar{\varphi} \in \Lambda_0(\hat{K})$, that is, $\Pi_0(\hat{K}) \subset \Lambda_0(\hat{K})$. We shall prove that $\Lambda_0(\hat{K}) \subset \Pi_0(\hat{K})$. Let $\bar{\varphi} \in \Lambda_0(\hat{K})$. We have to define $\bar{\varphi}_3 \in H^1(\Omega)$ such that $\hat{\varphi} = (\hat{\varphi}_1, \hat{\varphi}_2, \hat{\varphi}_3) = 0$ on $\Gamma_L$, $\hat{\varphi} \cdot n = 0$ on $\Gamma_F \cup \omega$. The first condition is just $\hat{\varphi}_3 = 0$ on $\Gamma_L$, the second is $\hat{\varphi}_3 = 0$ on $\omega$ and

$$\hat{\varphi}_1(x, h(x)) \frac{\partial h}{\partial x_1}(x) + \hat{\varphi}_2(x, h(x)) \frac{\partial h}{\partial x_2}(x) = \hat{\varphi}_3(x, h(x)) \quad \text{on} \quad \Gamma_F.$$

Thus, we have to show existence of $\hat{\varphi}_3 = \psi \in H^1(\Omega)$ such that $\psi = 0$ on $\Gamma_L \cup \omega$,

$$\psi = \hat{\varphi}_1(x, h(x)) \frac{\partial h}{\partial x_1}(x) + \hat{\varphi}_2(x, h(x)) \frac{\partial h}{\partial x_2}(x) = \psi_1 \quad \text{on} \quad \Gamma_F.$$

As $\bar{\varphi} = (\hat{\varphi}_1, \hat{\varphi}_2) \in (H^1(\Omega))^2$, $\psi_1$ is well defined on $\Gamma_F$ and belongs to $H^{\frac{1}{2}}(\Gamma_F)$ because $h \in C^2(\overline{\omega})$. Since $\hat{\varphi}_1 = \hat{\varphi}_2 = 0$ on $\Gamma_L$, we see that $\psi_1 = 0$ on $\partial\Omega$. Thus, there is no singularity at $\partial\omega$, and the function

$$\psi_b = \begin{cases} 0 & \text{on} \quad \Gamma_L \cup \omega, \\ \psi & \text{on} \quad \Gamma_F, \end{cases}$$

belongs to $H^{\frac{1}{2}}(\partial\Omega)$. Then, to prove the existence of $\psi \in H^1(\Omega)$ such that $\psi = \psi_b$ on $\partial\Omega$ we find $\psi$ as the solution of the Laplace problem

$$\Delta\psi = 0 \quad \text{in} \quad \Omega, \qquad \psi = \psi_b \quad \text{on} \quad \partial\Omega.$$

This proves that $\Lambda_0(\hat{K}) = \Pi_0(\hat{K})$, and we characterized $\Pi_0(\hat{K})$ in terms of the functions $\hat{\varphi}_1$ and $\hat{\varphi}_2$ only (and not $\hat{\varphi}_3$—the third component). Since $\Pi(\hat{K}) = \Pi_0(\hat{K}) + \bar{G}$, we have

$$\Pi(\hat{K}) = \{\bar{\varphi} = (\hat{\varphi}_1, \hat{\varphi}_2) \in (H^1(\Omega))^2 \; : \; \bar{\varphi} = \bar{G} \quad \text{on} \quad \Gamma_L\}.$$

**Lemma 6.8.** *Let $\bar{\varphi} = (\hat{\varphi}_1, \hat{\varphi}_2) \in \Pi(\hat{K})$ be such that the following condition, which will be denoted as condition* (D') *holds*

$$\int_\Omega \left( \hat{\varphi}_1(x, y) \frac{\partial \theta}{\partial x_1}(x) + \hat{\varphi}_2(x, y) \frac{\partial \theta}{\partial x_2}(x) \right) dxdy = 0 \quad \text{for all} \quad \theta \in C_0^1(\omega). \qquad (6.62)$$

*Then*

$$\int_{\Omega} p^{\star}(x)\left(\frac{\partial \hat{\varphi}_1}{\partial x_1}+\frac{\partial \hat{\varphi}_2}{\partial x_2}+\frac{\partial \hat{\varphi}_3}{\partial y}\right) dxdy = 0, \tag{6.63}$$

*which is equivalent to*

$$\int_{\Omega} p^{\star}(x)\left(\frac{\partial \hat{\varphi}_1}{\partial x_1}+\frac{\partial \hat{\varphi}_2}{\partial x_2}\right) dxdy + \int_{\omega} p^{\star}(x)\left(\hat{\varphi}_1(x,h(x))\frac{\partial h}{\partial x_1}+\hat{\varphi}_2(x,h(x))\frac{\partial h}{\partial x_2}\right) dx = 0. \tag{6.64}$$

*Proof.* Let $\theta_m \in C_0^1(\omega)$, $\theta_m \to p^{\star}$ in $L^2(\omega)$. Then

$$\int_{\Omega} \theta_m(x)\left(\frac{\partial \hat{\varphi}_1}{\partial x_1}+\frac{\partial \hat{\varphi}_2}{\partial x_2}+\frac{\partial \hat{\varphi}_3}{\partial y}\right) dxdy$$
$$= -\int_{\Omega}\left(\hat{\varphi}_1\frac{\partial \theta_m}{\partial x_1}+\hat{\varphi}_2\frac{\partial \theta_m}{\partial x_2}\right) dxdy + \int_{\partial\Omega} \theta_m(x)\left(\sum_{i=1}^{3}\hat{\varphi}_i n_i\right) dS,$$

as $\frac{\partial \theta_m}{\partial y} = 0$. Now, the surface integral equals zero since $\theta_m \in C_0^1(\omega)$ and $\hat{\varphi}\cdot n = 0$ on $\Gamma_F \cup \omega$. The first integral on the right-hand side equals zero by condition (D'). Thus, the integral on the left-hand side equals zero. We pass to the limit $\theta_m \to p^{\star}$ in $L^2(\omega)$ to get (6.63). As

$$\int_{\Omega} p^{\star}(x)\frac{\partial \hat{\varphi}_3}{\partial y}(x,y)\, dxdy = \int_{\omega} p^{\star}(x)\int_0^{h(x)}\frac{\partial \hat{\varphi}_3}{\partial y}(x,y)\, dy\, dx$$
$$= \int_{\omega} p^{\star}(x)\hat{\varphi}_3(x,h(x))dx = \int_{\omega} p^{\star}(x)\left(\sum_{i=1}^{2}\hat{\varphi}_i(x,h(x))\frac{\partial h}{\partial x_i}\right) dx,$$

we have the equivalence of (6.63) and (6.64). □

The above lemma tells us that if in the limit variational inequality (6.59) we choose a test function belonging to the set

$$\mathscr{F}_{\hat{K}} := \{\bar{\varphi} \in \Pi(\hat{K}) \quad \text{and} \quad \bar{\varphi} \quad \text{satisfies condition (D')}\}, \tag{6.65}$$

then this inequality will reduce to the following one,

$$\sum_{i=1}^{2}\frac{1}{2}\int_{\Omega}\frac{\partial u_i^{\star}}{\partial y}\frac{\partial(\hat{\varphi}_i - u_i^{\star})}{\partial y}\, dxdy + \hat{l}\int_{\omega} u_i^{\star}(x,h(x))(\hat{\varphi}_i - u_i^{\star})(x,h(x))\, dx + \hat{k}\int_{\omega}(|\bar{\varphi}-s| - |u^{\star}-s|)\, dx \geq 0 \quad \text{for all} \quad \bar{\varphi} \in \mathscr{F}_{\hat{K}}, \tag{6.66}$$

in which the pressure is not present.

Let

$$\mathscr{F} := \{u^\star \ : \ u^\star \text{ is a weak limit of } \hat{u} \text{ in the topology of } V_y \times V_y\}.$$

We prove that $\mathscr{F}$ is contained in the closure of $\Pi(\hat{K})$ in the topology of $V_y \times V_y$. This will allow us later to get useful conclusions from variational inequalities (6.59) and (6.66), by taking test functions from $\mathscr{F}$.

**Lemma 6.9.** *Each solution $u^\star$ of the limit variational inequality satisfies condition* (D′), *and $u^\star - \hat{G}$ satisfies the following stronger condition* (D)

$$\int_\Omega \left( (u_1^\star - \hat{G}_1)\frac{\partial \theta}{\partial x_1}(x) + (u_2^\star - \hat{G}_2)\frac{\partial \theta}{\partial x_2}(x) \right) dxdy = 0 \qquad \text{for all} \quad \theta \in C^1(\overline{\omega}). \tag{6.67}$$

*Proof.* From div $\hat{u} = 0$ in $\Omega$ we have, for $\theta \in C_0^1(\omega)$,

$$0 = \int_\Omega \theta(x) \left( \frac{\partial \hat{u}_1}{\partial x_1} + \frac{\partial \hat{u}_2}{\partial x_2} + \frac{\partial \hat{u}_3}{\partial y} \right) dxdy = -\int_\Omega \left( \hat{u}_1 \frac{\partial \theta}{\partial x_1} + \hat{u}_2 \frac{\partial \theta}{\partial x_2} \right) dxdy$$

(as $\hat{u} \cdot n = 0$ on $\Gamma_F \cup \omega$). As $\hat{u}_i \to u_i^\star$ weakly in $V_y$, $i = 1, 2$, we obtain condition (D') for $u^\star$, that is,

$$\int_\Omega \left( u_1^\star \frac{\partial \theta}{\partial x_1} + u_2^\star \frac{\partial \theta}{\partial x_2} \right) dxdy = 0 \qquad \text{for all} \quad \theta \in C_0^1(\omega).$$

Taking $\hat{u} - \hat{G}$ instead of $\hat{u}$ at the beginning of this proof, as $(\hat{u} - \hat{G}) \cdot n = 0$ on $\Gamma_F \cup \omega$ and $\hat{u} - \hat{G} = 0$ on $\Gamma_L$, we obtain (6.67) in the limit. □

From Lemmas 6.8 and 6.9 it follows that condition (D') is the right one to get rid of the pressure terms in the limit variational inequality. Note that in condition (D') we do not assume that $\bar{\varphi} \in (H^1(\Omega))^2$, that is, (6.62) has sense for $\bar{\varphi} \in (L^2(\Omega))^2$, for example. The solution $u^\star$ satisfies (D') and is not in $(H^1(\Omega))^2$.

We shall characterize the set $\mathscr{F}$ of solutions $u^\star$.

$$\mathscr{F} = \Big\{ u^\star = (u_1^\star, u_2^\star) \in (L^2(\Omega))^2 \ : \ \frac{\partial u_i^\star}{\partial y} \in L^2(\Omega),\ i = 1, 2, \quad u^\star \text{ satisfies (D}'), \\ u^\star \text{ satisfies condition } (C_{\Gamma_F}) \text{ coming from } \hat{u} \cdot n = 0 \text{ on } \Gamma_F \Big\}. \tag{6.68}$$

Our main problem now is to find relations between $\mathscr{F}_{\hat{K}}$ given by (6.65) and $\mathscr{F}$ given by (6.68). We do not know what is the condition $(C_{\Gamma_F})$ in (6.68) in the general case (note that if $\Gamma_F$ is given by $y = const.$ we have no condition on $u^\star$ coming from $\hat{u} \cdot n = 0$ on $\Gamma_F$). Thus we define

$$\mathscr{F}_1 = \left\{ u^\star = (u_1^\star, u_2^\star) \in (L^2(\Omega))^2 \;\; \frac{\partial u_i^\star}{\partial y} \in L^2(\Omega),\; i = 1,2,\;\; u^\star \text{ satisfies (D')} \right\}. \tag{6.69}$$

Then $\mathscr{F} \subset \mathscr{F}_1$ and it is sufficient to prove the following lemma.

**Lemma 6.10.** *$\mathscr{F}_1$ is contained in the closure of $\Pi(\hat{K})$ in the topology of $V_y \times V_y$.*

*Proof.* Let $v \in \mathscr{F}_1^o = \{v = u^\star - \hat{G} \,:\, u^\star \in \mathscr{F}_1\}$. We consider the map

$$\sqcap = \omega \times (0,1) \to \Omega \;:\quad (x,y) \mapsto (x, h(x)y) = (X,Y),$$

and define

$$v(X,Y) = v\,(x, h(x)y) = \bar{v}(x,y) \quad \text{and} \quad U(x,y) = \bar{v}(x,y)h(x).$$

Then

- $v$ and $U$ are of the same regularity,
- "$v$ satisfies (D') or (D) in $\Omega$" is equivalent to "$U$ satisfies (D') or (D) in $\sqcap$." Indeed for all $\theta \in C^1(\omega)$ we have

$$\int_\Omega v(X,Y)\nabla\theta(X)dXdY = \int_\sqcap U(x,y)\nabla\theta(x)dxdy,$$

- "$v = 0$ on $\Gamma_L$" is equivalent to "$U = 0$ on $\Gamma_L$."

and also

- If $U_n(x,y) \to U(x,y)$ in $V_y(\sqcap) \times V_y(\sqcap)$, then $v_n(X,Y) \to v(X,Y)$ in $V_y(\Omega) \times V_y(\Omega)$.

Indeed,

$$\begin{aligned}
\int_\Omega |v_n(X,Y) - v(X,Y)|^2\, dxdy &= \int_\sqcap |\bar{v}_n(x,y) - \bar{v}(x,y)|^2 h(x)\, dxdy \\
&= \int_\sqcap |\bar{v}_n(x,y)h(x) - \bar{v}(x,y)h(x)|^2 \frac{1}{h(x)}\, dxdy \\
&= \int_\sqcap |U_n(x,y) - U(x,y)|^2 \frac{1}{h(x)}\, dxdy \\
&\le \frac{1}{h_{min}} \int_\sqcap |U_n(x,y) - U(x,y)|^2\, dxdy,
\end{aligned}$$

and

$$\begin{aligned}
&\int_{\Omega} \left| \frac{\partial v_n}{\partial Y}(X,Y) - \frac{\partial v}{\partial Y}(X,Y) \right|^2 dXdY \\
&= \int_{\sqcap} \left| \frac{\partial \bar{v}_n}{\partial y}(x,y)\frac{1}{h(x)} - \frac{\partial \bar{v}}{\partial y}(x,y)\frac{1}{h(x)} \right|^2 h(x)\,dxdy \\
&= \int_{\sqcap} \left| h(x)\frac{\partial \bar{v}_n}{\partial y}(x,y) - h(x)\frac{\partial \bar{v}}{\partial y}(x,y) \right|^2 \frac{1}{h^3(x)}\,dxdy \\
&\le \frac{1}{h_{min}^3} \int_{\sqcap} \left| \frac{\partial U_n}{\partial y}(x,y) - \frac{\partial U}{\partial y}(x,y) \right|^2 dxdy.
\end{aligned}$$

Let $U_\sigma(x,y) = U(\sigma x, y), \sigma > 1$,

$$\int_{\sqcap} U(x,y)\nabla\theta(x)\,dxdy = 0 \quad \text{for all} \quad \theta \in C^1(\omega).$$

Then as $\eta(z) = \frac{1}{\sigma^2}\theta\left(\frac{z}{\sigma}\right) \in C^1(\omega)$, we have

$$\int_{\sqcap} U_\sigma(x,y)\nabla\theta(x)dxdy = \int_{\sqcap} U(\sigma x,y)\nabla\theta(x)dxdy = \int_{\sqcap} U(z,y)\nabla\left(\theta\left(\frac{z}{\sigma}\right)\right)\frac{dzdy}{\sigma^2} = 0.$$

Thus, $U_\sigma$ satisfies condition (D). Since it is not smooth enough in $x$, we take the usual mollification $\rho_\varepsilon \star U_\sigma$ with respect to $x$ variable only which is in $\Pi_0(\hat{K})$ for small $\varepsilon$, $\varepsilon \le \varepsilon(\sigma)$. We check condition (D).

$$\begin{aligned}
\int_{\sqcap} (\rho_\varepsilon \star U_\sigma)(x,y)\nabla\theta(x)dydx &= \int_\omega \int_0^1 \left( \int_\omega U_\sigma(t,y)\rho_\varepsilon(t-x)dt \right) \nabla\theta(x)dydx \\
&= \int_\omega \int_0^1 U_\sigma(t,y) \left( \int_\omega \rho_\varepsilon(x-t)\nabla\theta(x)dx \right) dydt \\
&= \int_{\sqcap} U_\sigma(t,y)\nabla(\rho_\varepsilon \star \theta)(t)dydt = 0.
\end{aligned}$$

As $\rho_\varepsilon\star\theta \in C^1(\omega)$ and $U_\sigma$ satisfies condition (D), then $\rho_\varepsilon \star U_\sigma$ also satisfies condition $(D)$ in $\sqcap$. Now,

$$\|U - \rho_\varepsilon \star U_\sigma\|_{V_y^2} \le \|U - U_\sigma\|_{V_y^2} + \|U_\sigma - \rho_\varepsilon \star U_\sigma\|_{V_y^2},$$

and we take $\sigma$ close to 1 first and then small $\varepsilon$.

Thus there exists $\rho_\varepsilon \star U_\sigma \in \Pi_0(\hat{K})$ such that $\|U - \rho_\varepsilon \star U_\sigma\|_{V_y^2(\sqcap)} \to 0$, which is equivalent to the existence of $\rho_\varepsilon \star v_\sigma \in \Pi_0(\hat{K})$ with $\|v - \rho_\varepsilon \star v_\sigma\|_{V_y^2(\Omega)} \to 0$. □

## 6.6 Strong Convergence of Velocities and the Limit Equation

From the Korn inequality (6.24) written in new variables we conclude that if $\hat{u}$ and $\hat{\varphi}$ in $\hat{K}$ are divergence free then, in particular

$$\sum_{i=1}^{2} \left\| \frac{\partial(\hat{u}_i - \hat{\varphi}_i)}{\partial y} \right\|_{L^2(\Omega)}^2 \leq \hat{a}(\hat{u} - \hat{\varphi}, \hat{u} - \hat{\varphi}) + C(\Gamma_F, h) \int_{\Gamma_F} |\hat{u} - \hat{\varphi}|^2 \, dS.$$

By (6.20),

$$\begin{aligned} \hat{a}(\hat{u} - \hat{\varphi}, \hat{u} - \hat{\varphi}) &= \hat{a}(\hat{\varphi}, \hat{\varphi} - \hat{u}) + \hat{a}(\hat{u}, \hat{u} - \hat{\varphi}) \\ &\leq \hat{a}(\hat{\varphi}, \hat{\varphi} - \hat{u}) + \hat{l} \int_{\Gamma_F} \hat{u}(\hat{\varphi} - \hat{u}) \, dS + \hat{k} \int_{\omega} (|\hat{\varphi} - s| - |\hat{u} - s|) \, dx, \end{aligned}$$

and we have

$$\begin{aligned} \sum_{i=1}^{2} \left\| \frac{\partial(\hat{u}_i - \hat{\varphi}_i)}{\partial y} \right\|_{L^2(\Omega)}^2 &\leq \hat{a}(\hat{\varphi}, \hat{\varphi} - \hat{u}) + \hat{l} \int_{\Gamma_F} \hat{u}(\hat{\varphi} - \hat{u}) \, dS \\ &\quad + \hat{k} \int_{\omega} (|\hat{\varphi} - s| - |\hat{u} - s|) \, dx \\ &\quad + C(\Gamma_F, h) \int_{\Gamma_F} |\hat{u} - \hat{\varphi}|^2 \, dS, \end{aligned}$$

where

$$\begin{aligned} \hat{a}(\hat{\varphi}, \hat{\varphi} - \hat{u}) = &\frac{1}{2} \sum_{i,j=1}^{2} \int_{\Omega} \left( \varepsilon \frac{\partial \hat{\varphi}_i}{\partial x_j} + \varepsilon \frac{\partial \hat{\varphi}_j}{\partial x_i} \right) \left( \varepsilon \frac{\partial}{\partial x_j} (\hat{\varphi}_i - \hat{u}_i) + \varepsilon \frac{\partial}{\partial x_i} (\hat{\varphi}_j - \hat{u}_j) \right) dxdy \\ &+ \frac{1}{2} \sum_{i=1}^{2} \int_{\Omega} \left( \frac{\partial \hat{\varphi}_i}{\partial y} + \varepsilon^2 \frac{\partial \hat{\varphi}_3}{\partial x_i} \right) \left( \frac{\partial}{\partial y} (\hat{\varphi}_i - \hat{u}_i) + \varepsilon^2 \frac{\partial}{\partial x_i} (\hat{\varphi}_3 - \hat{u}_3) \right) dxdy \\ &+ 2 \int_{\Omega} \varepsilon \frac{\partial \hat{\varphi}_3}{\partial y} \varepsilon \frac{\partial}{\partial y} (\hat{\varphi}_3 - \hat{u}_3) dxdy. \end{aligned}$$

Thus

$$\begin{aligned} \sum_{i=1}^{2} \left\| \frac{\partial(\hat{u}_i - \hat{\varphi}_i)}{\partial y} \right\|_{L^2(\Omega)}^2 &+ \hat{k} \int_{\omega} (|\hat{u} - s| - |\hat{\varphi} - s|) dx \\ &\leq \hat{a}(\hat{\varphi}, \hat{\varphi} - \hat{u}) + \hat{l} \int_{\Gamma_F} \hat{u}(\hat{\varphi} - \hat{u}) \, dS + C(\Gamma_F, h) \int_{\Gamma_F} |\hat{u} - \hat{\varphi}|^2 \, dS. \end{aligned}$$

Let $\bar{u} = (\hat{u}_1, \hat{u}_2)$, $u^\star = (u_1^\star, u_2^\star)$ as in (6.52) and let $\bar{\varphi} = (\hat{\varphi}_1, \hat{\varphi}_2) \in \Pi(\hat{K})$. Then

$$\limsup_{\varepsilon \to 0} \left\{ \sum_{i=1}^{2} \left\| \frac{\partial(\hat{u}_i - \hat{\varphi}_i)}{\partial y} \right\|^2_{L^2(\Omega)} + \hat{k} \int_\omega (|\bar{u} - s| - |\bar{\varphi} - s|) dx \right\}$$

$$\leq \int_\Omega \frac{\partial \bar{\varphi}}{\partial y} \frac{\partial(\bar{\varphi} - u^\star)}{\partial y} dxdy + \hat{l} \int_{\Gamma_F} u^\star (\bar{\varphi} - u^\star)\, dS + C(\Gamma_F, h) \int_{\Gamma_F} |u^\star - \bar{\varphi}|^2\, dS,$$

and

$$\sum_{i=1}^{2} \left\| \frac{\partial(\hat{u}_i - \hat{\varphi}_i)}{\partial y} \right\|^2_{L^2(\Omega)} + \hat{k} \int_\omega (|\bar{u} - s| - |\bar{\varphi} - s|) dx$$

$$\leq \int_\Omega \frac{\partial \bar{\varphi}}{\partial y} \frac{\partial(\bar{\varphi} - u^\star)}{\partial y} dxdy + \hat{l} \int_{\Gamma_F} u^\star (\bar{\varphi} - u^\star)\, dS$$

$$+ C(\Gamma_F, h) \int_{\Gamma_F} |u^\star - \bar{\varphi}|^2\, dS + \delta \quad \text{for all} \quad \delta < \varepsilon(\delta).$$

By Lemma 6.10, there exists a sequence of functions $\bar{\varphi}$ in $\Pi(\hat{K})$ which has $u^\star$ as a limit in $V_y^2$, whence

$$\sum_{i=1}^{2} \left\| \frac{\partial(\hat{u}_i - \hat{\varphi}_i)}{\partial y} \right\|^2_{L^2(\Omega)} + \hat{k} \int_\omega (|\bar{u} - s| - |u^\star - s|) dx \leq \delta \qquad \text{for all} \quad \delta < \varepsilon(\delta).$$

Since $\delta > 0$ is arbitrary, then by the weak lower semicontinuity of the functional $v \mapsto \int_\omega |v - s|\, dx$, we get

$$\int_\omega (|\bar{u} - s| - |u^\star - s|) dx \to 0.$$

and from Lemma 6.5 the strong convergence of $\bar{u}^\varepsilon$ to $u^\star$ in $\tilde{V}_y^2 \cap \mathscr{F}_1$ follows.

Now, we obtain from the limit inequality the classical relations for $u^\star$ and $p^\star$.

**Theorem 6.3.** *The limit functions $u^\star, p^\star$ satisfy*

$$p^\star(x_1, x_2, y) = p^\star(x_1, x_2) \quad \text{a.e. in} \quad \Omega, \quad p^\star \in H^1(\omega), \tag{6.70}$$

$$-\frac{1}{2} \frac{\partial^2 u_i^\star}{\partial y^2} + \frac{\partial p^\star}{\partial x_i} = 0 \quad (i = 1, 2) \quad \text{in} \quad L^2(\Omega). \tag{6.71}$$

*Proof.* Using Lemma 6.10 we first take $\hat{\varphi}_1 = u_1^\star \pm \psi_1$ with $\psi_1 \in H_0^1(\Omega)$ and $\hat{\varphi}_2 = u_2^\star$, and then $\hat{\varphi}_1 = u_1^\star$ and $\hat{\varphi}_2 = u_2^\star \pm \psi_2$ with $\psi_2 \in H_0^1(\Omega)$ in (6.59), to get

$$-\frac{1}{2}\frac{\partial^2 u_i^\star}{\partial y^2} + \frac{\partial p^\star}{\partial x_i} = 0 \quad \text{in} \quad H^{-1}(\Omega), \quad i = 1,2. \tag{6.72}$$

Multiplying (6.72) by $\psi_i(x,y) = y(y - h(x))\theta(x)$ with $\theta \in H_0^1(\omega)$ and using the Green formula we have

$$\begin{aligned}
\int_\Omega \frac{\partial u_i^\star}{\partial y}\frac{\partial \psi_i}{\partial y}dxdy &= \int_\Omega \frac{\partial u_i^\star}{\partial y}\frac{\partial [y^2 - yh(x)]\theta(x)}{\partial y}dxdy \\
&= \int_\Omega \frac{\partial u_i^\star}{\partial y}[2y - h(x)]\theta(x)dydx \\
&= -\int_\Omega 2u_i^\star(x,y)\theta(x)dydx + \int_{\partial\Omega} u_i^\star[2y - h(x)]\theta(x)n_3 dS \\
&= -\int_\omega 2h(x)\tilde{u}_i^\star(x)\theta(x)dx + \int_\omega h(x)u_i^\star(x,0)\theta(x)dx \\
&\quad + \int_\omega h(x)u_i^\star(x,h(x))\theta(x)dx,
\end{aligned}$$

where

$$h(x)\tilde{u}_i^\star(x) = \int_0^{h(x)} u_i^\star(x,y)dy,$$

and

$$\begin{aligned}
\int_\Omega p^\star(x)\frac{\partial \psi_i}{\partial x_i}dxdy &= \int_\omega p^\star(x)\left(\int_0^{h(x)} \frac{\partial [y^2 - yh(x)]\theta(x)}{\partial x_i}dy\right)dx \\
&= \int_\omega p^\star(x)\frac{\partial}{\partial x_i}\left(\theta(x)\int_0^{h(x)} [y^2 - yh(x)]dy\right)dx \\
&= -\frac{1}{6}\int_\omega p^\star(x)\frac{\partial h^3(x)\theta(x)}{\partial x_i}dx.
\end{aligned}$$

Thus

$$-h(x)\tilde{u}_i^\star(x) + \frac{h(x)}{2}u_i^\star(x,h(x)) + \frac{h(x)}{2}u_i^\star(x,0) = \frac{h^3(x)}{6}\frac{\partial p^\star}{\partial x_i}(x) \tag{6.73}$$

in $H^{-1}(\omega)$. As $u_i^\star \in V_y$, $\tilde{u}_i^\star \in L^2(\omega)$ and in consequence the left side of (6.73) belongs to $L^2(\omega)$. This gives (6.70) and (6.71). □

## 6.7 Reynolds Equation and the Limit Boundary Conditions

**Theorem 6.4.** *The pair $(u^\star, p^\star)$ satisfies the following weak form of the Reynolds equation*

$$\int_\omega \left(\frac{h^3}{12}\nabla p^\star - \frac{h}{2}s^\star - \frac{h}{2}s_h^\star\right)\nabla\varphi dx + \int_\omega s_h^\star \nabla h\varphi dx = -\int_{\partial\omega} \varphi\tilde{g}\cdot n\, dl \tag{6.74}$$

*for all $\varphi \in H^1(\omega)$, where*

$$s^\star(x) = u^\star(x,0), \quad s_h^\star(x) = u^\star(x,h(x)),$$

*and*

$$\tilde{g}(x) = \int_0^{h(x)} \hat{g}(x,y)dy \quad \text{for all} \quad x \in \partial\omega.$$

*Moreover, the traces $\tau^\star = \frac{\partial u^\star}{\partial y}(\cdot,0)$, $\tau_h^\star = \frac{\partial u^\star}{\partial y}(\cdot,h(x))$, $s_h^\star = u^\star(\cdot,h(x))$, and $s^\star = u^\star(\cdot,0)$ satisfy the following limit form of the Tresca boundary conditions (6.4) and the Fourier boundary condition (6.6)*

$$\left.\begin{array}{l} |\tau^\star| = \hat{k} \Longrightarrow \exists\lambda \geq 0 \quad u^\star = s + \lambda\tau^\star \\ |\tau^\star| < \hat{k} \Longrightarrow \qquad u^\star = s \end{array}\right\} \quad \text{a.e. in} \quad \omega, \tag{6.75}$$

$$\tau_h^\star \nabla h\cdot n + \hat{l}s_h^\star = 0 \quad \text{a.e. on} \quad \Gamma_F. \tag{6.76}$$

*Proof.* Integrating twice (6.71) between 0 and $y$ we obtain (for $1 \leq i \leq 2$)

$$u_i^\star(x,y) = \frac{y^2}{2}\frac{\partial p^\star(x)}{\partial x_i} + s_i^\star(x) + y\tau_i^\star(x), \tag{6.77}$$

and for $y = h(x)$ we get

$$s_h^\star(x) = \frac{h^2}{2}\nabla p^\star(x) + s^\star(x) + h\tau^\star(x) \quad \text{a.e. in} \quad \omega. \tag{6.78}$$

Now, integrating twice (6.71) between $y$ and $h(x)$ we obtain (for $1 \leq i \leq 2$)

$$u_i^\star(x,y) = \left(\frac{h^2}{2} - hy + \frac{y^2}{2}\right)\frac{\partial p^\star(x)}{\partial x_i} + s_{h,i}^\star(x) + (y-h)\tau_{h,i}^\star(x),$$

and for $y = 0$ we get

$$s^\star(x) = \frac{h^2}{2}\nabla p^\star(x) + s_h^\star(x) - h\tau_h^\star(x) \quad \text{a.e. in} \quad \omega. \tag{6.79}$$

Taking now the average

$$\tilde{v}(x) = \frac{1}{h(x)}\int_0^{h(x)} v(x,y)\,dy$$

of (6.77) we obtain

$$h\tilde{u}_i^\star(x) = \frac{h^3}{6}\frac{\partial p^\star(x)}{\partial x_i} + hs_i^\star(x) + \frac{h^2}{2}\tau_i^\star(x).$$

Moreover, for all $\varphi \in H^1(\omega)$, we have

$$\begin{aligned}0 = \int_\Omega \varphi \operatorname{div}\hat{u}\,dydx &= \int_\omega \varphi(x)\int_0^h \left(\sum_{i=1}^2 \frac{\partial \hat{u}_i}{\partial x_i} + \frac{\partial \hat{u}_3}{\partial y}\right) dydx\\ &= \int_\omega \varphi(x)\left(\sum_{i=1}^2 \frac{\partial (h\tilde{\hat{u}}_i)}{\partial x_i} + \hat{u}_3(x,h) - \hat{u}_3(x,0)\right) dx.\end{aligned}$$

Then as $\hat{u}\cdot n = \hat{u}_3(x,0) = 0$ on $\omega$, and $\Gamma_F$ is defined by $y = h(x)$, we obtain

$$\hat{u}_3(x,h(x)) = \hat{u}_1(x,h(x))\frac{\partial h}{\partial x_1} + \hat{u}_2(x,h(x))\frac{\partial h}{\partial x_2},$$

and

$$\int_\omega \varphi(x)\left(\sum_{i=1}^2 \frac{\partial (h\tilde{\hat{u}}_i)}{\partial x_i} + \hat{u}_i(x,h(x))\frac{\partial h}{\partial x_i}\right) dx = 0.$$

Using the Green formula we have

$$-\sum_{i=1}^2 \int_\omega h\tilde{\hat{u}}_i \frac{\partial \varphi}{\partial x_i} dx + \int_\omega \varphi(x)\left(\sum_{i=1}^2 \hat{u}_i(x,h(x))\frac{\partial h}{\partial x_i}\right) dx + \sum_{i=1}^2 \int_{\partial\omega} h\tilde{\hat{u}}_i n_i \varphi\,dl = 0.$$

As $\hat{u}_i \to u_i^\star$ weakly in $V_y$, $\tilde{\hat{u}}_i \to \tilde{u}_i^\star$ weakly in $L^2(\omega)$, and as $\partial\omega \subset \partial\Omega$, we obtain

$$\int_\omega h\tilde{\hat{u}}\nabla\varphi dx - \int_\omega \varphi(x)\,(\hat{u}(x,h(x))\nabla h)\,dx = \int_{\partial\omega} \tilde{g}(x)\cdot n\varphi\,dl \quad \text{for all} \quad \varphi \in H^1(\omega).$$

From (6.79) we have

$$\int_\omega \left(\frac{h^3}{6}\nabla p^\star + hs^\star + \frac{h^2}{2}\tau^\star\right)\nabla\varphi\, dx - \int_\omega \varphi s_h^\star \cdot \nabla h\, dx = \int_{\partial\omega} \varphi\tilde{g}\cdot n\, dl \tag{6.80}$$

for all $\varphi \in H^1(\omega)$. Then using (6.78) and (6.80), we obtain the weak formulation (6.74) of the Reynolds equation.

To prove (6.75) we transform (6.59). Using the Green formula, (6.71), conditions $\varphi_i = \hat{g}_i$ on $\Gamma_L$, $n_i = 0$ on $\omega$ for $i = 1, 2$, and $\varphi \cdot n = 0$ on $\omega \cup \Gamma_F$, as well as the density of $\mathscr{D}(\Omega)$ in $\Pi(\hat{K})$, and of $\mathscr{D}(\Omega)$ in $L^2(\Omega)$, we obtain

$$\begin{aligned}
&\int_\omega \hat{k}(|\varphi(x,0) - s| - |s^\star - s|)\, dx \\
&\quad + \int_\omega \hat{l} s_h^\star(\varphi(x,h(x)) - s_h^\star)\, dx - \int_\omega \tau^\star(\varphi(x,0) - s^\star)\, dx \\
&\quad + \int_\omega \tau_h^\star(\varphi(x,h(x)) - s_h^\star)\nabla h\cdot n\, dx \ge 0 \quad \text{for all} \quad \varphi \in (L^2(\Omega))^2.
\end{aligned} \tag{6.81}$$

We take first

$$\varphi(x,y) = \begin{cases} s_h^\star(x) & \text{if} \quad y = h(x), \\ s_i(x) & \text{if} \quad y = 0, \end{cases}$$

then

$$\varphi(x,y) = \begin{cases} s_h^\star(x) & \text{if} \quad y = h(x), \\ 2s_i^\star - s_i & \text{if} \quad y = 0, \end{cases}$$

in (6.81), to obtain

$$\int_\omega \left(\hat{k}|s^\star - s| - \tau^\star(s^\star - s)\right) dx = 0, \tag{6.82}$$

and from (6.81), with

$$\varphi(x,y) = \begin{cases} s_h^\star(x) & \text{if} \quad y = h(x), \\ \phi(x) + s & \text{if} \quad y = 0, \end{cases}$$

and $\phi \in (L^2(\omega))^2$, we obtain

$$\int_\omega \left(\hat{k}|\phi| - \tau^\star\phi\right) dx \ge \int_\omega \left(\hat{k}|s^\star - s| - \tau^\star(s^\star - s)\right) dx. \tag{6.83}$$

From (6.82) and (6.83) we deduce

$$\int_\omega \left(\hat{k}|\phi| - \tau^\star\phi\right) dx \ge 0 \quad \text{for all} \quad \phi \in (L^2(\omega))^2. \tag{6.84}$$

We will show that $|\tau^*| \leq \hat{k}$ a.e. on $\omega$. Suppose that there exists a set of positive measure $\omega_1 \subset \omega$ such that $|\tau^*| > \hat{k}$ on this set. Taking $\phi = 0$ on $\omega \setminus \omega_1$ and $\phi = \tau^*$ on $\omega$ in (6.84) we get

$$\int_{\omega_1} |\tau^*|(\hat{k} - |\tau^*|)\, dx \geq 0.$$

As the integrand must be a strictly negative function, we arrive at a contradiction. We have thus proved

$$|\tau^\star| \leq \hat{k} \quad \text{a.e. on} \quad \omega.$$

Then

$$\hat{k}|s^\star - s| \geq |\tau^\star||s^\star - s| \geq \tau^\star \cdot (s^\star - s) \quad \text{a.e. on} \quad \omega$$

and

$$\hat{k}|s^\star - s| - \tau^\star \cdot (s^\star - s) \geq 0 \quad \text{a.e. on} \quad \omega.$$

From (6.82) we deduce that

$$\hat{k}|s^\star - s| - \tau^\star \cdot (s^\star - s) = 0 \quad \text{a.e. on} \quad \omega. \tag{6.85}$$

If $|\tau^\star| = \hat{k}$, then from (6.85) we have $|\tau^\star||s^\star - s| = \tau^\star(s^\star - s)$ a.e. on $\omega$, which implies the existence of $\lambda \geq 0$ such that $s^\star - s = \lambda\tau^\star$.

If, in turn, $|\tau^\star| < \hat{k}$, then from (6.85) we have

$$0 = \hat{k}|s^\star - s| - \tau^\star(s^\star - s) \geq (\hat{k} - |\tau^\star|)|s^\star - s| \quad \text{a.e. on} \quad \omega \cap \{|\tau^*| < k\},$$

whence $s^\star - s = 0$ a.e. on $\omega \cap \{|\tau^*| < k\}$, and (6.75) follows.

We shall prove (6.76). First we take

$$\varphi(x, y) = \begin{cases} \phi(x) & \text{if} \quad y = h(x) \\ s^\star(x) & \text{if} \quad y = 0 \end{cases}$$

with $\phi \in (L^2(\omega))^2$ in (6.81), to get

$$\int_\omega \left(\hat{l}s_h^\star(\phi - s_h^\star) + \tau_h^\star(\phi - s_h^\star)\nabla h \cdot n\right) dx \geq 0 \quad \text{for all} \quad \phi \in (L^2(\omega))^2.$$

Then, taking $\phi = \psi + s_h^\star$ and $\phi = -\psi + s_h^\star$ we obtain

$$\int_\omega \left(\hat{l}s_h^\star + \tau_h^\star\nabla h \cdot n\right) \psi dx = 0 \quad \text{for all} \quad \psi \in (L^2(\omega))^2,$$

which implies (6.76). □

*Remark 6.5.* From formula (6.74) we obtain the following form of the Reynolds equation (for $\nu = 1$),

$$\operatorname{div}\left(\frac{h^3}{12}\nabla p^\star - \frac{h}{2}s^\star - \frac{h}{2}s_h^\star\right) + s_h^\star \nabla h = 0 \quad \text{in} \quad H^{-1}(\omega).$$

In the simplest classical one-dimensional case, with $\omega = (a,b)$, $s_h^\star = 0$, $s^\star = const$, we obtain the well-known equation describing the pressure distribution in the interval $(a,b)$,

$$-\frac{d}{dx}\left(\frac{h^3}{6}\frac{dp^\star}{dx}\right) + s^\star \frac{dh}{dx} = 0. \tag{6.86}$$

Observe how the Reynolds equation depends on the boundary conditions for the velocity field. When $\nu \neq 1$, the second term on the left-hand side of (6.86) should be multiplied by $\nu$.

## 6.8 Uniqueness

**Theorem 6.5.** *There exists a unique solution $(u^\star, p^\star)$ in $\tilde{V}_y \times (L_0^2(\omega) \cap H^1(\omega))$ of inequality (6.59).*

*Proof.* Let $(U^1, p^1)$ and $(U^2, p^2)$ be two solutions of (6.59). Taking $\varphi = U^2$ and $\varphi = U^1$, respectively, as test functions in (6.59) we reduce it to (6.66) and obtain

$$\frac{1}{2}\int_\Omega \left|\frac{\partial(U^1 - U^2)}{\partial y}\right|^2 dxdy + \hat{l}\int_\omega |U^1(x,h(x)) - U^2(x,h(x))|^2 dx \leq 0.$$

Using Lemma 6.5 we get

$$\|U^2 - U^1\|_{V_y^2} = 0 \quad \text{and} \quad |U^1(x,h(x)) - U^2(x,h(x))| = 0 \quad \text{a.e. in} \quad \omega. \tag{6.87}$$

Moreover, from (6.74), we have

$$\begin{aligned}
&\int_\omega \frac{h^3}{6}\nabla(p^2 - p^1)\nabla\varphi(x)dx \\
&\quad = \int_\omega h(x)[U^1(x,0) - U^2(x,0) + U^1(x,h(x)) - U^2(x,h(x))]\nabla\varphi(x)dx \\
&\qquad + \int_\omega h(x)(U^2(x,h(x)) - U^1(x,h(x)))\nabla h \cdot \varphi dx \quad \text{for all} \quad \varphi \in H^1(\omega).
\end{aligned}$$

Now, taking $\varphi = p^2 - p^1$, and by (6.87) we get

$$\|\nabla(p^2 - p^1)\|_{L^2(\omega)} = 0,$$

and from the Poincaré inequality, as $p^i \in L^2_0(\omega)$ we conclude that $\|p^2 - p^1\|_{L^2(\omega)} = 0$. The uniqueness of $u^\star$ follows then from (6.59). □

## 6.9 Comments and Bibliographical Notes

(1) *Remarks on boundary conditions.* The right choice of boundary conditions for the velocity field in hydrodynamical problems is fundamental for applications. In lubrication problems the widely assumed non-slip condition when the fluid has the same velocity as surrounding solid boundary is not respected any more when the shear rate becomes too high $10^7 \mathrm{s}^{-1}$. This phenomenon has been studied in a lot of mechanical papers, e.g., [120, 121, 214, 223]. Continuous experimental studies are conducted (e.g., [193]) but are still difficult due to thickness of the gap between the solid surfaces which can be as small as 50 nm.

In such operating conditions, non-slip condition is induced by chemical bounds between the lubricant and the surrounding surfaces and by the action of the normal stresses, which are linked to the pressure inside the flow. On the contrary, tangential stresses are so high that they tend to destroy the chemical bounds and induce slip phenomenon.

This is nothing else than a transposition of the well-known Coulomb law between two solids [90] to the fluid–solid interface.

Although being implicitly used in numerical procedures for lubrication problem, a Reynolds thin film equation taking account of such slip phenomenon have been hardly studied from the mathematical point of view. In [15, 21], respectively, the simplified Tresca and Coulomb interface conditions, posed on only one of the surrounding surfaces of a Newtonian fluid, were considered. This chapter is based on the results from [23]. Corresponding results for non-Newtonian fluids have been obtained in [20].

(2) *Comments on technical problems.* Two technical issues were crucial in the mathematical analysis of the problem. First, we could not use the usual Korn inequality as we did not assume that the velocity was equal to zero at one of the boundaries (on the top or the bottom) as is usually assumed in lubrication problems. We thus derived an analogue of the Korn inequality suitable for our boundary conditions and such that the constants could be controlled appropriately as the gap between the surfaces approached zero. This led us to the main uniform estimate of the velocity fields and to the limit variational inequality, in consequence. Second, to be able to make use of the latter, we had to characterize precisely the limit solution space and the set of admissible test functions. As the limit variational inequality was written in terms of the first two components of the velocity field, we had to characterize—in this very limit case—projections of the convex sets appearing in the weak form of the Stokes flow. This allowed us, in particular, to obtain a stronger convergence of the velocity fields than it is usually expected.

# 7
# Autonomous Two-Dimensional Navier–Stokes Equations

> *We put our faith in the tendency for dynamical systems with a large number of degrees of freedom, and with coupling between those degrees of freedom, to approach a statistical state which is independent (partially, if not wholly) of the initial conditions.*
>
> – George Keith Batchelor

This chapter contains some basic facts about solutions of nonstationary Navier–Stokes equations

$$\frac{\partial u}{\partial t} + (u \cdot \nabla)u = -\frac{1}{\rho}\nabla p + \nu \Delta u + f,$$

$$\operatorname{div} u = 0,$$

in two-dimensional spatial domains. We assume that the system is autonomous, that is, the external forces do not depend on time and that the domain of the flow is bounded. The boundary conditions are either periodic or homogeneous of Dirichlet type. In what follows, the viscosity $\nu$ is a positive constant and we set $\rho = 1$.

First, we prove the existence of a unique global in time weak solution. Then we analyze the time asymptotics of solutions proving the existence of the global attractor and showing its trivial structure in some cases of special external forces. Finally, we consider the problem of behavior of the kinetic energy of the fluid. To this end we introduce the notion of different scales of the flow, considered in the theory of turbulence, and corresponding in a way to different spatial dimensions of vortices of a given turbulent flow.

## 7.1 Navier–Stokes Equations with Periodic Boundary Conditions

Let $Q = [0, L]^2$ and define spaces

$$H = \{u = (u_1, u_2) \in \dot{L}_p^2(Q) \ : \ \operatorname{div} u = 0\},$$



G. Łukaszewicz, P. Kalita, *Navier–Stokes Equations*, Advances in Mechanics and Mathematics 34, DOI 10.1007/978-3-319-27760-8_7

and

$$V = \{u = (u_1, u_2) \in \dot{H}^1_p(Q) \ : \ \operatorname{div} u = 0\},$$

(cf. Sect. 3.4) with the scalar products

$$(u, v) = \int_Q u(x) \cdot v(x)\, dx \quad \text{in} \quad H, \quad ((u, v)) = \int_Q \nabla u(x) : \nabla v(x)\, dx \quad \text{in} \quad V,$$

and the norms

$$\|u\|_H = |u| = \sqrt{\int_Q |u(x)|^2 dx}, \quad |u|^2 = (u, u),$$

and

$$\|u\|_V = \|u\| = \sqrt{\int_Q |\nabla u(x)|^2 dx}, \quad \|u\|^2 = ((u, u)) = (\nabla u, \nabla u),$$

respectively. The duality between $V$ and its dual $V'$ is denoted by $\langle \cdot, \cdot \rangle$.

The weak formulation of the initial boundary value problem for the Navier–Stokes equations is given in Theorem 7.1 below. We adopt the notation from Chap. 4. In particular, the Stokes operator $A$ and the trilinear form $b$ as well as the associated nonlinear operator $B$ are defined in this chapter the same as in Chap. 4 (with the obvious difference that in Chap. 4 the stationary problem is three-dimensional and in present chapter we consider the two-dimensional problem).

Our aim is to prove the following existence and uniqueness result.

**Theorem 7.1.** *Let $u_0, f \in \dot{H}$. Then there exists a unique function $u : [0, \infty) \to H$ such that*

$$u \in C([0, T]; H) \cap L^2(0, T; V), \quad u(0) = u_0,$$

*for all $T > 0$, satisfying the following identity*

$$\frac{d}{dt}(u(t), v) + \nu((u(t), v)) + b(u(t), u(t), v) = (f, v) \quad \text{for all} \quad v \in V, \tag{7.1}$$

*in the sense of scalar distributions on $(0, \infty)$. Moreover, the function $u : [0, \infty) \to H$ is bounded and the map $u_0 \to u(t)$ is continuous as a map in $H$.*

*Proof.* Let $f(x) = \sum_{j=1}^{\infty} \hat{f}_j \omega_j(x)$, $\hat{f}_j = (f, \omega_j)$, where $\omega_j$, $j = 1, 2, \ldots$, is an orthonormal sequence in $H$, a sequence of eigenvectors of the Stokes operator. We seek our solution $u$ as a suitable limit of the sequence of *approximate solutions* $u_n$, $n = 1, 2, \ldots$, which are finite dimensional solutions of the following problems.

**Problem 7.1.** Let $n \in \mathbb{N}, f_n(x) = \sum_{j=1}^n \hat{f}_j \omega_j(x)$, $u_n(0) = u_{0n} = \sum_{j=1}^n (u_0, \omega_j)\omega_j(x)$. On the interval $(0, \infty)$ find $u_n(t) = u_n(\cdot, t)$ of the form

$$u_n(x,t) = \sum_{j=1}^n u_{nj}(t)\omega_j(x)$$

such that

$$\frac{d}{dt}(u_n(t), \omega_j) + \nu((u_n(t), \omega_j)) + b(u_n(t), u_n(t), \omega_j) = (f_n, \omega_j) \quad \text{for} \quad j = 1, \ldots, n,$$
$$u_n(0) = u_{0n}. \tag{7.2}$$

Observe that (7.2) is a system of $n$ ordinary differential equations for the vector functions

$$t \to (u_{n1}(t), \ldots, u_{nn}(t))$$

with initial condition $((u_0, \omega_1), \ldots, (u_0, \omega_n))$. From the classical theorem about existence and uniqueness of solutions of the system

$$\dot{x} = F(x), \quad x(0) = x_0,$$

with polynomial (hence locally Lipschitz) right-hand side it follows existence of a unique *local in time* solution $u_n(t)$ on some interval $[0, t_n)$, $t_n > 0$, for every positive integer $n$. We shall show that the solutions $u_n$ are defined in fact for all positive times (are global in time). This will follow from the *first energy inequality*. We multiply the $j$th equation in (7.2) by $u_{nj}(t)$ and add the resulting $n$ equations to get

$$\frac{1}{2}\frac{d}{dt}|u_n(t)|^2 + \nu\|u_n(t)\|^2 + b(u_n(t), u_n(t), u_n(t)) = (f_n, u_n(t)).$$

As $b(u_n(t), u_n(t), u_n(t)) = 0$, using the Cauchy inequality to the right-hand side as well as the Poincaré inequality, cf. (3.16),

$$|u| \leq C\|u\|,$$

we obtain the first energy inequality

$$\frac{d}{dt}|u_n(t)|^2 + \nu\|u_n(t)\|^2 \leq C(\nu)|f_n|^2, \tag{7.3}$$

and

$$\frac{d}{dt}|u_n(t)|^2 + k_1(\nu)|u_n(t)|^2 \leq C(\nu)|f_n|^2. \tag{7.4}$$

Integrating (7.3) in an interval $(0, t)$ we obtain

$$|u_n(t)|^2 + \nu \int_0^t \|u_n(t)\|^2 ds \leq C(\nu)t|f_n|^2 + |u_n(0)|^2 \leq C(\nu)t|f|^2 + |u(0)|^2.$$

Hence, for each $t > 0$, $|u_n(t)| < \infty$ and thus the local solution can be extended to the whole positive semiaxis. From this inequality we can see that for each $T > 0$ our approximate solutions are uniformly bounded in the space

$$C([0, T]; H) \cap L^2(0, T; V).$$

Moreover, the approximate solutions are infinitely differentiable functions in $x$ and in $t$.

**Exercise 7.1.** Prove that if $f = 0$ then any approximate solution, that is, solution starting from an arbitrary initial data $u_0 \in H$, converges at least exponentially fast to zero in $H$.

To get the boundedness in $H$ of our approximate solutions we shall use the Gronwall inequality (cf. Lemma 3.20). In our case, from the Gronwall inequality applied to (7.4), we have

$$\begin{aligned}|u_n(t)|^2 &\leq |u_n(0)|^2 e^{-k_1 t} + \int_0^t e^{k_1(s-t)} C(\nu)|f_n|^2 dt \\ &= |u_n(0)|^2 e^{-k_1 t} + \frac{C(\nu)}{k_1}|f_n|^2(1 - e^{-k_1 t}).\end{aligned}$$

Therefore, every approximate solution is bounded on $[0, \infty)$ as a function with values in $H$.

From the uniform boundedness of the sequence of approximate solutions in $C([0, T]; H) \cap L^2(0, T; V)$ it follows that there exists a subsequence (denoted again by $(u_n)$) and some $u$ in $L^\infty(0, T; H) \cap L^2(0, T; V)$ such that

$$u_n \to u \quad \text{weakly} - * \text{ in } \quad L^\infty(0, T; H), \tag{7.5}$$

which means that

$$\lim_{t \to \infty} \int_0^T \int_Q u_n(x, t)\varphi(x, t) dx dt = \int_0^T \int_Q u(x, t)\varphi(x, t) dx dt$$

for all $\varphi$ from $L^1(0, T; H)$, and

$$u_n \to u \quad \text{weakly in} \quad L^2(0, T; V), \tag{7.6}$$

which means that

$$\lim_{t\to\infty}\int_0^T\int_Q u_n(x,t)\varphi(x,t)dxdt=\int_0^T\int_Q u(x,t)\varphi(x,t)dxdt$$

for all $\varphi$ from $L^2(0,T;V)$. We shall prove that this limit $u$ is a solution of Problem 7.1. From (7.2) we have

$$-\int_0^T(u_n(t),v)\phi'(t)dt+\nu\int_0^T((u_n(t),v))\phi(t)dt$$
$$+\int_0^T b(u_n(t),u_n(t),v)\phi(t)dt=\int_0^T(f_n,v)\phi(t)dt$$

for all $v$ in $V_n=\text{span}\{\omega_1,\dots,\omega_n\}$ and $\phi$ in $C_0^\infty((0,T))$. We shall show that this identity is satisfied for the limit function $u$ and for all $v$ in $V$.

As for the linear terms, we have, for any $v\in V_m$, where $m$ is any given integer,

$$\lim_{n\to\infty}\int_0^T(u_n(t),v)\phi'(t)dt=\lim_{n\to\infty}\int_0^T\int_Q u_n(x,t)v(x)\phi'(t)dxdt$$
$$=\int_0^T\int_Q u(x,t)v(x)\phi'(t)dxdt=\int_0^T(u(t),v)\phi'(t)dt,$$

which follows from the definition of the weak-star convergence in $L^\infty(0,T;H)$, as $v(x)\phi'(t)\in L^1(0,T;H)$. Similarly,

$$\lim_{n\to\infty}\int_0^T((u_n(t),v))\phi(t)dt=\lim_{n\to\infty}\int_0^T\int_Q\nabla u_n(x,t)\nabla v(x)\phi(t)dxdt$$
$$=\int_0^T\int_Q\nabla u(x,t)\nabla v(x)\phi(t)dxdt=\int_0^T((u(t),v))\phi(t)dt,$$

as $u_n\to u$ weakly in $L^2(0,T;V)$ and $v(x)\phi(t)\in L^2(0,T;V)$. Moreover,

$$\lim_{n\to\infty}\int_0^T(f_n,v)\phi(t)dt=\int_0^T(f,v)\phi(t)dt,$$

as $f_n\to f$ strongly in $L^2(Q)$.

The only problem which remains is that with the convergence of the nonlinear term. Weak convergences of $u_n$ to $u$ as in (7.5) and (7.6) are not sufficient to pass to the limit in the nonlinear term. This is a typical problem when dealing with nonlinear terms.

**Exercise 7.2.** Prove that, in general, if $u_n\to u$ and $v_n\to v$ weakly in $L^2(Q)$ then it does not follow that $u_nv_n\to uv$ in the sense of distributions on $Q$.

However, to prove that

$$\lim_{n\to\infty}\int_0^T b(u_n(t), u_n(t), v)\phi(t)dt = \int_0^T b(u(t), u(t), v)\phi(t)dt \tag{7.7}$$

it suffices to know that

$$u_n \to u \quad \text{strongly in} \quad L^2(Q) \tag{7.8}$$

for the above subsequence of the sequence of approximate solutions (or for some of its subsequences).

**Lemma 7.1.** *Let $a(\cdot,\cdot)$ be a continuous bilinear form in the Banach space $X$ with norm $\|\cdot\|$, $u_n \to u$ strongly and $v_n \to v$ weakly in $X$. Then $\lim_{n\to\infty} a(u_n, v_n) = a(u, v)$.*

*Proof.* Since the sequence $u_n$ is strongly convergent and the sequence $v_n$ is bounded as weakly convergent, it is $a(u_n, v_n) - a(u, v) = a(u_n - u, v_n) + a(u, v_n - v)$ and $|a(u_n-u, v_n)| \le C\|u_n - u\|\|v_n\| \to 0$. As for the second term, there exists $u^* \in V'$, the dual of $V$ such that $a(u, v_n - v) = \langle u^*, v_n - v\rangle_{V'\times V} \to 0$ since the sequence $v_n$ is weakly convergent in $H$. □

**Exercise 7.3.** Prove (7.7) using (7.6), (7.8), and Lemma 7.1.

**Exercise 7.4.** Prove (7.8) by proving first that

$$\text{the sequence} \quad \{\partial_t u_n\} \quad \text{is bounded in} \quad L^2(0, T; V')$$

and then using Theorem 3.18 and Lemma 7.1.

We have thus

$$-\int_0^T (u(t), v)\phi'(t)dt + \nu\int_0^T ((u(t), v))\phi(t)dt + \int_0^T b(u(t), u(t), v)\phi(t)dt = \int_0^T (f, v)\phi(t)dt \tag{7.9}$$

for all $v$ in $V_m = \text{span}\{\omega_1, \dots, \omega_m\}$, where $m$ is any positive integer and $\phi$ is in $C_0^\infty(0, T)$.

**Exercise 7.5.** Prove that (7.9) holds for all $v \in V$ using the fact that the union of all the spaces $V_m$ is dense in $V$.

To prove that $u \in C([0, T]; H)$ for all $T > 0$ we apply Lemma 3.16.

**Exercise 7.6.** Prove, using the du Bois-Reymond lemma, that (7.9) implies (7.1).

**Exercise 7.7.** Prove that $u(0) = u_0$.

**Exercise 7.8.** Prove that we have the following equality in $V'$,

$$\frac{du(t)}{dt} + \nu Au(t) + B(u(t)) = f,$$

for a.e. $t \in (0, T)$. Use Proposition 3.2 and property (3.26) of the evolution triple $V \subset H \subset V'$.

To prove the uniqueness of a solution let us assume, on the contrary, that there exist two different solutions, $u$ and $v$, to our problem

$$\frac{du(t)}{dt} + \nu Au(t) + B(u(t)) = f, \quad u(0) = u_0$$

on the interval $[0, \infty)$. Setting $w(t) = u(t) - v(t)$ we have

$$\frac{dw(t)}{dt} + \nu Aw(t) + B(u(t)) - B(v(t)) = 0 \quad \text{in} \quad V'.$$

Thus, in particular,

$$\left\langle \frac{dw(t)}{dt} + \nu Aw(t) + B(u(t)) - B(v(t)), w(t) \right\rangle = 0,$$

that is, see Theorem 3.16,

$$\frac{1}{2}\frac{d}{dt}|w(t)|^2 + \nu \|w(t)\|^2 + b(u(t), u(t), w(t)) - b(v(t), v(t), w(t)) = 0.$$

Further, using a property of the trilinear form $b$, we obtain

$$\frac{1}{2}\frac{d}{dt}|w(t)|^2 + \nu \|w(t)\|^2 = -b(w(t), v(t), w(t)).$$

We shall estimate the right-hand side applying to it the Ladyzhenskaya inequality (cf. Exercise 4.8),

$$\|u\|_{L^4(Q)^2} \le C|u|^{1/2}\|u\|^{1/2} \quad \text{for} \quad u \in V.$$

We have

$$|b(u, v, w)| \le \|u\|_{L^4(Q)^2}\|v\|\|w\|_{L^4(Q)^2} \le c_1|u|^{1/2}\|u\|^{1/2}\|v\||w|^{1/2}\|w\|^{1/2}.$$

Therefore,

$$\begin{aligned} b(w(t), v(t), w(t)) &\le c_1|w(t)|\|v(t)\|\|w(t)\| \\ &\le \frac{\nu}{2}\|w(t)\|^2 + \frac{c_2}{\nu}|w(t)|^2\|v(t)\|^2. \end{aligned}$$

We have then

$$\frac{d}{dt}|w(t)|^2 + \nu\|w(t)\|^2 \leq \frac{c_2}{\nu}|w(t)|^2\|v(t)\|^2.$$

Now, in view of the Gronwall inequality we have, for $t \geq 0$,

$$|w(t)|^2 \leq |w(0)|^2 \exp\left\{\frac{c_2}{\nu}\int_0^t \|v(s)\|^2 ds\right\}.$$

As $w(0) = 0$, $u(t) = v(t)$ for all $t > 0$ which proves the uniqueness. From the uniqueness result together with our construction of a weak solution on an arbitrary interval $(0, T)$ follows the existence of a unique global solution defined on the interval $(0, \infty)$.

Observe that from the above inequality it follows also the continuous dependence of solutions on the initial data: the map $u_0 \to u(t)$ is continuous as a map in $H$ for $t > 0$. This completes the proof of Theorem 7.1. □

*Remark 7.1.* One can prove that under the assumptions of Theorem 7.1 the solution $u$ belongs to the space $L^2(\eta, T; D(A))$ for all $T > 0$ and $0 < \eta < T$, where $D(A)$ is the domain of the Stokes operator, cf. Sect. 4.1.

**Exercise 7.9.** Compute the constants $C(\nu)$, $k_1(\nu)$ in inequality (7.4).

**Exercise 7.10.** Prove an existence and uniqueness theorem analogous to Theorem 7.1 for the case of homogeneous Dirichlet boundary conditions $u = 0$ at $\partial\Omega \times (0, T)$, where $\Omega$ is a bounded open set in $\mathbb{R}^2$, and the spaces $H$ and $V$ in Theorem 7.1 are accordingly modified.

**Exercise 7.11.** Consider the nonhomogeneous Dirichlet boundary condition $u = a$ at $\partial\Omega \times (0, T)$, where $\Omega$ is a bounded open set in $\mathbb{R}^2$ with sufficiently smooth boundary, and where the function $a$ is defined in $\Omega$, $a \in H^1(\Omega)$, $\operatorname{div} a = 0$, so that $u - a \in V$. Formulate and prove an existence and uniqueness theorem analogous to Theorem 7.1 for this case.

**Exercise 7.12.** *Non-autonomous problem.* Using the argument of the proof of Theorem 7.1 prove the following result.

**Theorem 7.2.** *Let $V \subset H \subset V'$ be either given by the homogeneous Dirichlet boundary conditions on an open and bounded set $\Omega \subset \mathbb{R}^2$ with Lipschitz boundary or the periodic boundary conditions on a two-dimensional box. Let $u_0 \in H$ and $f \in L^2_{loc}(\mathbb{R}^+; V')$. Then there exists a unique function $u : \mathbb{R}^+ \to H$ such that*

$$u \in L^2(0, T; V) \cap C([0, T]; H), \quad u(0) = u_0,$$

*for all $T \geq 0$, satisfying the identity*

$$\frac{d}{dt}(u(t), v) + \nu((u(t), v)) + b(u(t), u(t), v) = \langle f(t), v\rangle \quad \textit{for all} \quad v \in V$$

*in the sense of scalar distributions on* $(0,\infty)$. *Moreover, the map* $u_0 \to u(t)$ *is continuous as a map in* $H$.

*Remark 7.2.* Consider either the two-dimensional periodic problem or the homogeneous Dirichlet problem on the bounded domain $\Omega \subset \mathbb{R}^2$ of class $C^2$. Let $f \in L^2_{loc}(\mathbb{R}^+;H)$ and $u_0 \in H$. Then the solution $u$ given in Theorem 7.2 belongs to $L^2(\eta, T; D(A))$ for all $0 < \eta < T < \infty$ (see, for example, Theorem 7.4 in [99] or Theorem III.3.10 in [219]).

*Remark 7.3.* For the case of homogeneous Dirichlet boundary conditions, the existence results of Theorems 7.1 and 7.2 can be extended to the case of *unbounded domains*, where the embedding $V \subset H$ is not compact. For the treatment of this case see, e.g., [201, 219].

## 7.2 Existence of the Global Attractor: Case of Periodic Boundary Conditions

We shall present first some notions from the theory of dynamical systems.

**Definition 7.1.** Let $X$ be a complete metric space. A family of maps $S(t) : X \to X$, $t \in [0,\infty)$, denoted as $\{S(t)\}_{t\geq 0}$ is called a *semigroup* in $X$ if $S(0)$ is the identity map and for all $s, t > 0$ it is $S(t+s) = S(t)S(s)$. A semigroup is *continuous* in $X$ if all the maps are continuous in $X$.

**Definition 7.2.** Let $\{S(t)\}_{t\geq 0}$ be a continuous semigroup in a complete metric space $X$. Then a subset $\mathscr{A}$ of $X$ is a *global attractor* for this semigroup in $X$ if the following conditions hold:

(*i*) *invariance*: $S(t)\mathscr{A} = \mathscr{A}$ for all $t > 0$,
(*ii*) *compactness*: $\mathscr{A}$ is a nonempty and compact subset of $X$,
(*iii*) *attraction*: for every bounded subset $B$ of $X$ and every $\varepsilon > 0$ there exists $t_0 = t_0(B,\varepsilon)$ such that for all $t \geq t_0$, $S(t)B \subset \mathscr{O}_\varepsilon(\mathscr{A})$, where $\mathscr{O}_\varepsilon(\mathscr{A})$ is the $\varepsilon$-neighborhood of $\mathscr{A}$,

$$\mathscr{O}_\varepsilon(\mathscr{A}) = \bigcup_{x\in\mathscr{A}} B(x,\varepsilon),$$

and $B(x,\varepsilon)$ is the open ball in $X$ with radius $\varepsilon$ and centered at $x$.

**Definition 7.3.** The set $A \subset X$ shall be called *invariant* with respect to the semigroup $\{S(t)\}_{t\geq 0}$ if $S(t)A = A$ for all $t \geq 0$.

**Definition 7.4.** The *Hausdorff semi-distance* in the metric space $X$ between two nonempty sets $A, B \subset X$ is defined as

$$\mathrm{dist}_X\,(A,B) = \sup_{x\in A}\inf_{y\in B} \rho(x,y).$$

**Exercise 7.13.** Prove that the last property in the definition of the global attractor (attraction) is equivalent to say that for every bounded set $B \subset X$ we have

$$\lim_{t\to\infty} \mathrm{dist}_X (S(t)B, \mathscr{A}) = 0.$$

**Definition 7.5.** A subset $B$ in $X$ is *absorbing* if for every bounded subset $G$ in $X$ there exists a time $t_0 = t_0(G)$ such that for all $t \geq t_0$, $S(t)G \subset B$.

Schematic illustration of a global attractor and an absorbing set is depicted in Fig. 7.1.

**Definition 7.6.** A semigroup $\{S(t)\}_{t\geq 0}$ in $X$ is *uniformly compact* if for every bounded subset $G$ in $X$ there exists a time $t_0 = t_0(G)$ such that the set $\bigcup_{t\geq t_0} S(t)G$ is precompact in $X$.

**Exercise 7.14.** By $\omega(B) = \bigcap_{s\geq 0} \overline{\bigcup_{t\geq s} S(t)B}$ we define the $\omega$-limit set $\omega(B)$ of a subset $B$ of $X$ (cf. Fig. 7.2 for an illustration of an $\omega$-limit set). Prove that $\phi \in \omega(B)$ if and only if there exists a sequence $\phi_n \in B$ and a sequence $t_n \to \infty$ such that $S(t_n)\phi_n \to \phi$ as $n \to \infty$.

We shall prove our main theorem about existence of a global attractor.

**Theorem 7.3.** *Let $\{S(t)\}_{t\geq 0}$ be a continuous semigroup in a complete metric space X. If there exists a bounded absorbing set $B \subset X$ and the semigroup is uniformly compact in X, then the $\omega$-limit set of B, $\mathscr{A} = \omega(B)$, is a global attractor for this semigroup in X.*

*Proof.* Let $B$ be an absorbing subset in $X$.

**Step 1.** $\omega(B)$ is nonempty and compact in $X$. As $\omega(B) = \bigcap_{s\geq 0} \overline{\bigcup_{t\geq s} S(t)B}$, where the sets $\overline{\bigcup_{t\geq s} S(t)B}$ are nonempty and compact for $s \geq t_0(B)$ (the semigroup is uniformly compact and $B$ is bounded) and constitute a nonincreasing family

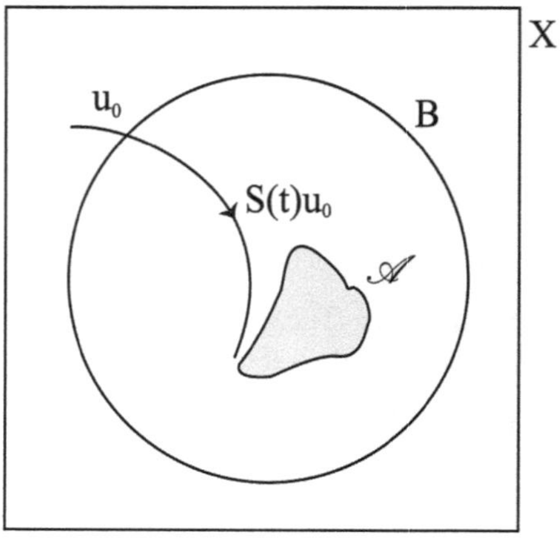


**Fig. 7.1** The absorbing set $B$ and the global attractor $\mathscr{A}$ for the semigroup $\{S(t)\}_{t\geq 0}$. All trajectories after some time (depending on the initial condition $u_0$) are in the set $B$. They become closer and closer to the attractor $\mathscr{A}$ as the time goes to infinity

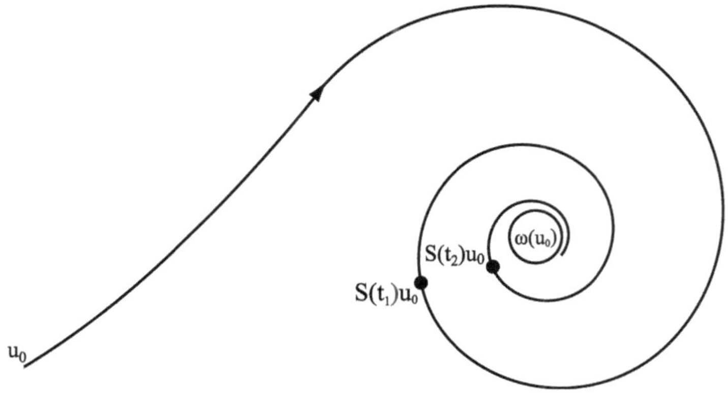


**Fig. 7.2** Illustration of the $\omega$-limit set of the point $u_0$. The set $\omega(\{u_0\})$ consists of all cluster points of all sequences $\{S(t_n)u_0\}$ for $t_n \to \infty$

of sets as $s$ grows, then $\omega(B)$ is not empty and compact in view of the Cantor theorem.
**Step 2.** $\omega(B)$ is invariant: $S(t)\omega(B) = \omega(B)$ for all $t > 0$.
($\Rightarrow$) $S(t)\omega(B) \subset \omega(B)$. Let $S(t)\psi = \phi$, $\psi \in \omega(B)$. We shall show that $\phi \in \omega(B)$. For $\psi$ there exist sequences: $\psi_n$ in $B$ and $t_n \to \infty$ such that $S(t)\psi_n \to \psi$. Hence, $S(t)S(t_n)\psi_n = S(t + t_n)\psi_n \to S(t)\psi = \phi$ (as the operator $S(t)$ is continuous). Moreover, $\phi \in \omega(B)$ by definition of the $\omega$-limit set, as the sequence $S(t + t_n)\psi_n$ converges to $\phi$, where $t + t_n \to \infty$.
($\Leftarrow$) $\omega(B) \subset S(t)\omega(B)$. Let $\phi \in \omega(B)$. We shall show that there exists $\psi \in \omega(B)$ such that $S(t)\psi = \phi$. Let $S(t_n)\phi_n \to \phi$, $t_n \to \infty$, $\phi_n \in B$. As the sets $\bigcup_{s \geq t_n - t} S(s)B$ are precompact for $t_n - t \geq t_0$ thus the sequence $S(t_n - t)\phi_n$ contains a convergent subsequence $S(t_\mu - t)\phi_\mu \to \psi$ for some $\psi \in X$. But $\psi \in \omega(B)$ as $\phi_\mu \in B$ and $t_\mu - t \to \infty$. Moreover, $S(t)S(t_\mu - t)\phi_\mu = S(t_\mu)\phi_\mu \to \phi = S(t)\psi$, from the continuity of $S(t)$.
**Step 3.** $\mathscr{A} = \omega(B)$ attracts bounded subsets $G$ of $X$.
Assume that this is not true. Then $\mathrm{dist}_X(S(t)G, \mathscr{A}) \geq \delta > 0$ for $t \to \infty$, that is, there exist sequences $t_n \to \infty$ and $\phi_n \in G$ such that $\mathrm{dist}_X(S(t_n)\phi_n, \mathscr{A}) \geq \delta > 0$ for all $n$. On the other hand, the sequence $S(t_n)\phi_n$ contains a convergent subsequence, $S(t_\mu)\phi_\mu \to \psi$, $\psi \in \mathscr{A} = \omega(B)$ as $S(t_\mu)\phi_\mu = S(t_\mu - t)S(t)\phi_\mu$ and for $t$ such that $S(t)G \subset B$, $S(t)\phi_\mu \in B$.

□

**Exercise 7.15.** Prove that $\mathscr{A}$ is a maximal, in the sense of the inclusion relation, bounded invariant set.

*Solution.* Assume that $\mathscr{A} \subset \mathscr{A}_1$, $\mathscr{A}_1$ different from $\mathscr{A}$, bounded and invariant. As $B$ absorbs $\mathscr{A}_1$, $S(t)\mathscr{A}_1 \subset B$ for large $t$, then $\mathscr{A}_1 \subset B$ as $S(t)\mathscr{A}_1 = \mathscr{A}_1$ from our assumption. Hence, $\omega(\mathscr{A}_1) = \overline{\mathscr{A}_1} \subset \omega(B) = \mathscr{A}$, and we have a contradiction.

**Exercise 7.16.** Prove that $\mathscr{A}$ is a connected set provided $X$ is connected.

*Solution.* Assume that this is not true. Then there exist open sets $U_1$, $U_2$ such that $U_i \cap \mathscr{A} \neq \emptyset$, $i = 1, 2$, $\mathscr{A} \subset U_1 \cup U_2$, $U_1 \cap U_2 = \emptyset$. Let $B$ be the closed convex hull of $\mathscr{A}$. Then $B$ is compact (and so bounded) and connected. From the continuity of $S(t)$, the sets $S(t)B$ are connected. The set $\mathscr{A}$ is a subset of each $S(t)B$, as $\mathscr{A} \subset B \Rightarrow \mathscr{A} = S(t)\mathscr{A} \subset S(t)B$. As $\mathscr{A}$ attracts bounded sets, $\text{dist}_X(S(t)B, \mathscr{A}) \to 0$ as $t \to \infty$. This means that $S(t)B \subset U_1 \cup U_2$—an open neighborhood of $\mathscr{A}$. But then we have $\emptyset \neq U_i \cap \mathscr{A} \subset U_i \cap S(t)B$, $i = 1, 2$, $S(t)B \subset U_1 \cup U_2$, $U_1 \cap U_2 = \emptyset$, which means that $S(t)B$ is not connected. We have come to a contradiction.

**Exercise 7.17.** Prove that $\mathscr{A}$ is a minimal closed set which attracts bounded sets of $X$.

*Solution.* Assume, to the contrary, that there exists a bounded set $F \subset \mathscr{A}$, $F \neq \mathscr{A}$, attracting bounded sets. As $\mathscr{A}$ is compact then $F$ is also compact. Let $y \in \mathscr{A} \setminus F$. For small $\varepsilon$, $\mathscr{O}_\varepsilon(y) \cap \mathscr{O}_\varepsilon(F) = \emptyset$. Let $B$ be an open absorbing set. As $F$ attracts $B$, $S(t)B \subset \mathscr{O}_\varepsilon(F)$ for large $t \geq t(\varepsilon)$. But $y \in \omega(B) = \mathscr{A}$ whence $y = \lim_{k\to\infty} S(t_k)x_k$, $x_k \in B$, $t_k \to \infty$, and we have $S(t_k)x_k \subset \mathscr{O}_\varepsilon(y)$ for large $k$. But we have also $S(t_k)x_k \in \mathscr{O}_\varepsilon(F)$ for large $k$. A contradiction as $\mathscr{O}_\varepsilon(y) \cap \mathscr{O}_\varepsilon(F) = \emptyset$.

*Remark 7.4.* The assumption of uniform compactness in Theorem 7.3 can be replaced by a weaker assumption of asymptotic compactness. However, we shall use the stronger version as in the case of the Navier–Stokes problem that we consider in this chapter the uniform compactness property holds and is easy to check. In the case of the Navier–Stokes equations in unbounded spatial domains the uniform compactness property does not hold but we can use the asymptotic compactness property [201].

**Definition 7.7.** We call a semigroup $\{S(t)\}_{t\geq 0}$ *asymptotically compact* if it has the following property: for any bounded sequence $\{u_n\}$, and any sequence $\{t_n\}$, $n \in \mathbb{N}$ such that $t_n \to \infty$ as $n \to \infty$, the sequence $\{S(t_n)u_n\}$ contains a convergent subsequence.

**Exercise 7.18.** Prove that uniform compactness implies asymptotic compactness. Consider the proof of Theorem 7.3 and show that the condition of uniform compactness in the hypothesis of the theorem can be replaced by the condition of asymptotic compactness.

**Exercise 7.19.** Prove that existence of a global attractor for a given semigroup implies existence of a bounded absorbing set and asymptotic compactness of the semigroup.

The global attractor for the semigroup governed by the solution map of the ODE $\dot{x} = |x|(1 - x)$ is presented in Fig. 7.3.

Now we can come back to our main considerations. From Theorem 7.1 it follows that the family of maps $S(t) : H \to H$ defined for all $t \geq 0$ by

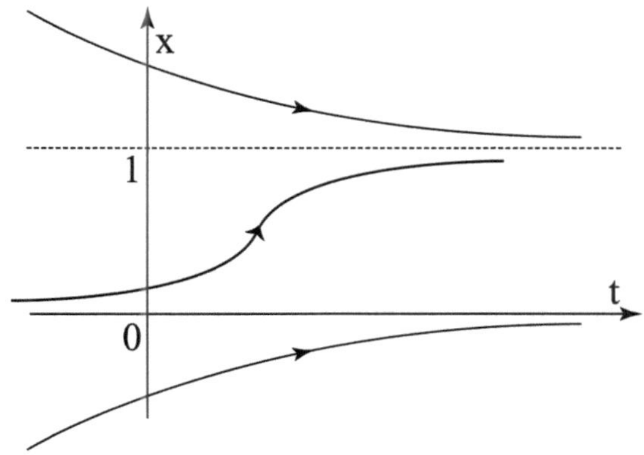


**Fig. 7.3** Example of a semigroup $S(t) : \mathbb{R} \to \mathbb{R}$ given by the solution map for the equation $\dot{x} = |x|(1 - x)$. The global attractor is given by $\mathscr{A} = [0, 1]$. It contains two stationary points $\{0, 1\}$ and all points belonging to a complete trajectory that connects them

$$S(t)u_0 = u(t), \tag{7.10}$$

where $u(\cdot)$ is the unique weak solution of the considered problem with $u(0) = u_0$, is a continuous semigroup in $X$. Our aim is to prove the following

**Theorem 7.4 (Existence of a Global Attractor in** $H$**).** *Under the assumptions of Theorem 7.1 there exists a global attractor* $\mathscr{A}$ *for the semigroup* $\{S(t)\}_{t\geq 0}$ *in* $H$ *associated with the two-dimensional Navier–Stokes equations with periodic boundary conditions.*

*Proof.* **Step 1.** *Existence of a bounded absorbing set in* $H$. From the first energy inequality, cf. inequality (7.3),

$$\frac{d}{dt}|u(t)|^2 + \nu\|u(t)\|^2 \leq C(\nu)|f|^2, \tag{7.11}$$

we have as an immediate consequence, cf. inequality (7.4),

$$\frac{d}{dt}|u(t)|^2 + k_1(\nu)|u(t)|^2 \leq C(\nu)|f|^2.$$

Applying the Gronwall inequality to the last one we have,

$$|u(t)|^2 \leq |u_0|^2 e^{-k_1 t} + \int_0^t e^{k_1(s-t)} C(\nu)|f|^2 dt = |u_0|^2 e^{-k_1 t} + \frac{C(\nu)}{k_1}|f|^2(1 - e^{-k_1 t}), \tag{7.12}$$

from which it follows that for $\rho^2 > \rho_0^2 = \frac{C(\nu)}{k_1}|f|^2$, balls $B(0, \rho)$ in the phase space $H$ absorb bounded sets in $H$.

Let $u_0 \in B_1$ be a bounded subset in $H$. Then, $|u(t)|^2 \leq \rho^2$ for $t \geq t_0(B_1)$.

**Exercise 7.20.** Let $u_0 \in B(0, r)$. Compute $t_0 = t_0(B(0, r))$ such that $|u(t)|^2 \leq \rho^2$ for $t \geq t_0$.

**Step 2.** *Existence of a bounded absorbing set in V and uniform compactness of the semigroup.* To prove these properties we need the *second energy inequality*. Taking the scalar product of the Navier–Stokes equation with $Au$ (for $u \in D(A), Au = -\Delta u$, cf. Remark 7.1) we obtain

$$\frac{d}{dt}\|u\|^2 + 2\nu|Au|^2 = 2(f, Au), \tag{7.13}$$

as $b(u, u, Au) = 0$. The relation (7.13) is called the *enstrophy equation*.

**Exercise 7.21.** Prove that for the two-dimensional Navier–Stokes flow with periodic boundary conditions we have not only $b(u, u, u) = 0$ but also $b(u, u, Au) = 0$, cf. [197, 220]. The latter property makes the analysis of the problem much easier.

From (7.13) we obtain

$$\frac{d}{dt}\|u\|^2 + \nu|Au|^2 \leq C|f|^2,$$

and, as $\|u\| \leq C_1|Au|$ for $u \in D(A)$, we have also

$$\frac{d}{dt}\|u\|^2 + k_1\|u\|^2 \leq C|f|^2. \tag{7.14}$$

**Exercise 7.22.** Prove that $\|u\| \leq C_1|Au|$ for $u \in D(A)$ and compute the constants $C$, $C_1$, and $k_1$ above.

Now, to estimate the norm $\|u(t)\|$ we cannot use the Gronwall inequality since our initial condition $u_0$ is only in $H$ and not in $V$. Instead, we shall use the *uniform Gronwall inequality* (cf. Lemma 3.21). To apply the uniform Gronwall lemma, we proceed as follows. First, from inequality (7.11), after integration in the interval $(t, t+r)$, where $t \geq t_0(u_0)$, we obtain in particular,

$$\nu \int_t^{t+r} \|u(t)\|^2 ds \leq C(\nu)r|f|^2 + \rho^2 = a_3\nu \tag{7.15}$$

for all $t \geq t_0(B_1)$. Then, from inequality (7.14) and the uniform Gronwall lemma we get directly

$$\|u(t)\|^2 \leq \left(\frac{a_3}{r} + a_2\right) \quad \text{for all} \quad t \geq t_0 + r, \tag{7.16}$$

where

$$a_2 = \int_t^{t+r} C|f|^2 dt = rC|f|^2, \quad a_3 = \frac{1}{\nu}(rC|f|^2 + \rho^2),$$

[cf. (7.15)]. Estimate (7.16) means that our bounded set $B_1$ is moved by $S(t)$, for $t \geq t_0 + r$, into a ball in $V$,

$$\|S(t)x\|^2 \leq \left(\frac{a_3}{r} + a_2\right) \quad \text{for all} \quad t \geq t_0 + r, \ x \in B_1,$$

which is a precompact set in $H$; due to the Rellich lemma, $V$ is compactly embedded in $H$. Thus, the whole trajectory of the set $B_1$ in $H$, starting from $t_0 + r$,

$$\bigcup_{t \geq t_0 + r} S(t)B_1 \quad \text{is precompact in} \quad H.$$

This is the condition of the *uniform compactness* of the semigroup $S(t)$ in $H$.

From the above considerations and Theorem 7.3 it follows the existence of the global attractor $\mathscr{A} \subset H$ for the two-dimensional Navier–Stokes equations with periodic boundary conditions. □

**Exercise 7.23.** Prove the existence of a global attractor in the case of homogeneous Dirichlet boundary conditions. To deal with the trilinear form $b(u, u, Au)$ apply the *Agmon inequality*

$$\|u\|_\infty \leq c_2 |u|^{1/2} |Au|^{1/2} \quad \text{for} \quad u \in D(A).$$

**Exercise 7.24.** Prove that from the uniform compactness of the semigroup $S(t)$ in $H$ it follows the existence of a *compact* absorbing set in $H$.

**Exercise 7.25.** Prove that if $f \in L^\infty(\mathbb{R}^+; H)$, then the bounded absorbing set in $V$ exists for the corresponding non-autonomous problem.

## 7.3 Convergence to the Stationary Solution: The Simplest Case

Let us recall the stationary problem:

$$-\nu \Delta u + (u \cdot \nabla)u + \nabla p = f \quad \text{in} \quad Q,$$
$$\operatorname{div} u = 0 \quad \text{in} \quad Q,$$

with one of the boundary conditions:

1. $Q$ is a square $[0, L]^2$ in $\mathbb{R}^2$ and we assume periodic boundary conditions, or
2. $Q$ is a bounded domain in $\mathbb{R}^2$, with smooth boundary, and we assume the homogeneous boundary condition $u = 0$ on $\partial Q$.

Let $H$ and $V$ be the corresponding function spaces, cf. Sect. 4.1. Then the weak formulation of the above problem is as follows.

**Problem 7.2.** For $f \in H$ (or $f \in V'$) find $u \in V$ such that

$$\nu((u, v)) + b(u, u, v) = \langle f, v \rangle \quad \text{for all} \quad v \in V$$

or, equivalently,

$$\nu Au + Bu = f \quad \text{in} \quad H \quad \text{or} \quad V'.$$

Assuming that the viscosity $\nu$ is large enough with respect to the volume force $f$ we shall prove convergence of solutions $u(t)$ of the nonstationary problem to the unique stationary solution $u_\infty$ as time goes to infinity.

**Theorem 7.5.** *There exists a constant $c_1(Q)$ such that if $\nu^2 > c_1(Q)\|f\|_{V'}$ then the stationary solution $u_\infty$ is unique. Moreover, for each $u_0 \in H$ the solution $u$ of the problem:*

$$u \in C([0,\infty); H) \cap L^2_{loc}(0,\infty; V), \quad u(0) = u_0,$$
$$\frac{d}{dt}(u(t), v) + \nu((u(t), v)) + b(u(t), u(t), v) = \langle f, v\rangle \quad \text{for all} \quad v \in V,$$

*converges to $u_\infty$ in $H$ as time goes to infinity,*

$$|u(t) - u_\infty| \to 0 \quad \text{as} \quad t \to \infty.$$

*Proof.* Let $w(t) = u(t) - u_\infty$. Since $\nu Au_\infty + B(u_\infty) = f$, we have

$$\frac{dw(t)}{dt} + \nu Aw(t) + B(u(t)) - B(u_\infty) = 0 \quad \text{in} \quad V'.$$

Thus, in particular,

$$\left\langle \frac{dw(t)}{dt} + \nu Aw(t) + B(u(t)) - B(u_\infty), w(t)\right\rangle = 0,$$

that is,

$$\frac{1}{2}\frac{d}{dt}|w(t)|^2 + \nu\|w(t)\|^2 + b(u(t), u(t), w(t)) - b(u_\infty, u_\infty, w(t)) = 0.$$

Further, using a property of the trilinear form $b$, we obtain

$$\frac{1}{2}\frac{d}{dt}|w(t)|^2 + \nu\|w(t)\|^2 = -b(w(t), u_\infty, w(t)). \tag{7.17}$$

We shall estimate the right-hand side. Using the Ladyzhenskaya inequality we obtain

$$b(u, v, w) \le \|u\|_{L^4(Q)^2}\|v\|\|w\|_{L^4(Q)^2} \le c_1|u|^{1/2}\|u\|^{1/2}\|v\||w|^{1/2}\|w\|^{1/2}.$$

Thus,

$$\begin{aligned} b(w(t), u_\infty, w(t)) &\le c_1 |w(t)| \|u_\infty\| \|w(t)\| \\ &\le \frac{\nu}{2}\|w(t)\|^2 + \frac{c_2}{\nu}|w(t)|^2\|u_\infty\|^2. \end{aligned}$$

We have then

$$\frac{d}{dt}|w(t)|^2 + \nu\|w(t)\|^2 \le \frac{c_2}{\nu}|w(t)|^2\|u_\infty\|^2,$$

and further, as $|w(t)| \le C\|w(t)\|$, we have

$$\frac{d}{dt}|w(t)|^2 + \left(\nu C - \frac{c_2}{\nu}\|u_\infty\|^2\right)|w(t)|^2 \le 0. \tag{7.18}$$

Now, we know that $\|u_\infty\| \le \frac{1}{\nu}\|f\|_{V'}$, whence

$$\nu C - \frac{c_2}{\nu}\|u_\infty\|^2 \ge \nu C - \frac{c_2}{\nu^3}\|f\|_{V'}^2 > 0$$

for large $\nu$ [satisfying $\nu^2 > c_1(Q)\|f\|_{V'}$ for some constant $c_1(Q)$]. Thus, from inequality (7.18) it follows (for these $\nu$) the uniqueness of the stationary solution $u_\infty$ and also the exponential decay in $H$ of $u(t)$ to $u_\infty$ as $t \to \infty$, and this ends the proof. □

**Exercise 7.26.** Prove the above theorem without using the Ladyzhenskaya inequality, by estimating in a different way the right-hand side of Eq. (7.17). Compare the constants $c_1(Q)$ in both cases for the problem with periodic boundary conditions.

## 7.4 Convergence to the Stationary Solution for Large Forces

We consider two-dimensional flows on $Q = [0, L]^2$ as in the preceding sections and prove that for any force of the form $f = \left(0, \alpha \sin\left(\frac{2\pi}{L}x_1\right)\right)$, where $\alpha > 0$ can be arbitrarily large, all nonstationary solutions of the Navier–Stokes problem converge to the unique solution $\bar{u} = \frac{1}{\nu\lambda_1}f$ of the stationary Navier–Stokes problem, where $\lambda_1 = \frac{4\pi^2}{L^2}$ is the first eigenvalue of the Stokes operator.

*Remark 7.5.* Observe that for our force $f$ the *Grashof number*

$$G = \frac{|f|}{\nu^2\lambda_1} = \frac{L|\alpha|}{\sqrt{2}\nu^2\lambda_1},$$

which is a dimensionless quantity which measures the strength of the forcing relative to viscosity, is large for large $\alpha$.

It is easy to see that $f$ has the following properties,

$$\operatorname{div} f = 0, \quad Af = \lambda_1 f, \quad B(f,f) = 0. \tag{7.19}$$

Let us write the energy equation

$$\frac{1}{2}\frac{d}{dt}|u|^2 + \nu\|u\|^2 = (f,u),$$

and the enstrophy equation

$$\frac{1}{2}\frac{d}{dt}\|u\|^2 + \nu|Au|^2 = (f,Au) = (Af,u) = \lambda_1(f,u).$$

We have used the property, cf. Exercise 7.21,

$$b(u,u,Au) = 0.$$

Multiplying the energy equation by $\lambda_1$ and subtracting it from the enstrophy equation we have

$$\frac{1}{2}\frac{d}{dt}(\|u\|^2 - \lambda_1|u|^2) + \nu(|Au|^2 - \lambda_1\|u\|^2) = 0.$$

Observe that $\|u\|^2 - \lambda_1|u|^2 \geq 0$ in view of the Poincaré inequality. More precisely, recalling that

$$u = \sum_{k=1}^{\infty} \hat{u}_k \omega_k$$

we have,

$$\|u\|^2 - \lambda_1|u|^2 = \sum_{k=5}^{\infty}(\lambda_k - \lambda_1)|\hat{u}_k|^2,$$

and also

$$\begin{aligned}|Au|^2 - \lambda_1\|u\|^2 &= \sum_{k=5}^{\infty}\lambda_k(\lambda_k - \lambda_1)|\hat{u}_k|^2 \\ &\geq \lambda_5\sum_{k=5}^{\infty}(\lambda_k - \lambda_1)|\hat{u}_k|^2 = \lambda_5(\|u\|^2 - \lambda_1|u|^2).\end{aligned}$$

Thus, we have

$$\frac{1}{2}\frac{d}{dt}(\|u\|^2 - \lambda_1|u|^2) + \nu\lambda_5(\|u\|^2 - \lambda_1|u|^2) \leq 0,$$

and

$$\lim_{t\to\infty}(\|u(t)\|^2 - \lambda_1|u(t)|^2) = 0. \tag{7.20}$$

Now, let $u = P_4u + Q_4u$. Then

$$\begin{aligned}|Q_4u|^2 = \sum_{k=5}^{\infty}|\hat{u}_k|^2 &\leq \frac{1}{\lambda_5 - \lambda_1}\sum_{k=5}^{\infty}(\lambda_k - \lambda_1)|\hat{u}_k|^2 \\ &= \frac{1}{\lambda_5 - \lambda_1}(\|u\|^2 - \lambda_1|u|^2). \end{aligned} \tag{7.21}$$

Therefore, from (7.20) and (7.21) we conclude that

$$\lim_{t\to\infty}|Q_4u(t)|^2 = 0.$$

We have shown that all higher harmonics go to zero as time goes to infinity. Let $\bar{u} = \frac{1}{\nu\lambda_1}f$ be the stationary solution. We define $v(t) = u(t) - \bar{u}$, the difference between some nonstationary solution and the stationary solution $\bar{u}$. As $P_4\bar{u} = \bar{u}$ then $Q_4v(t) = Q_4u(t) \to 0$ as $t \to \infty$. We shall show that also $P_4v(t) \to 0$ as $t \to \infty$, so that $v(t) \to 0$ and hence $u(t) \to \bar{u}$ as $t \to \infty$. We have

$$\frac{dv}{dt} + \nu Av + B(v, v) + B(v, \bar{u}) + B(\bar{u}, v) = 0.$$

We multiply this equation by $P_4v$ in $H$ and use $\|P_4v\|^2 = \lambda_1|P_4v|^2$ to get

$$\frac{1}{2}\frac{d}{dt}|P_4v|^2 + \nu\lambda_1|P_4v|^2 + b(v, v, P_4v) + b(v, \bar{u}, P_4v) + b(\bar{u}, v, P_4v) = 0.$$

Let us write $v = P_4v + Q_4v$ and use $|Q_4v(t)| \to 0$ as $t \to \infty$. We can write the above equation in the form

$$\frac{1}{2}\frac{d}{dt}|P_4v|^2 + \nu\lambda_1|P_4v|^2 + b(P_4v, P_4v, P_4v) + b(P_4v, \bar{u}, P_4v) + b(\bar{u}, P_4v, P_4v) = \beta(t),$$

where $\beta(t) \to 0$ as $t \to \infty$ since all terms in $\beta(t)$ are of the form $b(x, y, z)$ where at least one variable is equal to $Q_4v(t)$. Moreover $b(P_4v, P_4v, P_4v) = 0$ and

$$b(P_4v, \bar{u}, P_4v) + b(\bar{u}, P_4v, P_4v) = -b(P_4v, P_4v, \bar{u}) = 0.$$

**Exercise 7.27.** Write $P_4 v$ as a sum of the first four eigenvectors to check directly that $b(P_4 v, P_4 v, \bar{u}) = 0$.

In this way we have obtained

$$\frac{1}{2}\frac{d}{dt}|P_4 v|^2 + \nu\lambda_1 |P_4 v|^2 \leq \beta(t), \quad \beta(t) \to 0, \quad t \to \infty,$$

whence

$$\lim_{t\to\infty} |P_4 u(t)|^2 = 0,$$

and

$$\lim_{t\to\infty} |u(t) - \bar{u}| = 0.$$

This relation proves also the uniqueness of the stationary solution.

**Exercise 7.28.** Write $\beta(t)$ explicitly and prove that $\beta(t) \to 0$ as $t \to \infty$.

**Exercise 7.29.** Assume that the external force $f$ satisfies: $B(f,f) = 0$ and $Af = \lambda_m f$ for $\lambda_m$ large enough. Prove that the stationary solution $\bar{u} = \frac{1}{\nu\lambda_m} f$ is unique and globally stable, that is, all nonstationary solutions $u(t)$ converge to $\bar{u}$ in $H$ as $t \to \infty$.

**Exercise 7.30.** Assume that the external force $f = f_m + f_n$ satisfies: $B(f_m, f_m) = 0$, $B(f_n, f_n) = 0$, $B(f_n, f_m) = 0$ and $Af_m = \lambda_m f_m$, $Af_n = \lambda_n f_n$ for $\lambda_m, \lambda_n$ large enough. Find the unique stationary solution $\bar{u}$ and prove that it is globally stable, that is, all nonstationary solutions $u(t)$ converge to $\bar{u}$ in $H$ as $t \to \infty$. Generalize the result to any finite combination $f = f_{m_1} + \ldots + f_{m_k}$.

*Solution to Exercise 7.29.* Multiplying the equation

$$\frac{du}{dt} + \nu Au + B(u,u) = f$$

by $\lambda_m$ and using $\lambda_m f = Af$ we obtain

$$\lambda_m \frac{du}{dt} + A(\lambda_m \nu u - f) + \lambda_m B(u,u) = 0. \tag{7.22}$$

Let $v(t) = \lambda_m \nu u(t) - f$ (then $u = \frac{1}{\lambda_m \nu}(v + f)$). We shall show that $v(t) \to 0$ as $t \to \infty$. Observe that

$$\lambda_m \frac{du}{dt} = \frac{1}{\nu}\frac{dv}{dt},$$

hence, multiplying Eq. (7.22) by $\nu$ we have the following equation for $v(t)$,

$$\frac{dv}{dt} + \nu Av + B(u,u) = 0, \quad \text{where} \quad u = \frac{1}{\lambda_m \nu}(v + f).$$

Writing the nonlinear term in terms of $v$, and using $B(f,f) = 0$ we obtain

$$\frac{dv}{dt} + \nu Av + \frac{1}{\lambda_m \nu}(B(v,v) + B(v,f) + B(f,v)) = 0.$$

Now, multiplying both sides by $v$ in $H$ we get

$$\frac{1}{2}\frac{d}{dt}|v|^2 + \nu\|v\|^2 + \frac{1}{\lambda_m \nu} b(v,f,v) = 0.$$

Since $Af = \lambda_m f$ then $\|f\|^2 = \lambda_m |f|^2$, and

$$\left|\frac{1}{\lambda_m \nu} b(v,f,v)\right| \leq \frac{c_1}{\lambda_m \nu}\|v\|^2\|f\| = \frac{c_1}{\lambda_m \nu}\|v\|^2\sqrt{\lambda_m}|f| = \frac{c_1}{\sqrt{\lambda_m}\nu}\|v\|^2|f|,$$

whence,

$$\frac{1}{2}\frac{d}{dt}|v|^2 + \left(\nu - \frac{c_1}{\sqrt{\lambda_m}\nu}|f|\right)\|v\|^2 \leq 0. \tag{7.23}$$

For any fixed $f$ satisfying (7.19) there exists a natural number $m_0$ such that for all $m \geq m_0$, $\sigma_m = \nu - \frac{c_1}{\sqrt{\lambda_m}\nu}|f| > 0$ and then all solutions $v(t)$ converge to zero in $H$ at least exponentially fast as $t \to \infty$. In fact, using $|v|^2 \leq \frac{1}{\lambda_1}\|v\|^2$ we obtain from (7.23),

$$|v(t)|^2 \leq |v(0)|^2 \exp(-2\lambda_1 \sigma_m t) \quad \text{for} \quad t \geq 0.$$

This means that $u(t) \to \bar{u} = \frac{1}{\nu\lambda_m} f$ in $H$ as $t \to \infty$ and proves also that $\bar{u}$ is the unique solution of the stationary problem.

*Remark 7.6.* Observe that though the norm of our $f$ in $H$ may be large, its norm in $H^{-1}$ is equal to $\frac{1}{\sqrt{\lambda_m}}|f|$ and for $m$ large is small.

## 7.5 Average Transfer of Energy

We shall consider the Navier–Stokes system as in Sect. 7.1.

Let $e(u) = \frac{1}{2}|u|^2$ be the kinetic energy of the flow. From the Navier–Stokes equation we obtain

$$\frac{d}{dt} e(u(t)) = -\nu\|u(t)\|^2 + (f, u(t))$$

and we can see that, thanks to the identity $b(u,u,u) = 0$, the global balance of the kinetic energy of the flow does not depend on the nonlinear term $(u \cdot \nabla)u$. It is the same as in the case of the linear (Stokes) flow. In this section we shall study more precisely the transport of energy in the fluid. We know (cf. Sect. 7.4) that the velocity $u(x,t)$ can be represented as a Fourier series

$$u(x,t) = \sum_{k=1}^{\infty} \hat{u}_k(t)\omega_k(x),$$

where $\omega_k$, $k = 1, 2, 3, \ldots$ are eigenfunctions of the Stokes operator constituting an orthonormal family in the phase space $H$. We have thus,

$$e(u) = \frac{1}{2}\sum_{k=1}^{\infty} |\hat{u}_k|^2.$$

Let $u = y + z$, where

$$y = \sum_{k=1}^{m} \hat{u}_k\omega_k, \quad z = \sum_{k=m+1}^{\infty} \hat{u}_k\omega_k.$$

We have,

$$e(u) = e(y) + e(z),$$

where

$$e(y) = \frac{1}{2}\sum_{k=1}^{m} |\hat{u}_k|^2, \quad e(z) = \frac{1}{2}\sum_{k=m+1}^{\infty} |\hat{u}_k|^2$$

are kinetic energies of the *large scale* component $y$ of the flow and the *small scale* component $z$. Now, in the balance of the kinetic energies of components $y$ and $z$ the nonlinear term will not reduce to zero. We shall show how the kinetic energy of particular components of the flow is distributed in time among the Fourier modes. Let us assume that the external force is given by

$$f = \sum_{k=m_1}^{m_2} \hat{f}_k\omega_k, \quad 1 \le m_1 \le m_2 < \infty.$$

Let $P_m : u \to y$, $Q_m : u \to z$ be orthogonal projections, with $Q_m = I - P_m$. Applying these projections to the Navier–Stokes system

$$\frac{du}{dt} + \nu Au + B(u,u) = f,$$

and assuming that

$$m \geq m_2,$$

we obtain

$$\frac{dy}{dt} + \nu Ay + P_m B(y+z, y+z) = P_m f$$

and

$$\frac{dz}{dt} + \nu Az + Q_m B(y+z, y+z) = 0,$$

as $Q_m z = 0$ $(m \geq m_2)$, $P_m A = AP_m$, $Q_m A = AQ_m$. The corresponding energy equations are

$$\frac{1}{2}\frac{d}{dt}|y|^2 + \nu\|y\|^2 + b(y+z, y+z, y) = (f, y),$$

and

$$\frac{1}{2}\frac{d}{dt}|z|^2 + \nu\|z\|^2 + b(y+z, y+z, z) = 0.$$

Define $E(u) = \|u\|^2$—the *enstrophy*. Using the property $b(u, v, w) = -b(u, w, v)$ we can write

$$\frac{d}{dt}e(y) + \nu E(y) = \Phi_z(y) + (f, y), \tag{7.24}$$

and

$$\frac{d}{dt}e(z) + \nu E(z) = \Phi_y(z), \tag{7.25}$$

where

$$\Phi_z(y) = -b(z, z, y) + b(y, y, z), \quad \Phi_y(z) = -b(y, y, z) + b(z, z, y).$$

Observe that $\Phi_z(y) + \Phi_y(z) = 0$. In (7.24) $\Phi_z(y)$ is the net amount of kinetic energy per unit time transferred from small length scales $z$ to large length scales $y$ (similarly, $\Phi_y(z)$ is the net amount of kinetic energy per unit time transferred from large length scales $y$ to small length scales $z$), while the term $\nu E(y)$ represents the rate of energy dissipation by viscous effects per unit time for the large length scale and $(f, y)$ is the energy flow injected into the large length scales by the forcing term.

We shall show that on average, over time and space, $\Phi_y(z) \geq 0$, that is,

if the external forces act in the range of large length scales $y$ then, on average, the net transfer of kinetic energy occurs only into the small scales.

This transfer of energy is called *direct energy transfer*.

To prove the above statement, we integrate Eq. (7.25) in $t$ to get

$$e(z(T)) + \nu \int_0^T E(z(s))ds = \int_0^T \Phi_y(z(s))ds + e(z(0)).$$

As the function $t \to e(z(t))$ is bounded for $t \geq 0$, dividing both sides by $T$ and passing with $T$ to infinity we obtain, in particular,

$$0 \leq \liminf_{T\to\infty} \frac{1}{T}\nu \int_0^T E(z(s))ds = \liminf_{T\to\infty} \frac{1}{T} \int_0^T \Phi_y(z(s))ds.$$

Assuming that the limit

$$\lim_{T\to\infty} \frac{1}{T} \int_0^T \Phi_y(z(s))ds$$

exists, we have

$$\frac{1}{|\Omega|} \lim_{T\to\infty} \frac{1}{T} \int_0^T \Phi_y(z(s))ds \geq 0.$$

**Exercise 7.31.** Let us assume that the external force is given by

$$f = \sum_{k=m_1}^{m_2} \hat{f}_k \omega_k,$$

where now $1 \leq m \leq m_1$, $m_1 > 1$. Thus, the force acts in the range of small length scales. Prove that the transfer of energy is then from higher modes (small length scales) to lower modes (large length scales), which is called an *inverse energy transfer*. The inverse energy transfer in the large scales may be responsible for the observed persistent low-wave number coherent structures in turbulent flows.

## 7.6 Comments and Bibliographical Notes

In this chapter we introduced and applied some basic tools and techniques of dealing with the nonstationary Navier–Stokes equations. As in the case of the stationary Navier–Stokes equations, the problem of existence of solutions was resolved by introducing the appropriate notion of a weak solution. We proved the uniqueness of a weak solution and considered, in the simplest context, a problem of regularity of the weak solution. We saw the utility of the Galerkin method of approximate solutions, the importance of compactness theorems, here, the Aubin–Lions compactness theorem, in particular. The usefulness of basic functional inequalities: the Poincaré inequality, the Ladyzhenskaya inequality, and the Agmon inequality, as well as the role of the Gronwall and the uniform Gronwall inequalities were demonstrated.

Further considerations and details concerning the existence of solutions and global attractors for autonomous two-dimensional Navier–Stokes flows in a bounded domain can be found, e.g., in [88, 197, 220]. For the study of three-dimensional case see, e.g., [76]. For the existence and regularity of solutions for problems on unbounded domains which do not satisfy the Poincaré inequality see [237, 238] and for the problems on cylindrical domains see [192, 196, 233]. In Sects. 7.4 and 7.5 we followed closely the argument from [99].

In Chap. 8 we shall show the connections of the global attractor with statistical solutions of the Navier–Stokes equations.

In Chap. 9 we shall apply the presented techniques to a problem of a shear flow taken from the theory of lubrication and shall prove that the global attractor has a finite fractal dimension.

Non-autonomous system of the Navier–Stokes equations and flows in unbounded domains will be considered in Chaps. 11 and 13, respectively.

# 8
# Invariant Measures and Statistical Solutions

In this chapter we prove the existence of invariant measures associated with two-dimensional autonomous Navier–Stokes equations. Then we introduce the notion of a stationary statistical solution and prove that every invariant measure is also such a solution.

## 8.1 Existence of Invariant Measures

We shall consider flows with homogeneous Dirichlet boundary conditions, in a smooth bounded domain $\Omega$.

Let $f \in H$. Then there exists a semigroup $\{S(t)\}_{t\geq 0}$ in $H$ such that the solution of the problem

$$u_t + \nu Au + B(u, u) = f, \quad u(0) = u_0$$

is given by $u(t) = S(t)u_0$ for $t \geq 0$. The map $(t, u) \to S(t)u$ is continuous from $[0, \infty) \times H$ to $H$.

Our first aim is to prove the following theorem.

**Theorem 8.1.** *Let* $\mathrm{LIM}_{T\to\infty}$ *be a fixed Banach generalized limit. Then for every* $u_0 \in H$ *there exists a probability measure* $\mu$ *on* $H$ *such that*

$$\mathrm{LIM}_{T\to\infty} \frac{1}{T} \int_0^T \phi(S(t)u_0)dt = \int_H \phi(u)d\mu(u) \tag{8.1}$$

*for every continuous function* $\phi : H \to \mathbb{R}$ $(\phi \in C(H))$*. Moreover, the measure* $\mu$ *is regular, invariant with respect to the semigroup* $\{S(t)\}_{t\geq 0}$*, that is, for every* $\mu$*-measurable set* $E \subset H$ *it holds* $\mu(S(t)^{-1}E) = \mu(E)$*, and its support is contained in the global attractor* $\mathscr{A}$*.*



G. Łukaszewicz, P. Kalita, *Navier–Stokes Equations*, Advances in Mechanics and Mathematics 34, DOI 10.1007/978-3-319-27760-8_8

We shall explain the notion of a Banach generalized limit $\mathrm{LIM}_{T\to\infty}$.

**Definition 8.1.** $\mathrm{LIM}_{T\to\infty}$—a Banach generalized limit is any linear functional on the set of all bounded functions $\psi : [0,\infty) \to \mathbb{R}$ ($\psi \in B([0,\infty))$), such that

- $\mathrm{LIM}_{T\to\infty}\psi(T) \geq 0 \quad \text{for} \quad \psi \geq 0,$
- $\mathrm{LIM}_{T\to\infty}\psi(T) = \lim_{t\to\infty}\psi(T)$ if the limit on the right-hand side exists.

*Remark 8.1.* Such functionals exist, there may be many of them, hence in Eq. (8.1) the measure $\mu$ depends not only on the initial condition $u_0$ but also on the choice of $\mathrm{LIM}_{T\to\infty}$ [note that in (8.1) the right-hand side is defined by the left-hand side].

To prove the existence of $\mathrm{LIM}_{T\to\infty}$ we shall use the Hahn–Banach theorem (cf. Theorem 8.2). Let

$$G_0 = \{\psi \in B([0,\infty)) \ : \ \lim_{T\to\infty}\psi(T) \quad \text{exists}\}.$$

The map $\Lambda_0(\psi) = \lim_{T\to\infty}\psi(T)$ is a linear functional defined on the subspace $G_0$ of $B([0,\infty))$. We can extend it to a linear functional $\Lambda$ defined on the whole space $B([0,\infty))$. In fact, let $p(\psi) = \limsup_{t\to\infty}\psi(t)$. Then

$$\Lambda_0(\psi) \leq p(\psi) \quad \text{on} \quad G_0,$$

and

$$p(\psi+\phi) \leq p(\psi) + p(\phi), \quad p(\alpha\psi) = \alpha p(\psi), \quad \alpha \geq 0$$

for $\psi, \phi \in B([0,\infty))$. In view of the Hahn–Banach theorem there exists an extension $\Lambda$ of $\Lambda_0$ to the whole space, satisfying $\Lambda(\psi) \leq p(\psi) = \limsup_{t\to\infty}\psi(t)$ for all $\psi \in B([0,\infty))$. It then follows (cf. also Exercise 8.1) that $\Lambda(\psi) \geq 0$ for all nonnegative $\psi \in B([0,\infty))$. We denote $\Lambda(\psi) = \mathrm{LIM}_{T\to\infty}\psi(T)$.

**Exercise 8.1.** Prove that for all $\psi \in B([0,\infty))$,

$$\liminf_{T\to\infty}\psi(T) \leq \mathrm{LIM}_{T\to\infty}\psi(T) \leq \limsup_{T\to\infty}\psi(T), 1$$

whence

$$|\mathrm{LIM}_{T\to\infty}\psi(T)| \leq \limsup_{T\to\infty}\psi(T) \leq \sup_{T\geq 0}|\psi(T)|.$$

We are going now to prove Theorem 8.1. At first, we assume that there exists a measure $\mu$ as in (8.1) for some fixed $u_0 \in H$ and $\mathrm{LIM}_{T\to\infty}$, and derive its basic properties which follow from this formula.

1. *$\mu$ is a probability measure, i.e., $\mu(H) = 1$.* Take $\phi \equiv 1$ on $H$ to get

$$1 = \mathrm{LIM}_{T\to\infty}\frac{1}{T}\int_0^T 1dt = \int_H 1d\mu(u) = \mu(H).$$

2. *$\mu$ is an invariant measure for the semigroup* $\{S(t)\}_{t\geq 0}$. We shall prove that

$$\int_H \phi(S(t)u)d\mu(u) = \int_H \phi(u)d\mu(u) \tag{8.2}$$

for all $\phi \in C(H)$. To this end observe that by the uniform compactness of the semigroup (cf. Sect. 7.2) we have

$$\begin{aligned}
\int_H \phi(S(\tau)u)d\mu(u) &= \mathrm{LIM}_{T\to\infty}\frac{1}{T}\int_0^T \phi(S(t+\tau)u_0)dt \\
&= \mathrm{LIM}_{T\to\infty}\frac{1}{T}\int_\tau^{T+\tau} \phi(S(t)u_0)dt \\
&= \mathrm{LIM}_{T\to\infty}\Big\{\frac{1}{T}\int_0^T \phi(S(t)u_0)dt + \frac{1}{T}\int_T^{T+\tau} \phi(S(t)u_0)dt \\
&\quad -\frac{1}{T}\int_0^\tau \phi(S(t)u_0)dt\Big\} \\
&= \mathrm{LIM}_{T\to\infty}\frac{1}{T}\int_0^T \phi(S(t)u_0)dt = \int_H \phi(u)d\mu(u).
\end{aligned}$$

This identity can be extended to all functions $\phi$ which are $\mu$-integrable on $H$. Setting $\phi = \chi_E$—the characteristic function of $E \subset H$, we obtain

$$\int_H \chi_E(S(t)u)d\mu(u) = \int_H \chi_E(u)d\mu(u),$$

which can be written as $\mu(S(t)^{-1}E) = \mu(E)$. Thus, $\mu$ is invariant.

**Exercise 8.2.** Let identity (8.2) hold for all $\phi \in C(H)$, $\mu$ be a regular measure [in the sense of Theorem 8.3 (c)] and $C(H)$ be a dense subset of $L^1(H, d\mu)$. Prove that then (8.2) holds for all functions $\phi$ which are $\mu$-integrable on $H$. Prove that the latter property is equivalent to the property of invariance of $\mu$ given by $\mu(S(t)^{-1}E) = \mu(E)$.

*Remark 8.2.* To prove invariance of $\mu$, the uniform compactness property is not needed. In the argument above we need only the boundedness of the function $t \to \phi(S(t)u_0)$, which follows directly from the existence of the global attractor. In fact, if $\phi(S(t)u_0)$ is not bounded on the positive semiaxis $t \geq 0$, then there exists a sequence $s_n \to \infty$ such that

$$|\phi(S(s_n)u_0)| \geq n \quad \text{for each} \quad n \geq 1. \tag{8.3}$$

But then, as $\mathrm{dist}_H(S(t)u_0, \mathscr{A}) \to 0$ as $t \to \infty$, we can find a subsequence $s_n'$ and a sequence $v_n$, $v_n \in \mathscr{A}$, such that $\mathrm{dist}_H(S(s_n')u_0, v_n) \leq 1/n$ for each $n \geq 1$. Then, as $\mathscr{A}$ is compact, there is a subsequence of the $v_n$ (which we relabel) such that

$v_n \to v \in \mathscr{A}$. But then also $S(s'_n)u_0 \to v$, and therefore $\phi(S(s'_n)u_0) \to \phi(v)$ as $m \to \infty$, which contradicts (8.3).

3. *The support of $\mu$ is contained in the global attractor $\mathscr{A}$*. We shall make use of the following lemma.

**Lemma 8.1.** *The support of the measure $\mu$ contains only the nonwandering points, that is, points $u \in H$ such that for every open neighborhood $A$ of $u$ in $H$ there exists a sequence of times $t_n \to \infty$ for which $S(t_n)A \cap A \neq \emptyset$ for every $n$.*

*Remark 8.3.* The set $M_1$ of nonwandering points is closed and invariant. If the phase space is compact, then $M_1$ is a nonempty and compact set. Then, every trajectory converges to $M_1$ [186].

*Proof (of Lemma 8.1).* Let $F$ be the support of the measure $\mu$, $F = \operatorname{supp}\mu$—the smallest closed set such that $\mu(F) = 1$.

**Step 1.** First we shall prove that for $u \in F$ and a set $A$ containing $u$ which is relatively open in $F$, $\mu(A) > 0$. Assuming that $\mu(A) = 0$ we obtain $\mu(F \backslash A) = 1$, $F \setminus A$ is closed, compact in $F$, and different from $F$, hence $F$ is not the support of $\mu$ which gives a contradiction.

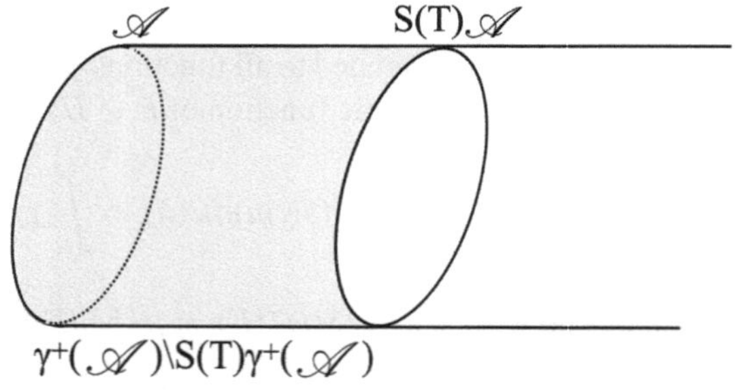


**Fig. 8.1** Illustration of the set $\gamma^+(\mathscr{A}) \setminus S(T)\gamma^+(\mathscr{A})$ used in the proof of Lemma 8.1

**Step 2.** To prove that $u \in F$ is nonwandering, let us assume to the contrary, that for some $T$ and all $t \geq T$, $S(t)A \cap A = \emptyset$. Consider the positive trajectory $\gamma^+(A) = \bigcup_{t \geq 0} S(t)A$ of the set $A$. By assumption, $A \cap S(T)\gamma^+(A) = \emptyset$. Moreover, $S(T)\gamma^+(A) \subset \gamma^+(A)$, $A \subset \gamma^+(A) \setminus S(T)\gamma^+(A)$, (see Fig. 8.1) and using step 1 we get

$$0 < \mu(A) < \mu(\gamma^+(A) \setminus S(T)\gamma^+(A)) = \mu(\gamma^+(A)) - \mu(S(T)\gamma^+(A)).$$

On the other hand,

$$\mu(S(T)\gamma^+(A)) = \mu(S(T)^{-1}S(T)\gamma^+(A)) \geq \mu(\gamma^+(A)),$$

as always $S(T)^{-1}S(T)B \supset B$. Thus, we have come to a contradiction. The proof of Lemma 8.1 is complete. □

**Corollary 8.1.** *The support of the measure* $\mu$ *is contained in the global attractor* $\mathscr{A}$ *for the semigroup* $\{S(t)\}_{t\geq 0}$.

*Proof.* Every point outside of the global attractor is *wandering*, that is, it is not a nonwandering point. □

We have also the following property of the measure $\mu$.

**Corollary 8.2.** $\mu(S(t)E) = \mu(E)$ *for all* $\mu$*-measurable* $E \subset \mathscr{A}$ *and* $t \in \mathbb{R}$.

*Proof.* The semigroup $\{S(t)\}_{t\geq 0}$ reduced to $\mathscr{A}$ is a complete dynamical system (we have a result about the backward uniqueness on the attractor for the two-dimensional Navier–Stokes equations, see [197], Theorem 12.8), hence, in view of the invariance of $\mu$,

$$\mu(E) = \mu(S(t)^{-1}S(t)E) = \mu(S(t)E),$$

which ends the proof. □

**Proof of the existence of the measure $\mu$ in (8.1)**

**Step 1.** We shall prove that for every $u_0 \in H$ and for every continuous function $\phi : H \to \mathbb{R}$ there exists a limit

$$L(\phi) = \mathrm{LIM}_{T\to\infty}\frac{1}{T}\int_0^T \phi(S(t)u_0)dt.$$

Assume we have chosen the functional $\mathrm{LIM}_{T\to\infty}$. It suffices to show that the function

$$f(T) = \frac{1}{T}\int_0^T \phi(S(t)u_0)dt$$

is bounded on the interval $[0,\infty)$. We know that there exists a compact absorbing set $X$ for the semigroup $\{S(t)\}_{t\geq 0}$ (cf. Sect. 7.2). Hence, the function $t \to \phi(S(t)u_0)$ is continuous and bounded on the compact set $\overline{\gamma^+(u_0)}$. Therefore, the function $T \to \phi(T)$ is bounded (and also continuous) on $[0,\infty)$.

**Step 2.** $L(\phi)$ depends only on values of $\phi$ on any compact absorbing set $X$. It means that if $\phi_1$ is another function in $C(H)$ which is equal to $\phi$ on $X$ then $L(\phi) = L(\phi_1)$.

To prove this property observe that there exists $T_0$ such that

$$\phi(S(t)u_0) - \phi_1(S(t)u_0) = 0$$

for $t \geq T_0$, as $S(t)u_0 \in X$ for large times. Thus,

$$\begin{aligned} L(\phi - \phi_1) &= \mathrm{LIM}_{T\to\infty}\frac{1}{T}\int_0^T (\phi(S(t)u_0) - \phi_1(S(t)u_0))dt \\ &= \mathrm{LIM}_{T\to\infty}\frac{1}{T}\int_0^{T_0} (\phi(S(t)u_0) - \phi_1(S(t)u_0))dt = 0. \end{aligned}$$

**Step 3.** Consider a compact absorbing set $X$ as above. From the Riesz–Kakutani representation theorem (Theorem 8.3) we know that for every positive linear continuous functional $G$ on the space of continuous functions on $X$ ($G \in C(X)'$), there exists a Borel measure $\mu$ on $X$ such that

$$G(\psi) = \int_X \psi(u)d\mu(u) \quad \text{for all} \quad \psi \in C(X)'.$$

We can extend this measure to the whole phase space setting $\tilde{\mu}(E) = \mu(E \cap X)$ for $E \subset H$. We shall denote this extended measure simply by $\mu$ in what follows. Then, in particular, $\mu(H \setminus X) = 0$. Further, from the Tietze extension theorem (Theorem 8.4) we know that every function $\psi \in C(X)$ can be extended to a continuous function on the whole space without enlarging the norm. Thus,

$$\begin{aligned}
\mathrm{LIM}_{T\to\infty}\frac{1}{T}\int_0^T \phi(S(t)u_0)dt &= L(\phi)\\
&= G(\phi|_X) \quad (L(\phi) \text{ depends only on } \phi|_X)\\
&= G(\psi) \quad (\phi|_X = \psi)\\
&= \int_X \psi(u)d\mu(u) \quad (\text{Riesz–Kakutani theorem})\\
&= \int_H \phi(u)d\mu(u) \quad (\text{support of } \mu \text{ is contained in } X).
\end{aligned}$$

Since $\mu$ is a time averaged measure, it is invariant. We have shown that every such measure is supported on the global attractor. Thus, $\mu(\mathscr{A}) = 1$ and

$$\mathrm{LIM}_{T\to\infty}\frac{1}{T}\int_0^T \phi(S(t)u_0)dt = \frac{1}{\mu(\mathscr{A})}\int_{\mathscr{A}} \phi(u)d\mu(u) \quad \text{for all} \quad \phi \in C(H).$$

The proof of Theorem 8.1 is complete. □

*Remark 8.4.* The existence of a compact absorbing set is not needed to prove the existence of $\mu$. We shall provide a simple proof that works in uniformly convex Banach spaces.

Let $K$ be the closed convex hull of the global attractor. As $\mathscr{A}$ is compact, it is known that $K$ is also a compact subset of $H$ (see [3], Theorem 5.35). As $K$ is compact and $H$ is uniformly convex, for each $u \in H$ there exists a unique $k_u \in K$ such that $|u - k_u| = \inf_{k\in K} |u - k|$ and the projection operator $P : u \in H \to k_u \in K$ is continuous.

Given $u_0 \in H$ consider $t \to P(S(t)u_0)$, the projection onto $K$ of the trajectory $t \to S(t)u_0$. Since $K$ is compact, the function $[0, \infty) \ni t \to \phi(P(S(t)u_0)) \in \mathbb{R}$ is continuous and bounded for $\phi \in C(H)$.

Moreover

$$|\phi(S(t)u_0) - \phi(P(S(t)u_0))| \to 0 \quad \text{as} \quad s \to \infty. \tag{8.4}$$

We then conclude that

$$\text{LIM}_{T\to\infty}\frac{1}{T}\int_0^T \phi(S(t)u_0)dt = \text{LIM}_{T\to\infty}\frac{1}{T}\int_0^T \phi(P(S(t)u_0))dt. \tag{8.5}$$

The right-hand side defines a linear positive functional on $C(K)$. The rest of the proof is similar to that in step 3 above, where now $K$ plays the role of the compact absorbing set $X$.

**Exercise 8.3.** Prove that the projection operator $P : u \in H \to k_u \in K$ is continuous.

**Exercise 8.4.** Prove (8.4) and (8.5).

Below we present the basic theorems which have been used above.

**Theorem 8.2 (Hahn–Banach).** *Let $X$ be a real linear space, $p : X \to \mathbb{R}$ a function such that*

$$p(x + y) \leq p(x) + p(y), \quad p(\alpha x) = \alpha p(x)$$

*for $\alpha \geq 0$, $x, y \in X$. Let $f$ be a real-valued linear functional on a subspace $Y$ of $X$ such that $f(x) \leq p(x)$ for $x \in Y$. Then there exists a real linear functional $F$ on $X$ such that*

$$F(x) = f(x), \quad x \in Y, \quad F(x) \leq p(x), \quad x \in X.$$

**Theorem 8.3 (Riesz–Kakutani).** *Let $X$ be a locally compact Hausdorff space. Let $\Lambda$ be a positive linear functional on $C_c(X)$, the space of real-valued continuous functions of compact support on $X$. Then there exist a $\sigma$-algebra $\mathscr{M}$ on $X$ containing all Borel sets in $X$ and a unique positive measure $\mu$ on $\mathscr{M}$ such that:*

*(a)* $\Lambda f = \int_X f(u)d\mu(u)$ *for every* $f \in C_c(X)$,
*(b)* $\mu(K) < \infty$ *for every compact set* $K \subset X$,
*(c)* $\mu$ *is regular in this sense that for every* $E \subset \mathscr{M}$,

$$\mu(E) = \inf\{\mu(\Theta) \ : \ E \subset \Theta, \quad \Theta \text{ is open}\},$$

*and, for every* $E \in \mathscr{M}$ *with* $\mu(E) < \infty$,

$$\mu(E) = \sup\{\mu(K) \ : \ K \subset E, \quad K \text{ is compact}\}.$$

*(d)* $\mu$ *is complete, i.e., if* $E \in \mathscr{M}$, $A \subset E$ *and* $\mu(A) = 0$, *then* $A \in \mathscr{M}$.

**Theorem 8.4 (Tietze).** *Let $X$ be a normal topological space and $A$ be a closed subset of $X$. If $f$ is a continuous and bounded real-valued function defined on $A$, then there exists a continuous real-valued function $F$ defined on $X$ such that:*

*(i)* $F(x) = f(x)$ *for* $x \in A$,
*(ii)* $\sup_{x\in X} |F(x)| = \sup_{x\in X} |f(x)|$.

## 8.2 Stationary Statistical Solutions

Let $u(t) = S(t)u$, $t \in \mathbb{R}^+$, be a family of maps in $(X, \mu_0)$. If $\mu_t(E) = \mu_0(S(t)^{-1}E)$ for all $\mu_0$-measurable sets $E$, then

$$\int_X \Phi(u)d\mu_t(u) = \int_X \Phi(S(t)u)d\mu_0(u) \quad \text{for all} \quad \Phi \in L^1(X, \mu_t). \tag{8.6}$$

If

$$\frac{du(t)}{dt} = F(u(t)), \tag{8.7}$$

then from (8.6) by formal differentiation with respect to $t$ and from (8.7) we obtain

$$\begin{aligned}\frac{d}{dt}\int_X \Phi(u)d\mu_t(u) &= \frac{d}{dt}\int_X \Phi(S(t)u)d\mu_0(u)\\ &= \int_X \Phi'(S(t)u)\frac{d}{dt}(S(t)u)d\mu_0(u) = \int_X \Phi'(S(t)u)F(S(t)u)d\mu_0(u)\\ &= \int_X \Phi'(u)F(u)d\mu_t(u).\end{aligned}$$

Hence,

$$\frac{d}{dt}\int_X \Phi(u)d\mu_t(u) = \int_X \Phi'(u)F(u)d\mu_t(u). \tag{8.8}$$

The above equation we shall call the *Liouville equation.*

If measure $\mu_0$ is invariant with respect to the semigroup of transformations $\{S(t)\}_{t\geq 0}$, then

$$\mu_0(T^{-1}E) = \mu_0(E) = \mu_t(E),$$

that is, all measures $\mu_t$, $t \geq 0$, coincide with $\mu_0$. In this case the Liouville equation (8.8) reduces to the *stationary Liouville equation.*

$$\int_X \Phi'(u)F(u)d\mu(u) = 0, \tag{8.9}$$

where we denoted by $\mu$ the measure $\mu_t = \mu_0$.

Let us consider the two-dimensional Navier–Stokes equations

$$\frac{du(t)}{dt} = F(u(t)), \quad F(u) = f - \nu Au - B(u),$$
$$u(0) = u_0.$$

We proved (in Sect. 8.1) existence of probability measures invariant with respect to the semigroup $\{S(t)\}_{t\geq 0}$ associated with the above Navier–Stokes system. Thus, there exists a Borel measure $\mu$ on the phase space $H$ for which $\mu(S(t)^{-1}E) = \mu(E)$ for Borel sets $E$ in $H$. Moreover, $\mu(H) = 1$. For such measure formula (8.9) holds true. We shall give the precise meaning to this formula in the case of the two-dimensional Navier–Stokes dynamical system. Let us define the class of the test functions $\Phi$.

**Definition 8.2.** To the class $\mathcal{T}$ of test functions belong real-valued functionals $\Phi = \Phi(u)$ defined on $H$, which are bounded on bounded subsets of $H$ and such that

(i) For any $u \in V$, the Fréchet derivative $\Phi'(u)$ taken in $H$ along $V$ exists. More precisely, for each $u \in V$, there exists an element in $H$ denoted $\Phi'(u)$ such that

$$\frac{|\Phi(u+v) - \Phi(u) - (\Phi'(u), v)|}{|v|} \to 0 \quad \text{as} \quad |v| \to 0, \quad v \in V,$$

(ii) $\Phi'(u) \in V$ for all $u \in V$, and $u \to \Phi'(u)$ is continuous and bounded as a function from $V$ to $V$.

For example, let $\Phi(u) = \psi((u, g_1), \ldots, (u, g_n))$, where $\psi \in C_0^1(\mathbb{R}^n)$, $g_j \in V$. Then $\Phi'(u) = \sum_{j=1}^n \partial_j \psi((u, g_1), \ldots, (u, g_n)) g_j \in V$.

Let $\mu$ denote a probability measure in $H$ such that

$$\int_H \|u\|^2 d\mu(u) < \infty, \tag{8.10}$$

where $\|u\| = \infty$ for $u \in H \setminus V$. We thus have $\mu(H \setminus V) = 0$, that is, the support of $\mu$ is included in $V$, and the map $u \to (\nu Au + B(u) - f, \Phi'(u))$ is continuous in $V$ for every test function $\Phi \in \mathcal{T}$. Moreover,

$$\begin{aligned} |(\nu Au + B(u) - f, \Phi'(u))| &= |\nu((u, \Phi'(u))) + b(u, u, \Phi'(u)) - (f, \Phi'(u))| \\ &\leq \nu\|u\|\|\Phi'(u)\| + c\|u\|^2\|\Phi'(u)\| + |f||\Phi'(u)| \\ &\leq (\nu\|u\| + c\|u\|^2 + \lambda_1^{1/2}|f|) \sup_{u\in V} \|\Phi'(u)\|. \end{aligned} \tag{8.11}$$

As $\mu(H\setminus V) = 0$, from (8.10) and (8.11) it follows [cf. condition (ii) of the definition of the class $\mathcal{T}$] that the integral

$$\int_H (\nu Au + B(u) - f, \Phi'(u))d\mu(u)$$

has sense and is finite.

Now we can define the *stationary statistical solution* of the two-dimensional Navier–Stokes system.

**Definition 8.3.** Stationary statistical solution of the Navier–Stokes system is a probability measure $\mu$ on $H$ such that

(i) $\int_H \|u\|^2 d\mu(u) < \infty$,
(ii) $\int_H (F(u), \Phi'(u))d\mu(u) = 0$ for all $\Phi \in \mathcal{T}$, where $F(u) = \nu Au + B(u) - f$, $f \in H$,
(iii) $\int_{E_1 \le |u|^2 < E_2} \{\nu\|u\|^2 - (f,u)\}d\mu(u) \le 0$ for all $0 \le E_1 < E_2 \le \infty$.

Inequality in (iii) is a weak form of the energy inequality. It gives an estimate of the integral in (i) and an estimate of the support of the measure $\mu$ in $H$.

Let $E = \{E_1 \le |u|^2 < E_2\}$. Then, from (i),

$$\int_E (f,u)d\mu(u) \le |f| \int_E |u|d\mu(u) \le |f| \left(\int_E |u|^2 d\mu(u)\right)^{\frac{1}{2}} \mu(E)^{\frac{1}{2}}$$

$$\le \frac{|f|}{\lambda_1^{\frac{1}{2}}} \left(\int_E \|u\|^2 d\mu(u)\right)^{\frac{1}{2}} < \infty.$$

Then from (iii),

$$\nu \int_E \|u\|^2 d\mu(u) \le \int_E (f,u)d\mu(u) \le \frac{|f|}{\lambda_1^{\frac{1}{2}}} \left(\int_E \|u\|^2 d\mu(u)\right)^{\frac{1}{2}} < \infty,$$

and we have,

$$\int_E \|u\|^2 d\mu(u) \le \frac{|f|^2}{\nu^2 \lambda_1},$$

and

$$\int_E |u|^2 d\mu(u) \le \frac{|f|^2}{\nu^2 \lambda_1^2}.$$

The last inequality we can write in the form

$$\int_E \left(|u|^2 - \frac{|f|^2}{\nu^2 \lambda_1^2}\right) d\mu(u) \le 0.$$

For $E_1 = \frac{|f|^2}{\nu^2\lambda_1^2}$ and $E_2 = \infty$ we have

$$|u|^2 - \frac{|f|^2}{\nu^2\lambda_1^2} \leq 0$$

for all $u \in E$, whence $\mu(E) = 0$. Thus

$$\mu\left(\left\{u \in H : |u| \leq \frac{|f|}{\nu\lambda_1}\right\}\right) = 1,$$

and

$$\operatorname{supp}\mu \subset \left\{u \in H : |u| \leq \frac{|f|}{\nu\lambda_1}\right\} \cap V.$$

Now, we shall prove existence of a stationary statistical solution for the two-dimensional Navier–Stokes system.

**Theorem 8.5.** *Let $\mu$ be a probability measure on $H$ invariant with respect to the semigroup $\{S(t)\}_{t\geq 0}$ associated with the two-dimensional Navier–Stokes system. Then $\mu$ is a stationary statistical solution of the system.*

*Proof.* **Step 1.** Since $\mu$ is invariant, it is concentrated on the global attractor. Thus

$$\int_H \|u\|^2 d\mu(u) = \int_{\mathscr{A}} \|u\|^2 d\mu(u) < \infty,$$

and we have the first property of the stationary statistical solution.
**Step 2.** We shall show that

$$\int_H (F(u), \Phi'(u))d\mu(u) = 0, \tag{8.12}$$

where $F(u) = \nu Au + B(u) - f$, $F(u) = u_t$, $\Phi \in \mathscr{T}$. From the invariance of $\mu$, for every $\Phi \in \mathscr{T}$,

$$\int_H (F(u), \Phi'(u))d\mu(u) = \int_H (F(S(t)u), \Phi'(S(t)u))d\mu(u),$$

and the right-hand side is independent of $t \geq 0$. Therefore, from the Fubini theorem,

$$\begin{aligned}\int_H (F(S(t)u), \Phi'(S(t)u))d\mu(u) &= \frac{1}{T}\int_0^T \int_H (F(S(t)u), \Phi'(S(t)u))d\mu(u)dt \\ &= \int_H \frac{1}{T}\int_0^T (F(S(t)u), \Phi'(S(t)u))dtd\mu(u).\end{aligned}$$

Thus,

$$\int_H (F(u), \Phi'(u))d\mu(u) = \int_H \frac{1}{T}\{\Phi(S(t)u) - \Phi(u)\}d\mu(u), \tag{8.13}$$

as

$$\frac{d}{dt}\Phi(u) = (F(u), \Phi'(u)).$$

The right-hand side of (8.13) tends to 0 as $T \to \infty$, as the function $t \to \Phi(S(t)u)$ is bounded. This proves (8.12).

**Step 3.** Now we shall prove that

$$\int_{E_1 \le |u|^2 < E_2} \{\nu\|u\|^2 - (f, u)\}d\mu(u) = 0 \tag{8.14}$$

for all $0 \le E_1 < E_2 \le \infty$. In particular the left-hand side is not positive which gives the third property of the stationary statistical solution. From the energy equation

$$\frac{1}{2}|S(t)u|^2 + \nu \int_0^t \|S(s)u\|^2 ds = \frac{1}{2}|u|^2 + \int_0^t (f, S(s)u)ds,$$

we obtain, by regrouping the terms,

$$\int_0^t \{\nu\|S(s)u\|^2 - (f, S(s)u)\}ds = \frac{1}{2}\{|u|^2 - |S(t)u|^2\} \tag{8.15}$$

for $t \ge 0$. From the invariance of $\mu$, the Fubini theorem, and (8.15) we get

$$\begin{aligned}
\int_{E_1 \le |u|^2 < E_2} \{\nu\|u\|^2 - (f, u)\}d\mu(u) &= \int_{E_1 \le |u|^2 < E_2} \{\nu\|S(s)u\|^2 - (f, S(s)u)\}d\mu(u) \\
&= \frac{1}{T}\int_0^T \int_{E_1 \le |u|^2 < E_2} \{\nu\|S(s)u\|^2 - (f, S(s)u)\}d\mu(u) \\
&= \int_{E_1 \le |u|^2 < E_2} \frac{1}{T}\int_0^T \{\nu\|S(s)u\|^2 - (f, S(s)u)\}d\mu(u) \\
&= \frac{1}{2}\int_{E_1 \le |u|^2 < E_2} \frac{1}{T}\{|u|^2 - |S(t)u|^2\}d\mu(u)
\end{aligned}$$

for every $T \ge 0$. Passing with $T$ to $\infty$ we obtain zero on the right-hand side. This proves (8.14).

□

*Remark 8.5.* In fact, the converse is also true. If $\mu$ is a stationary statistical solution, then $\mu$ is a probability measure invariant with respect to the semigroup $\{S(t)\}_{t\geq 0}$ (cf. Chap. 4 in [99]).

## 8.3 Comments and Bibliographical Notes

Our presentation follows that in [99], where this subject is treated in greater detail. The alternative proof of Theorem 8.1 and its generalization to noncompact semigroups presented in Remarks 8.2 and 8.4 come from [165]. Theorem 8.5 follows also from the results of Chap. 12, cf. [160].

For statistical solutions for the Navier–Stokes equations, see [96, 97, 99, 101, 116, 195, 203, 228]. Some recent results for three-dimensional problems can be found in [29, 104].

# 9 Global Attractors and a Lubrication Problem

We start this chapter from necessary background on the theory of fractal dimension. Next, we formulate and study a problem which models the two-dimensional boundary driven shear flow in lubrication theory. After the derivation of the energy dissipation rate estimate and a version of Lieb–Thirring inequality we provide an estimate from above on the global attractor fractal dimension.

## 9.1 Fractal Dimension

We start this section from the definition and some properties of a *fractal dimension* of a relatively compact set $K \subset X$, where $X$ is a Banach space. By $N_\varepsilon^X(K)$ we denote the minimal number of closed balls in $X$ of radius $\varepsilon$ needed to cover the set $K$. From the relative compactness of $K$ it follows that this number is finite. Obviously, we expect the quantity $N_\varepsilon^X(K)$ to increase as $\varepsilon \to 0$. The fractal dimension, given by the following definition, measures how fast is this increase.

**Definition 9.1.** Let $K \subset X$ be a relatively compact set. The *fractal dimension* of $K$, denoted by $d_f^X(K)$ is a number defined by

$$d_f^X(K) = \limsup_{\varepsilon \to 0} \frac{\log N_\varepsilon^X(K)}{\log \frac{1}{\varepsilon}}.$$

The value $d_f^X(K)$ can be equal to $\infty$ even if $K$ is a compact set. If the set $K$ is not relatively compact, we set $d_f^K(K) = \infty$. In the next result we recall several properties of $d_f^X$.



G. Łukaszewicz, P. Kalita, *Navier–Stokes Equations*, Advances in Mechanics and Mathematics 34, DOI 10.1007/978-3-319-27760-8_9

**Theorem 9.1 (Cf. [197, Proposition 13.2]).**

*(i) if $K_k \subset X$ for $k = 1, \dots, N$ are relatively compact sets, then for every $k = 1, \dots, N$*

$$d_f^X(K_k) \leq d_f^X\left(\bigcup_{k=1}^{N} K_k\right) \leq \max_{k=1,\dots,N} d_f^X(K_k).$$

*(ii) If $f : X \to X$ is Hölder continuous with the exponent $\theta \in (0, 1]$, i.e.,*

$$\|f(x) - f(y)\|_X \leq L\|x - y\|_X^\theta \quad \textit{for all} \quad x, y \in X,$$

*then*

$$d_f^X(f(K)) \leq \frac{d_f^X(K)}{\theta} \quad \textit{for every relatively compact} \quad K \subset X.$$

*(iii) We have $d_f^X(\overline{K}) = d_f^X(K)$ for every relatively compact $K \subset X$, where $\overline{K}$ is the closure of $K$ in $X$.*

Note that property (i) does not hold for countable unions. Indeed, let $X = \mathbb{R}$ and $K = [0, 1] \cap \mathbb{Q}$. From (iii) it follows that $d_f^X(K) = d_f^X([0, 1])$, and it is easy to verify that $d_f^X([0, 1]) = 1$, while for every point $x \in K$ we have $d_f^X(\{x\}) = 0$.

If we know that an infinite dimensional semiflow $\{S(t)\}_{t\geq 0}$ is defined by the solution maps of an initial and boundary value problem and it has a global attractor $\mathscr{A}$, then we can restrict the dynamical system to the attractor, an invariant set, i.e., treat the family $\{S(t)\}_{t\geq 0}$ as the mappings $S(t) : \mathscr{A} \to \mathscr{A}$, thus interpreting the global attractor as the image of the flow after a long time of evolution (keeping in mind that all trajectories are arbitrarily close to the attractor after a long time, but each trajectory does not necessarily have one corresponding trajectory on the attractor, to which it becomes attracted in the norm, such trajectories exist, but only on finite time intervals, the length of which goes to infinity as time goes to infinity, see [197] Proposition 10.14 and Corollary 10.15).

An important question that arises is the following: can the dynamics on the attractor be described by means of the time evolution of a finite number of variables without loss of information? From the fact that the global attractor exists, we only know that it is compact, and hence it must have an empty interior in an infinite dimensional phase space $H$, but such compact sets can be still "large." Indeed, if a Banach space $V$ embeds compactly in the phase space $H$, then any set, that is bounded in $V$, is relatively compact in $H$, so, potentially, a ball in $V$ is an admissible candidate for the attractor. In certain cases, however, it is possible to obtain much more knowledge on the attractor than only its compactness: namely it is possible to show that it has *finite fractal dimension.* If we could prove that the attractor is a finite dimensional manifold, then it would be natural to use its parameterization, and reduce the dynamics to the finite dimensional one. Unfortunately, the set of finite fractal dimension can be very complicated and it does not have to be a manifold. Still, due to the Hölder–Mañé Theorem, if the fractal dimension of the attractor is finite we can reduce the dynamics on the attractor to the finite dimensional one.

**Theorem 9.2 (Hölder–Mañé, cf. [117]).** *Let $H$ be a Banach space and let $\mathscr{A} \subset H$ be a compact set such that $d_f^H(\mathscr{A}) \leq \frac{m}{2}$. Then, almost every bounded linear function $\pi : H \to \mathbb{R}^m$ is a bijection as a map $\pi : \mathscr{A} \to \pi(\mathscr{A})$ and its inverse $\pi^{-1}|_{\pi(\mathscr{A})}$ is Hölder continuous with a certain exponent $\alpha \in (0,1)$ depending only on $m$ and $d_f^H(\mathscr{A})$.*

The expression "almost every" in the last theorem is understood in the sense of prevalence, a generalization on the notion "almost everywhere in the sense of Lebesgue measure" to the infinite dimensional Banach spaces. Here we consider the prevalence in the space of linear and bounded operators $\mathscr{L}(H;\mathbb{R}^m)$.

**Definition 9.2.** The Borel set $E \subset X$, where $X$ is a Banach space, is said to be prevalent if there exists a compactly supported probability measure $\mu$ such that $\mu(E+x) = 1$ for all $x \in X$. A non-Borel set that contains a prevalent set is also prevalent.

Using a linear map $\pi$ that satisfies the assumptions of Theorem 9.2 it is possible to construct the finite dimensional semiflow $S_\pi(t) : \pi(\mathscr{A}) \to \pi(\mathscr{A})$ for $t \geq 0$ conjugate with the original one according to the formula $S_\pi(t)x = \pi(S(t)\pi^{-1}x)$. Unfortunately, this new semiflow is not necessarily Lipschitz continuous, even if $S(t)$ are Lipschitz, since $\pi^{-1}$ is only Hölder. Therefore it can have more than one solution. Still, this result plays an important role in the theory of finite dimensional reduction of infinite dimensional dynamical systems, and, clearly, the finite dimensionality of the global attractor is a strong evidence that the semiflow is in fact finite dimensional. More information about this theory can be found in the monograph [198], and, in context of two-dimensional Navier–Stokes equations, in the review article [199].

## 9.2 Abstract Theorem on Finite Dimensionality and an Algorithm

In this section we present two classes of methods that can be used to prove the finite dimensionality of the global attractor.

The first of this two approaches was introduced by Constantin and Foiaş in the article [74] (also see [57, 58, 220]) and it relies on the linearization of the underlying PDEs and the fact that if the linearized system contracts $m$-dimensional volumes uniformly on the global attractor, then its fractal dimension must not exceed $m$. Below we present an algorithm, based on this idea, that can be used to obtain the estimate of the fractal dimension of the global attractor for the Navier–Stokes equations. The exposition is based on [57, 220].

Let $\{S(t)\}_{t\geq 0}$ be a semiflow on a Hilbert space $X$, that has a global attractor $\mathscr{A}$, being a compact, invariant, and attracting set in $X$.

**Definition 9.3.** We say that the semiflow $\{S(t)\}_{t\geq 0}$ is uniformly quasidifferentiable on $\mathscr{A}$ if for every $t \geq 0$ and $u \in \mathscr{A}$ there exists a linear and continuous operator $\Lambda(t,u) \in \mathscr{L}(X;X)$ such that for all $t \geq 0$

$$\sup_{u,v\in\mathscr{A}} \sup_{0<\|u-v\|_X\leq\epsilon} \frac{\|S(t)v - S(t)u - \Lambda(t,u)(v-u)\|_X}{\|v-u\|_X} \to 0 \quad \text{as} \quad \epsilon \to 0. \tag{9.1}$$

Note (see Sect. V.2.2 in [220]) that the mapping $\Lambda(t,u)$ does not have to be defined uniquely.

We assume that the semiflow is given by a solution map of the following abstract evolution problem.

$$\begin{aligned} \frac{du(t)}{dt} &= F(u(t)) \quad \text{in} \quad X \quad \text{for} \quad t > 0, \\ u(0) &= u_0 \quad \text{in} \quad X, \end{aligned}$$

where $F : W \to X$, $W$ being a certain Banach space continuously embedded in $X$. Then we have $u(t) = S(t)u_0$. We construct the following *linearized* problem

$$\frac{dU(t)}{dt} = F'(S(t)u_0)U(t) \quad \text{in} \quad X \quad \text{for} \quad t > 0, \tag{9.2}$$

$$U(0) = \xi \quad \text{in} \quad X, \tag{9.3}$$

where $F'$ is the Fréchet derivative of $F$, and we make the following assumption.

**Assumption A.** *The semiflow $\{S(t)\}_{t\geq 0}$ is uniformly quasidifferentiable on $\mathscr{A}$, with*

$$\sup_{t\in[0,1]} \sup_{u\in\mathscr{A}} \|\Lambda(t,u)\|_{\mathscr{L}(X;X)} < \infty. \tag{9.4}$$

*Moreover, for $u_0 \in \mathscr{A}$ the mapping $X \times [0,\infty) \ni (\xi,t) \to \Lambda(t,u_0)\xi = U(t) \in X$ is a solution map of the linearized system (9.2)–(9.3).*

In practice, the verification of this, technical, assumption depends on the structure of the studied problem and relies on the construction of the linearized problem (9.2)–(9.3) and proving that its solution map is a quasidifferential, i.e., it satisfies (9.1), and that (9.4) holds (see, for instance, examples in Sects. VI.2.1 and VI.3.1 in [197]).

In addition to **Assumption A**, for the fractal dimension of a global attractor to be finite (and equal to $d$), we need the linearized semiflow to *contract uniformly* the $d$-dimensional volumes. To understand this notion properly we must first recall several useful ideas.

If $\phi_1, \ldots, \phi_m \in X$, then the tensor product

$$\phi_1 \otimes \phi_2 \otimes \ldots \otimes \phi_m$$

is an $m$-linear form over $X$ defined as

$$(\phi_1 \otimes \phi_2 \otimes \ldots \otimes \phi_m)(\psi_1, \ldots, \psi_m) = (\phi_1, \psi_1)_X(\phi_2, \psi_2)_X \ldots (\phi_m, \psi_m)_X,$$

and the space spanned by all tensor products $\phi_1 \otimes \phi_2 \otimes \ldots \otimes \phi_m$ of $m$ elements from $X$ is denoted by $\bigotimes^m X$. We will be interested in the subspace of $\bigotimes^m X$ denoted by $\bigwedge^m X = X \wedge X \wedge \ldots \wedge X$ ($X$ appears $m$ times on the right-hand side). This subspace, containing the $m$-linear antisymmetric forms, is spanned by the forms

$$\phi_1 \wedge \phi_2 \wedge \ldots \wedge \phi_m$$

being the *exterior products* of the elements from $X$, defined by

$$\phi_1 \wedge \phi_2 \wedge \ldots \wedge \phi_m = \sum_{\sigma} \operatorname{sgn}(\sigma)\phi_{\sigma(1)} \otimes \phi_{\sigma(2)} \otimes \ldots \otimes \phi_{\sigma(m)},$$

where the sum is taken over all permutations $\sigma$ of the set $\{1, \ldots, m\}$, and $\operatorname{sgn}(\sigma)$ is the sign of the permutation $\sigma$. We define on $\bigwedge^m X$ the inner product given, for exterior products of elements from $X$, by the formula

$$(\phi_1 \wedge \phi_2 \wedge \ldots \wedge \phi_m, \psi_1 \wedge \psi_2 \wedge \ldots \wedge \psi_m)_{\bigwedge^k X} = \det\{(\phi_i, \psi_j)_X\}_{i,j=1,\ldots,m},$$

and we extend it to the whole $\bigwedge^k X$ by linearity. The associated norm is given, for exterior products of elements from $X$, by

$$\|\phi_1 \wedge \phi_2 \wedge \ldots \wedge \phi_m\|^2_{\bigwedge^k X} = \det\{(\phi_i, \phi_j)_X\}_{i,j=1,\ldots,m}.$$

The importance of the norm of the exterior product $\|\phi_1 \wedge \phi_2 \wedge \ldots \wedge \phi_m\|_{\bigwedge^m X}$ lies in the fact that this quantity is the volume of the $m$-dimensional parallelepiped with the edges given by $\phi_1, \ldots, \phi_m$. We will start the evolution from the volume spanned by the vectors which are the initial conditions of the linearized system (9.2)–(9.3) and we will derive the condition which guarantees that this volume is uniformly contracted on the attractor by the linearized semiflow. To this end, for a linear and bounded operator $L \in \mathscr{L}(X; X)$ let us introduce the operator $L_m : X^m \to \bigwedge^m X$ by the formula

$$L_m(\xi_1, \ldots, \xi_m) = L\xi_1 \wedge \xi_2 \wedge \ldots \wedge \xi_m + \xi_1 \wedge L\xi_2 \wedge \ldots \wedge \xi_m$$
$$+ \ldots + \xi_1 \wedge \xi_2 \wedge \ldots \wedge L\xi_m.$$

We have the following lemma.

**Lemma 9.1 (Cf. [220, Lemma V.1.2]).** *For any* $\xi_1, \ldots, \xi_m \in X$ *we have*

$$(L_m(\xi_1, \ldots, \xi_m), \xi_1 \wedge \ldots \xi_m)_{\bigwedge^m X} = \|\xi_1 \wedge \xi_2 \wedge \ldots \wedge \xi_m\|^2_{\bigwedge^m X} Tr(L \circ Q),$$

*where* $Q$ *is the orthogonal projection on the space spanned by* $\{\xi_i\}_{i=1}^m$ *and* $Tr$ *is the trace of the operator of the finite rank, given by*

$$Tr(L \circ Q) = \sum_{i=1}^{m} \left( L\frac{\theta_i}{\|\theta_i\|_X}, \frac{\theta_i}{\|\theta_i\|_X} \right)_X,$$

*where* $\{\theta_i\}_{1=1}^m$ *is the orthogonal basis of the space spanned by* $\{\xi_i\}_{i=1}^m$*, obtained, for example, by the Gram–Schmidt procedure.*

We fix $u_0 \in \mathscr{A}$ and we take the linearly independent vectors $\xi_1, \ldots, \xi_m \in H$. Next, we solve (9.2)–(9.3) for $U_1(t), \ldots, U_m(t)$ and calculate how the volume spanned by $U_1(t), \ldots, U_m(t)$ is related to the volume spanned by $\xi_1, \ldots, \xi_m$ (see Fig. 9.1). We have

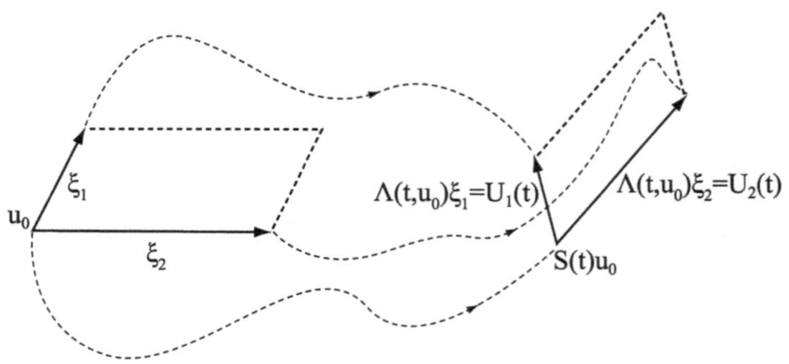


**Fig. 9.1** Transformation by a linearized semiflow $\Lambda(t, u_0)$ of a two-dimensional volume spanned by $\xi_1, \xi_2$

$$\begin{aligned}
&\frac{1}{2}\frac{d}{dt}\|U_1(t) \wedge \ldots \wedge U_m(t)\|^2_{\bigwedge^m X} \\
&\quad = \left( \frac{d}{dt}(U_1(t) \wedge \ldots \wedge U_m(t)), U_1(t) \wedge \ldots \wedge U_m(t) \right)_{\bigwedge^m X} \\
&\quad = (U_1'(t) \wedge \ldots \wedge U_m(t), U_1(t) \wedge \ldots \wedge U_m(t))_{\bigwedge^m X} + \ldots \\
&\qquad + (U_1(t) \wedge \ldots \wedge U_{m-1}(t) \wedge U_m'(t), U_1(t) \wedge \ldots \wedge U_m(t))_{\bigwedge^m X} \\
&\quad = (F'(S(t)u_0)U_1(t) \wedge \ldots \wedge U_m(t), U_1(t) \wedge \ldots \wedge U_m(t))_{\bigwedge^m X} + \ldots
\end{aligned}$$

$$+ (U_1(t) \wedge \ldots \wedge U_{m-1}(t) \wedge F'(S(t)u_0)U_m(t), U_1(t) \wedge \ldots \wedge U_m(t))_{\bigwedge^m X}$$
$$= ((F'(S(t)u_0))_m(U_1(t), \ldots, U_m(t)), U_1(t) \wedge \ldots \wedge U_m(t))_{\bigwedge^m X}$$
$$= \|U_1(t) \wedge \ldots \wedge U_m(t)\|^2_{\bigwedge^m X} \mathrm{Tr}(F'(S(t)u_0) \circ Q(t)),$$

where $Q(t) = Q(t, u_0; \xi_0, \ldots, \xi_m)$ is the orthogonal projection onto the space spanned by $U_1(t), \ldots, U_m(t)$. It follows that

$$\|U_1(t) \wedge \ldots \wedge U_m(t)\|_{\bigwedge^m X} = \|\xi_1 \wedge \ldots \wedge \xi_m\|_{\bigwedge^m X} e^{\left(\int_0^t \mathrm{Tr}(F'(S(t)u_0)\circ Q(t))\, dt\right)}.$$

We introduce the quantity

$$\omega_m(S(t)u_0) = \sup_{\xi_1,\ldots,\xi_m \in X, \|\xi_i\|_X \le 1} \|U_1(t) \wedge \ldots \wedge U_m(t)\|_{\bigwedge^m X},$$

being the largest volume obtained from the $m$-dimensional parallelepiped with sides no greater than one, by the linearized semiflow. We also introduce auxiliary quantities

$$\overline{\omega}_m(t) = \sup_{u_0 \in \mathscr{A}} \omega_m(S(t)u_0),$$

$$q_m(t) = \sup_{u_0 \in \mathscr{A}} \sup_{\xi_1,\ldots,\xi_m \in X, \|\xi_i\|_X \le 1} \frac{1}{t} \int_0^t \mathrm{Tr}(F'(S(t)u_0) \circ Q(t))\, dt,$$

whence we have $\frac{1}{t} \log \overline{\omega}_m(t) \le q_m(t)$. Finally, we introduce the numbers

$$q_m = \limsup_{m \to \infty} q_m(t).$$

It is easy to deduce the following result

**Proposition 9.1.** *If, for certain $m \in \mathbb{N}^+$,*

$$q_m < 0,$$

*then, for some $t_0 > 0$ and all $t \ge t_0$ we have $q_m(t) \le -\delta < 0$, and the volume element $\|U_1(t) \wedge \ldots \wedge U_m(t)\|_{\bigwedge^m X}$ decays exponentially as $t \to \infty$ uniformly for $u_0 \in \mathscr{A}$ and $\xi_1, \ldots, \xi_m \in X$, i.e.,*

$$\|U_1(t) \wedge \ldots \wedge U_m(t)\|_{\bigwedge^m X} \le c e^{-\delta t} \quad \text{for} \quad t \ge t_0.$$

We conclude the presentation of the method with a result that links the values $q_m$ with the fractal dimension of the global attractor.

**Theorem 9.3 (Cf. [57, Corollary 2.2]).** *Let* ***Assumption A*** *hold and let $f : [0, \infty) \to \mathbb{R}$ be a concave function such that for every positive natural number $m$ we have $q_m \le f(m)$ and let $d^* > 1$ be such that $f(d^*) = 0$. Then*

$$d_f^X(\mathscr{A}) \le d^*.$$

The above result gives rise to an algorithm. We need to estimate from above the quantities

$$\mathrm{Tr}(F'(S(t)u_0) \circ Q(t)),$$

and by taking the supremum over $\xi_1, \ldots, \xi_m$ and over the points $u_0$ in the attractor, after the integration over time, we obtain the bound on $q_m$ by the concave function of $m$. Then, it is enough to find a zero of this function to get the bound on the global attractor fractal dimension. We conclude with two remarks: firstly, the algorithm works only if the phase space is a Hilbert space, and secondly, the semiflow must be smooth, i.e., it must be possible to linearize the underlying initial and boundary value problem and the solution map of the linearized problem must be a quasidifferential of the original semiflow in accordance with Definition 9.3. Further in this chapter we will apply the described algorithm to estimate the attractor dimension for a shear flow in the lubrication theory. Before we move to this problem, however, we give an account of other techniques useful for the fractal dimension estimation.

The second class of methods useful for proving the attractor finite dimensionality relies on obtaining the estimates on the *difference* of two trajectories. We present several results from this class of methods.

**Theorem 9.4 (Cf. [168, 236]).** *Let $E$ be a bounded subset of a Banach space $H$. Let moreover $V$ be another Banach space compactly embedded in $H$ and let $L : H \to H$ be a mapping such that $E \subset L(E)$, and we have*

$$\|Lu - Lv\|_V \le c\|u - v\|_H \quad \textit{for all} \quad u, v \in E \quad \textit{with} \quad c > 0. \tag{9.5}$$

*Then, the fractal dimension of $E$ is finite and*

$$d_f^H(E) \le \frac{\log N^H_{\frac{1}{4c}}(\overline{B}_V(0,1))}{\log 2},$$

*where $\overline{B}_V(0,1)$ is the closed unit ball in $V$.*

*Proof.* We choose $R > 0$ and $u \in E$ such that $E \subset \overline{B}_H(u,R)$ (we use the notation $\overline{B}_H(x,R) = \{u \in H : \|u - x\|_H \le R\}$ and $\overline{B}_V(x,R) = \{u \in V : \|u - x\|_V \le R\}$). We have

$$E \subset L(E) \subset \overline{B}_V(Lu, cR) \subset \bigcup_{i=1}^N \overline{B}_H\left(v_i, \frac{R}{4}\right),$$

where $v_i \in H$ and $N$ is the number of balls in $H$ having the radius $\frac{R}{4}$ needed to cover the ball in $V$ centered at zero having the radius $cR$. Note that $N = N^H_{\frac{1}{4c}}(\overline{B}_V(0,1))$. We can assume that all balls from the last sum have a nonempty intersection with $E$, and we denote the arbitrary points from these intersections by $z_i$, whence

$$E \subset \bigcup_{i=1}^{N} \overline{B}_H\left(z_i, \frac{R}{2}\right),$$

with $z_i \in E$. It follows that

$$E = \bigcup_{i=1}^{N} \left(\overline{B}_H\left(z_i, \frac{R}{2}\right) \cap E\right).$$

Now

$$\overline{B}_H\left(z_i, \frac{R}{2}\right) \cap E \subset \overline{B}_V\left(Lz_i, c\frac{R}{2}\right).$$

The ball in $V$ having the radius $c\frac{R}{2}$ can be covered by $N$ balls in $H$ having the radius $\frac{R}{8}$, whence

$$\overline{B}_H\left(z_i, \frac{R}{2}\right) \cap E \subset \bigcup_{j=1}^{N} \overline{B}_H\left(v_{ij}, \frac{R}{8}\right),$$

and, if we want the centers to be the points of $E$, we have

$$\overline{B}_H\left(z_i, \frac{R}{2}\right) \cap E \subset \bigcup_{j=1}^{N} \overline{B}_H\left(z_{ij}, \frac{R}{4}\right),$$

with $z_{ij} \in E$. Repeating this procedure inductively, we can construct a cover of $E$ by $N^k$ balls of radius $\frac{R}{2^k}$. For $\varepsilon > 0$ we can always find $k \in \mathbb{N}$ such that $\frac{R}{2^{k+1}} < \varepsilon \leq \frac{R}{2^k}$, whence $N_\varepsilon(E) \leq N_{\frac{R}{2^{k+1}}}(E) \leq N^{k+1}$, and

$$\frac{\log N_\varepsilon(E)}{\log \frac{1}{\varepsilon}} \leq \frac{(k+1)\log N}{k\log 2 - \log R}.$$

The assertion follows. □

To get the finite dimensionality of the global attractor by the above theorem, it would seem easiest to apply it for $L = S(t)$ for certain $t > 0$ and $E = \mathscr{A}$. It is then enough to derive the estimate (9.5), which is known as the *smoothing estimate* as it implies that in fact we have $\mathscr{A} \subset V$. While this estimate may be difficult to obtain for $L = S(t)$, in some cases it is easier to get it for $L$ being the translation operator defined on the trajectories starting from the attractor (and staying in it, by invariance), and $V, H$ being the appropriately chosen spaces of time-dependent functions. This method, known as the method of $l$-trajectories, was introduced in [168] and will be used for the two-dimensional Navier–Stokes equations with nonmonotone friction in Sect. 10.

The next result, very useful for proving finite dimensionality of global attractors of infinite dimensional dynamical systems, proved originally by Ladyzhenskaya (see, for example, [147], where, however, the notion of the Hausdorff dimension is used), uses the so-called *squeezing property*. This property involves the estimates which are not, as in Theorem 9.4, in the space compactly embedded in the state space, but which involve the finite dimensional projectors. The variant of Ladyzhenskaya result, that we present here, comes from [70].

**Theorem 9.5 (Cf. [70, Theorem 8.1]).** *Let $H$ be the Hilbert space and let $E \subset H$ be a compact set. Let moreover $L : H \to H$ be a continuous mapping such that $E \subset L(E)$ and there exists an orthogonal projector $P_N : H \to H_N$ onto the subspace $H_N \subset H$ of dimension $N$, such that*

$$\|P(Lu - Lv)\|_H \leq l\|u - v\|_H \quad \textit{for all} \quad u, v \in E,$$

*and*

$$\|(I - P)(Lu - Lv)\|_H \leq \delta\|u - v\|_H \quad \textit{for all} \quad u, v \in E,$$

*where $\delta < 1$. Then the fractal dimension $d_f^H(E)$ is finite and*

$$d_f^H(E) \leq N \log \frac{9l}{1-\delta} \left(\log \frac{2}{1-\delta}\right)^{-1}.$$

The above theorem has also the non-autonomous version (see Theorem 13.5) which will be used later in Chap. 13 to prove the finite dimensionality of the pullback attractor for two-dimensional boundary driven Navier–Stokes flow.

**Comments** We refer the reader to [198] for more information about dimensions. Fractal dimension is also known as upper box counting dimension (see [198] Definition 3.1). Sometimes another notion of dimension, namely, the above mentioned Hausdorff dimension, is used to study the global attractors for infinite dimensional dynamical systems. To define it, one needs to use the $s$-dimensional Hausdorff measure valid for $s \geq 0$ and a subset $K$ of a Banach space $X$, given by

$$\mathscr{H}_s^X(K) = \lim_{\delta \to 0} \inf \left\{ \sum_{i=1}^{\infty} |U_i|^s \; : \; K \subset \bigcup_{i=1}^{\infty} U_i \text{ with } |U_i| \leq \delta \right\},$$

where $|A| = \sup_{x,y \in A} \|x - y\|_X$. We define the Hausdorff dimension as

$$d_H^X(K) = \inf\{s \geq 0 \ : \ \mathscr{H}_s^K(K) = 0\}.$$

For the properties of $d_H^X$ see Chap. 2 in [198]. Here we only note that, in contrast to the fractal dimension, the Hausdorff dimension is stable under the countable unions (see Proposition 2.8 in [198]), and for a set $K \subset X$, we have

$$d_H^X(K) \leq d_f^X(K),$$

whence the ability to estimate from above the fractal dimension gives us also the corresponding estimate on the Hausdorff dimension, while the opposite implication does not hold (in particular it is possible to construct sets with zero Hausdorff dimension and infinite fractal dimension, see [198], Exercise 13.3). We also note that both the notions of fractal and Hausdorff dimensions remain valid in metric spaces.

## 9.3 An Application to a Shear Flow in Lubrication Theory

In this section we consider a two-dimensional Navier–Stokes shear flow for which there exists a unique global in time solution as well as the global attractor for the associated semigroup. Our aim is to estimate from above the fractal dimension of the attractor in terms of given data and geometry of the domain of the flow.

The problem we consider is motivated by a typical problem from lubrication theory where the domain of the flow is usually very thin and the Reynolds number of the flow is large.

### *9.3.1 Formulation of the Problem*

We consider two-dimensional Navier–Stokes equations,

$$u_t - \nu \Delta u + (u \cdot \nabla)u + \nabla p = 0, \tag{9.6}$$

$$\operatorname{div} u = 0, \tag{9.7}$$

in the channel

$$\Omega_\infty = \{x = (x_1, x_2) \ : \ -\infty < x_1 < \infty, \ \ 0 < x_2 < h(x_1)\},$$

where $h$ is a function, positive, smooth, and $L$-periodic in $x_1$.

Let

$$\Omega = \{x = (x_1, x_2) \ : \ 0 < x_1 < L, \ 0 < x_2 < h(x_1)\},$$

and $\partial\Omega = \overline{\Gamma_C} \cup \overline{\Gamma_L} \cup \overline{\Gamma_D}$, where $\Gamma_C$ and $\Gamma_D$ are the bottom and the top, and $\Gamma_L$ is the lateral part of the boundary of $\Omega$.

We are interested in solutions of (9.6)–(9.7) in $\Omega$ which are $L$-periodic with respect to $x_1$. Moreover, we assume

$$u = 0 \quad \text{on} \quad \Gamma_D, \tag{9.8}$$

$$u = U_0 e_1 = (U_0, 0) \quad \text{on} \quad \Gamma_C, \tag{9.9}$$

where $U_0$ is a positive constant, and the initial condition

$$u(x, 0) = u_0(x) \quad \text{for} \quad x \in \Omega. \tag{9.10}$$

The problem is motivated by a flow in an infinite (rectified) journal bearing $\Omega \times (-\infty, +\infty)$, where $\Gamma_D \times (-\infty, +\infty)$ represents the outer cylinder, and $\Gamma_C \times (-\infty, +\infty)$ represents the inner, rotating cylinder. In the lubrication problems the gap $h$ between cylinders is never constant. We can assume that the rectification does not change the equations as the gap between cylinders is very small with respect to their radii.

*Remark 9.1.* When $h = const$, problem (9.6)–(9.10) was intensively studied in several contexts, cf. Sect. 9.4.

In our considerations we use the background flow method and transform the boundary condition (9.9) to the homogeneous ones by defining a smooth background flow, a simple version of the Hopf construction, which we shall be described in detail. Let

$$u(x_1, x_2, t) = U(x_2)e_1 + v(x_1, x_2, t), \tag{9.11}$$

with

$$U(0) = U_0, \quad U(h(x_1)) = 0, \quad x_1 \in (0, L). \tag{9.12}$$

Then $v$ is $L$-periodic in $x_1$ and satisfies

$$v_t - \nu \Delta v + (v \cdot \nabla)v + U v_{,x_1} + (v)_2 U' e_1 + \nabla p = \nu U'' e_1, \tag{9.13}$$

$$\operatorname{div} v = 0, \tag{9.14}$$

$$v = 0 \quad \text{on} \quad \Gamma_D \cup \Gamma_C, \tag{9.15}$$

and the initial condition

$$v(x,0) = v_0(x) = u_0(x) - U(x_2)e_1. \tag{9.16}$$

By $(v)_2$ we denoted the second component of $v$.

Now, we define a weak form of the transformed problem above. To this end we need some notations. Let

$$\tilde{V} = \{v \in C^\infty(\overline{\Omega})^2 \ : \ v \text{ is } L\text{-periodic in } x_1, \ \operatorname{div} v = 0, \ v = 0 \text{ on } \Gamma_D \cup \Gamma_C\},$$

$$V = \text{ closure of } \tilde{V} \text{ in } H^1(\Omega)^2,$$

$$H = \text{ closure of } \tilde{V} \text{ in } L^2(\Omega)^2.$$

We define the scalar products

$$(u,v) = \int_\Omega u(x) \cdot v(x) dx,$$

and

$$((u,v)) = (\nabla u, \nabla v)$$

in $H$ and $V$, respectively, and the corresponding norms

$$|v| = (v,v)^{\frac{1}{2}} \quad \text{and} \quad \|v\| = ((v,v))^{\frac{1}{2}}.$$

Let

$$b(u,v,w) = ((u \cdot \nabla)v, w),$$

and

$$a(u,v) = \nu(\nabla u, \nabla v).$$

Then the natural weak formulation of the transformed problem (9.13)–(9.16) is as follows.

**Problem 9.1.** Find

$$v \in C([0,T];H) \cap L^2(0,T;V)$$

for each $T > 0$, such that

$$\frac{d}{dt}(v(t), \Theta) + a(v(t), \Theta) + b(v(t), v(t), \Theta) = F(v(t), \Theta) \tag{9.17}$$

for all $\Theta \in V$, and

$$v(x,0) = v_0(x),$$

where

$$F(v,\Theta) = -a(\xi,\Theta) - b(\xi,v,\Theta) - b(v,\xi,\Theta), \tag{9.18}$$

and $\xi = Ue_1$ is a suitable background flow.

We have the following existence theorem.

**Theorem 9.6.** *There exists a unique weak solution of Problem 9.1 such that for all $\eta$, $T$, $0 < \eta < T$, $v \in L^2(\eta, T; H^2(\Omega))$, and for each $t > 0$ the map $v_0 \mapsto v(t)$ is continuous as a map in $H$. Moreover, there exists a global attractor for the associated semigroup $\{S(t)\}_{t\geq 0}$ in the phase space $H$.*

*Proof.* The standard existence part of the proof is based on an energy inequality, the Galerkin approximations, and the compactness method [220]. We shall not reproduce the standard argument, however, we stop for a moment to mention the properties of the Stokes operator and the regularity problem. As usual, we define the Stokes operator as the self-adjoint operator from $V$ to its dual $V'$ and also as an unbounded strictly positive operator in $H$ with compact inverse $A^{-1}$, by

$$\langle Au, v\rangle = a(u,v) \quad \text{for all} \quad u, v \in V.$$

Then, there exists a complete orthonormal family $w_j, j \in \mathbb{N}$ in $H$, with $w_j \in D(A) = \{u \in V : Au \in H\} \subset H^2(\Omega)$, such that $Aw_j = \lambda_j w_j$, $0 < \lambda_1 \leq \lambda_2 \leq \ldots, \lambda_j \to \infty$ as $j \to \infty$. The inclusion $D(A) \subset H^2(\Omega)$ can be proved in the usual way by using the regularity theory. First, we extend $u, p$, and $f$ by periodicity to some smooth set $\Omega_3$ such that $\Omega \subset \Omega_3 \subset \Omega_\infty$, and then use the classical Cattabriga–Solonnikov estimate [219] to the solution $(\varphi u, \varphi p)$ of the related generalized Stokes problem in the domain $\Omega_3$, with $\varphi$ being a suitable cut-off function that equals one on $\Omega$. This argument gives us also the equivalence of the norms $\|u\|_{H^2(\Omega)}$ and $|Au|$ on $D(A)$.

Now, defining the Galerkin approximations by $v_m(x,t) = \sum_{j=1}^m \tilde{v}(t)w_j(x)$ we can see that they belong to $H^2(\Omega)$ for $t > \tau$. Further estimates performed on the Galerkin approximations lead, in a standard way, to $v \in L^2(\tau, T; H^2(\Omega))$. □

**Exercise 9.1.** Provide the details of the proof of Theorem 9.6.

### *9.3.2 Energy Dissipation Rate Estimate*

Our aim now is to estimate the time averaged energy dissipation rate per unit mass $\epsilon$ of the flow $u$—the weak solution of problem (9.6)–(9.10). We define

$$\epsilon = \frac{\nu}{|\Omega|} < \|u\|^2 > = \limsup_{T \to +\infty} \frac{\nu}{|\Omega|} \frac{1}{T} \int_0^T \|u(t)\|^2 dt. \tag{9.19}$$

We estimate first the averaged energy dissipation rate of the flow $v$, and then use the relation, cf. (9.11),

$$\|u(t)\|^2 = \|v(t)\|^2 + 2 \int_\Omega U'(v)_{1,x_2}\, dx + \int_\Omega |U'|^2 dx. \tag{9.20}$$

To estimate the right-hand side of (9.20) we use Eq. (9.17). Taking $\Theta = v$ in (9.17), we obtain

$$\frac{1}{2}\frac{d}{dt}|v|^2 + a(v,v) + b(v,v,v) = F(v,v).$$

Since $v = 0$ on $\Gamma_D \cup \Gamma_C$, is $L$-periodic in $x_1$, and $\operatorname{div} v = 0$, we have $b(v,v,v) = 0$, and

$$\frac{1}{2}\frac{d}{dt}|v|^2 + \nu\|v\|^2 = F(v,v). \tag{9.21}$$

Integrating the above equation in t on the interval $(0,T)$, dividing by $T$, and taking lim sup of both sides we estimate the averaged energy dissipation rate of $v$. Before, however, we have to estimate carefully the term $F(v,v)$ on the right-hand side of (9.21). By (9.18),

$$F(v,v) = -a(\xi,v) - b(v,\xi,v). \tag{9.22}$$

We have

$$|a(\xi,v)| \le \nu\|\xi\|\|v\| \le \nu\|\xi\|^2 + \frac{\nu}{4}\|v\|^2. \tag{9.23}$$

To estimate the last term in (9.22) we use the following lemma.

**Lemma 9.2.** *For any $\eta > 0$ there exists a smooth extension*

$$\xi = \xi(x_2) = U(x_2)e_1 = (U(x_2), 0)$$

*of the boundary condition for u, such that*

$$|b(v,\xi,v)| \le \eta\|v\|^2 \quad \text{for all} \quad v \in V.$$

*Proof.* Let $\rho : [0,\infty) \to [0,1]$ be a smooth function such that

$$\rho(0) = 1, \quad \operatorname{supp}\rho \subset [0,1/2], \quad \max|\rho'(s)| \le \sqrt{8},$$

and let, for $0 < \varepsilon \le 1$,

$$U(x_2) = U_0\rho(x_2/(h_0\varepsilon)), \quad h_0 = \min_{0\le x_1\le L} h(x_1).$$

Then the function $U$ matches the boundary conditions (9.12), is $L$-periodic in $x_1$, and $\operatorname{div} Ue_1 = 0$. Moreover, for $Q_\varepsilon = (0, L) \times (0, h_0\varepsilon/2)$,

$$|b(v, Ue_1, v)| = |b(v, v, Ue_1)| \le \left\| \frac{v}{x_2} \right\|_{L^2(Q_\varepsilon)} \|\nabla v\|_{L^2(Q_\varepsilon)} \|x_2 U(x_2)\|_{L^\infty(Q_\varepsilon)}.$$

Now,

$$\|x_2 U(x_2)\|_{L^\infty(Q_\varepsilon)} \le \frac{h_0\varepsilon}{2} U_0,$$

and

$$\left\| \frac{v}{x_2} \right\|^2_{L^2(Q_\varepsilon)} \le \int_0^L \int_0^{h(x_1)} \left| \frac{v(x_1, x_2)}{x_2} \right|^2 dx_2 dx_1 \le 4 \int_0^L \int_0^{h(x_1)} \left| \frac{\partial v(x_1, x_2)}{\partial x_2} \right|^2 dx_2 dx_1$$

by the Hardy inequality (cf. Theorem 3.12 and Exercise 3.12). Thus, we obtain

$$|b(v, Ue_1, v)| \le \eta\|v\|^2 \quad \text{with} \quad \eta = 2h_0 U_0 \varepsilon. \quad \square$$

In particular,

$$|b(v, \xi, v)| \le \frac{\nu}{4}\|v\|^2 \quad \text{for} \quad \varepsilon = \min\{\nu/(8h_0U_0), 1\}. \tag{9.24}$$

In view of estimates (9.23) and (9.24),

$$|F(v, v)| \le \nu\|\xi\|^2 + \frac{\nu}{2}\|v\|^2,$$

and, by (9.21),

$$\frac{1}{2}\frac{d}{dt}|v|^2 + \frac{\nu}{2}\|v\|^2 \le \nu\|\xi\|^2. \tag{9.25}$$

To estimate the right-hand side in terms of the data, we use

**Lemma 9.3.** *Let $U$ be as in Lemma 9.2. Then*

$$\int_\Omega |U(x_2)|^2 dx_1 dx_2 \le \frac{1}{2} L h_0 U_0^2 \varepsilon, \tag{9.26}$$

*and*

$$\int_{\Omega} |U'(x_2)|^2 dx_1 dx_2 \leq \frac{4LU_0^2}{h_0} \frac{1}{\varepsilon}. \tag{9.27}$$

*Proof.* We have $|U(x_2)| \leq U_0$ and $\operatorname{supp} U \subset Q_\varepsilon$, which gives (9.26). Moreover,

$$\begin{aligned}\int_{\Omega} |U'(x_2)|^2 dx &= \frac{U_0^2}{h_0^2 \varepsilon^2} \int_{\Omega} \left|\rho'\left(\frac{x_2}{h_0 \varepsilon}\right)\right|^2 dx \\ &= \frac{U_0^2}{h_0^2 \varepsilon^2} L \int_0^{\frac{h_0 \varepsilon}{2}} \left|\rho'\left(\frac{x_2}{h_0 \varepsilon}\right)\right|^2 dx_2 \leq \frac{4LU_0^2}{h_0} \frac{1}{\varepsilon},\end{aligned}$$

as $|\rho'| \leq \sqrt{8}$, which gives the second inequality of the lemma. □

Applying Lemma 9.3 to inequality (9.25) we obtain

$$\frac{1}{2} \frac{d}{dt} |v|^2 + \frac{\nu}{2} \|v\|^2 \leq \nu \frac{4LU_0^2}{h_0} \frac{1}{\varepsilon}. \tag{9.28}$$

Let $h_M = \max_{0 \leq x_1 \leq L} h(x_1)$ and $h_M/h_0 \simeq 1$. Then we can define the Reynolds number of the flow $u$ by $Re = (h_0 U_0)/\nu$. Observe that for a turbulent flow $u$ (of some large Reynolds number) $\varepsilon$ in (9.24) is equal to $\nu/(8h_0 U_0)$. Therefore, from (9.28) we obtain, as indicated above,

**Lemma 9.4.** *If $Re \gg 1$ then the time averaged energy dissipation rate per unit mass $\epsilon(v)$ for the flow $v$ can be estimated as follows:*

$$\epsilon(v) \leq \frac{64L}{|\Omega|} U_0^3 \leq \frac{64}{h_0} U_0^3.$$

In the end, we have the following theorem.

**Theorem 9.7.** *For the Navier–Stokes turbulent flow $u$ with $Re \gg 1$ the time averaged energy dissipation rate per unit mass $\epsilon$ defined in (9.19) can be estimated as follows:*

$$\epsilon \leq C \frac{U_0^3}{h_0}, \tag{9.29}$$

*where $C$ is a numeric constant.*

*Proof.* By (9.20) we have

$$< \|u\|^2 > \; \leq \; 2 < \|v\|^2 > +2 \int_{\Omega} |U'|^2 dx,$$

and then we use (9.27) to estimate the second term on the right-hand side. □

Estimate (9.29) has the same form as the usual in turbulence theory estimate of the averaged energy dissipation rate for the Navier–Stokes boundary driven flow in a rectangular domain [87, 88, 99].

In the sequel we shall use the estimates of $\epsilon$ to find an upper bound of the dimension of the global attractor.

### 9.3.3 A Version of the Lieb–Thirring Inequality

Let

$$\tilde{H}^1 = \{v \in C^\infty(\overline{\Omega})^2 \ : \ v \text{ is } L\text{-periodic in } x_1, \ \ v = 0 \text{ on } \Gamma_D \cup \Gamma_C\}$$

and

$$H^1 = \text{closure of } \tilde{H}^1 \text{ in } H^1(\Omega)^2.$$

**Lemma 9.5.** *Let $\varphi_j \in H^1, j = 1, \dots, m$ be an orthonormal family in $L^2(\Omega)$ and let $h_M = \max_{0 \le x_1 \le L} h(x_1)$. Then the following inequality holds:*

$$\int_\Omega \left( \sum_{j=1}^m \varphi_j^2 \right)^2 dx \le \sigma \left( 1 + \left( \frac{h_M}{L} \right)^2 \right) \sum_{j=1}^m \int_\Omega |\nabla \varphi_j|^2 dx, \tag{9.30}$$

*where $\sigma$ is an absolute constant.*

*Proof.* Let $\Omega_1 = (0, L) \times (0, h_0)$, and let $\psi_j \in H^1(\Omega_1), j = 1, \dots, m$, be a family of functions that are orthonormal in $L^2(\Omega_1)$, with $\psi_j(0, x_2) = \psi_j(L, x_2)$ and $\psi_j = 0$ for $x_2 = 0, x_2 = h_0$. We know [89] that for this family there exist absolute constants $C_0'$ and $C_1'$, such that

$$\int_{\Omega_1} \left( \sum_{j=1}^m \psi_j^2 \right)^2 dx \le C_0' \left( \sum_{j=1}^m \int_{\Omega_1} \left| \frac{\partial \psi_j}{\partial x_1} \right|^2 dx + \frac{C_1' m}{L^2} \right)^{\frac{1}{2}} \left( \sum_{j=1}^m \int_{\Omega_1} \left| \frac{\partial \psi_j}{\partial x_2} \right|^2 dx \right)^{\frac{1}{2}}.$$

We have also the Poincaré inequality

$$\int_{\Omega_1} \psi_j^2 dx \le h_M^2 \int_{\Omega_1} \left| \frac{\partial \psi_j}{\partial x_2} \right|^2 dx.$$

From the above two inequalities and the observation that

$$m = \sum_{j=1}^m \int_{\Omega_1} \psi_j^2 dx,$$

(9.30) follows. □

### 9.3.4 Dimension Estimate of the Global Attractor

We rewrite Eq. (9.17) for the transformed flow $v$ in the short form

$$\frac{dv}{dt} = L(v), \quad v(0) = v_0,$$

where, for all $\Theta \in V$,

$$\langle L(v(t)), \Theta\rangle = -a(v(t), \Theta) - b(v(t), v(t), \Theta) + F(v(t), \Theta),$$

and $F$ is defined in (9.18). Now, to estimate from above the dimension of the global attractor we follow the procedure described in Sect. 9.2. First, for an integer $m > 1$ and vectors $\xi_j \in H, j = 1, \dots, m$, we consider the corresponding linearized around the orbit $v(t)$ problems

$$\frac{d}{dt}U_j = L'(v)U_j \quad \text{and} \quad U_j(0) = \xi_j \quad \text{for} \quad j = 1, \dots, m, \tag{9.31}$$

where $L'(v)$ is the Fréchet derivative of $L$ at $v = v(t)$, with

$$\begin{aligned}(L'(v)U_j, \Theta) &= -a(U_j, \Theta) - b(v, U_j, \Theta) - b(U_j, v, \Theta) - b(\xi, U_j, \Theta) - b(U_j, \xi, \Theta) \\ &= -a(U_j, \Theta) - b(u, U_j, \Theta) - b(U_j, u, \Theta). \end{aligned} \tag{9.32}$$

Let, for a particular time $\tau$, $\Theta_j = \Theta_j(\tau), j \in \mathbb{N}$ be an orthonormal basis of $H$ with $\Theta_1(\tau), \dots, \Theta_m(\tau)$ spanning

$$Q_m(\tau)H = Q_m(\tau, v_0, \xi_1, \dots, \xi_m)H = \text{span}\{U_1(\tau), \dots, U_m(\tau)\},$$

$Q_m(\tau)$ being the orthogonal projector of $H$ onto the space spanned by $U_j(\tau)$ for $j = 1, \dots, m$, solutions of (9.31). We then have $\Theta_j(\tau) \in V, j = 1, \dots, m$, for a.e. $\tau \in \mathbb{R}_+$.

The trace of $L'(v(\tau)) \circ Q_m(\tau)$ is

$$Tr\left(L'(v(\tau)) \circ Q_m(\tau)\right) = \sum_{j=1}^{m} \left(L'(v(\tau))\Theta_j(\tau), \Theta_j(\tau)\right). \tag{9.33}$$

Let $\mathscr{A}$ be the global attractor for the transformed flow $v(\tau) = S(\tau)v_0$, and let

$$q_m = \limsup_{T\to\infty} q_m(T),$$

where

$$q_m(T) = \sup_{v_0 \in \mathscr{A}} \sup_{\substack{\xi_j \in H, \\ |\xi_j| \leq 1, j=1,\dots,m}} \left\{ \frac{1}{T} \int_0^T Tr\left(L'(v(\tau)) \circ Q_m(\tau)\right) d\tau \right\}.$$

Now, our aim is to obtain estimate (9.37) and then inequality (9.38).

**Lemma 9.6.** *The following estimate holds:*

$$Tr\left(L'(v) \circ Q_m\right) \leq -\nu \sum_{j=1}^{m} \|\Theta_j\|^2 + |\rho|_{L^2(\Omega)} \|u\|, \tag{9.34}$$

*where*

$$\rho(x) = \sum_{j=1}^{m} |\Theta_j(x)|^2.$$

*Proof.* From (9.32) and (9.33) we obtain

$$\begin{aligned}\sum_{j=1}^{m} \left(L'(v)\Theta_j, \Theta_j\right) &\leq -\nu \sum_{j=1}^{m} \|\Theta_j\|^2 - \sum_{j=1}^{m} b(\Theta_j, u, \Theta_j) \\ &\leq -\nu \sum_{j=1}^{m} \|\Theta_j\|^2 + \int_{\Omega} |\nabla u| \left( \sum_{j=1}^{m} |\Theta_j(x)|^2 \right) dx,\end{aligned}$$

whence (9.34) follows. □

Now, in our estimates of the trace we use Lemma 9.5. By this lemma, taking into account that $h_M \ll L$ for a thin domain $\Omega$, and from

$$\int_{\Omega} \rho(x) dx = m$$

we have

$$\frac{m^2}{|\Omega|} \leq |\rho|^2_{L^2(\Omega)} \leq \sigma_1 \sum_{j=1}^{m} \|\Theta_j\|^2, \tag{9.35}$$

where $\sigma_1 = 2\sigma$. Moreover,

$$|\rho|_{L^2(\Omega)} \|u\| \leq \frac{\nu}{2\sigma_1} |\rho|^2_{L^2(\Omega)} + \frac{\sigma_1}{2\nu} \|u\|^2 \leq \frac{\nu}{2} \sum_{j=1}^{m} \|\Theta_j\|^2 + \frac{\sigma_1}{2\nu} \|u\|^2 \tag{9.36}$$

by the second inequality in (9.35). Thus, by (9.34), (9.36)

$$Tr\left(L'(v) \circ Q_m\right) \leq -\frac{\nu}{2} \sum_{j=1}^{m} \|\Theta_j\|^2 + \frac{\sigma_1}{2\nu} \|u\|^2,$$

and by (9.35) again, we obtain in the end

$$Tr\left(L'(v(\tau)) \circ Q_m(\tau)\right) \leq -\frac{\nu}{2\sigma_1 |\Omega|} m^2 + \frac{\sigma_1}{2\nu} \|u\|^2. \tag{9.37}$$

From (9.37), recalling $\epsilon$ defined in (9.19), we have

$$q_m \leq -\frac{\nu}{2\sigma_1 |\Omega|} m^2 + \frac{\sigma_1 |\Omega|}{2\nu^2} \epsilon. \tag{9.38}$$

Define

$$a = \frac{\nu}{2\sigma_1 |\Omega|}, \qquad c = \frac{\sigma_1 |\Omega|}{2\nu^2} \epsilon.$$

We can write

$$q_m \leq -am^2 + c.$$

By Theorem 9.3 we have

$$d_f^H(\mathscr{A}) \leq p,$$

where $p > 0$ is such that $-ap^2 + c = 0$. Obviously, we have

$$p = \sqrt{\frac{c}{a}}.$$

Further, taking into account definitions of $a$ and $c$, as well as estimate (9.29) of $\epsilon$, we obtain

$$p \leq \sigma_1 \frac{|\Omega|}{h_0^2} (Re)^{3/2}. \tag{9.39}$$

From the above estimates we conclude the following theorem about strongly turbulent flows in thin (or elongated) domains.

**Theorem 9.8.** *Consider the Navier–Stokes flow u as described in this section. Assume that the domain $\Omega$ is thin and that the flow is strongly turbulent, namely*

$$\frac{h_M}{L} \ll 1 \quad \text{and} \quad Re \gg 1. \tag{9.40}$$

*Then the fractal dimension of the global attractor $\mathscr{A}$ for this flow can be estimated as follows:*

$$d_f^H(\mathscr{A}) \leq \kappa \frac{|\Omega|}{h_0^2}(Re)^{3/2}, \tag{9.41}$$

*where $\kappa$ is some absolute constant. For a rectangular domain $\Omega = (0, L) \times (0, h_0)$ we obtain, in particular,*

$$d_f^H(\mathscr{A}) \leq \kappa \frac{L}{h_0}(Re)^{3/2}.$$

*Proof.* By (9.39), (9.40), and by the definition of $m_0$, (9.41) follows. □

## 9.4 Comments and Bibliographical Notes

The standard procedure of estimating the global attractor dimension based on the dynamical system theory [88, 99, 220], we used in Sect. 9.3, involves two important ingredients: estimate of the time averaged energy dissipation rate $\epsilon$ and a Lieb–Thirring-like inequality. The precision and physical soundness of an estimate of the number of degrees of freedom of a given flow (expressed by an estimate of its global attractor) depends directly on the quality of the estimate of $\epsilon$ and a good choice of the Lieb–Thirring-like inequality which depends, in particular, on the geometry of the domain and on the boundary conditions of the flow.

There is a quickly growing literature devoted to better and better estimates of averaged parameters and attractor dimension of a variety of flows. We mention only a few positions which are related to the problem considered in Sect. 9.3 and some other to give a larger context. Estimate (9.41) is a direct generalization of that in [89], where the domain of the flow is an elongated rectangle $\Omega = (0, L) \times (0, h)$, $L \gg h$. In this case the attractor dimension can be estimated from above by $c\frac{L}{h}Re^{3/2}$, where $c$ is a universal constant, and $Re = \frac{Uh}{\nu}$ is the Reynolds number. Here, the dependence on the geometry of the domain is explicit. (To obtain such result the authors prove a suitable anisotropic form of the Lieb–Thirring inequality. For a variety of forms of this inequality cf. [220]. Moreover, a discussion on agreement of their estimates with Kolmogorov and Kraichnan scaling in turbulence theory is provided in [89].)

In [241] optimal bounds of the attractor dimension are given for a flow in a rectangle $(0, 2\pi L) \times (0, 2\pi L/\alpha)$, with periodic boundary conditions and given external forcing. The estimates are of the form $c_0/\alpha \le d_f^H(\mathscr{A}) \le c_1/\alpha$, see also [180]. A free boundary conditions are considered in [242], see also [222], and an upper bound on the attractor dimension established with the use of a suitable anisotropic version of the Lieb–Thirring inequality.

Estimates of the energy dissipation rate for boundary driven flows, essential in turbulence theory, are investigated, e.g., in [230] for $\Omega = (0, L_x) \times (0, L_y) \times (0, h)$, and in [229] for a smooth and bounded three-dimensional domains. The best estimates come from variational methods, cf. [88], and, in particular [134]. Other contexts and problems can be found, e.g., in monographs [62, 88, 99, 197, 220], and the literature quoted there.

Boundary driven flows in smooth and bounded two-dimensional domains for a non-autonomous Navier–Stokes system are considered in [178], by using an approach of Chepyzhov and Vishik, cf. [62]. In Chap. 13 we shall consider a shear flow with time-dependent boundary driving.

# 10
# Exponential Attractors in Contact Problems

In this chapter we consider two examples of contact problems. First, we study the problem of time asymptotics for a class of two-dimensional turbulent boundary driven flows subject to the Tresca friction law which naturally appears in lubrication theory. Then we analyze the problem with the generalized Tresca law, where the friction coefficient can depend on the tangential slip rate.

While the first problem is governed by a variational inequality, since the underlying potential is convex, the problem with the generalized Tresca law due to the lack of potential convexity leads to a differential inclusion where the multivalued term has the form of the Clarke subdifferential.

We prove that in both cases the global attractors exist and they have finite fractal dimensions. Moreover, there exists an object, called exponential attractor. It contains the global attractor, has finite fractal dimension and attracts the trajectories exponentially fast in time. This property allows, among other things, to locate the exponential and thus the global attractor in the phase space by using numerical analysis.

We start the chapter with the section in which we remind a general method of proving the existence of exponential attractors.

## 10.1 Exponential Attractors and Fractal Dimension

Let us consider an abstract autonomous evolutionary problem

$$\frac{dv(t)}{dt} = F(v(t)) \quad \text{in} \quad Z \quad \text{for a.e.} \quad t \in \mathbb{R}^+, \tag{10.1}$$
$$v(0) = v_0,$$



G. Łukaszewicz, P. Kalita, *Navier–Stokes Equations*, Advances in Mechanics and Mathematics 34, DOI 10.1007/978-3-319-27760-8_10

where $Z$ is a Banach space, $F$ is a nonlinear operator, and $v_0$ is in a Banach space $X$ that embeds in $Z$. We assume that the above problem has a global in time unique solution $\mathbb{R}^+ \ni t \to v(t) \in X$ for every $v_0 \in X$. In this case one can associate with the problem a semigroup $\{S(t)\}_{t\geq 0}$ of (nonlinear) operators $S(t) : X \to X$ setting $S(t)v_0 = v(t)$, where $v(t), t > 0$, is the unique solution of (10.1).

Below we recall the notion of the global attractor and its properties (cf. Sect. 7.2).

We know that from the properties of the semigroup of operators $\{S(t)\}_{t\geq 0}$ we can derive the information on the behavior of solutions of problem (10.1), in particular, on their time asymptotics. One of the objects, the existence of which characterizes the asymptotic behavior of solutions, is the global attractor. It is a compact and invariant with respect to operators $S(t)$ subset of the phase space $X$ (in general, a metric space) that uniformly attracts all bounded subsets of $X$.

**Definition 10.1.** A global attractor for a semigroup $\{S(t)\}_{t\geq 0}$ in a Banach space $X$ is a subset $\mathscr{A}$ of $X$ such that:

- $\mathscr{A}$ is compact in X,
- $\mathscr{A}$ is invariant, i.e., $S(t)\mathscr{A} = \mathscr{A}$ for every $t \geq 0$,
- for every $\varepsilon > 0$ and every bounded set $B$ in $X$ there exists $t_0 = t_0(B, \varepsilon)$ such that $\bigcup_{t\geq t_0} S(t)B$ is a subset of the $\varepsilon$-neighborhood of the attractor $\mathscr{A}$ (uniform attraction property).

The global attractor defined above is uniquely determined by the semigroup. Moreover, it is connected and also has the following properties: it is the maximal compact invariant set and the minimal closed set that attracts all bounded sets. The global attractor may have a very complex structure. However, as a compact set (in an infinite dimensional Banach space) its interior is empty.

In Sect. 7.2 we proved the following theorem that guarantees the global attractor existence (see Theorem 7.3 and Remark 7.4).

**Theorem 10.1.** *Let $\{S(t)\}_{t\geq 0}$ be a semigroup of operators in a Banach space $X$ such that:*

- *for all $t \geq 0$ the operators $S(t) : X \to X$ are continuous,*
- *$\{S(t)\}_{t\geq 0}$ is dissipative, i.e., there exists a bounded set $B_0 \subset X$ such that for every bounded set $B \subset X$ there exists $t_0 = t_0(B)$ such that $\bigcup_{t\geq t_0} S(t)B \subset B_0$,*
- *$\{S(t)\}_{t\geq 0}$ is asymptotically compact, i.e., for every bounded set $B \subset X$ and all sequences $t_k \to \infty$ and $y_k \in S(t_k)B$, the sequence $\{y_k\}$ is relatively compact in $X$.*

*Then $\{S(t)\}_{t\geq 0}$ has a global attractor $\mathscr{A}$.*

In this chapter, however, we shall make use of another general theorem on the existence of the global attractor, cf. Theorem 10.2.

As we have seen in Chap. 9, for some dissipative dynamical systems the global attractor has a finite fractal dimension. This fact has a number of important consequences for the behavior of the flow generated by the semigroup [197, 198].

**Definition 10.2.** The *fractal dimension* (also known as *upper box counting dimension*) of a compact set $K$ in a Banach space $X$ is defined as

$$d_f^X(K) = \limsup_{\varepsilon \to 0} \frac{\log N_\varepsilon^X(K)}{\log \frac{1}{\varepsilon}},$$

where $N_\varepsilon^X(K)$ is the minimal number of balls of radius $\varepsilon$ in $X$ needed to cover $K$.

Another important property that holds for many dynamical systems is the existence of an exponential attractor.

**Definition 10.3.** An *exponential attractor* for a semigroup $\{S(t)\}_{t \geq 0}$ in a Banach space $X$ is a subset $\mathscr{M}$ of $X$ such that:

- $\mathscr{M}$ is compact in X,
- $\mathscr{M}$ is positively invariant, i.e., $S(t)\mathscr{M} \subset \mathscr{M}$ for every $t \geq 0$,
- fractal dimension of $\mathscr{M}$ is finite, i.e., $d_f^X(\mathscr{M}) < \infty$,
- $\mathscr{M}$ attracts exponentially all bounded subsets of $X$, i.e., there exist a universal constant $c_1$ and a monotone function $\Phi$ such that for every bounded set $B$ in $X$, its image $S(t)B$ is a subset of the $\varepsilon(t)$-neighborhood of $\mathscr{M}$ for all $t \geq t_0$, where $\varepsilon(t) = \Phi(\|B\|_X)e^{-c_1 t}$ (exponential attraction property).

Since the global attractor $\mathscr{A}$ is the minimal compact attracting set it follows that if both global and exponential attractors exist, then $\mathscr{A} \subset \mathscr{M}$ and the fractal dimension of $\mathscr{A}$ must be finite. Moreover, in contrast to the global attractor, an exponential attractor does not have to be unique.

In the following we will remind the abstract framework of [168] (see also [162]) that uses the so-called method of $l$-trajectories (or short trajectories) to prove the existence of global and exponential attractors.

Let $X$, $Y$, and $Z$ be three Banach spaces such that

$$Y \subset X \quad \text{with compact imbedding} \quad \text{and} \quad X \subset Z.$$

We assume, moreover, that $X$ is reflexive and separable.

For $\tau > 0$, let

$$X_\tau = L^2(0, \tau; X),$$

and

$$Y_\tau = \left\{ u \in L^2(0, \tau; Y) : \frac{du}{dt} \in L^2(0, \tau; Z) \right\}.$$

By $C_w([0, \tau]; X)$ we denote the space of weakly continuous functions from the interval $[0, \tau]$ to the Banach space $X$, and we assume that the solutions of (10.1) are at least in $C_w([0, T]; X)$ for all $T > 0$. Then by an $l$-trajectory we mean the restriction of any solution to the time interval $[0, l]$. If $v = v(t), t \geq 0$, is the solution of (10.1) then both $\chi = v|_{[0,l]}$ and all shifts $L_t(\chi) = v|_{[t,l+t]}$ for $t > 0$ are $l$-trajectories. Note that the mapping $L_t$ is defined as $L_t(\chi)(\tau) = v(\tau + t)$ for $t > 0$ and $\tau \in [0, l]$, where $v$ is a unique solution such that $v|_{[0,l]} = \chi$.

We can now formulate a theorem which gives criteria for the existence of a global attractor $\mathscr{A}$ for the semigroup $\{S(t)\}_{t\geq 0}$ in $X$ and its finite dimensionality. The following criteria are stated as assumptions $(A1)$–$(A8)$ in [168]:

$(A1)$ for any $v_0 \in X$ and arbitrary $T > 0$ there exists (not necessarily unique) $v \in C_w([0, T]; X) \cap Y_T$, a solution of the evolutionary problem on $[0, T]$ with $v(0) = v_0$; moreover, for any solution $v$, the estimates of $\|v\|_{Y_T}$ are uniform with respect to $\|v_0\|_X$,
$(A2)$ there exists a bounded set $B^0 \subset X$ with the following properties: if $v$ is an arbitrary solution with initial condition $v_0 \in X$ then

1. there exists $t_0 = t_0(\|v_0\|_X)$ such that $v(t) \in B^0$ for all $t \geq t_0$,
2. if $v_0 \in B^0$ then $v(t) \in B^0$ for all $t \geq 0$,

$(A3)$ each $l$-trajectory has a unique continuation among all solutions of the evolution problem which means that from an endpoint of an $l$-trajectory there starts at most one solution,
$(A4)$ for all $t > 0$, $L_t : X_l \to X_l$ is continuous on $\mathscr{B}_l^0$—the set of all $l$-trajectories starting at any point of $B^0$ from (A2),
$(A5)$ for some $\tau > 0$, the closure in $X_l$ of the set $L_\tau(\mathscr{B}_l^0)$ is contained in $\mathscr{B}_l^0$,
$(A6)$ there exists a space $W_l$ such that $W_l \subset X_l$ with compact embedding, and $\tau > 0$ such that $L_\tau : X_l \to W_l$ is Lipschitz continuous on $\mathscr{B}_l^1$—the closure of $L_\tau(\mathscr{B}_l^0)$ in $X_l$,
$(A7)$ the map $e : X_l \to X$, $e(\chi) = \chi(l)$, is continuous on $\mathscr{B}_l^1$,
$(A8)$ the map $e : X_l \to X$ is Hölder-continuous on $\mathscr{B}_l^1$.

**Theorem 10.2.** *Let the assumptions* (A1)–(A5), (A7) *hold. Then there exists a global attractor $\mathscr{A}$ for the semigroup $\{S(t)\}_{t\geq 0}$ in $X$. Moreover, if the assumptions* (A6), (A8) *are satisfied then the fractal dimension of the attractor is finite.*

For the existence of an exponential attractor we need two additional properties, where now $X$ is a Hilbert space (cf. [168]):

$(A9)$ for all $\tau > 0$ the operators $L_t : X_l \to X_l$ are (uniformly with respect to $t \in [0, \tau]$) Lipschitz continuous on $\mathscr{B}_l^1$,
$(A10)$ for all $\tau > 0$ there exists $c > 0$ and $\beta \in (0, 1]$ such that for all $\chi \in \mathscr{B}_l^1$ and $t_1, t_2 \in [0, \tau]$ it holds that

$$\|L_{t_1}\chi - L_{t_2}\chi\|_{X_l} \leq c|t_1 - t_2|^\beta. \tag{10.2}$$

**Theorem 10.3.** *Let $X$ be a separable Hilbert space and let the assumptions* $(A1)$–$(A6)$ *and* $(A8)$–$(A10)$ *hold. Then there exists an exponential attractor $\mathscr{M}$ for the semigroup $\{S(t)\}_{t\geq 0}$ in $X$.*

For the proofs of Theorems 10.2 and 10.3 we refer the readers to corresponding theorems in [168].

## 10.2 Planar Shear Flows with the Tresca Friction Condition

### 10.2.1 Problem Formulation

We start this section from the description of the problem for which we will later prove the exponential attractor existence. The flow of an incompressible fluid in a two-dimensional domain $\Omega$ is described by the equation of motion

$$u_t - \nu\Delta u + (u\cdot\nabla)u + \nabla p = 0 \quad \text{in} \quad \Omega\times\mathbb{R}^+, \tag{10.3}$$

and the incompressibility condition

$$\operatorname{div} u = 0 \quad \text{in} \quad \Omega\times\mathbb{R}^+. \tag{10.4}$$

To define the domain $\Omega$ of the flow, let $\Omega_\infty$ be the channel

$$\Omega_\infty = \{x = (x_1, x_2) : \; -\infty < x_1 < \infty, \; 0 < x_2 < h(x_1)\},$$

where $h$ is a positive function, smooth, and L-periodic in $x_1$. Then we set

$$\Omega = \{x = (x_1, x_2) : \; 0 < x_1 < L, \; 0 < x_2 < h(x_1)\},$$

and $\partial\Omega = \overline{\Gamma_C}\cup\overline{\Gamma_L}\cup\overline{\Gamma_D}$, where $\Gamma_C$ and $\Gamma_D$ are the bottom and the top, and $\Gamma_L$ is the lateral part of the boundary of $\Omega$. By $n$ we will denote the unit outward normal to $\partial\Omega$ and by $T = T(u,p)$ the stress tensor given by $T(u,p) = (T_{ij}(u,p))_{i,j=1}^2$, where $T_{ij}(u,p) = -p\delta_{ij} + \nu\left(u_{i,j} + u_{j,i}\right)$.

We are interested in solutions of (10.3)–(10.4), such that the velocity $u$ and the Cauchy stress vector $T(u,p)n$ on $\Gamma_L$ are $L$-periodic with respect to $x_1$. Note that if both velocity and pressure are smooth functions defined on $\Omega_\infty$ and $L$-periodic with respect to $x_1$, then the periodicity of the Cauchy stress vector on $\Gamma_L$ follows automatically from the constitutive law.

We assume that

$$u = 0 \quad \text{on} \quad \Gamma_D\times\mathbb{R}^+. \tag{10.5}$$

Moreover, we assume the no flux condition across $\Gamma_C$ so that the normal component of the velocity on $\Gamma_C$ satisfies

$$u_n = u\cdot n = 0 \quad \text{on} \quad \Gamma_C\times\mathbb{R}^+, \tag{10.6}$$

and that the tangential component of the velocity $u_\tau$ on $\Gamma_C$ is unknown and satisfies the Tresca friction law with a constant and positive maximal friction coefficient $k$. This means that, cf., e.g., [91, 215],

$$\left.\begin{array}{l}|T_\tau(u,p)| \le k \\[2ex] |T_\tau(u,p)| < k \Rightarrow u_\tau = U_0 e_1 \\[2ex] |T_\tau(u,p)| = k \Rightarrow \exists \lambda \ge 0 \text{ such that } u_\tau = U_0 e_1 - \lambda T_\tau(u,p)\end{array}\right\} \quad \text{on} \quad \Gamma_C \times \mathbb{R}^+, \tag{10.7}$$

where $U_0 e_1 = (U_0, 0)$, $U_0 \in \mathbb{R}$, is the velocity of the lower surface producing the driving force of the flow and $T_\tau$ is the tangential component of the stress vector on $\Gamma_C$ given by

$$T_\tau(u,p) = T(u,p)n - ((T(u,p)n) \cdot n)n, \tag{10.8}$$

and $u_\tau$ is the tangential velocity on $\Gamma_C$ given by $u_\tau = u - u_n n$. From (10.6) it follows that $u_\tau = u$.

Finally, the initial condition for the velocity field is

$$u(x,0) = u_0(x) \quad \text{for} \quad x \in \Omega.$$

The problem is motivated by the examination of a certain two-dimensional flow in an infinite (rectified) journal bearing $\Omega \times (-\infty, +\infty)$, where $\Gamma_D \times (-\infty, +\infty)$ represents the outer cylinder, and $\Gamma_C \times (-\infty, +\infty)$ represents the inner, rotating cylinder. In the lubrication problems the gap $h$ between cylinders is never constant. We can assume that the rectification does not change the equations as the gap between cylinders is very small with respect to their radii.

Since the widely used no-slip boundary condition when the fluid has the same velocity as surrounding solid boundary is not respected if the shear rate becomes too high we can model such situation by a transposition of the well-known friction laws between two solids [215] to the fluid–solid interface. The Tresca friction law is one of them.

To work with the above problem it is convenient to transform the boundary condition (10.7) to the homogeneous one. To this end, let

$$u(x_1, x_2, t) = U(x_2)e_1 + v(x_1, x_2, t), \tag{10.9}$$

with

$$U(0) = U_0, \quad U(h(x_1)) = 0, \quad x_1 \in (0, L). \tag{10.10}$$

The new vector field $v$ is $L$-periodic in $x_1$ and satisfies the equation of motion

$$v_t - \nu \Delta v + (v \cdot \nabla)v + \nabla p = G(v), \tag{10.11}$$

with

$$G(v) = -U v_{,x_1} - (v)_2 \, U_{,x_2} \, e_1 + \nu U_{,x_2 x_2} \, e_1,$$

where by $(v)_2$ we denoted the second component of $v$. As $\operatorname{div}(Ue_1) = 0$ we get

$$\operatorname{div} v = 0 \quad \text{in} \quad \Omega \times \mathbb{R}^+. \tag{10.12}$$

From (10.9)–(10.10) we obtain

$$v = 0 \quad \text{on} \quad \Gamma_D \times \mathbb{R}^+ \tag{10.13}$$

and

$$v_n = v \cdot n = 0 \quad \text{on} \quad \Gamma_C \times \mathbb{R}^+. \tag{10.14}$$

Moreover,

$$T_\tau(v,p) = T_\tau(u,p) + \left( \nu \frac{\partial U(x_2)}{\partial x_2}\bigg|_{x_2=0}, 0 \right) \quad \text{on} \quad \Gamma_C \times \mathbb{R}^+.$$

Since we can define the extension $U$ in such a way that

$$\frac{\partial U(x_2)}{\partial x_2}\bigg|_{x_2=0} = 0,$$

the Tresca condition (10.7) transforms to

$$\left.\begin{aligned} &|T_\tau(v,p)| \le k \\ &|T_\tau(v,p)| < k \Rightarrow v_\tau = 0 \\ &|T_\tau(v,p)| = k \Rightarrow \exists \lambda \ge 0 \text{ such that } v_\tau = -\lambda T_\tau(v,p) \end{aligned}\right\} \quad \text{on} \quad \Gamma_C \times \mathbb{R}^+. \tag{10.15}$$

Finally, the initial condition becomes

$$v(x,0) = v_0(x) = u_0(x) - U(x_2)e_1 \quad \text{for} \quad x \in \Omega. \tag{10.16}$$

The Tresca condition (10.15) is a particular case of an important in contact mechanics class of subdifferential boundary conditions of the form, cf., e.g., [188],

$$\varphi(\Theta_\tau(x)) - \varphi(v_\tau(x,t)) \ge -T_\tau(x,t)(\Theta_\tau(x) - v_\tau(x,t)) \quad \text{on} \quad \Gamma_C \times \mathbb{R}^+, \tag{10.17}$$

where $\Theta_\tau$ belongs to a certain set of admissible functions (in our case the tangential components of the traces on $\Gamma_C$ of functions from the space $V$ defined below) and $\varphi$ is a convex functional. For $\varphi(v_\tau) = k|v_\tau|$ the last condition is equivalent to (10.15).

We pass to the variational formulation of the homogeneous problem (10.11)–(10.16). Then, for the convenience of the reader, we describe the relations between the classical and the weak formulations.

We begin with some basic definitions. Let

$$\tilde{V} = \{v \in C^\infty(\overline{\Omega})^2 : \ \operatorname{div} v = 0 \ \text{in} \ \Omega, \quad v \text{ is } L\text{-periodic in } x_1, \\ v = 0 \ \text{on} \ \Gamma_D, \quad v \cdot n = 0 \ \text{on} \ \Gamma_C\},$$

and

$$V = \text{closure of} \ \ \tilde{V} \ \text{in} \ H^1(\Omega)^2, \qquad H = \text{closure of} \ \ \tilde{V} \ \text{in} \ L^2(\Omega)^2.$$

We define the scalar products in $H$ and $V$, respectively, by

$$(u\,,\,v) = \int_\Omega u(x) \cdot v(x)\,dx \quad \text{and} \quad (\nabla u\,,\,\nabla v) = \int_\Omega \nabla u(x) \colon \nabla v(x)\,dx,$$

and their associated norms by

$$|v| = (v, v)^{\frac{1}{2}} \quad \text{and} \quad \|v\| = (\nabla v\,,\,\nabla v)^{\frac{1}{2}}.$$

Let, for $u$, $v$, and $w$ in $V$,

$$a(u\,,\,v) = (\nabla u\,,\,\nabla v) \quad \text{and} \quad b(u\,,\,v\,,\,w) = ((u \cdot \nabla)v\,,\,w).$$

We denote the trace operator from $V$ to $L^2(\Gamma_C)^2$ by $\gamma$. Its norm is denoted by $\|\gamma\| = \|\gamma\|_{\mathscr{L}(V;L^2(\Gamma_C)^2)}$. Finally, let us define the convex and continuous functional $\Phi : V \to \mathbb{R}$ by

$$\Phi(u) = \int_{\Gamma_C} k|u(x_1, 0)|dx_1 = \int_{\Gamma_C} \varphi(u_\tau)\,dS.$$

The variational formulation of the homogeneous problem (10.11)–(10.16) is as follows:

**Problem 10.1.** Given $v_0 \in H$, find $v : [0, \infty) \to H$ such that:

(*i*) for all $T > 0$,

$$v \in C([0, T]; H) \cap L^2(0, T; V) \quad \text{with} \quad v_t \in L^2(0, T; V'),$$

where $V'$ is the dual space to $V$,

(*ii*) for all $\Theta$ in $V$, and for almost all $t$ in the interval $(0, \infty)$, the following variational inequality holds

$$\langle v_t(t), \Theta - v(t)\rangle + \nu a(v(t), \Theta - v(t)) + b(v(t), v(t), \Theta - v(t)) + \Phi(\Theta) - \Phi(v(t)) \geq \langle \mathscr{L}(v(t)), \Theta - v(t)\rangle, \quad (10.18)$$

(*iii*) the following initial condition holds

$$v(x, 0) = v_0(x) \quad \text{for a.e.} \quad x \in \Omega. \quad (10.19)$$

In (10.18) the operator $\mathscr{L} : V \to V'$ is defined by

$$\langle \mathscr{L}(v), \Theta\rangle = -\nu a(\xi, \Theta) - b(\xi, v, \Theta) - b(v, \xi, \Theta),$$

where $\xi = Ue_1$ is a suitable smooth background flow.

We have the following relations between the classical and weak formulations.

**Proposition 10.1.** *Every classical solution of Problem* (10.11)–(10.16) *is also a solution of Problem* 10.1. *On the other hand, every solution of Problem* 10.1 *which is smooth enough is also a classical solution of Problem* (10.11)–(10.16).

*Proof.* As the proof is quite technical it is only sketched, and the details are left to the reader. Let $v$ be a classical solution of Problem (10.11)–(10.16). As it is (by assumption) sufficiently regular, we have to check only (10.18). Remark first that (10.11) can be written as

$$v_t(t) - \operatorname{div} T(v(t), p(t)) + (v(t) \cdot \nabla)v(t) = G(v(t)). \quad (10.20)$$

Let $\Theta \in V$. Multiplying (10.20) by $\Theta - v(t)$ and using the Green formula we obtain

$$\int_\Omega v_t(t) \cdot (\Theta - v(t))\, dx + \int_\Omega T_{ij}(v(t), p(t))(\Theta - v(t))_{i,j}\, dx + b(v(t), v(t), \Theta - v(t)) = \int_{\partial\Omega} T_{ij}(v(t), p(t))n_j(\Theta - v(t))_i\, dS + \int_\Omega G(v(t)) \cdot (\Theta - v(t))\, dx \quad (10.21)$$

for $t \in (0, T)$. As $v(t)$ and $\Theta$ are in $V$, after some calculations we obtain

$$\int_\Omega T_{ij}(v(t), p(t))(\Theta - v(t))_{i,j}\, dx = \nu a(v(t)\, , \Theta - v(t)). \quad (10.22)$$

By (10.17) with $\varphi(v_\tau) = k|v_\tau|$, taking into account the boundary conditions, we get

$$\int_{\partial\Omega} T_{ij}(v(t), p(t))n_j(\Theta - v(t))_i\, dS = \int_{\Gamma_C} T_\tau(v(t), p(t)) \cdot (\Theta_\tau - v_\tau(t))\, dS \geq -\int_{\Gamma_C} k(|\Theta_\tau| - |v_\tau(t)|)\, dS. \quad (10.23)$$

Finally,

$$\int_{\Omega} G(v(t)) \cdot (\Theta - v(t))dx = \langle \mathscr{L}(v(t)), \Theta - v(t) \rangle. \tag{10.24}$$

From (10.22)–(10.24) we see that (10.21) yields (10.18), and (10.19) is the same as (10.16).

Conversely, suppose that $v$ is a sufficiently smooth solution to Problem 10.1. We have immediately (10.12)–(10.14) and (10.16). Let $\varphi$ be in the space

$$(H^1_{div}(\Omega))^2 = \{\varphi \in V : \varphi = 0 \text{ on } \partial\Omega\}.$$

We take $\Theta = v(t) \pm \varphi$ in (10.18) to get

$$\langle v_t(t) - \nu \Delta v(t) + (v(t) \cdot \nabla)v(t) - G(v(t)), \varphi \rangle = 0 \quad \text{for every} \quad \varphi \in (H^1_{div}(\Omega))^2.$$

Thus, there exists a distribution $p(t)$ on $\Omega$ such that

$$v_t(t) - \nu \Delta v(t) + (v(t) \cdot \nabla)v(t) - G(v(t)) = \nabla p(t) \quad \text{in} \quad \Omega, \tag{10.25}$$

so that (10.11) holds. It follows that $p$ is smooth and without loss of generality we may assume that it is also $L$-periodic with respect to $x_1$. Now, we shall derive the Tresca boundary condition (10.15) from the weak formulation. We have

$$\int_{\Omega} T_{ij}(v(t), p(t))(\Theta - v(t))_{i,j}\, dx = -\int_{\Omega} \operatorname{div} T(v(t), p(t)) \cdot (\Theta - v(t))\, dx + \int_{\partial\Omega} T(v(t), p(t))n \cdot (\Theta - v(t))\, dS. \tag{10.26}$$

Applying (10.22) and (10.26) to (10.18) we get

$$\int_{\Omega} (v_t(t) - \operatorname{div} T(v(t), p(t)) + (v(t) \cdot \nabla)v(t) - G(v(t))) \cdot (\Theta - v(t))\, dx + \int_{\partial\Omega} T(v(t), p(t))n \cdot (\Theta - v(t))\, dS \geq \Phi(v(t)) - \Phi(\Theta).$$

By (10.25) we have (10.20) and so the first integral on the left-hand side vanishes. Thus we obtain condition (10.23). Due to the fact that $\Theta_n - v_n(t) = 0$ on $\Gamma_C$, we conclude

$$\int_{\partial\Omega} T(v(t), p(t))n \cdot (\Theta - v(t))_i\, dS = \int_{\Gamma_C} T_\tau(v(t), p(t)) \cdot (\Theta_\tau - v_\tau(t))\, dS + \int_{\Gamma_L} T(v(t), p(t))n \cdot (\Theta - v(t))\, dS,$$

and then,

$$\int_{\Gamma_L} T(v(t), p(t))n \cdot (\Theta - v(t))\, dS + \int_{\Gamma_C} T_\tau(v(t), p(t)) \cdot (\Theta_\tau - v_\tau(t))\, dS$$
$$\geq - \int_{\Gamma_C} k(|\Theta_\tau| - |v_\tau(t)|)\, dS,$$

where $\Theta$ is any element of $V$. We assume that the weak solution $v$ is a sufficiently smooth and $L$-periodic with respect to $x_1$ function defined on the infinite strip $\Omega_\infty$. Then, as the associated pressure is also $L$-periodic, it follows that the Cauchy stress vector $T(v(t), p(t))n$ must be $L$-periodic with respect to $x_1$ and the integral on $\Gamma_L$ vanishes, whereas

$$\int_{\Gamma_C} T_\tau(v(t), p(t)) \cdot (\Theta_\tau - v_\tau(t))\, dS \geq - \int_{\Gamma_C} k(|\Theta_\tau| - |v_\tau(t)|)\, dS. \tag{10.27}$$

Taking $\Theta = 0$ and $\Theta = 2v$, respectively, we get

$$\int_{\Gamma_C} T_\tau \cdot v_\tau\, dS = - \int_{\Gamma_C} k|v_\tau|\, dS, \tag{10.28}$$

whence it follows that

$$\int_{\Gamma_C} T_\tau \cdot \Theta_\tau\, dS \geq - \int_{\Gamma_C} k|\Theta_\tau|\, dS \quad \text{for every} \quad \Theta \in V. \tag{10.29}$$

Note that only the second coordinate of $T_\tau$ and $v_\tau$ is nonzero on $\Gamma_C$, so by a slight abuse of notation we can consider these functions as scalars. If $\Theta_\tau$ is smooth on $\Gamma_C$, using the methods of Sect. 1.2 in [146] it is possible to construct $\Theta \in V$ such that its tangential part is equal to $\Theta_\tau$ on $\Gamma_C$. From density of smooth functions in $L^2(\Gamma_C)$ it follows that the following inequality analogous to (10.29)

$$\int_{\Gamma_C} T_\tau \Psi\, dS \geq - \int_{\Gamma_C} k|\Psi|\, dS$$

holds for all $\Psi \in L^2(\Gamma_C)$. From the above inequality it follows that the pointwise inequality $|T_\tau| \leq k$ must hold a.e. on $\Gamma_C$. Indeed, if the opposite inequality holds on a set $S$ of positive measure then we arrive at contradiction by taking $\Psi = 0$ outside $S$ and $\Psi = -\frac{T_\tau}{|T_\tau|}$ on $S$. Now, (10.28) implies that the Tresca condition (10.15) holds and the proof is complete. □

### 10.2.2 Existence and Uniqueness of a Global in Time Solution

In this subsection we establish the existence and uniqueness of a global in time solution for Problem 10.1. First, we present two lemmas.

**Lemma 10.1 (cf. [28]).** *There exists a smooth extension*

$$\xi(x_1, x_2) = U(x_2)e_1$$

*of $U_0 e_1$ from $\Gamma_C$ to $\Omega$ satisfying: (10.10),*

$$\left.\frac{\partial U(x_2)}{\partial x_2}\right|_{x_2=0} = 0,$$

*and such that*

$$|b(v\,,\,\xi,\,v)| \leq \frac{\nu}{4}\|v\|^2 \quad \text{for all} \quad v \in V.$$

*Moreover,*

$$|\xi|^2 + |\nabla\xi|^2 = \int_\Omega |U(x_2)|^2 dx_1 dx_2 + \int_\Omega |U_{,x_2}(x_2)|^2 dx_1 dx_2 \leq F,$$

*where $F$ is a constant depending on $\nu$, $\Omega$, and $U_0$.*

**Exercise 10.1.** Provide a proof of Lemma 10.1, cf. Lemma 9.2.

**Lemma 10.2.** *For all $v$ in $V$ we have the Ladyzhenskaya inequality*

$$\|v\|_{L^4(\Omega)} \leq C(\Omega)|v|^{\frac{1}{2}}\|v\|^{\frac{1}{2}}. \tag{10.30}$$

*Proof.* Let $v \in V$ and $r \in C^1([-L, L])$ such that $r = 1$ on $[0\,,\,L]$ and $r = 0$ at $x_1 = -L$. Define $\varphi = rv$ (keep in mind that $v$ can be periodically extended to whole $\Omega_\infty$), and extend $\varphi$ by 0 to $\Omega_1 = (-L\,,\,L) \times (0\,,\,h)$, where $h = \max_{0\leq x_1\leq L} h(x_1)$. We obtain

$$\varphi^2(x_1, x_2) = 2\int_{-L}^{x_1} \varphi(t_1, x_2)\frac{\partial\varphi}{\partial t_1}(t_1, x_2)\,dt_1 \leq 2\int_{-L}^{L} |\varphi(x_1, x_2)|\left|\frac{\partial\varphi}{\partial x_1}(x_1, x_2)\right| dx_1$$

and

$$\varphi^2(x_1, x_2) = -2\int_{x_2}^{h} \varphi(x_1, t_2)\frac{\partial\varphi}{\partial t_2}(x_1, t_2)\,dt_2 \leq 2\int_0^h |\varphi(x_1, x_2)|\left|\frac{\partial\varphi}{\partial x_2}(x_1, x_2)\right| dx_2,$$

whence

$$\begin{aligned}\|\varphi\|^4_{L^4(\Omega_1)} &= \int_{\Omega_1} \varphi^2(x_1,x_2)\varphi^2(x_1,x_2)\,dx_1dx_2\\ &\le \left(\int_0^h \sup_{-L\le x_1\le L} \varphi^2(x_1,x_2)\,dx_2\right)\left(\int_{-L}^L \sup_{0\le x_2\le h} \varphi^2(x_1,x_2)\,dx_1\right)\\ &\le 4\left(\int_0^h\int_{-L}^L |\varphi|\left|\frac{\partial\varphi}{\partial x_1}\right| dx_1dx_2\right)\times\left(\int_{-L}^L\int_0^h |\varphi|\left|\frac{\partial\varphi}{\partial x_2}\right| dx_2dx_1\right).\end{aligned}$$

By the Cauchy–Schwartz inequality,

$$\begin{aligned}\|\varphi\|^4_{L^4(\Omega_1)} &\le 4|\varphi|^2_{L^2(\Omega_1)}\left|\frac{\partial\varphi}{\partial x_1}\right|_{L^2(\Omega_1)}\left|\frac{\partial\varphi}{\partial x_2}\right|_{L^2(\Omega_1)}\\ &\le 2|\varphi|^2_{L^2(\Omega_1)}\left(\left|\frac{\partial\varphi}{\partial x_1}\right|^2_{L^2(\Omega_1)} + \left|\frac{\partial\varphi}{\partial x_2}\right|^2_{L^2(\Omega_1)}\right)\\ &\le 2|\varphi|^2_{L^2(\Omega_1)}|\nabla\varphi|^2_{L^2(\Omega_1)}.\end{aligned}$$

We use the fact that $|r|\le 1$ and the Poincaré inequality to get

$$\|v\|_{L^4(\Omega)} \le \|\varphi\|_{L^4(\Omega_1)},\quad |\varphi|_{L^2(\Omega_1)}\le 2|v|,\quad \text{and}\quad |\nabla\varphi|_{L^2(\Omega_1)}\le C\|v\|$$

for some constant $C$, whence (10.30) holds. □

**Theorem 10.4.** *For any $v_0\in H$ and $U_0\in\mathbb{R}$ there exists a solution of Problem 10.1.*

*Proof.* We provide only the main steps of the proof as it is quite standard and, on the other hand, long. The estimates we obtain will be used further in the section.

Observe that the functional $\Phi$ is convex, lower semicontinuous (in fact it is continuous) but nondifferentiable at zero. To overcome this difficulty we use the following approach (see, e.g., [91, 113, 188]). For $\delta>0$ let $\Phi_\delta : V\to\mathbb{R}$ be a functional defined by

$$V\ni w\mapsto \Phi_\delta(w) = \frac{1}{1+\delta}\int_{\Gamma_C} k|w_\tau|^{1+\delta}\,dx.$$

This functional is convex and lower semicontinuous on $V$. Moreover, for $v_\delta\to v$ weakly in $L^2(0,T;V)$,

$$\liminf_{\delta\to0+}\int_0^T \Phi_\delta(v_\delta(t))\,dt \ge \int_0^T \Phi(v(t))\,dt,$$

and

$$\lim_{\delta\to0+}\Phi_\delta(w) = \Phi(w)$$

for all $w \in V$. The functional $\Phi_\delta$ is Gâteaux differentiable in $V$, with

$$\langle \Phi_\delta'(v)\,,\,\Theta\rangle = \int_{\Gamma_C} k|v_\tau|^{\delta-1}\, v_\tau \cdot \Theta_\tau \quad \text{for all} \quad \Theta \in V.$$

Let us consider the following equation,

$$\left\langle \frac{dv_\delta(t)}{dt}, \Theta \right\rangle + \nu a(v_\delta(t), \Theta) + b(v_\delta(t), v_\delta(t), \Theta) + \langle \Phi_\delta'(v_\delta(t)), \Theta\rangle$$
$$= -\nu a(\xi, \Theta) - b(\xi, v_\delta(t), \Theta) - b(v_\delta(t), \xi, \Theta) \tag{10.31}$$

for any test function $\Theta \in V$, with the initial condition

$$v_\delta(0) = v_0. \tag{10.32}$$

For $\delta > 0$, we establish a priori estimates of $v_\delta$. Since $\langle \Phi_\delta'(v_\delta), v_\delta\rangle \geq 0$, $v_\delta \in V$, and $b(v_\delta, v_\delta, v_\delta) = b(\xi, v_\delta, v_\delta) = 0$, taking $\Theta = v_\delta(t)$ in (10.31) we get

$$\frac{1}{2}\frac{d}{dt}|v_\delta(t)|^2 + \nu\|v_\delta(t)\|^2 \leq -\nu a(\xi, v_\delta(t)) - b(v_\delta(t), \xi, v_\delta(t)).$$

In view of Lemma 10.1 we obtain

$$\frac{1}{2}\frac{d}{dt}|v_\delta(t)|^2 + \frac{\nu}{2}\|v_\delta(t)\|^2 \leq \nu\|\xi\|^2.$$

We estimate the right-hand side in terms of the data using Lemma 10.1 to get

$$\frac{1}{2}\frac{d}{dt}|v_\delta(t)|^2 + \frac{\nu}{2}\|v_\delta(t)\|^2 \leq F, \tag{10.33}$$

with $F = F(\nu, \Omega, U_0)$. From (10.33) we conclude that

$$|v_\delta(t)|^2 + \nu\int_0^t \|v_\delta(s)\|^2 ds \leq |v(0)|^2 + 2tF, \tag{10.34}$$

whence

$$v_\delta \text{ is bounded in } L^2(0, T; V) \cap L^\infty(0, T; H) \text{ independently of } \delta. \tag{10.35}$$

The existence of $v_\delta$ satisfying (10.31)–(10.32) is based on inequality (10.33), the Galerkin approximations, and the compactness method. Moreover, from (10.34) we can deduce that

$$\frac{dv_\delta}{dt} \quad \text{is bounded in} \quad L^2(0, T; V'). \tag{10.36}$$

From (10.35) and (10.36) we conclude that there exists $v$ such that (possibly for a subsequence)

$$v_\delta \to v \text{ weakly in } L^2(0,T;V), \qquad \frac{dv_\delta}{dt} \to \frac{dv}{dt} \text{ weakly in } L^2(0,T;V'). \tag{10.37}$$

In view of (10.37), $v \in C([0,T];H)$, and

$$v_\delta \to v \quad \text{strongly in} \quad L^2(0,T;H).$$

We can now pass to the limit $\delta \to 0$ in (10.31)–(10.32) as in [91] to obtain the variational inequality (10.18) for almost every $t \in (0,T)$. Thus the existence of a solution of Problem 10.1 is established. □

**Theorem 10.5.** *Under the hypotheses of Theorem 10.4, the solution $v$ of Problem 10.1 is unique and the map $v(\tau) \to v(t)$, for $t > \tau \geq 0$, is Lipschitz continuous in $H$.*

*Proof.* Let $v$ and $w$ be two solutions of Problem 10.1. Set $\Theta = w$ in the variational inequality for $v$, $\Theta = v$ in the variational inequality for $w$, and add the two thus obtained inequalities. The terms with the boundary functionals reduce and for $u(t) = w(t) - v(t)$ we obtain

$$\frac{1}{2}\frac{d}{dt}|u(t)|^2 + \nu\|u(t)\|^2 \leq b(u(t), w(t), u(t)) + b(u(t), \xi, u(t)).$$

By Lemma 10.1 and the Ladyzhenskaya inequality (10.30) we obtain

$$\frac{d}{dt}|u(t)|^2 + \frac{\nu}{2}\|u(t)\|^2 \leq \frac{2}{\nu}C(\Omega)^4\|w(t)\|^2|u(t)|^2, \tag{10.38}$$

and then in view of the Poincaré inequality we conclude

$$\frac{d}{dt}|u(t)|^2 + \frac{\sigma}{2}|u(t)|^2 \leq \frac{2}{\nu}C(\Omega)^4\|w(t)\|^2|u(t)|^2,$$

where $\sigma > 0$ is a constant. Using the Gronwall inequality, we obtain

$$|u(t)|^2 \leq |u(\tau)|^2 \exp\left\{-\int_\tau^t \left(\frac{\sigma}{2} - \frac{2}{\nu}C(\Omega)^4\|w(s)\|^2\right) ds\right\}. \tag{10.39}$$

From (10.37) it follows that the solution $w$ of Problem 10.1 belongs to $L^2(\tau,t;V)$. By (10.39) the map $v(\tau) \to v(t)$, $t > \tau \geq 0$, in $H$ is Lipschitz continuous, with

$$|w(t) - v(t)| \leq C|w(\tau) - v(\tau)| \tag{10.40}$$

uniformly for $t, \tau$ in a given interval $[0,T]$ and initial conditions $w(0), v(0)$ in a given bounded set $B$ in $H$.

In particular, as $u(0) = w(0) - v(0) = 0$, the solution $v$ of Problem 10.1 is unique. This ends the proof of Theorem 10.5. □

### 10.2.3 Existence of Finite Dimensional Global Attractor

In this subsection we prove the following theorem:

**Theorem 10.6.** *There exists a global attractor of a finite fractal dimension for the semigroup $\{S(t)\}_{t\geq 0}$ associated with Problem 10.1.*

*Proof.* From the considerations in the previous section it follows that to prove the theorem it suffices to check assumptions $(A1)$–$(A6)$, $(A8)$. For the convenience of the reader we repeat their statements in appropriate places.

*Assumption* $(A1)$. For any $v_0 \in X$ and arbitrary $T > 0$ there exists (not necessarily unique) $v \in C_w([0,T];X) \cap Y_T$, a solution of the evolutionary problem on $[0,T]$ with $v(0) = v_0$. Moreover, for any solution the estimates of $\|v\|_{Y_T}$ are uniform with respect to $\|v(0)\|_X$.
In our case, set $X = H$, $Y = V$, and $Y_T = \{u \in L^2(0,T;V) : u' \in L^2(0,T;V')\}$. From Theorems 10.4 and 10.5 we know that for any $v_0 \in H$ and arbitrary $T > 0$ there exists a unique $v \in C([0,T];H) \cap Y_T$, solution of Problem 10.1. We shall obtain the needed estimates directly from the variational inequality, cf. (10.18),

$$\langle v_t(t), \Theta - v(t)\rangle + \nu a(v(t), \Theta - v(t)) + b(v(t), v(t), \Theta - v(t)) \\ + \Phi(\Theta) - \Phi(v(t)) \geq \langle \mathscr{L}(v(t)), \Theta - v(t)\rangle. \qquad (10.41)$$

Set $\Theta = 0$ in (10.41) to get

$$\frac{1}{2}\frac{d}{dt}|v|^2 + \nu\|v\|^2 + \Phi(v) \leq \langle \mathscr{L}(v), v\rangle$$

as $\Phi(0) = 0$. Since, by Lemma 10.1,

$$\langle \mathscr{L}(v), v\rangle \leq \frac{\nu}{2}\|v\|^2 + \nu\|\xi\|^2$$

we obtain

$$\frac{d}{dt}|v|^2 + \nu\|v\|^2 + 2\Phi(v) \leq 2\nu\|\xi\|^2 = F. \qquad (10.42)$$

Integration in $t$ gives

$$|v(t)|^2 + \nu \int_0^t \|v(s)\|^2 ds \le |v(0)|^2 + 2tF, \tag{10.43}$$

and we deduce that

$$v \quad \text{is bounded in} \quad L^2(0,T;V) \cap L^\infty(0,T;H)$$

uniformly with respect to $|v(0)|$.
To get a uniform with respect to $|v(0)|$ estimate of $v'$ in $L^2(0,T;V')$ set $\Theta = v - \psi$, $\psi \in V$, in (10.41). We then have

$$\langle v', \psi\rangle \;\le\; \langle \mathscr{L}(v), \psi\rangle - \nu a(v,\psi) - b(v,v,\psi) + \Phi(v-\psi) - \Phi(v). \tag{10.44}$$

Using the Poincaré and the trace inequalities we obtain $\|\gamma(v)\|_{L^2(\partial\Omega)} \le C\|v\|$, and

$$\Phi(v-\psi) - \Phi(v) = k\int_{\Gamma_C} (|v_\tau - \psi_\tau| - |v_\tau|)\, dS \le k\int_{\Gamma_C} |\psi_\tau|\, dS \le C(\Gamma_C)\|\psi\|. \tag{10.45}$$

Moreover, using the Ladyzhenskaya inequality (10.30) to the nonlinear term, we have

$$\langle \mathscr{L}(v), \psi\rangle - \nu a(v,\psi) - b(v,v,\psi) \le C_1(\|v\| + |v|\,\|v\| + 1)\|\psi\|. \tag{10.46}$$

From (10.44)–(10.46) we obtain

$$\|v'\|_{V'} \le C_2(\|v\| + |v|\,\|v\| + 1),$$

and

$$\begin{aligned}\|v'\|^2_{L^2(0,T;V')} &= \int_0^T \|v'(t)\|^2_{V'}\, dt \\ &\le C_2\left(\int_0^T \|v(t)\|^2\, dt + \|v\|^2_{L^\infty(0,T;H)}\int_0^T \|v(t)\|^2\, dt + T\right) \le C(|v(0)|).\end{aligned}$$

Thus, $(A1)$ holds true.

*Assumption* $(A2)$. There exists a bounded set $B^0 \subset X$ with the following properties: if $v$ is an arbitrary solution with initial condition $v_0 \in X$ then

(i) there exists $t_0 = t_0(\|v_0\|_X)$ such that $v(t) \in B^0$ for all $t \ge t_0$,
(ii) if $v_0 \in B^0$ then $v(t) \in B^0$ for all $t \ge 0$.

From (10.42) and the Poincaré inequality,

$$\frac{d}{dt}|v|^2 + \nu\lambda_1 |v|^2 \le F,$$

and then by the Gronwall inequality (cf. Lemma 3.20)

$$|v(t)|^2 \le |v(0)|^2 e^{-\nu\lambda_1 t} + \frac{F}{\nu\lambda_1}.$$

Thus, there exists a bounded absorbing set (e.g., the ball $B_H(0,\rho)$ with $\rho^2 = 2\frac{F}{\nu\lambda_1}$) in $H$. Let $B_0$ be defined as $B_0 = \overline{\bigcup_{t\ge0} S(t)B_H(0,\rho)}^H$. If $v_0 \in B^0$ then there exist sequences $t_k$ and $v_k \to v_0$ in $H$ such that $v_k \in S(t_k)B_H(0,\rho)$. From the continuity of $S(t)$ it follows that $S(t+t_k)B_H(0,\rho) \ni S(t)v_k \to S(t)v_0$ in $H$ and hence $S(t)v_0 \in B^0$ for all $t \ge 0$. Thus, $(A2)$ holds true. Note that we need $B^0$ to be closed in $H$ to be able to satisfy assumption $(A5)$ below.

*Assumption* $(A3)$. Each $l$-trajectory has among all solutions a unique continuation. We recall that by the $l$-trajectory we mean any solution on the time interval $[0,l]$, and the unique continuation means that from an endpoint of an $l$-trajectory there starts at most one solution. In our case $(A3)$ is satisfied as the solutions are unique (Theorem 10.5).

*Assumption* $(A4)$. For all $t > 0$, $L_t : X_l \to X_l$ is continuous on $\mathscr{B}_l^0$.
$\mathscr{B}_l^0$ is defined as the set of all $l$-trajectories starting at any point of $B^0$ from $(A2)$, and $X_l = L^2(0,l;X)$. The semigroup $\{L_t\}_{t\ge0}$ acts on the set of $l$-trajectories as the shifts operators: $(L_t\chi)(\tau) = v(t+\tau)$ for $0 \le \tau \le l$, where $v$ is the unique solution on $[0, l+\tau]$ such that $v|_{[0,l]} = \chi$.
In our case, $X = H$, hence $X_l = H_l = L^2(0,l;H)$. In view of inequality (10.40) the map $S(t) : B^0 \to B^0$ is Lipschitz continuous for every $t > 0$, whence for any two $\chi_1, \chi_2$ in $\mathscr{B}_l^0$,

$$\int_0^l |L_t\chi_1(s) - L_t\chi_2(s)|^2 ds \le C^2(t) \int_0^l |\chi_1(s) - \chi_2(s)|^2 ds. \tag{10.47}$$

Thus, $(A4)$ holds true.

*Assumption* $(A5)$. For some $\tau > 0$, the closure in $X_l$ of the set $L_\tau(\mathscr{B}_l^0)$ is included in $\mathscr{B}_l^0$.
As $L_\tau(\mathscr{B}_l^0) \subset \mathscr{B}_l^0$, it is enough to check that the set $\mathscr{B}_l^0$ is closed in $X_l$. We have to prove that if $\{\chi_k\}$ is a sequence in $\mathscr{B}_l^0$ converging to some $\chi$ in $X_l$ then $\chi$ is also a trajectory and that $\chi(0) \in B^0$.

From assumption $(A1)$ it follows that the sequence $\{\chi_k\}$ is bounded in $Y_l$ and thus contains a subsequence (relabeled $\{\chi_k\}$) such that

$$\chi_k \to \chi \text{ weakly in } L^2(0,l;V) \text{ and } \frac{d\chi_k}{dt} \to \frac{d\chi}{dt} \text{ weakly in } L^2(0,l;V'). \tag{10.48}$$

Moreover, by the Aubin–Lions compactness theorem,

$$\chi_k \to \chi \quad \text{strongly in} \quad L^2(0,l;H). \tag{10.49}$$

We have

$$\langle \chi_k'(t), \Theta - \chi_k(t)\rangle \; + \; \nu a(\chi_k(t), \Theta - \chi_k(t)) \; + \; b(\chi_k(t), \chi_k(t), \Theta - \chi_k(t)) \\ + \; \Phi(\Theta) - \Phi(\chi_k(t)) \geq \langle \mathscr{L}(\chi_k(t)) \, , \, \Theta - \chi_k(t)\rangle.$$

We multiply both sides by a nonnegative smooth function $\eta = \eta(t)$ with support in the interval $(0, l)$ and integrate with respect to $t$ in this interval. We shall prove that taking $\liminf_{k\to\infty}$ of both sides and using (10.48) and (10.49) we obtain

$$\int_0^l \langle \chi'(t), \Theta - \chi(t)\rangle \eta(t)\, dt + \nu \int_0^l a(\chi(t), \Theta - \chi(t))\eta(t) dt \\ + \int_0^l b(\chi(t), \chi(t), \Theta - \chi(t))\eta(t)\, dt + \int_0^l \Phi(\Theta)\eta(t)\, dt - \int_0^l \Phi(\chi(t))\eta(t)\, dt \\ \geq \int_0^l \langle \mathscr{L}(\chi(t)), \Theta - \chi(t)\rangle \eta(t)\, dt. \tag{10.50}$$

First we shall show (without using the convexity argument) that

$$\lim_{n\to\infty} \int_0^l \Phi(\chi_k(t))\eta(t) dt = \int_0^l \Phi(\chi(t))\eta(t) dt. \tag{10.51}$$

Let $\epsilon \in \left(0, \frac{1}{2}\right)$. From the fact that $H^1(\Omega)$ embeds in $H^{1-\epsilon}(\Omega)$ compactly, the trace operator $\gamma_{1-\epsilon} : H^{1-\epsilon}(\Omega) \to L^2(\partial\Omega)$ is linear and continuous, and from the Ehrling lemma (cf. Lemma 3.17) for the spaces $H^1(\Omega), H^{1-\epsilon}(\Omega)$, and $L^2(\Omega)$ it follows that for every $\varepsilon > 0$ there exists $C_\varepsilon > 0$ such that for all $v \in H^1(\Omega)$,

$$\|\gamma(v)\|_{L^2(\partial\Omega)} \leq \varepsilon|\nabla v| + C_\varepsilon |v|.$$

We have thus

$$\|\gamma(\chi_k) - \gamma(\chi)\|^2_{L^2(\Gamma_C)} \leq \varepsilon\|\chi_k - \chi\|^2 + C'_\varepsilon|\chi_k - \chi|^2,$$

and

$$\int_0^l \|\gamma(\chi_k) - \gamma(\chi)\|^2_{L^2(\Gamma_C)}\, dt \leq \varepsilon \int_0^l \|\chi_k - \chi\|^2\, dt + C'_\varepsilon \int_0^l |\chi_k - \chi|^2\, dt.$$

As, in view of (10.48), there exists $M > 0$ such that for all $n$,

$$\int_0^l \|\chi_k(t) - \chi(t)\|^2\, dt \leq M,$$

we obtain, using (10.49),

$$\limsup_{k\to\infty} \int_0^l \|\gamma(\chi_k) - \gamma(\chi)\|^2_{L^2(\Gamma_C)}\, dt \le \varepsilon M.$$

Now, as $\varepsilon$ is any positive number, we obtain

$$\lim_{k\to\infty} \int_0^l \|\gamma(\chi_k) - \gamma(\chi)\|^2_{L^2(\Gamma_C)}\, dt = 0. \tag{10.52}$$

From (10.52), (10.51) easily follows.

We have also

$$\begin{aligned}\lim_{k\to\infty} \int_0^l \langle \chi_k', \chi_k\rangle \eta(t)\, dt &= \lim_{k\to\infty} \int_0^l \frac{1}{2}\frac{d}{dt}|\chi_k|^2 \eta(t)\, dt \\ &= -\lim_{k\to\infty} \int_0^l \frac{1}{2}|\chi_k|^2 \eta(t)'\, dt = -\int_0^l \frac{1}{2}|\chi|^2\eta(t)'\, dt \\ &= \int_0^l \langle \chi', \chi\rangle \eta(t)\, dt.\end{aligned}$$

In view of (10.48) and (10.49), there are no problems to get the other terms in (10.50) and finally, (10.50) itself. As inequality (10.50) is equivalent to the inequality

$$\begin{aligned}\langle \chi(t)', \Theta - \chi\rangle \,+\, \nu a(\chi(t), \Theta - \chi(t)) + b(\chi(t), \chi(t), \Theta - \chi(t)) \\ + \Phi(\Theta) - \Phi(\chi(t)) \ge \langle \mathscr{L}(\chi(t)), \Theta - \chi(t)\rangle,\end{aligned}$$

satisfied for almost all $t \in (0,l)$, $\chi$ is a solution with $\chi(0)$ in $H$. To end the proof we have to show that $\chi(0)$ belongs to the set $B^0$. We have $\chi_k(t) \in B^0$ for all $t \in [0,l]$ and, by (10.49), for a subsequence, $\chi_k(t) \to \chi(t)$ for almost all $t \in (0,l)$. As $B^0$ is closed, $\chi(t) \in B^0$ for almost all $t \in (0,l)$. Now, from the continuity of $\chi : [0,l] \to H$ and the closedness of $B^0$ it follows that $\chi(0)$ is in $B^0$. Thus, assumption (A5) holds.

*Assumption* (A6). There exists a space $W_l$ such that $W_l \subset X_l$ with compact embedding, and $\tau > 0$ such that $L_\tau : X_l \to W_l$ is Lipschitz continuous on $\mathscr{B}_l^1$—the closure in $X_l$ of $L_\tau(\mathscr{B}_l^0)$.

Define

$$W_l = \{u \in L^2(0,l;V) : \ u' \in L^1(0,l;U')\},$$

where $U = \{\psi \in V : \psi = 0 \text{ on } \Gamma_C\}$. We have, $W_l \subset X_l$, with compact embedding and we shall prove that $L_l : X_l \to W_l$ is Lipschitz continuous on $\mathscr{B}_l^1$.

Let $w$ and $v$ be two solutions of Problem 10.1 starting from $B^0$ and let $u = w - v$. Then, cf. (10.38),

$$\frac{d}{dt}|u(t)|^2 + \frac{\nu}{2}\|u(t)\|^2 \leq \frac{2}{\nu}C(\Omega)^4\|w(t)\|^2|u(t)|^2.$$

Take $s \in (0, l)$ and integrate this inequality over $\tau \in (s, 2l)$ to get

$$|u(2l)|^2 + \frac{\nu}{2}\int_s^{2l} \|u(\tau)\|^2 d\tau \leq \frac{2}{\nu}C(\Omega)^4 \int_s^{2l} \|u(\tau)\|^2|u(\tau)|^2\, d\tau + |u(s)|^2. \tag{10.53}$$

From (10.40) we conclude that

$$|u(\tau)|^2 \leq C_1|u(s)|^2$$

for $\tau \in (s, 2l)$ and from (10.43) we have

$$\int_s^{2l} \|u(\tau)\|^2\, d\tau \leq \int_0^{2l} (\|v(\tau)\|^2 + \|w(\tau)\|^2)\, d\tau \leq \frac{1}{\nu}(|v(0)|^2 + |w(0)|^2 + 8lF),$$

whence

$$\int_s^{2l} \|u(\tau)\|^2\, d\tau \leq C_2$$

uniformly for $w(s), v(s) \in B^0$. Therefore, from (10.53) we obtain

$$\int_l^{2l} \|u(\tau)\|^2\, d\tau \leq C_3|u(s)|^2.$$

Integrating over $s \in (0, l)$ we obtain

$$\int_l^{2l} \|u(\tau)\|^2\, d\tau \leq \frac{C_3}{l}\int_0^l |u(s)|^2\, ds,$$

and therefore

$$\|L_l\chi_1 - L_l\chi_2\|_{L^2(0,l;V)} \leq \sqrt{\frac{C_3}{l}}\|\chi_1 - \chi_2\|_{L^2(0,l;H)} \tag{10.54}$$

for any $\chi_1, \chi_2$ in $\mathscr{B}_l^0$.

In order to prove that

$$\|(L_l\chi_1 - L_l\chi_2)'\|_{L^1(0,l;U')} \leq C\|\chi_1 - \chi_2\|_{L^2(0,l;H)}$$

for some $C > 0$, it is sufficient, in view of (10.54), to prove that

$$\|(\chi_1 - \chi_2)'\|_{L^1(0,l;U')} \leq C'\|\chi_1 - \chi_2\|_{L^2(0,l;V)} \tag{10.55}$$

with some $C' > 0$.

We have

$$\langle v', \Theta - v\rangle + \nu a(v, \Theta - v) + b(v, v, \Theta - v) + \Phi(\Theta) - \Phi(v) \geq \langle \mathscr{L}(v), \Theta - v\rangle,$$

and

$$\langle w', \Theta - w\rangle + \nu a(w, \Theta - w) + b(w, w, \Theta - w) + \Phi(\Theta) - \Phi(w) \geq \langle \mathscr{L}(w), \Theta - w\rangle.$$

Setting $\Theta = v - \psi$ in the first inequality and $\Theta = w + \psi$ in the second one, where $\psi \in U$, $\|\psi\| \leq 1$, and adding the obtained two inequalities we get

$$\langle u', -\psi\rangle \leq \mathscr{N}(u, w, v; \psi), \tag{10.56}$$

where

$$\mathscr{N}(u, w, v; \psi) = b(\xi, u, \psi) + b(u, \xi, \psi) + \nu a(u, \psi) + b(w, u, \psi) + b(u, v, \psi).$$

Estimating the right-hand side of (10.56) we get

$$\langle u', -\psi\rangle \;\leq\; C''(1 + \|v\| + \|w\|)\|u\|\,\|\psi\|,$$

whence

$$\|u'(t)\|_{U'} = \sup\{\langle u'(t), -\psi\rangle \,:\, \|\psi\| \leq 1\} \;\leq\; C''(1 + \|v(t)\| + \|w(t)\|)\|u(t)\|.$$

Finally, integration over $t \in (0, l)$ gives

$$\int_0^l \|u'(t)\|_{U'}\, dt \leq C''' \left(\int_0^l \|u(t)\|^2\, dt\right)^{1/2}$$

with

$$C''' = C_0 \left(\int_0^l (1 + \|v(t)\|^2 + \|w(t)\|^2)\, dt\right)^{1/2} \quad \text{with a constant} \quad C_0 > 0,$$

uniformly for trajectories starting from $B^0$. This proves (10.55) and ends the proof of the Lipschitz continuity of the map $L_l : X_l \to W_l$. Assumption (*A*6) holds true.

*Assumption* (*A*8). The map $e : X_l \to X$ is Hölder-continuous on $\mathscr{B}_l^1$.

*Assumption* (*A*8) follows directly from the Lipschitz continuity of the map $e : X_l \to X$, $e(\chi) = \chi(l)$. To check the latter, let $w, v$ be two solutions as above, starting from $B^0$. From (10.40) we have, in particular,

$$|w(l) - v(l)| \leq C|w(\tau) - v(\tau)|$$

for $\tau \in (0,l)$. Integrating this inequality in $\tau$ on the interval $(0,l)$ we obtain

$$|w(l) - v(l)| \leq \frac{C}{\sqrt{l}} \|w - v\|_{L^2(0,l;H)}.$$

This ends the proof of Theorem 10.6. □

### 10.2.4 Existence of an Exponential Attractor

Now we can prove the main theorem of Sect. 10.2.

**Theorem 10.7.** *There exists an exponential attractor for the semigroup* $\{S(t)\}_{t\geq 0}$ *associated with Problem 10.1.*

*Proof.* In view of Theorem 10.3 and the considerations in the previous section it suffices to check conditions $(A9)$ and $(A10)$. The first one follows immediately from inequality (10.47). Thus it remains to prove that

$$\|L_{t_1}\chi - L_{t_2}\chi\|_{X_l} \leq c|t_1 - t_2|^{\beta} \tag{10.57}$$

holds for all $\tau > 0$, $0 \leq t_2 \leq t_1 \leq \tau$, $\chi \in \mathscr{B}_l^1$, some $c > 0$ and some $\beta \in (0,1]$, where, in our case, $X_l = H_l$.

To obtain (10.57) it suffices to know that $\chi'$, the time derivatives of $\chi \in \mathscr{B}_l^1$, are uniformly bounded in $L^q(0,l;H)$ for some $1 < q \leq \infty$, cf. [168]. In fact, we have then

$$\begin{aligned} |L_{t_1}\chi(s) - L_{t_2}\chi(s)| &= |u(t_1 + s) - u(t_2 + s)| \\ &= \left\| \int_{t_2+s}^{t_1+s} u'(\eta)\, d\eta \right\| \\ &\leq |t_1 - t_2|^{1-\frac{1}{q}} \|u'\|_{L^q(t_2+s,t_1+s;H)}. \end{aligned}$$

and integration with respect to $s$ in the interval $[0,l]$ gives (10.57) with $c$ depending on $\tau$ and $l$.

We shall prove that there exists $M > 0$ such that

$$\|\chi'\|_{L^\infty(0,l;H)} \leq M \quad \text{for all} \quad \chi \in \mathscr{B}_l^1.$$

The formal a priori estimates that follow can be performed on the smooth in the time variable Galerkin approximations $v_\delta^n(t)$, $n = 1,2,3,\ldots$, of the regularized problem (10.31)–(10.32), as in [148]. The obtained estimates are preserved by solutions $v_\delta(t)$ of problem (10.31)–(10.32), with bounds independent of $\delta$ when $\delta \to 0$, and in the end, by solutions $v(t)$ of Problem 10.1.

Let us consider the solution $v = v(t)$ of problem (10.31)–(10.32) (we drop the subscript $\delta$ for short),

$$\left\langle \frac{dv(t)}{dt}, \Theta \right\rangle + \nu a(v(t), \Theta) + b(v(t), v(t), \Theta) + \langle \Phi'_\delta(v(t)), \Theta \rangle$$
$$= -\nu a(\xi, \Theta) - b(\xi, v(t), \Theta) - b(v(t), \xi, \Theta), \qquad (10.58)$$

with the initial condition

$$v(0) = v_0.$$

Our aim is to derive, following the method used in [148], two a priori estimates following from (10.58). To get the first one, set $\Theta = v_t$ ($v_t = v'$) in (10.58). We obtain

$$|v_t|^2 + \frac{d}{dt}\frac{\nu}{2}\|v\|^2 + \frac{d}{dt}\Phi_\delta(v) = \langle \mathscr{L}(v), v_t \rangle - b(v, v, v_t). \qquad (10.59)$$

Using the Ladyzhenskaya inequality (10.30) we have

$$\langle \mathscr{L}(v), v_t \rangle \le c_1(\|v_t\| + \|v\||v_t|^{1/2}\|v_t\|^{1/2} + |v|^{1/2}\|v\|^{1/2}|v_t|^{1/2}\|v_t\|^{1/2}). \quad (10.60)$$

Now, we differentiate (10.58) with respect to the time variable and set $\Theta = v_t$, to get

$$\frac{1}{2}\frac{d}{dt}|v_t|^2 + \nu\|v_t\|^2 \le -b(v_t, \xi, v_t) - b(v_t, v, v_t), \qquad (10.61)$$

as $b(\xi, v_t, v_t) = 0$, $b(v, v_t, v_t) = 0$, and, cf. [91, 188],

$$\left\langle \frac{d}{dt}\Phi'_\delta(v), v_t \right\rangle \ge 0.$$

Using Lemma 10.1 and the Ladyzhenskaya inequality (10.30) to estimate the right-hand side of (10.61) we obtain

$$\frac{d}{dt}|v_t|^2 + \nu\|v_t\|^2 \le c_2\|v\|^2|v_t|^2. \qquad (10.62)$$

Now, we multiply (10.62) by $t^2$ to get

$$\frac{d}{dt}(t^2|v_t|^2) + t^2\nu\|v_t\|^2 \le c_2\|v\|^2(t^2|v_t|^2) + 2t|v_t|^2. \qquad (10.63)$$

To get rid of the last term on the right-hand side we add to (10.63) Eq. (10.59) multiplied by $2t$. After simple calculations and using (10.60) we obtain

$$\begin{aligned}\frac{d}{dt}&(t^2|v_t|^2 + t\nu\|v\|^2 + 2t\Phi_\delta(v)) + t^2\nu\|v_t\|^2\\ &\leq 2tc_1(\|v_t\| + \|v\||v_t|^{1/2}\|v_t\|^{1/2} + |v|^{1/2}\|v\|^{1/2}|v_t|^{1/2}\|v_t\|^{1/2})\\ &\qquad + c_2\|v\|^2(t^2|v_t|^2) + \nu\|v\|^2 + 2\Phi_\delta(v)\\ &\qquad + c_3|v|^{1/2}\|v\|^{3/2}(t|v_t|)^{1/2}(t\|v_t\|)^{1/2}.\end{aligned} \tag{10.64}$$

Define $y = t^2|v_t|^2 + t\nu\|v\|^2 + 2t\Phi_\delta(v)$. Using the Young inequality to the right-hand side of (10.64) and observing that $\Phi_\delta(v) \leq C(\|v\|^2 + 1)$ for $0 < \delta \leq 1$, we obtain the inequality of the form

$$\frac{d}{dt}y(t) + \frac{t^2\nu}{2}\|v_t\|^2 \leq C_1(t)y(t) + C_2(t),$$

where the coefficients $C_i(\cdot)$, $i = 1, 2$, are locally integrable and do not depend on $\delta$. They are also independent of the initial conditions for $v$ in a given bounded sets in $H$. This proves that the time derivative of solutions of the regularized problems is uniformly bounded with respect to $\delta$ in $L^\infty(\eta, T; H) \cap L^2(\eta, T; V)$ for all intervals $[\eta, T]$, $0 < \eta < T$. As a consequence, this property holds for all $\chi \in \mathscr{B}^1_l$. In view of the above considerations this ends the proof of the existence of an exponential attractor. □

## 10.3 Planar Shear Flows with Generalized Tresca Type Friction Law

The problem considered in this section is based on the results of Sect. 10.2, cf. also [162]. As we show, the arguments used there can be generalized to a class of problems with the friction coefficient dependent on the slip rate.

### *10.3.1 Classical Formulation of the Problem*

We will consider the planar flow of an incompressible viscous fluid governed by the equation of momentum

$$v_t - \nu\Delta v + (v \cdot \nabla)v + \nabla p = f \quad \text{in} \quad \Omega \times \mathbb{R}^+, \tag{10.65}$$

and the incompressibility condition

$$\operatorname{div} v = 0 \quad \text{in} \quad \Omega \times \mathbb{R}^+ \tag{10.66}$$

in domain $\Omega$ given by

$$\Omega = \{x = (x_1, x_2) : \ 0 < x_1 < L, \ 0 < x_2 < h(x_1)\},$$

with boundary $\partial\Omega = \overline{\Gamma_D} \cup \overline{\Gamma_C} \cup \overline{\Gamma_L}$, where $\Gamma_D$ is the top part of the boundary given by $\Gamma_D = \{(x_1, h(x_1)) : x_1 \in (0, L)\}$, $\Gamma_C$ is the bottom part of the boundary given by $\Gamma_C = (0, L) \times \{0\}$, and $\Gamma_L$ is the lateral one given by $\Gamma_L = \{0, L\} \times (0, h(0))$. The function $h$ is smooth and $L$-periodic such that $h(x_1) \geq \varepsilon > 0$ for all $x_1 \in \mathbb{R}$ with a constant $\varepsilon > 0$. We will use the notation $e_1 = (1, 0)$ and $e_2 = (0, 1)$ for the canonical basis of $\mathbb{R}^2$. Note that on $\Gamma_C$ the outer normal unit vector is given by $n = -e_2$.

The unknowns are the velocity field $v : \Omega \times \mathbb{R}^+ \to \mathbb{R}^2$ and the pressure field $p : \Omega \times \mathbb{R}^+ \to \mathbb{R}$, $\nu > 0$ is the viscosity coefficient and $f : \Omega \to \mathbb{R}^2$ is the mass force density. The stress tensor $T$ is related to the velocity and pressure through the linear constitutive law $T_{ij} = -p\delta_{ij} + \nu(v_{i,j} + v_{j,i})$.

We are looking for the solutions of (10.65)–(10.66) which are $L$-periodic in the sense that $v(0, x_2, t) = v(L, x_2, t)$, $\frac{\partial v_2(0,x_2,t)}{\partial x_1} = \frac{\partial v_2(L,x_2,t)}{\partial x_1}$, and $p(0, x_2, t) = p(L, x_2, t)$ for $x_2 \in [0, h(0)]$ and $t \in \mathbb{R}^+$. The first condition represents the $L$-periodicity of velocities, while the latter two ones, the $L$-periodicity of the Cauchy stress vector $Tn$ on $\Gamma_L$ in the space of divergence free functions. Moreover we assume that

$$v = 0 \quad \text{on} \quad \Gamma_D \times \mathbb{R}^+. \tag{10.67}$$

On the contact boundary $\Gamma_C$ we decompose the velocity into the normal component $v_n = v \cdot n$ and the tangential one $v_\tau = v \cdot e_1$. Note that since the domain $\Omega$ is two-dimensional it is possible to consider the tangential components as scalars, for the sake of the ease of notation. Likewise, we decompose the stress on the boundary $\Gamma_C$ into its normal component $T_n = Tn \cdot n$ and the tangential one $T_\tau = Tn \cdot e_1$.

We assume that there is no flux across $\Gamma_C$ and hence

$$v_n = 0 \quad \text{on} \quad \Gamma_C \times \mathbb{R}^+. \tag{10.68}$$

The boundary $\Gamma_C$ is assumed to be moving with the constant velocity $U_0 e_1 = (U_0, 0)$ which, together with the mass force, produces the driving force of the flow. The friction coefficient $k$ is assumed to be related to the slip rate through the relation $k = k(|v_\tau - U_0|)$, where $k : \mathbb{R}^+ \to \mathbb{R}^+$. If there is no slip between the fluid and the boundary then the friction force is bounded by the friction threshold $k(0)$

$$v_\tau = U_0 \Rightarrow |T_\tau| \leq k(0) \quad \text{on} \quad \Gamma_C \times \mathbb{R}^+, \tag{10.69}$$

while if there is a slip, then the friction force is given by the expression

$$v_\tau \neq U_0 \Rightarrow -T_\tau = k(|v_\tau - U_0|)\frac{v_\tau - U_0}{|v_\tau - U_0|} \quad \text{on} \quad \Gamma_C \times \mathbb{R}^+. \tag{10.70}$$

Note that (10.69)–(10.70) generalize the Tresca law considered in Sect. 10.2, where $k$ was assumed to be a constant. Here $k$ depends on the slip rate, this dependence represents the fact that the kinetic friction is less than the static one, which holds if $k$ is a decreasing function. Similar friction law is used, for example, in the study of the motion of tectonic plates, see [119, 190, 207] and the references therein. We make the following assumptions on the friction coefficient $k$:

($k1$) $k \in C(\mathbb{R}^+; \mathbb{R}^+)$,
($k2$) $k(s) \le \alpha(1 + s)$ for all $s \in \mathbb{R}^+$ with $\alpha > 0$,
($k3$) $k(s) - k(r) \ge -\mu(s - r)$ for all $s > r \ge 0$ with $\mu > 0$.

Note that the assumption that $k$ has values in $\mathbb{R}^+$ has a clear physical interpretation, namely it means that the friction force is dissipative.

Finally, the initial condition for the velocity field is

$$v(x, 0) = v_0(x) \quad \text{for} \quad x \in \Omega. \tag{10.71}$$

In the next section we present a weak formulation of the problem in the form of an evolutionary differential inclusion with a suitable potential corresponding to the generalized Tresca condition.

### 10.3.2 Weak Formulation of the Problem

To be able to define a weak solution of problem (10.65)–(10.71) we introduce some notation. Let

$$\tilde{V} = \{v \in C^\infty(\overline{\Omega})^2 : \ \text{div}\, v = 0 \ \text{in} \ \Omega, \quad v \text{ is } L\text{-periodic in } x_1, \\ v = 0 \ \text{on} \ \Gamma_D, \quad v_n = 0 \ \text{on} \ \Gamma_C\},$$

and

$$V = \text{ closure of } \tilde{V} \text{ in } H^1(\Omega)^2, \qquad H = \text{ closure of } \tilde{V} \text{ in } L^2(\Omega)^2.$$

We define scalar products in $H$ and $V$, respectively, by

$$(u\,,\, v) = \int_\Omega u(x)v(x)\,dx \quad \text{and} \quad (\nabla u\,,\, \nabla v) = \int_\Omega \nabla u(x) \colon \nabla v(x)\,dx,$$

and their associated norms by

$$|v| = (v, v)^{\frac{1}{2}} \quad \text{and} \quad \|v\| = (\nabla v\,,\,\nabla v)^{\frac{1}{2}}.$$

We denote the trace operator from $V$ to $L^2(\Gamma_C)^2$ by $\gamma$ and its norm by $\|\gamma\| = \|\gamma\|_{\mathscr{L}(V;L^2(\Gamma_C)^2)}$. Moreover, let, for $u$, $v$, and $w$ in $V$,

$$a(u\,,\,v) = (\nabla u\,,\,\nabla v) \quad \text{and} \quad b(u\,,\,v\,,\,w) = ((u\cdot\nabla)v\,,\,w).$$

Finally, we define the functional $j : \mathbb{R} \to \mathbb{R}$ corresponding to the generalized Tresca condition (10.69)–(10.70) by

$$j(v) = \int_0^{|v-U_0|} k(s)\,ds. \tag{10.72}$$

Note that the functional $j$ is not necessarily convex, but it is locally Lipschitz. By $\partial j$ we will denote its Clarke subdifferential (see Sect. 3.7). The variational formulation of problem (10.65)–(10.71) can be derived by the calculation analogous to the proof of Proposition 10.1.

**Problem 10.2.** Given $v_0 \in H$ and $f \in H$, find $v : \mathbb{R}^+ \to H$ such that:

(*i*) $v \in C(\mathbb{R}^+; H) \cap L^2_{\rm loc}(\mathbb{R}^+; V)$, with $v_t \in L^2_{\rm loc}(\mathbb{R}^+; V')$,

(*ii*) for all $\Theta$ in $V$ and for almost all $t \in \mathbb{R}^+$, the following equality holds

$$\langle v_t(t), \Theta\rangle + \nu a(v(t), \Theta) + b(v(t), v(t), \Theta) + (\xi(t), \Theta_\tau)_{L^2(\Gamma_C)} = (f, \Theta), \tag{10.73}$$

with $\xi(t) \in S^2_{\partial j(v_\tau(t))}$ for a.e. $t \in \mathbb{R}^+$, where for $w \in L^2(\Gamma_C)$ we denote by $S^2_{\partial j(w)}$ the set of all functions $\eta \in L^2(\Gamma_C)$ such that $\eta(x) \in \partial j(w(x))$ for a.e. $x \in \Gamma_C$,

(*iii*) the following initial condition holds:

$$v(0) = v_0. \tag{10.74}$$

The above definition is justified by the assertion (*i*) of the following lemma.

**Lemma 10.3.** *Under assumptions* (*k*1)–(*k*3) *the functional j defined by (10.72) satisfies the following properties:*

(*i*) *$j$ is locally Lipschitz and conditions (10.69)–(10.70) are equivalent to*

$$-\,T_\tau \in \partial j(v_\tau), \tag{10.75}$$

(*ii*) $|\xi| \le \tilde{\alpha}(1 + |w|)$ *for all* $w \in \mathbb{R}$ *and* $\xi \in \partial j(w)$, $\tilde{\alpha} > 0$, *(iii)* $m(\partial j) \geq -\mu$, *where* $m(\partial j)$ *is the one sided Lipschitz constant defined by*

$$m(\partial j) = \inf_{\substack{u,v\in\mathbb{R},u\neq v,\\ \xi\in\partial j(u),\eta\in\partial j(v)}} \frac{(\xi-\eta)\cdot(u-v)}{|u-v|^2},$$

*(iv) for all* $w \in \mathbb{R}$ *and* $\xi \in \partial j(w)$, $\xi \cdot w \geq -\beta(1+|w|)$, *with a constant* $\beta > 0$.

*Proof.* (*i*) The fact that $j$ is locally Lipschitz follows from the continuity of $k$ and the direct calculation. To see that (10.69)–(10.70) is equivalent to (10.75) observe that for $v \neq U_0$ the functional $j$ is continuously differentiable on a neighborhood of $v$ and its derivative is given as $j'(v) = k(|v-U_0|)\frac{v-U_0}{|v-U_0|}$, whence (10.70) is equivalent to (10.75) by the characterization of the Clarke subdifferential given in Theorem 3.19. The same characterization implies that (10.69) is equivalent to (10.75) if $v = U_0$.

Assertion (*ii*) follows in a straightforward way from ($k$2).

Assertion (*iii*) can be obtained by a following computation. Let $\xi \in \partial j(u)$, $\eta \in \partial j(v)$, where $u \neq U_0$ and $v \neq U_0$. Then

$$\begin{aligned}(\xi-\eta)\cdot(u-v) &= \left(k(|u-U_0|)\frac{u-U_0}{|u-U_0|} - k(|v-U_0|)\frac{v-U_0}{|v-U_0|}\right)\cdot(u-v)\\ &= k(|u-U_0|)|u-U_0| - k(|u-U_0|)\frac{(u-U_0)\cdot(v-U_0)}{|u-U_0|}\\ &\quad -k(|v-U_0|)\frac{(u-U_0)\cdot(v-U_0)}{|v-U_0|} + k(|v-U_0|)|v-U_0|\\ &\geq (k(|u-U_0|)-k(|v-U_0|))\,(|u-U_0|-|v-U_0|)\\ &\geq -\mu(|u-U_0|-|v-U_0|)^2 \geq -\mu|u-v|^2.\end{aligned}$$

The calculation in the case, when either $u = U_0$ or $v = U_0$ is similar.

Proof of assertion (*iv*) is also straightforward. Indeed, for $w = 0$ the assertion obviously holds. Let $w \neq 0$ and $\xi \in \partial j(w)$. We have, by ($k$1)–($k$2),

$$\begin{aligned}\xi\cdot w &= k(|w-U_0|)\frac{w-U_0}{|w-U_0|}(w-U_0+U_0)\\ &\geq k(|w-U_0|)|w-U_0| - k(|w-U_0|)|U_0| \geq -\alpha(1+|w|+|U_0|)|U_0|,\end{aligned}$$

and (*iv*) follows. The proof is complete. □

*Remark 10.1.* Observe that assertion (*iii*) of Lemma 10.3 is equivalent to the claim that the functional $j$ is semiconvex, i.e., the functional $s \to j(s) + \frac{\mu s^2}{2}$ is in fact convex (see Definition 10 in [18]).

### 10.3.3 Existence and Properties of Solutions

We begin with some estimates that are satisfied by the solutions of Problem 10.2. We define the auxiliary operators $A : V \to V'$ and $B : V \to V'$ by $\langle Au, v\rangle = a(u, v)$ and $\langle Bu, v\rangle = b(u, u, v)$.

**Lemma 10.4.** *Let $v$ be a solution of Problem 10.2. Then, for all $t \geq 0$,*

$$\max_{s\in[0,t]} |v(s)|^2 + \int_0^t \|v(s)\|^2\, ds \leq C(t, |v_0|), \tag{10.76}$$

$$\int_0^t \|v'(s)\|_{V'}^2\, ds \leq C(t, |v_0|), \tag{10.77}$$

*where $C(t, |v_0|)$ is a nondecreasing function of $t$ and $|v_0|$.*

*Proof.* Take $\Theta = v(t)$ in (10.73). Since, for $v \in V$ it is $b(v, v, v) = 0$, we get, with an arbitrary $\varepsilon > 0$,

$$\begin{aligned}\frac{1}{2}\frac{d}{dt}|v(t)|^2 + \nu\|v(t)\|^2 &\leq |f||v(t)| + \|\xi(t)\|_{L^2(\Gamma_C)}\|v_\tau(t)\|_{L^2(\Gamma_C)}\\ &\leq \varepsilon|v(t)|^2 + \frac{|f|^2}{4\varepsilon} + \frac{3}{2}\tilde{\alpha}\|v(t)\|^2_{L^2(\Gamma_C)} + \tilde{\alpha}m_1(\Gamma_C),\end{aligned}$$

where we used (ii) of Lemma 10.3 and $m_1(\Gamma_C) = L$ is the one-dimensional boundary measure of $\Gamma_C$.

Denoting $Z = V \cap H^{1-\delta}(\Omega)^2$ with some small $\delta \in \left(0, \frac{1}{2}\right)$ endowed with the norm $H^{1-\delta}(\Omega)^2$, observe that the embedding $V \subset Z$ is compact and the trace map $\gamma_C : Z \to L^2(\Gamma_C)^2$ is continuous. We denote $\|\gamma_C\| = \|\gamma_C\|_{\mathscr{L}(Z;L^2(\Gamma_C)^2)}$. From the Ehrling lemma (cf. Lemma 3.17) we get, for an arbitrary $\varepsilon > 0$ independent on $v \in V$ and $C(\varepsilon) > 0$ that

$$\|v\|_Z^2 \leq \varepsilon\|v\|^2 + C(\varepsilon)|v|^2.$$

Hence, noting that the Poincaré inequality

$$\lambda_1|v|^2 \leq \|v\|^2$$

is valid for $v \in V$, we get

$$\begin{aligned}\frac{d}{dt}|v(t)|^2 + 2\nu\|v(t)\|^2 &\leq \frac{2\varepsilon}{\lambda_1}\|v(t)\|^2 + \frac{|f|}{2\varepsilon}\\ &\quad + 3\tilde{\alpha}\|\gamma_C\|^2\varepsilon\|v(t)\|^2 + 3\tilde{\alpha}\|\gamma_C\|^2C(\varepsilon)|v(t)|^2 + 2\tilde{\alpha}L.\end{aligned}$$

It is clear that we can choose $\varepsilon$ small enough such that

$$\frac{d}{dt}|v(t)\|^2 + \nu\|v(t)\|^2 \leq C_1 + C_2|v(t)|^2,$$

with $C_1, C_2 > 0$. Using the Gronwall inequality (cf. Lemma 3.20) we get (10.76).

For the proof of (10.77) observe that for $u \in V$, $\|Bu\|_{V'} \leq C|u|\|u\|$. Indeed, by the Hölder inequality and the Ladyzhenskaya inequality,

$$\begin{aligned} |\langle Bu, z\rangle| &= |b(u,u,z)| = |b(u,z,u)| \\ &\leq \|u\|^2_{L^4(\Omega)^2}|\nabla z| \leq C\|u\||u|\|z\| \quad \text{for} \quad u,z \in V. \end{aligned} \tag{10.78}$$

Hence (10.77) follows by (10.76) and the straightforward computation that uses the assertion *(ii)* of Lemma 10.3 to estimate the multivalued term. □

We formulate and prove the theorem on existence of solutions to Problem 10.2.

**Theorem 10.8.** *For any $u_0 \in H$ Problem 10.2 has at least one solution.*

*Proof.* Since the proof of solution existence, based on the Galerkin method, is standard and quite long, we provide only its main steps.

Let $\varrho \in C_0^\infty((-1,1))$ be a mollifier such that $\int_{-1}^1 \varrho(s)\,ds = 1$ and $\varrho(s) \geq 0$. We define $\varrho_k : \mathbb{R} \to \mathbb{R}$ by $\varrho_k(x) = k\varrho(kx)$ for $k \in \mathbb{N}$ and $x \in \mathbb{R}$. Then supp $\varrho_k \subset \left(-\frac{1}{k}, \frac{1}{k}\right)$. We consider $j_k : \mathbb{R} \to \mathbb{R}$ defined by the convolution

$$j_k(r) = \int_{\mathbb{R}} \varrho_k(s) j(r-s)\,ds \quad \text{for} \quad r \in \mathbb{R}. \tag{10.79}$$

Note that $j_k \in C^\infty(\mathbb{R})$. From the assertion *(B)* of Theorem 3.21 it follows that for all $n \in \mathbb{N}$ and $r \in \mathbb{R}$ we have $|j_k'(r)| \leq \overline{\alpha}(1+|r|)$, where $\overline{\alpha}$ is different from $\tilde{\alpha}$ in *(ii)* of Lemma 10.3 above but independent of $k$.

Let us furthermore take the sequence $V_k$ of finite dimensional spaces such that $V_k = \text{span}\{z_1, \ldots, z_k\}$, where $\{z_i\}$ are orthonormal in $H$ eigenfunctions of the Stokes operator with the Dirichlet and periodic boundary conditions given in the definition of the space $V$. Then $\{V_k\}_{k=1}^\infty$ approximate $V$ from inside, i.e., $\overline{\bigcup_{k=1}^\infty V_k} = V$. We denote by $P_k : V' \to V_k$ the orthogonal projection on $V_k$ defined as $P_k u = \sum_{i=1}^k \langle u, z_i\rangle z_i$. Using Theorem 3.17 we conclude that $P_k v \to v$ strongly in $V'$ for any $v \in V'$.

We formulate the regularized Galerkin problems for $k \in \mathbb{N}$.

Find $v_k \in C(\mathbb{R}^+; V_k)$ such that for a.e. $t \in \mathbb{R}^+$, $v_k$ is differentiable and

$$\begin{aligned} \langle v_k'(t), \Theta\rangle + \nu a(v_k(t), \Theta) + b(v_k(t), v_k(t), \Theta) \\ + (j_k'(v_{n\tau}(t)(\cdot)), \Theta_\tau)_{L^2(\Gamma_C)} = (f, \Theta), \end{aligned} \tag{10.80}$$

$$v_k(0) = P_k v_0 \tag{10.81}$$

for a.e. $t \in \mathbb{R}^+$ and all $\Theta \in V_k$.

Note that due to the fact that $j_k', a, b$ are smooth, the solution of (10.80)–(10.81), if it exists, is a continuously differentiable function of time variable, with values in $V_k$.

We first show that if $v_k$ solves (10.80) then estimates analogous to the ones in Lemma 10.4 hold. Taking $v_k(t)$ in place of $\Theta$ in (10.80) and using the argument similar to that in the proof of (10.76) in Lemma 10.4 we obtain that, for a given $T > 0$,

$$\|v_k\|_{L^\infty(0,T;H)} \leq \text{const}, \tag{10.82}$$

$$\|v_k\|_{L^2(0,T;V)} \leq \text{const}. \tag{10.83}$$

Now, denoting by $\iota : V \to L^2(\Gamma_C)$ the map $\iota v = (\gamma v)_\tau$ and by $\iota^*$ the adjoint to $\iota$, we observe that (10.80) is equivalent to the following equation in $V'$,

$$v_k'(t) + \nu A v_k(t) + P_k B(v_k(t)) + P_k \iota^* j_k'(v_{k\tau}(t)(\cdot)) = P_k f. \tag{10.84}$$

Since, by Theorem 3.17, for $w \in V'$ we have $\|P_k w\|_{V'} \leq \|w\|_{V'}$, then, using the estimate $\|Bw\|_{V'} \leq C|w|\|w\|$ valid for $w \in V$, the growth condition on $j_k'$ as well as (10.82) and (10.83), from Eq. (10.84) we get

$$\|v_k'\|_{L^2(0,T;V')} \leq \text{const}. \tag{10.85}$$

The existence of the solution to the Galerkin problem (10.80) follows by the Carathéodory theorem and estimates (10.82)–(10.83).

Since the bounds (10.82), (10.83), and (10.85) are independent of $k$ then, for a subsequence, we get

$$v_k \to v \quad \text{weakly} - * \text{ in} \quad L^\infty(0,T;H),$$

$$v_k \to v \quad \text{weakly in} \quad L^2(0,T;V),$$

$$v_k' \to v' \quad \text{weakly in} \quad L^2(0,T;V'),$$

$$v_{k\tau} \to v_\tau \quad \text{strongly in} \quad L^2(0,T;L^2(\Gamma_C)), \tag{10.86}$$

$$v_k \to v \quad \text{strongly in} \quad L^2(0,T;H), \tag{10.87}$$

where (10.86) is a consequence of the Aubin–Lions compactness theorem used for the spaces $V \subset Z \subset V'$ and (10.87) is a consequence of the Aubin–Lions compactness theorem used for the spaces $V \subset H \subset V'$.

From (10.83) and the growth condition on $j_k'$ it follows that

$$\|j_k'(v_{n\tau})\|_{L^2(0,T;L^2(\Gamma_C))} \leq \text{const}.$$

Hence, perhaps for another subsequence,

$$j_k'(v_{k\tau}) \to \xi \quad \text{weakly in} \quad L^2(0,T;L^2(\Gamma_C)).$$

In a standard way (see, for example, Sects. 7.4.3 and 9.4 in [197]) we can pass to the limit in (10.84) and obtain that $v$ and $\xi$ satisfy (10.73) for a.e. $t \in (0, T)$. The assertion $(C)$ of Theorem 3.21 implies that $\xi(t) \in S^2_{\partial j(v_\tau(t))}$ for a.e. $t \in (0, T)$.

We shall show that $v(0) = v_0$ in $H$. We have, for $t \in (0, T)$,

$$v_k(t) = P_k v_0 + \int_0^t v_k'(s)\, ds, \quad v(t) = v(0) + \int_0^t v'(s)\, ds,$$

where the equalities hold in $V'$. Take $\phi \in V$. Then

$$\begin{aligned}\int_0^T &\langle v_k(t) - v(t), \phi\rangle\, dt \\ &= \int_0^T \langle P_k v_0 - v(0), \phi\rangle\, dt + \int_0^T \left\langle \int_0^t (v_k'(s) - v'(s))\, ds, \phi \right\rangle dt \\ &= T\langle P_k v_0 - v(0), \phi\rangle + \int_0^T \langle v_k'(s) - v'(s), (T-s)\phi\rangle\, ds.\end{aligned}$$

Since $v_k \to v$ weakly in $L^2(0, T; V)$ and $v_k' \to v'$ weakly in $L^2(0, T; V')$, it follows that $P_k v_0 \to v(0)$ weakly in $V'$. We know that $P_k v_0 \to v_0$ strongly in $V'$ and hence $v(0) = v_0$.

Finally, the solution can be extended from $[0, T]$ to the whole $\mathbb{R}^+$. Indeed, we can take the value of the solution at the time point $T$ as an initial condition and solve the resulting problem, obtaining another solution on $[0, T]$. Then, we concatenate the two solutions, obtaining the solution defined on the interval $[0, 2T]$. Repeating this procedure we get the solution on the whole $\mathbb{R}^+$ (see, for example, [93], Theorem 2 in Sect. 9.2.1). □

*Remark 10.2.* In the proof of Theorem 10.8 we used only assertions $(i)$ and $(ii)$ of Lemma 10.3 and therefore the theorem holds for a class of potentials $j$ that satisfy these two conditions and are not necessarily given by (10.72). This is a new result of independent interest since it weakens the conditions required for the existence of solutions provided, for example, in [174]. Indeed, the sign condition (see H(j)(iv) p. 583 in [174]) is not needed here for the existence proof.

Next lemma shows that assertion $(iii)$ of Lemma 10.3 gives the Lipschitz continuity of the solution map on bounded sets as well as the solution uniqueness.

**Lemma 10.5.** *The solution of Problem 10.2 is unique and if $v$, $w$ are two solutions of Problem 10.2 with the initial conditions $v_0, w_0$, respectively, then for any $t > s \geq 0$ we have*

$$|v(t) - w(t)| \leq D(t, |w_0|)|v(s) - w(s)|, \tag{10.88}$$

*where $D(t, |w_0|) > 0$ is a nondecreasing function of $t, |w_0|$.*

*Proof.* Let $v$ and $w$ be two solutions of Problem 10.2 with the initial conditions $v_0$ and $w_0$. We denote $u(t) = v(t) - w(t)$. Subtracting (10.73) for $v$ and $w$, respectively, and taking $u(t)$ as a test function we get, for a.e. $t \in \mathbb{R}^+$,

$$\frac{1}{2}\frac{d}{dt}|u(t)|^2 + \nu\|u(t)\|^2 + b(u(t), w(t), u(t)) + (\xi(t) - \eta(t), v(t)_\tau - w(t)_\tau)_{L^2(\Gamma_C)} = 0,$$

where $\xi(t) \in S^2_{\partial j(v_\tau(t))}$ and $\eta(t) \in S^2_{\partial j(w_\tau(t))}$.

Using estimate (10.78) to estimate the convective term and also assertion (*iii*) in Lemma 10.3 to estimate the multivalued one we get, for a.e. $t \in \mathbb{R}^+$,

$$\frac{1}{2}\frac{d}{dt}|u(t)|^2 + \nu\|u(t)\|^2 - \mu\|u_\tau(t)\|^2_{L^2(\Gamma_C)} \leq C|u(t)|\|u(t)\|\|w(t)\|. \tag{10.89}$$

As in the proof of Lemma 10.4 we use the fact that the embedding $V \subset Z$ is compact and the trace $\gamma_C : Z \to L^2(\Gamma_C)^2$ is continuous. We get, for a.e. $t \in \mathbb{R}^+$,

$$\begin{aligned} \frac{1}{2}\frac{d}{dt}|u(t)|^2 + \nu\|u(t)\|^2 &\leq \varepsilon\|u(t)\|^2 + C(\varepsilon)|u(t)|^2\|w(t)\|^2 \\ &\quad + \mu\|\gamma_C\|^2(\varepsilon\|u(t)\|^2 + C(\varepsilon)|u(t)|^2) \end{aligned} \tag{10.90}$$

with an arbitrary $\varepsilon > 0$ and the constant $C(\varepsilon) > 0$ independent on $u$, $w$, and $t$. By choosing $\varepsilon > 0$ small enough we get

$$\frac{d}{dt}|u(t)|^2 \leq |u(t)|^2(C_1\|w(t)\|^2 + C_2) \quad \text{for a.e.} \quad t \in \mathbb{R}^+,$$

with the constants $C_1, C_2 > 0$. The Gronwall inequality (cf. Lemma 3.20) gives

$$|u(t)|^2 \leq |u(s)|^2 e^{C_2(t-s)+C_1\int_s^t \|w(\tau)\|^2\,d\tau} \leq |u(s)|^2 e^{C_2 t+C_1\int_0^t \|w(\tau)\|^2\,d\tau}.$$

Using (10.76) we get

$$|v(t) - w(t)| \leq |v(s) - w(s)| e^{\frac{1}{2}C_2 t+\frac{1}{2}C_1 C(t,|w_0|)}, \tag{10.91}$$

and the proof of (10.88) is complete. Taking $s = 0$ and $v(0) = w(0)$ in (10.88) we obtain the uniqueness. □

Now we shall prove that assertion (*iv*) of Lemma 10.3 gives additional, dissipative, a priori estimates.

**Lemma 10.6.** *If $v$ is a solution of Problem 10.2 with the initial condition $v_0$, then,*

$$|v(t)| \leq |v_0|e^{-C_1 t} + C_2 \quad \textit{for all} \quad t \geq 0,$$

*with positive constants $C_1, C_2$ independent of $t$, $v_0$.*

*Proof.* We proceed as in the proof of (10.76) in Lemma 10.3. Taking the test function $\Theta = v(t)$ in (10.73) we get

$$\frac{1}{2}\frac{d}{dt}|v(t)|^2 + \nu\|v(t)\|^2 + (\xi(t), v_\tau(t))_{L^2(\Gamma_C)} \leq |f||v(t)|$$

for a.e. $t \in \mathbb{R}^+$. We use assertion $(iv)$ of Lemma 10.3 and the Poincaré inequality $\lambda_1|v|^2 \leq \|v\|^2$ to get

$$\frac{1}{2}\frac{d}{dt}|v(t)|^2 + \nu\|v(t)\|^2 - \beta L\left(1 + \frac{1}{4\varepsilon_1}\right) - \beta\varepsilon_1\|v_\tau(t)\|^2_{L^2(\Gamma_C)} \leq \frac{|f|^2}{4\varepsilon_2} + \frac{\varepsilon_2}{\lambda_1}\|v(t)\|^2$$

for a.e. $t \in \mathbb{R}^+$, with an arbitrary $\varepsilon_1 > 0$ and $\varepsilon_2 > 0$. Using the trace inequality and the Poincaré inequality again, we can choose $\varepsilon_1$ and $\varepsilon_2$ small enough to get

$$\frac{d}{dt}|v(t)|^2 + \nu\lambda_1|v(t)|^2 \leq C$$

for a.e. $t \in \mathbb{R}^+$, where $C > 0$ is a constant. Finally, the assertion follows from the Gronwall inequality (cf. Lemma 3.20). □

### *10.3.4 Existence of Finite Dimensional Global Attractor*

In this subsection we prove the following theorem.

**Theorem 10.9.** *The semigroup $\{S(t)\}_{t\geq 0}$ associated with Problem 10.2 has a global attractor $\mathscr{A} \subset H$ of finite fractal dimension in the space $H$.*

*Proof.* In view of Theorem 10.2 we need to verify assumptions $(A1)$–$(A8)$. We set $Y = V, X = H, Z = V', Y_T = \{u \in L^2(0,T;V) \ : \ u' \in L^2(0,T;V')\}$ for $T > 0$, and $l > 0, X_l = L^2(0,l;H)$.

*Assumption* $(A1)$. From Theorem 10.8 it follows that for any $u_0 \in H$ there exists at least one solution of Problem 10.2. This solution restricted to the interval $(0,T)$ belongs to $Y_T$ and $C([0,T];H)$. Uniform bounds in $Y_T$ follow directly from Lemma 10.4.

*Assumption* $(A2)$. In view of Lemma 10.6 the closed ball $B_H(0, C_2 + 1)$ absorbs in finite time all bounded sets in $H$. Let $B^0 = \overline{\bigcup_{t\geq 0} S(t)B_H(0, C_2 + 1)}^H$. Obviously this set is closed in $H$ (we will need this fact to verify the assumption $(A5)$ below). From Lemma 10.6 it follows that $B^0$ is bounded. Let us assume that $u \in B^0$ and $v = S(t)u$. Then, for a sequence $u_k \in S(t_k)B_H(0, C_2 + 1)$ with certain $t_k$ we have

$u_k \to u$ in $H$. From Lemma 10.5 it follows that $S(t)u_k \to S(t)u = v$ in $H$. But $S(t)u_k \in S(t+t_k)B_H(0, C_2+1) \subset B^0$. Hence, from the closedness of $B^0$ it follows that $v \in B^0$ and the assertion is proved.

In the sequel of the proof the generic constant depending on $B^0$, $l$ and the problem data will be denoted by $C_{l,B^0}$.

*Assumption* $(A3)$. Recall that the $l$-trajectory is a restriction of any solution of Problem 10.2 to the time interval $[0,l]$. The fact that assumption $(A3)$ holds follows immediately from Lemma 10.5.

*Assumption* $(A4)$. Define the shift operator $L_t : X_l \to X_l$ by $(L_t\chi)(s) = u(s+t)$ for $s \in [0,l]$, where $u$ is the unique solution on $[0, l+t]$ such that $u|_{[0,l]} = \chi$. We will prove that this operator is continuous on $\mathscr{B}_l^0$ for all $t \geq 0$ and $l > 0$, where $\mathscr{B}_l^0$ is the set of all $l$-trajectories with the initial condition in the set $B^0$ from assumption $(A2)$. Let $u, v$ be two solutions with the initial conditions $u_0, v_0 \in B^0$. By Lemma 10.5 we have

$$\int_0^l |v(s+t) - u(s+t)|^2\, ds \leq D^2(t+l, |B^0|) \int_0^l |v(s) - u(s)|^2\, ds,$$

where $|B^0| = \sup_{v \in B_0} |v|$. Hence, denoting $\chi_1 = u|_{[0,l]}$ and $\chi_2 = v|_{[0,l]}$, we get

$$\|L_t\chi_1 - L_t\chi_2\|_{X_l}^2 \leq D^2(t+l, |B^0|)\|\chi_1 - \chi_2\|_{X_l}^2, \tag{10.92}$$

and the proof of $(A4)$ is complete.

*Assumption* $(A5)$. We need to prove that, for some $\tau > 0$, $\overline{L_\tau(\mathscr{B}_l^0)}^{X_l} \subset \mathscr{B}_l^0$. Let $u_0 \in B^0$. From the fact that for all $\tau > 0$, $S(\tau)u_0 \in B^0$, it follows that $L_\tau(\mathscr{B}_l^0) \subset \mathscr{B}_l^0$. To obtain the assertion we must show that $\mathscr{B}_l^0$ is closed in $X_l$. To this end assume that $u_k$ is a sequence of solutions to Problem 10.2 such that $u_k(0) \in B^0$ and that

$$\chi_k \to \chi \text{ strongly in } L^2(0,l;H), \tag{10.93}$$

where $\chi_k = u_k|_{[0,l]}$. From $(A1)$ it follows that

$$\chi_k \to \chi \quad \text{weakly in} \quad L^2(0,l;V), \tag{10.94}$$

$$\chi_k' \to \chi' \quad \text{weakly in} \quad L^2(0,l;V'). \tag{10.95}$$

We shall show that $\chi \in \mathscr{B}_l^0$. Let us first prove that $\chi(0) \in B^0$. Indeed, from (10.93) it follows that, for a subsequence, $\chi_k(t) \to \chi(t)$ strongly in $H$ for a.e. $t \in (0,l)$. Since $\chi_k(t) \in B^0$ for all $k \in \mathbb{N}$ and all $t \in [0,l]$, then $\chi(t)$ belongs to the closed set $B^0$ for a.e. $t \in (0,l)$. The closedness of $B^0$ and the fact that $\chi \in C([0,l];H)$ imply that $\chi(t) \in B^0$ for all $t \in [0,l]$ and, in particular, $\chi(0) \in B^0$. We need to prove that $\chi$ satisfies (10.73) for a.e. $t \in (0,l)$. It is enough to show that for all $w \in L^2(0,l;V')$,

$$\int_0^l \langle \chi'(t), w(t)\rangle \, dt + \nu \int_0^l \langle A\chi(t), w(t)\rangle \, dt$$
$$+ \int_0^l \langle B\chi(t), w(t)\rangle \, dt + \int_0^l (\xi(t), w_\tau(t))_{L^2(\Gamma_C)} \, dt = \int_0^l (f, w(t)) \, dt \qquad (10.96)$$

with $\xi(t) \in S^2_{\partial j(\chi_\tau(t))}$ for a.e. $t \in (0, l)$. From the fact that $\chi_k$ satisfies (10.73) for a.e. $t \in (0, l)$ we have

$$\int_0^l \langle \chi_k'(t), w(t)\rangle \, dt + \nu \int_0^l \langle A\chi_k(t), w(t)\rangle \, dt$$
$$+ \int_0^l \langle B\chi_k(t), w(t)\rangle \, dt + \int_0^l (\xi_k(t), w_\tau(t))_{L^2(\Gamma_C)} \, dt = \int_0^l (f, w(t)) \, dt, \qquad (10.97)$$

with $\xi_k(t) \in S^2_{\partial j(\chi_{k\tau}(t))}$ for a.e. $t \in (0, l)$. Then, from the growth condition *(ii)* in Lemma 10.3 it follows that $\|\xi_k\|_{L^2(0,l;L^2(\Gamma_C))} \le \sqrt{2lL}\tilde{\alpha} + \sqrt{2}\tilde{\alpha}\|\chi_{k\tau}\|_{L^2(0,l;L^2(\Gamma_C))}$. Since $\chi_k$ is bounded in $L^2(0, l; V)$, then $\chi_{k\tau}$ is bounded in $L^2(0, l; L^2(\Gamma_C))$ and hence $\xi_k$ is bounded in the same space. Therefore, for a subsequence,

$$\xi_k \to \xi \quad \text{weakly in} \quad L^2(0, l; L^2(\Gamma_C)). \qquad (10.98)$$

Now (10.93), (10.94), and (10.98) are sufficient to pass to the limit in all the terms in (10.97) and thus to prove (10.96). It remains to show that $\xi(t) \in S^2_{\partial j(\chi_{n\tau}(t))}$ for a.e. $t \in (0, l)$. Let $Z = V \cap H^{1-\delta}(\Omega)^2$ be equipped with $H^{1-\delta}$ topology, where $\delta \in \left(0, \frac{1}{2}\right)$. Then for the triple $V \subset Z \subset V'$ we can use the Aubin–Lions compactness theorem and conclude from (10.93) and (10.94) that $\chi_k \to \chi$ strongly in $L^2(0, l; Z)$. Now, since the trace operator $\gamma_C : Z \to L^2(\Gamma_C)^2$ is linear and continuous, and so is the corresponding Nemytskii operator,

$$\chi_{k\tau} \to \chi_\tau \quad \text{strongly in} \quad L^2(0, l; L^2(\Gamma_C)). \qquad (10.99)$$

We use assertion $(H2)$ in Theorem 3.24 to deduce that $\xi(x, t) \in \partial j(\chi_\tau(x, t))$ a.e. in $\Gamma_C \times (0, l)$. This completes the proof of $(A5)$.

*Assumption* $(A6)$. We define $W = H_0^1(\Omega)^2 \cap V$, where we equip $W$ with the norm of $V$. Then $V' \subset W'$. By the Aubin–Lions compactness theorem the space

$$W_l = \{\chi \in L^2(0, l; V) : \ \chi' \in L^1(0, T; W')\}$$

is embedded compactly in $X_l$. We must show that the shift operator $L_\tau : X_l \to W_l$ is Lipschitz continuous on $\mathscr{B}_l^1 = \overline{L_\tau(\mathscr{B}_l^0)}^{X_l}$ for some $\tau > 0$. In fact, we shall prove that $L_\tau$ is Lipschitz continuous on the larger set $\mathscr{B}_l^0$ for $\tau = l$. Indeed, let $w, v$ be two solutions of Problem 10.2 starting from $B^0$. Denote $u = v - w$. Then, by (10.90),

$$\frac{d}{dt}|u(t)|^2 + \nu\|u(t)\|^2 \leq C|u(t)|^2(1 + \|w(t)\|^2) \quad \text{for a.e.} \quad t \in \mathbb{R}^+,$$

where the constant $C$ depends only on the problem data. We fix $s \in (0, l)$ and integrate the last inequality over interval $(s, 2l)$ to obtain

$$|u(2l)|^2 + \nu \int_s^{2l} \|u(t)\|^2 \, dt \leq C \int_s^{2l} |u(t)|^2(1 + \|w(t)\|^2) \, dt + |u(s)|^2.$$

Using (10.91) we get

$$\nu \int_l^{2l} \|u(t)\|^2 \leq |u(s)|^2 \left( 2C_{l,B^0} l + C_{l,B^0} \int_0^{2l} \|w(t)\|^2 \, dt + 1 \right).$$

Since, by $(A1)$,

$$\int_0^{2l} \|w(t)\|^2 \, dt \leq C_{l,B^0},$$

it follows that

$$\int_l^{2l} \|u(t)\|^2 \, dt \leq C_{l,B^0} |u(s)|^2.$$

Integrating this inequality over interval $(0, l)$ with respect to the variable $s$ it follows that

$$l \int_l^{2l} \|u(t)\|^2 \, dt \leq C_{l,B^0} \int_0^l |u(s)|^2 \, ds,$$

and therefore

$$\|L_l v - L_l w\|_{L^2(0,l;V)} \leq \sqrt{\frac{C_{l,B^0}}{l}} \|v - w\|_{X_l}. \tag{10.100}$$

It remains to prove that

$$\|(L_l v - L_l w)'\|_{L^2(0,l;W')} \leq C_{l,B^0} \|v - w\|_{X_l}.$$

In view of (10.100) it is sufficient to prove that

$$\|(v - w)'\|_{L^2(0,l;W')} \leq C_{l,B^0} \|v - w\|_{L^2(0,l;V)}.$$

To this end, we subtract (10.73) for $w$ and $v$, respectively, and take $\Theta \in W$ as a test function. We get

$$\langle u'(t), \Theta\rangle + \nu\langle Au(t), \Theta\rangle + \langle Bv(t) - Bw(t), \Theta\rangle = 0.$$

We have, for a constant $k > 0$ dependent only on the problem data,

$$\begin{aligned}\langle Bv(t) - Bw(t), \Theta\rangle &= b(u(t), w(t), \Theta) + b(w(t), u(t), \Theta) + b(u(t), u(t), \Theta)\\ &\le k(\|u(t)\|\,\|w(t)\| + \|u(t)\|^2)\|\Theta\|\\ &\le k(2\|w(t)\| + \|v(t)\|)\|u(t)\|\,\|\Theta\|,\end{aligned}$$

whence, for a constant $C > 0$ dependent only on the problem data,

$$\begin{aligned}\|u'(t)\|_{W'} &= \sup_{\Theta\in W, \|\Theta\|=1} |\nu\langle Au(t), \Theta\rangle + \langle Bv(t) - Bw(t), \Theta\rangle|\\ &\le \nu\|Au(t)\|_{V'} + \|Bv(t) - Bw(t)\|_{V'}\\ &\le C(1 + \|w(t)\| + \|v(t)\|)\|u(t)\|.\end{aligned}$$

It follows that

$$\|u'\|_{L^1(0,l;W')} \le C\|u\|_{L^2(0,l;V)}\sqrt{\int_0^l (1 + \|w(t)\| + \|v(t)\|)^2\,dt}.$$

The assertion follows from $(A1)$.

*Assumptions* $(A7)$ *and* $(A8)$. Since $(A8)$ implies $(A7)$ we prove only $(A8)$. We shall show that the mapping $e : X_l \to X$ defined as $e(\chi) = \chi(l)$ is Lipschitz on $\mathscr{B}_l^0$. Indeed, by (10.91), for any two solutions $v, w$ of Problem 10.2 such that their initial conditions belong to $B^0$, we get

$$|w(l) - v(l)| \le C_{l,B^0}|w(s) - v(s)| \quad \text{for all} \quad s \in (0, l).$$

Integrating the square of this inequality over $s \in (0, l)$ we obtain

$$l|w(l) - v(l)|^2 \le C_{l,B^0}^2\|w - v\|_{X_l}^2,$$

whence

$$|w(l) - v(l)| \le \frac{C_{l,B^0}}{\sqrt{l}}\|w - v\|_{X_l},$$

and $(A8)$ holds.

This completes the proof Theorem 10.9 about the existence of the global attractor of a finite fractal dimension. □

In the following subsection we shall prove the existence of an exponential attractor.

### 10.3.5 Existence of an Exponential Attractor

Our goal is to prove the following:

**Theorem 10.10.** *The semigroup $\{S(t)\}_{t\geq 0}$ associated with Problem 10.2 has an exponential attractor in the space H.*

*Proof.* Since we have shown, in the proof of Theorem 10.9, that assumptions $(A1)$–$(A8)$ hold, in view of Theorem 10.3 it is enough to prove $(A9)$ and $(A10)$.

*Assumption* $(A9)$. Lipschitz continuity of $L_t : X_l \to X_l$, uniform with respect to $t \in [0, \tau]$ for all $\tau > 0$, follows immediately from inequality (10.92).

*Assumption* $(A10)$. It remains to prove that the inequality

$$\|L_{t_1}\chi - L_{t_2}\chi\|_{X_l} \leq c|t_1 - t_2|^{\delta}$$

holds for all $t_0 > 0$, $0 \leq t_2 \leq t_1 \leq t_0$ and $\chi \in \mathscr{B}_l^1$ with $\delta \in (0, 1]$ and $c > 0$.

Since $\mathscr{B}_l^1 = \overline{L_\tau(\mathscr{B}_l^0)}^{X_l}$ for certain $\tau > 0$, it is enough to prove the assertion for $\chi \in L_\tau(\mathscr{B}_l^0)$. Let $v$ be the unique solution of Problem 10.2 on $[0, \tau + l + t_0]$ with $v(0) \in B^0$ such that $v|_{[\tau,\tau+l]} = \chi$. It is enough to obtain the uniform bound on $v'$ in the space $L^\infty(\tau, \tau + l + t_0; H)$. Indeed,

$$\begin{aligned}\|L_{t_1}\chi - L_{t_2}\chi\|_{X_l} &= \sqrt{\int_0^l |v(\tau + s + t_2) - v(\tau + s + t_1)|^2\, ds} \\ &\leq \sqrt{\int_0^l \left(\int_{\tau+s+t_1}^{\tau+s+t_2} |v'(r)|\, dr\right)^2 ds} \\ &\leq |t_1 - t_2|\sqrt{l}\|v'\|_{L^\infty(\tau,\tau+l+t_0;H)}.\end{aligned}$$

The a priori estimate will be computed for the smooth solutions of the regularized Galerkin problem (10.80)–(10.81) formulated in the proof of Theorem 10.8. We shall show that, for the Galerkin approximations $v_k$, for any $T > \tau$, the functions $v_k'$ are bounded in $L^\infty(\tau, T; H)$ independently of $k$ and, in consequence, the desired estimate will be preserved by their limit $v$, a solution of Problem 10.2.

The argument follows the lines of the proof of Theorem 10.7 (compare [148]). First we take $\Theta = v_k'(t)$ in (10.80) which gives us

$$|v_k'(t)|^2 + \nu \frac{1}{2}\frac{d}{dt}\|v_k(t)\|^2 + b(v_k(t), v_k(t), v_k'(t)) \\ + \frac{d}{dt}\int_{\Gamma_C} j_k(v_{k\tau}(x,t))\, dS = (f, v_k'(t)).$$

Estimating the convective term by using the Ladyzhenskaya inequality (10.30) we get

$$|v_k'(t)|^2 + \nu \frac{d}{dt}\|v_k(t)\|^2 + 2\frac{d}{dt}\int_{\Gamma_C} j_k(v_{k\tau}(x,t))\, dS \\ \leq |f|^2 + C|v_k(t)|^{\frac{1}{2}}\|v_k(t)\|^{\frac{3}{2}}|v_k'(t)|^{\frac{1}{2}}\|v_k'(t)\|^{\frac{1}{2}}, \tag{10.101}$$

where $C > 0$ is a constant.

Now, a straightforward and technical calculation that uses the Fatou lemma, the Lebourg mean value theorem (cf. Theorem 3.20) and the conditions (*ii*) and (*iii*) in Lemma 10.3, shows, that for all $r_1, r_2 \in \mathbb{R}$,

$$(j_k'(r_2) - j_k'(r_1))(r_2 - r_1) \geq -\mu|r_2 - r_2|^2,$$

where the constant $\mu$ is the same as in (*iii*) of Lemma 10.3. Hence, since $v_k$ is a continuously differentiable function of time variable,

$$\left(\frac{d}{dt}j_k'(v_{k\tau}(x,t))\right) v_{k\tau}'(x,t) \geq -\mu|v_{k\tau}'(x,t)|^2 \quad \text{for all} \quad (x,t) \in \Omega \times (0,T). \tag{10.102}$$

We differentiate both sides of (10.80) with respect to time and take $\Theta = v_k'(t)$, which gives us

$$\frac{1}{2}\frac{d}{dt}|v_k'(t)|^2 + \nu\|v_k'(t)\|^2 + \int_{\Gamma_C}\left(\frac{d}{dt}j_k'(v_{k\tau}(x,t))\right) v_{k\tau}'(x,t)\, dS \\ + b(v_k'(t), v_k(t), v_k'(t)) + b(v_k(t), v_k'(t), v_k'(t)) = 0.$$

Using (10.102) and the fact that $b(v_k(t), v_k'(t), v_k'(t)) = 0$ we get

$$\frac{1}{2}\frac{d}{dt}|v_k'(t)|^2 + \nu\|v_k'(t)\|^2 - \mu\|v_{k\tau}'(t)\|^2_{L^2(\Gamma_C)} + b(v_k'(t), v_k(t), v_k'(t)) \leq 0.$$

Proceeding exactly as in the proof of (10.89) we get, for an arbitrary $\varepsilon > 0$ and a constant $C(\varepsilon) > 0$,

$$\frac{1}{2}\frac{d}{dt}|v_k'(t)|^2 + \nu\|v_k'(t)\|^2 \leq C(\varepsilon)\|v_k(t)\|^2|v_k'(t)|^2 \\ + \varepsilon\|v_k'(t)\|^2 + C(\varepsilon)|v_k'(t)|^2,$$

whence

$$\frac{d}{dt}|v_k'(t)|^2 + \nu\|v_k'(t)\|^2 \leq C|v_k'(t)|^2(\|v_k(t)\|^2 + 1).$$

Multiplying this inequality by $t^2$ we get

$$\frac{d}{dt}(t^2|v_k'(t)|^2) + \nu t^2\|v_k'(t)\|^2 \leq Ct^2|v_k'(t)|^2(\|v_k(t)\|^2 + 1) + 2t|v_k'(t)|^2.$$

We add this inequality to inequality (10.101) multiplied by $2t$ to obtain, after simple calculations that use the fact that $\int_{\Gamma_C} j_k(v_{k\tau}(x,t))\,dS \leq C(1 + \|v_k(t)\|^2)$,

$$\begin{aligned}
&\frac{d}{dt}\left(t^2|v_k'(t)|^2 + 2t\|v_k'(t)\|^2 + 4t\int_{\Gamma_C} j_k(v_{k\tau}(x,t))\,dS\right) + \nu t^2\|v_k'(t)\|^2 \\
&\quad\leq C_1\|v_k(t)\|^2(t|v_k'(t)|)^2 + C_2(t|v_k'(t)|)^2 + C_3 + C_4\|v_k(t)\|^2 \\
&\quad+ C_5|v_k(t)|^{\frac{1}{2}}\|v_k(t)\|^{\frac{3}{2}}(t|v_k'(t)|)^{\frac{1}{2}}(t\|v_k'(t)\|)^{\frac{1}{2}},
\end{aligned}$$

with $C_i > 0$ for $i = 1, \ldots, 5$ independent on $k$ and initial data. Denoting

$$y_k(t) = t^2|v_k'(t)|^2 + 2t\|v_k'(t)\|^2 + 4t\int_{\Gamma_C} j_k(v_{k\tau}(x,t))\,dS,$$

using the bound of Lemma 10.6 for $|v_k(t)|$, Young inequality, and the fact that $j_k$ assumes nonnegative values, we get

$$\frac{d}{dt}y_k(t) + \frac{\nu}{2}t^2\|v_k'(t)\|^2 \leq \left(C_6\|v_k(t)\|^2 + C_2\right)y_k(t) + C_3 + C_7\|v_k(t)\|^2,$$

where $C_6, C_7 > 0$. Finally, the Gronwall inequality (cf. Lemma 3.20) and the bound in (10.76) of Lemma 10.4 for $\int_0^T \|v_k(s)\|^2\,ds$ yield the uniform boundedness of $v_k'$ in $L^\infty(\eta, T; H) \cap L^2(\eta, T; V)$ for all intervals $[\eta, T]$, $0 < \eta < T$. This completes the proof of $(A10)$ and thus the existence of an exponential attractor for the semigroup $\{S(t)\}_{t\geq 0}$ associated with Problem 10.2. □

## 10.4 Comments and Bibliographical Notes

Large time behavior of solutions of problems in contact mechanics is an important though somewhat neglected part of the theory. In fact, remarking on future directions of research in the field of contact mechanics, in their book [215], the authors wrote: "The infinite-dimensional dynamical systems approach to contact problems is virtually nonexistent. (...) This topic certainly deserves further consideration."

From the mathematical point of view, a considerable difficulty in analysis of contact mechanics problems, and dynamical problems in particular, comes from the presence of involved boundary constraints which are often modeled by multivalued boundary conditions of a subdifferential type and lead to a formulation of the considered problem in terms of a variational inequality or differential inclusion with, frequently, nondifferentiable boundary functionals.

To establish the existence of the global attractor of a finite fractal dimension we used the method of $l$-trajectories as presented in [168] (cf. also [167]). This method appears very useful when one deals with variational inequalities, cf. [208], as it overcomes obstacles coming from the usual methods presented in [62, 69, 111, 197, 220]. One does not need to check directly neither compactness of the dynamics which results from the second energy inequality nor asymptotic compactness, cf., e.g., [28, 220], which results from the energy equation. In the case of variational inequalities it is sometimes not possible to get the second energy inequality and the differentiability of the associated semigroup due to the presence of nondifferentiable boundary functionals.

While there are other methods to establish the existence of the global attractor, where the problem of the lack of regularity appears, such as, e.g., the approach based on the notion of the Kuratowski measure of noncompactness of bounded sets, where we do not need even the strong continuity of the semigroup associated with a given dynamical problem, cf., e.g., [240], the problem of a finite dimensionality of the attractor is more involved.

The method of $l$-trajectories allows to prove the existence of an even more desirable object, the exponential attractor, for many problems for which there exists a finite dimensional global attractor [179]. It contains the global attractor and thus its existence implies the finite dimensionality of the global attractor itself. Its crucial property is an exponential rate of attraction of solution trajectories [94, 179]. The proof of the existence of an exponential attractor requires the solution to be regular enough to ensure the Hölder continuity of the semigroup in the time variable [168]. We established this property by providing additional a priori estimates of solutions.

In the end we would like to mention some related problems. First, there is a question of the existence of global and exponential attractors for other contact problems with multivalued boundary conditions in form of Clarke subdifferentials. As concerns exponential attractors, the main difficulty seems to be produced by assumption $(A10)$. Usually (10.2) follows from some better regularity of the time derivative of the solution, e.g., it easily follows if we know that $u' \in L^q(\tau, T; X)$ for $T > \tau > 0$ and some $q > 1$. While some regularity results for contact problems are known [91, 215], such a property is not known to hold for most cases, and hence a question of the solution regularity for contact problems still demand further study. In particular we needed the convexity of underlying potential in the Tresca case, and $(k3)$, the one sided Lipschitz condition, in the generalized Tresca case. These properties guaranteed us the solution uniqueness. It is, to our knowledge, an open problem, to obtain results on the global attractor fractal dimension for the problems without the solution uniqueness, for example, if we do not assume $(k3)$.

For some visco-plastic flows (e.g., Bingham flows) governed by variational inequalities (however, only with homogeneous or periodic boundary conditions) the regularity problem in question was solved, e.g., in [148, 211] and used to study the time asymptotics of solutions.

In this context, there are other important open problems, namely these of the time asymptotics of solutions of the Navier–Stokes equations, visco-plastic, and other fluid models governed by evolution variational or even hemivariational inequalities (cf., e.g., [176]) that take into account involved boundary conditions coming from a variety of applications in mechanics.

Note that for the dissipative second order (in time) problems with multivalued boundary conditions the situation can be totally different from that for the first order problems. Namely, even for the case of a very simple convex potential and the gradient system, it is possible to show that the global attractor can have infinite fractal dimension (see [126]).

# 11
# Non-autonomous Navier–Stokes Equations and Pullback Attractors

*Running water has within itself an infinite number of movements which are greater or less than its principal course.*

– Leonardo da Vinci

In this chapter we study the time asymptotics of solutions to the two-dimensional Navier–Stokes equations. In the first two sections we prove two properties of the equations in a bounded domain, concerning the existence of determining modes and nodes. Then we study the equations in an unbounded domain, in the framework of the theory of infinite dimensional non-autonomous dynamical systems and pullback attractors.

## 11.1 Determining Modes

Our aim in this section is to prove that if a number of Fourier modes of two different solutions of the Navier–Stokes equations have the same asymptotic behavior as $t \to \infty$ then, under the condition that the corresponding external forces have the same asymptotic behavior as $t \to \infty$, the remaining infinite number of modes also have the same asymptotic behavior. This means that both solutions have the same time asymptotics.

We shall consider first two-dimensional flows with homogeneous Dirichlet boundary conditions, in an open, bounded, and connected domain $\Omega$ of class $C^2$.

To be more precise, let

$$\frac{\partial u}{\partial t} + \nu\Delta u + (u \cdot \nabla)u + \nabla p = f, \qquad \operatorname{div} u = 0,$$

and

$$\frac{\partial v}{\partial t} + \nu\Delta v + (v \cdot \nabla)v + \nabla q = g, \qquad \operatorname{div} v = 0,$$


G. Łukaszewicz, P. Kalita, *Navier–Stokes Equations*, Advances in Mechanics and Mathematics 34, DOI 10.1007/978-3-319-27760-8_11

and let

$$u(x,t) = \sum_{k=1}^{\infty} \hat{u}_k(t)\omega_k(x), \qquad P_m u(x,t) = \sum_{k=1}^{m} \hat{u}_k(t)\omega_k(x),$$

and

$$v(x,t) = \sum_{k=1}^{\infty} \hat{v}_k(t)\omega_k(x), \qquad P_m v(x,t) = \sum_{k=1}^{m} \hat{v}_k(t)\omega_k(x),$$

where the functions $\omega_k$, $k = 1, 2, 3, \ldots$, are the eigenfunctions of the Stokes operator $A$, $A\omega_k = \lambda_k \omega_k$, cf. Chap. 4.

We assume that the forcing terms $f$ and $g$, $f, g \in L^\infty(0,\infty; H)$, have the same asymptotic behavior as time goes to infinity,

$$\int_\Omega |f(x,t) - g(x,t)|^2 dx \to 0 \quad \text{as} \quad t \to \infty. \tag{11.1}$$

**Definition 11.1.** Let (11.1) hold. Then the first $m$ modes associated with $P_m$ are called the determining modes if the condition

$$\int_\Omega |P_m u(x,t) - P_m v(x,t)|^2 dx \to 0 \quad \text{as} \quad t \to \infty$$

implies

$$\int_\Omega |u(x,t) - v(x,t)|^2 dx \to 0 \quad \text{as} \quad t \to \infty. \tag{11.2}$$

*Remark 11.1.* Relation (11.2) implies further a similar convergence in the norm of $V$.

Let

$$\limsup_{t\to\infty} \left( \int_\Omega |f(x,t)|^2 dx \right)^{1/2} = F.$$

We shall give the explicit estimate of the number $m$ of determining modes in terms of the Grashof number $G$, which is a nondimensional parameter, defined here as

$$G = \frac{F}{\lambda_1 \nu^2} \quad \text{for the space dimension} \quad n = 2.$$

We shall prove the following

**Theorem 11.1.** *Suppose that $m \in \mathbb{N}$ is such that $m \geq cG^2$ where $G$ is the Grashof number, $G = \frac{F}{\lambda_1 \nu^2}$, and $c$ is a constant depending only on the shape of $\Omega$. Then the first $m$ modes are determining (in the sense defined above) for the two-dimensional Navier–Stokes system with non-slip boundary conditions.*

*Proof.* The proof is based on the generalized Gronwall lemma (Lemma 3.23), cf. [99]. We shall apply the lemma to $\xi(t) = |Q_m w(t)|^2$, where $w(t) = u(t) - v(t)$ is the difference of the two solutions and $Q_m = I - P_m$, showing that if $m$ is sufficiently large and $|P_m w(t)| \to 0$ together with $|f(t) - g(t)| \to 0$ as $t \to \infty$ then (3.36) holds.

We have the following equation for the difference of solutions,

$$\frac{dw}{dt} + \nu A w + B(w, u) + B(v, w) = f(t) - g(t).$$

Taking the inner product with $Q_m w$ in $H$ we obtain

$$\frac{1}{2}\frac{d}{dt}|Q_m w|^2 + \nu \|Q_m w\|^2 + b(w, u, Q_m w) + b(v, w, Q_m w) = (f(t) - g(t), Q_m w). \tag{11.3}$$

We estimate the nonlinear terms as follows. Observe that

$$b(w, u, Q_m w) = b(P_m w, u, Q_m w) + b(Q_m w, u, Q_m w). \tag{11.4}$$

We estimate the first term on the right-hand side using the Ladyzhenskaya inequality (3.8) to get

$$\begin{aligned} |b(P_m w, u, Q_m w)| &= |b(P_m w, Q_m w, u)| \\ &\leq c_1 |P_m w|^{1/2} \|P_m w\|^{1/2} |u|^{1/2} \|u\|^{1/2} \|Q_m w\|. \end{aligned}$$

The second term on the right-hand side of (11.4) is estimated similarly to the first term, obtaining

$$\begin{aligned} |b(Q_m w, u, Q_m w)| &\leq c_1 |Q_m w| \|Q_m w\| \|u\| \\ &\leq \frac{c_1^2}{2\nu} |Q_m w|^2 \|u\|^2 + \frac{\nu}{2} \|Q_m w\|^2. \end{aligned}$$

Observe that

$$\begin{aligned} |b(v, w, Q_m w)| &= |b(v, P_m w, Q_m w)| \\ &\leq c_1 |v|^{1/2} \|v\|^{1/2} |P_m w|^{1/2} \|P_m w\|^{1/2} \|Q_m w\|. \end{aligned}$$

For the right-hand side of (11.3) we have

$$|(f(t) - g(t), Q_m w)| \le |f(t) - g(t)||Q_m w|.$$

From the above estimates we conclude that

$$\begin{aligned}\frac{1}{2}\frac{d}{dt}|Q_m w|^2 + \frac{\nu}{2}\|Q_m w\|^2 - \frac{c_1^2}{2\nu}\|u\|^2|Q_m w|^2 \\ \le c_1|P_m w|^{1/2}\|P_m w\|^{1/2}|u|^{1/2}\|u\|^{1/2}\|Q_m w\| \qquad (11.5) \\ +c_1|v|^{1/2}\|v\|^{1/2}|P_m w|^{1/2}\|P_m w\|^{1/2}\|Q_m w\| + |f(t) - g(t)||Q_m w|.\end{aligned}$$

We use

$$\lambda_{m+1}|Q_m w|^2 \le \|Q_m w\|^2 \qquad (11.6)$$

in the second term on the left-hand side of (11.5) to obtain

$$\frac{d\xi(t)}{dt} + \alpha(t)\xi(t) \le \beta(t),$$

with

$$\alpha(t) = \nu\lambda_{m+1} - \frac{c_1^2}{2\nu}\|u(t)\|^2,$$

and

$$\beta = 2 \times \{\text{the right-hand side of } (11.5)\}.$$

We shall check the conditions of the generalized Gronwall lemma (i.e., Lemma 3.23). The existence of uniquely determined $u$ and $v$ follows from Theorem 7.2. Remark 7.2 implies that both $u$ and $v$ belong to $L^2(\eta, T; D(A))$ for all $0 < \eta < T < \infty$. Now, as $f, g \in L^\infty(\mathbb{R}^+; H)$ we use an argument based on the enstrophy estimate and the uniform Gronwall inequality, similar as in the proof of Theorem 7.4, to deduce that the solutions $u(t)$ and $v(t)$ are uniformly bounded in $t$ away from zero in both $H$ and $V$ norms (for the initial conditions in $H$), see Exercise 7.25. Therefore, if $|P_m w(t)| \to 0$ and $|f(t) - g(t)| \to 0$ as $t \to \infty$ then from inequality (11.5) it follows that $\beta(t) \to 0$ as $t \to \infty$, which implies condition (3.35).

Second, from the energy equation

$$\frac{1}{2}|u(t)|^2 + \nu\int_{t_0}^t \|u(s)\|^2 ds = \frac{1}{2}|u(t_0)|^2 + \int_{t_0}^t (f(s), u(s))ds,$$

the boundedness of $|u(t)|$, and the Poincaré inequality we have

$$\frac{1}{T}\int_t^{t+T} \|u(s)\|^2 ds \leq \frac{2}{\nu^2\lambda_1}\|f\|^2_{L^\infty(t,t+T;H)} \quad \text{for large} \quad T. \tag{11.7}$$

By (11.7), condition (3.34) is satisfied. We have also

$$\begin{aligned}\liminf_{t\to\infty} \frac{1}{T}\int_t^{t+T} \alpha(\tau)d\tau &\geq \nu\lambda_m - \frac{c_1^2}{\nu}\limsup_{t\to\infty}\frac{1}{T}\int_t^{t+T} \|u(\tau)\|^2 d\tau \\ &\geq \nu\lambda_m - \frac{2c_1^2F^2}{\nu^3\lambda_1}.\end{aligned}$$

Thus, for large $m$,

$$\lambda_{m+1} > \frac{2c_1^2F^2}{\nu^4\lambda_1}, \tag{11.8}$$

and (3.33) is satisfied. In view of Lemma 3.23, $\xi(t) \to 0$ as $t \to \infty$ for sufficiently large $m$. For $m \to \infty$ we have $\lambda_m \sim c_1'\lambda_1 m$ where $c_1'$ is a nondimensional constant (cf. [99]), and we can see that (11.8) is satisfied for $m \geq cG^2$, where $c$ is a nondimensional number depending only on the shape of $\Omega$. □

**Exercise 11.1.** Prove inequality (11.6).

**Exercise 11.2.** Prove (11.7).

**Determining Modes in the Space-Periodic Case** In this case we can prove in a similar way a stronger result due to better properties of the trilinear form $b$.

**Theorem 11.2.** *Suppose that $m \in \mathbb{N}$ is such that $m \geq cG$ where $G$ is the Grashof number, $G = \frac{F}{\lambda_1\nu^2}$, and $c$ is a constant depending only on the shape of $\Omega$ (i.e., the ratio between the periods in the two space directions). Then the first $m$ modes are determining in the sense that*

$$\int_\Omega |\nabla u(x,t) - \nabla v(x,t)|^2 dx \to 0 \quad \text{as} \quad t \to \infty.$$

*for the two-dimensional Navier–Stokes system with periodic boundary conditions and vanishing space average.*

*Proof.* The proof is more involved than that of Theorem 11.1 and can be found in [99]. The idea is again to use the generalized Gronwall lemma to the function $\xi(t) = Q_m w(t)$. We start with equation

$$\frac{dw}{dt} + \nu Aw + B(w,u) + B(v,w) = f(t) - g(t),$$

take the inner product with $AQ_m w$ in $H$ (we know that $u(t) \in D(A)$ for $t > 0$) to obtain

$$\frac{1}{2}\frac{d}{dt}\|Q_m w\|^2 + \nu\|Q_m Aw\|^2 + b(w,u,AQ_m w) + b(v,w,AQ_m w) \\ = (f(t) - g(t), AQ_m w),$$

and then proceed as above using further properties (as these in Exercises 11.3 and 11.4) of solutions of two-dimensional Navier–Stokes equations with space-periodic boundary conditions. □

**Exercise 11.3.** Prove that for $u, v \in D(A)$

$$b(v,u,Au) + b(u,v,Au) + b(u,u,Av) = 0.$$

**Exercise 11.4.** Prove that

$$\frac{1}{T}\int_t^{t+T} |Au(s)|^2 ds \leq \frac{2}{\nu^2}\|f\|^2_{L^\infty(t,t+T;H)} \quad \text{for large } T.$$

(cf. Exercise 11.6).

## 11.2 Determining Nodes

In many practical situations the experimental data is collected from measurements at a finite number of nodal points in the physical space.

A set of points in the physical space (in the domain filled by the fluid) is called a set of determining nodes if, whenever the difference between the measurements at those points of the velocity field of any two flows goes to zero as time goes to infinity, then the difference between those velocity fields goes to zero uniformly on the domain (in a certain norm).

We shall prove that there exists a finite number of determining nodes and shall estimate their lowest number.

Let $\Omega$ be an open, bounded, and connected two-dimensional domain of class $C^2$. We consider two velocity fields $u = u(x,t)$ and $v = v(x,t)$ satisfying in $\Omega$ the Navier–Stokes equations,

$$\frac{\partial u}{\partial t} + \nu\Delta u + (u\cdot\nabla)u + \nabla p = f, \qquad \operatorname{div} u = 0,$$

and

$$\frac{\partial v}{\partial t} + \nu\Delta v + (v\cdot\nabla)v + \nabla q = g, \qquad \operatorname{div} v = 0,$$

with the same homogeneous Dirichlet boundary conditions; $p, q$ and $f, g$ are the corresponding pressures and external forces, respectively.

We assume that the forcing terms $f$ and $g, f, g \in L^\infty(0, \infty; H)$, have the same asymptotic behavior as time goes to infinity,

$$\int_\Omega |f(x,t) - g(x,t)|^2 dx \to 0 \quad \text{as} \quad t \to \infty.$$

Let $\Omega$ be covered by $N$ identical squares $Q_1, \ldots, Q_N$ (for example, by a regular mesh) and let each square $Q_i$ contain exactly one measurement point $x^i$, $i = 1, \ldots, N$. We thus have a set of $N$ measurement points $\mathscr{E} = \{x^1, \ldots, x^N\}$. Let

$$\max_{j=1,\ldots,N} |u(x^j,t) - v(x^j,t)| \to 0 \quad \text{as} \quad t \to \infty. \tag{11.9}$$

The set $\mathscr{E}$ is then called a set of determining nodes if (11.9) implies

$$\int_\Omega |u(x,t) - v(x,t)|^2 dx \to 0 \quad \text{as} \quad t \to \infty.$$

We shall show that if (11.9) holds for some $N$ then also

$$\int_\Omega |\nabla u(x,t) - \nabla v(x,t)|^2 dx \to 0 \quad \text{as} \quad t \to \infty.$$

Denote

$$\eta(w) = \max_{1 \le j \le N} |w(x^j)|.$$

To prove the existence of a finite number of determining nodes we shall use the generalized Gronwall lemma (cf. Lemma 3.23) as well as inequality (11.10) of the following lemma.

**Lemma 11.1.** *Let the domain $\Omega$ be covered by N identical squares. Consider the set $\mathscr{E} = \{x^1, \ldots, x^N\}$ of points in $\Omega$, distributed one in each square. Then for each vector field w in $D(A)$ the following inequalities hold,*

$$\begin{aligned} |w|^2 &\le \frac{c}{\lambda_1}\eta(w)^2 + \frac{c}{\lambda_1^2 N^2}|Aw|^2, \\ \|w\|^2 &\le cN\eta(w)^2 + \frac{c}{\lambda_1 N}|Aw|^2, \\ \|w\|^2_{L^\infty(\Omega)} &\le cN\eta(w)^2 + \frac{c}{\lambda_1 N}|Aw|^2, \end{aligned} \tag{11.10}$$

*where the constant c depends only on the shape of the domain $\Omega$.*

A proof of the lemma can be found in [99].

**Theorem 11.3.** *Let the domain $\Omega$ be covered by N identical squares. Consider the set $\mathscr{E} = \{x^1, \ldots, x^N\}$ of points in $\Omega$, distributed one in each square. Let f and g be two forcing terms in $L^\infty(0,\infty;H)$ that satisfy*

$$\int_\Omega |f(x,t) - g(x,t)|^2 dx \to 0 \quad \text{as} \quad t \to \infty,$$

*and let*

$$F = \limsup_{t\to\infty} |f(t)| = \limsup_{t\to\infty} |g(t)| \quad \text{as} \quad t \to \infty.$$

*Then there exists a constant $C = C(F,\nu,\Omega)$ such that if*

$$N \geq C = C(F,\nu,\Omega),$$

*then $\mathscr{E}$ is a set of determining nodes in the sense defined above for the two-dimensional Navier–Stokes equations with homogeneous Dirichlet boundary conditions.*

*Proof.* Let $w = u - v$. We know that $\eta(w(t)) \to \infty$ as $t \to \infty$. We have to show that $|w(t)| \to 0$ as well. Actually, we will show that

$$\|w(t)\| \to 0 \quad \text{as} \quad t \to \infty.$$

We have, from the Navier–Stokes equations,

$$\frac{dw(t)}{dt} + \nu Aw(t) + B(w(t),u(t)) + B(v(t),w(t)) = f(t) - g(t).$$

Taking the inner product with $Aw(t)$ in $H$ we obtain

$$\frac{1}{2}\frac{d}{dt}\|w(t)\|^2 + \nu|Aw(t)|^2 + b(w(t),u(t),Aw(t)) + b(v(t),w(t),Aw(t)) \\ = (f(t) - g(t), Aw(t)). \tag{11.11}$$

By $C$ we shall denote a generic constant that might depend on $\Omega, f, g, \nu$, but not on the initial conditions $u_0, v_0$. By $c$ we denote constants that might depend only on $\Omega$.

We estimate the nonlinear terms in the following way:

$$|b(w,u,Aw)| \leq c_1|w|^{1/2}\|u\||Aw|^{3/2} \leq \frac{\nu}{6}|Aw|^2 + \frac{c}{\nu^3}\|u\|^4|w|^2,$$

$$|b(v,w,Aw)| \leq c_1|v|^{1/2}|Av|^{1/2}\|w\||Aw| \leq \frac{\nu}{6}|Aw|^2 + \frac{c}{\nu}|v||Av|\|w\|^2.$$

**Exercise 11.5.** Prove the above two inequalities by using the following Agmon inequality:

$$\|u\|_\infty \le c_2|u|^{1/2}|Au|^{1/2} \quad \text{for} \quad u \in D(A). \tag{11.12}$$

The right-hand side of (11.11) is estimated as

$$|(f-g, Aw)| \le \frac{\nu}{6}|Au|^2 + \frac{3}{2\nu}|f-g|^2.$$

Thus, (11.11) yields

$$\frac{d}{dt}\|w\|^2 + \nu|Aw|^2 \le \frac{c}{\nu^3}\|u\|^4|w|^2 + \frac{c}{\nu}|v||Av|\|w\|^2 + \frac{3}{\nu}|f-g|^2. \tag{11.13}$$

From (11.10) we have

$$|Aw|^2 \ge c^{-1}\lambda_1 N\|w\|^2 - \lambda_1 N^2 \eta(w)^2.$$

Using the Poincaré inequality in the first term on the right-hand side of (11.13) we get, in the end,

$$\frac{d}{dt}\|w\|^2 + \left(c^{-1}\lambda_1 N\nu - \frac{c}{\nu^3\lambda_1}\|u\|^4 - \frac{c}{\nu}|v||Av|\right)\|w\|^2 \le \frac{3}{\nu}|f-g|^2 + \lambda_1 N^2 \nu \eta(w)^2.$$

This inequality can be written in the form

$$\frac{d\xi}{dt} + \alpha\xi \le \beta \quad \text{where} \quad \xi = \|w(t)\|^2. \tag{11.14}$$

We see that

$$\beta(t) \to 0 \quad \text{as} \quad t \to \infty,$$

and hence

$$\lim_{t\to\infty} \frac{1}{T}\int_t^{t+T} \beta^+(\tau)d\tau = 0. \tag{11.15}$$

As the functions

$$t \to \|u(t)\| \quad \text{and} \quad t \to |v(t)| \tag{11.16}$$

are bounded on $[t_0, \infty)$ for $t_0 > 0$, and

$$\frac{1}{T}\int_t^{t+T} |Av(\tau)|d\tau \leq C(|g|^2_{L^\infty(t,t+T;H)}, \nu, \Omega), \tag{11.17}$$

we deduce that

$$\liminf_{t\to\infty} \frac{1}{T}\int_t^{t+T} \alpha(\tau)d\tau \geq c^{-1}\nu\lambda_1 N - C(f, g, \nu, \Omega).$$

For $N$ sufficiently large we have

$$\liminf_{t\to\infty} \frac{1}{T}\int_t^{t+T} \alpha(\tau)d\tau > 0. \tag{11.18}$$

Also

$$\limsup_{t\to\infty} \frac{1}{T}\int_t^{t+T} \alpha^-(\tau)d\tau < \infty. \tag{11.19}$$

In view of (11.14) together with (11.15), (11.18), and (11.19) we deduce that

$$\xi(t) = \|w(t)\| \to 0 \quad \text{as} \quad t \to \infty.$$

Hence, the set $\mathscr{E}$ is a set of determining nodes. □

**Exercise 11.6.** Prove (11.17) from

$$\frac{1}{2}\frac{d}{dt}\|v(t)\|^2 + \nu|Av(t)|^2 + b(v(t), v(t), Av(t)) = (g(t), Av(t)),$$

knowing that the functions in (11.16) are bounded.

## 11.3 Pullback Attractors for Asymptotically Compact Non-autonomous Dynamical Systems

Let $\Omega$ be a nonempty set and let $\{\theta_t\}_{t\in\mathbb{R}}$ be a family of mappings $\theta_t : \Omega \to \Omega$ satisfying:

1. $\theta_0\omega = \omega$ for all $\omega \in \Omega$,
2. $\theta_t(\theta_\tau\omega) = \theta_{t+\tau}\omega$ for all $\omega \in \Omega$ and all $t, \tau \in \mathbb{R}$.

The mappings $\theta_t$ are called the shift operators.

Let moreover $X$ be a metric space with the distance $d(\cdot,\cdot)$, and let $\phi$ be a $\theta$-cocycle on $X$, i.e., a mapping $\phi : \mathbb{R}_+ \times \Omega \times X \to X$, which satisfies:

(a) $\phi(0,\omega,x) = x$ for all $(\omega,x) \in \Omega \times X$,
(b) $\phi(t+\tau,\omega,x) = \phi(t,\theta_\tau\omega,\phi(\tau,\omega,x))$ for all $t,\tau \in \mathbb{R}_+$ and $(\omega,x) \in \Omega\times X$.

The $\theta$-cocycle $\phi$ is said to be continuous if for all $(t,\omega) \in \mathbb{R}_+ \times \Omega$, the mapping $\phi(t,\omega,\cdot) : X \to X$ is continuous.

The sets parameterized by the index $\omega \in \Omega$ will be called the parameterized sets and denoted $\widehat{D} = \{D(\omega)\}_{\omega\in\Omega}$, while their particular case, the sets parameterized by time $t \in \mathbb{R}$ will be called the non-autonomous sets and denoted $\mathbb{D} = \{D(t)\}_{t\in\mathbb{R}}$. This different notation stresses the special case $\Omega = \mathbb{R}$ and $\theta_t s = s+t$. Let $\mathscr{P}(X)$ denote the family of all nonempty subsets of $X$, and $\mathscr{S}$ the class of all parameterized sets $\widehat{D} = \{D(\omega)\}_{\omega\in\Omega}$ such that $D(\omega) \in \mathscr{P}(X)$ for all $\omega \in \Omega$. We say that the fibers $D(\omega)$ of the parameterized set $\widehat{D}$ are nonempty.

We consider a given nonempty subclass $\mathscr{D} \subset \mathscr{S}$. This class will be called an *attraction universe*.

**Definition 11.2.** The $\theta$-cocycle $\phi$ is said to be pullback $\mathscr{D}$-asymptotically compact ($\mathscr{D}$-a.c.) if for any $\omega \in \Omega$, any $\widehat{D} \in \mathscr{D}$, and any sequences $t_n \to +\infty$, $x_n \in D(\theta_{-t_n}\omega)$, the sequence $\phi(t_n,\theta_{-t_n}\omega,x_n)$ has a convergent subsequence.

**Definition 11.3.** A parameterized set $\widehat{B} = \{B(\omega)\}_{\omega\in\Omega} \in \mathscr{S}$ is said to be pullback $\mathscr{D}$-absorbing if for each $\omega \in \Omega$ and $\widehat{D} \in \mathscr{D}$, there exists $t_0(\omega,\widehat{D}) \geq 0$ such that

$$\phi(t,\theta_{-t}\omega,D(\theta_{-t}\omega)) \subset B(\omega) \quad \text{for all} \quad t \geq t_0(\omega,\widehat{D}).$$

We denote by $\operatorname{dist}_X(C_1,C_2)$ the Hausdorff semi-distance between $C_1$ and $C_2$, defined as

$$\operatorname{dist}_X(C_1,C_2) = \sup_{x\in C_1}\inf_{y\in C_2} d(x,y) \quad \text{for} \quad C_1,C_2 \subset X.$$

**Definition 11.4.** A parameterized set $\widehat{C} = \{C(\omega)\}_{\omega\in\Omega} \in \mathscr{S}$ is said to be pullback $\mathscr{D}$-attracting if

$$\lim_{t\to+\infty} \operatorname{dist}_X(\phi(t,\theta_{-t}\omega,D(\theta_{-t}\omega)),C(\omega)) = 0 \quad \text{for all} \quad \widehat{D} \in \mathscr{D},\ \omega \in \Omega.$$

**Definition 11.5.** A parameterized set $\widehat{A} = \{A(\omega)\}_{\omega\in\Omega} \in \mathscr{S}$ is said to be a global pullback $\mathscr{D}$-attractor if it satisfies

(i) $A(\omega)$ is compact for any $\omega \in \Omega$,
(ii) $\widehat{A}$ is pullback $\mathscr{D}$-attracting,
(iii) $\widehat{A}$ is invariant, i.e.,

$$\phi(t,\omega,A(\omega)) = A(\theta_t\omega) \quad \text{for any} \quad (t,\omega) \in \mathbb{R}_+ \times \Omega.$$

*Remark 11.2.* Observe that Definition 11.5 does not guarantee the uniqueness of a pullback $\mathscr{D}$-attractor (see [36] for a discussion on this point). In order to ensure the uniqueness one needs to impose additional conditions as, for instance, the condition that the attractor belongs to the same attraction universe $\mathscr{D}$, or some kind of minimality. However, we do not include this stronger assumptions in the definition since, as we will show in Theorem 11.4, under very general hypotheses, it is possible to ensure the existence of a global pullback $\mathscr{D}$-attractor which is minimal in an appropriate sense.

**Definition 11.6.** For each $\widehat{D} \in \mathscr{S}$ and $\omega \in \Omega$, we define the $\omega$-limit set of $\widehat{D}$ at $\omega$ as

$$\Lambda(\widehat{D}, \omega) = \bigcap_{s \geq 0} \left( \overline{\bigcup_{t \geq s} \phi(t, \theta_{-t}\omega, D(\theta_{-t}\omega))} \right).$$

Obviously, $\Lambda(\widehat{D}, \omega)$ is a closed subset of $X$ that can be possibly empty. It is easy to see that for each $y \in X$, one has that $y \in \Lambda(\widehat{D}, \omega)$ if and only if there exist a sequence $t_n \to +\infty$ and a sequence $x_n \in D(\theta_{-t_n}\omega)$ such that

$$\lim_{t_n \to +\infty} d(\phi(t_n, \theta_{-t_n}\omega, x_n), y) = 0.$$

Our first aim is to prove the following result on the existence of global pullback $\mathscr{D}$-attractor.

**Theorem 11.4.** *Suppose the $\theta$-cocycle $\phi$ is continuous and pullback $\mathscr{D}$-asymptotically compact, and there exists $\widehat{B} \in \mathscr{D}$ which is pullback $\mathscr{D}$-absorbing. Then, the parameterized set $\widehat{A}$ defined by*

$$A(\omega) = \Lambda(\widehat{B}, \omega) \quad \textit{for} \quad \omega \in \Omega,$$

*is a global pullback $\mathscr{D}$-attractor which is minimal in the sense that if $\widehat{C} \in \mathscr{S}$ is a parameterized set such that $C(\omega)$ is closed and*

$$\lim_{t \to +\infty} \operatorname{dist}_X(\phi(t, \theta_{-t}\omega, B(\theta_{-t}\omega)), C(\omega)) = 0,$$

*then $A(\omega) \subset C(\omega)$.*

In order to prove Theorem 11.4, we first need the following results.

**Proposition 11.1.** *If $\widehat{B} \in \mathscr{S}$ is a pullback $\mathscr{D}$-absorbing parameterized set, then*

$$\Lambda(\widehat{D}, \omega) \subset \Lambda(\widehat{B}, \omega) \quad \text{for all} \quad \widehat{D} \in \mathscr{D} \quad \text{and} \quad \omega \in \Omega.$$

*If in addition $\widehat{B} \in \mathscr{D}$, then*

$$\Lambda(\widehat{D}, \omega) \subset \Lambda(\widehat{B}, \omega) \subset \overline{B(\omega)} \quad \text{for all} \quad \widehat{D} \in \mathscr{D} \quad \text{and} \quad \omega \in \Omega.$$

*Proof.* Let us fix $\widehat{D} \in \mathscr{D}$, $\omega \in \Omega$, and $y \in \Lambda(\widehat{D}, \omega)$. There exist a sequence $t_n \to +\infty$ and a sequence $x_n \in D(\theta_{-t_n}\omega)$ such that

$$\phi(t_n, \theta_{-t_n}\omega, x_n) \to y. \tag{11.20}$$

As $\widehat{B}$ is pullback $\mathscr{D}$-absorbing, for each integer $k \geq 1$ there exists an index $n_k$ such that $t_{n_k} \geq k$ and

$$\phi(t_{n_k} - k, \theta_{-(t_{n_k}-k)}(\theta_{-k}\omega), D(\theta_{-(t_{n_k}-k)}(\theta_{-k}\omega))) \subset B(\theta_{-k}\omega). \tag{11.21}$$

In particular, as $x_{n_k} \in D(\theta_{-t_{n_k}}\omega) = D(\theta_{-(t_{n_k}-k)}(\theta_{-k}\omega))$, we obtain from (11.21)

$$y_k = \phi(t_{n_k} - k, \theta_{-t_{n_k}}\omega, x_{n_k}) \in B(\theta_{-k}\omega).$$

But then

$$\begin{aligned}
\phi(t_{n_k}, \theta_{-t_{n_k}}\omega, x_{n_k}) &= \phi((t_{n_k} - k) + k, \theta_{-t_{n_k}}\omega, x_{n_k}) \\
&= \phi(k, \theta_{-k}\omega, \phi(t_{n_k} - k, \theta_{-t_{n_k}}\omega, x_{n_k})) \\
&= \phi(k, \theta_{-k}\omega, y_k),
\end{aligned}$$

and consequently, by (11.20),

$$\phi(k, \theta_{-k}\omega, y_k) \to y \quad \text{with} \quad y_k \in B(\theta_{-k}\omega).$$

Thus, $y$ belongs to $\Lambda(\widehat{B}, \omega)$.

Finally, observe that if $z$ belongs to $\Lambda(\widehat{B}, \omega)$, then there exist a sequence $\tau_n \to +\infty$ and a sequence $w_n \in B(\theta_{-\tau_n}\omega)$ such that $\phi(\tau_n, \theta_{-\tau_n}\omega, w_n) \to z$. But, as $\widehat{B}$ is pullback $\mathscr{D}$-absorbing, there exists $t_0(\widehat{B}, \omega) \geq 0$ such that $\phi(\tau_n, \theta_{-\tau_n}\omega, w_n) \in B(\omega)$ for all $\tau_n \geq t_0(\widehat{B}, \omega)$, and consequently $z$ belongs to $\overline{B(\omega)}$. □

*Remark 11.3.* As a straightforward consequence of Proposition 11.1, we have that if $\widehat{B} \in \mathscr{D}$ is pullback $\mathscr{D}$-absorbing, then

$$\overline{\bigcup_{\widehat{D} \in \mathscr{D}} \Lambda(\widehat{D}, \omega)} = \Lambda(\widehat{B}, \omega),$$

so that, under the assumptions in Theorem 11.4, we have

$$\mathcal{A}(\omega) = \Lambda(\widehat{B}, \omega) = \overline{\bigcup_{\widehat{D} \in \mathscr{D}} \Lambda(\widehat{D}, \omega)} \quad \text{for} \quad \omega \in \Omega.$$

**Proposition 11.2.** *If $\phi$ is pullback $\mathscr{D}$-asymptotically compact, then for $\widehat{D} \in \mathscr{D}$ and $\omega \in \Omega$, the set $\Lambda(\widehat{D}, \omega)$ is nonempty, compact, and*

$$\lim_{t \to +\infty} \text{dist}_X(\phi(t, \theta_{-t}\omega, D(\theta_{-t}\omega)), \Lambda(\widehat{D}, \omega)) = 0. \tag{11.22}$$

*Proof.* Let us fix $\widehat{D} \in \mathscr{D}$ and $\omega \in \Omega$. If we consider a sequence $t_n \to +\infty$, and a sequence $x_n \in D(\theta_{-t_n}\omega)$, then from the sequence $\phi(t_n, \theta_{-t_n}\omega, x_n)$ we can extract a convergent subsequence $\phi(t_\mu, \theta_{-t_\mu}\omega, x_\mu) \to y$. By its construction, $y$ belongs to $\Lambda(\widehat{D}, \omega)$, and thus this set is nonempty.

We know that $\Lambda(\widehat{D}, \omega)$ is closed, and consequently in order to prove its compactness it is enough to see that for any given sequence $\{y_n\} \subset \Lambda(\widehat{D}, \omega)$, we can extract a convergent subsequence. First, observe that as $y_n \in \Lambda(\widehat{D}, \omega)$, there exist $t_n \geq n$ and $x_n \in D(\theta_{-t_n}\omega)$ such that

$$d(\phi(t_n, \theta_{-t_n}\omega, x_n), y_n) \leq \frac{1}{n}. \tag{11.23}$$

As $\phi$ is pullback $\mathscr{D}$-a.c., from the sequence $\phi(t_n, \theta_{-t_n}\omega, x_n)$ we can extract a convergent subsequence $\phi(t_\mu, \theta_{-t_\mu}\omega, x_\mu) \to y$, and consequently, from (11.23), we also obtain $y_\mu \to y$.

Finally, if (11.22) does not hold, there exist $\varepsilon > 0$, a sequence $t_n \to +\infty$, and a sequence $x_n \in D(\theta_{-t_n}\omega)$, such that

$$d(\phi(t_n, \theta_{-t_n}\omega, x_n), y) \geq \varepsilon \quad \text{for all} \quad y \in \Lambda(\widehat{D}, \omega). \tag{11.24}$$

But from the sequence $\phi(t_n, \theta_{-t_n}\omega, x_n)$ we can extract a convergent subsequence $\phi(t_\mu, \theta_{-t_\mu}\omega, x_\mu) \to y \in \Lambda(\widehat{D}, \omega)$, which contradicts (11.24). □

**Proposition 11.3.** *If the $\theta$-cocycle $\phi$ is continuous and pullback $\mathscr{D}$-asymptotically compact, then for any $(t, \omega) \in \mathbb{R}_+ \times \Omega$ and any $\widehat{D} \in \mathscr{D}$, one has*

$$\phi(t, \omega, \Lambda(\widehat{D}, \omega)) = \Lambda(\widehat{D}, \theta_t\omega).$$

*Proof.* Let $(t, \omega) \in \mathbb{R}_+ \times \Omega$ and $\widehat{D} \in \mathscr{D}$ be fixed, and let $y \in \Lambda(\widehat{D}, \omega)$. We show that $\phi(t, \omega, y) \in \Lambda(\widehat{D}, \theta_t\omega)$, and consequently we obtain

$$\phi(t, \omega, \Lambda(\widehat{D}, \omega)) \subset \Lambda(\widehat{D}, \theta_t\omega).$$

We already know that there exist a sequence $t_n \to +\infty$ and a sequence $x_n \in D(\theta_{-t_n}\omega)$ such that

$$\phi(t_n, \theta_{-t_n}\omega, x_n) \to y.$$

But

$$\phi(t,\omega,\phi(t_n,\theta_{-t_n}\omega,x_n)) = \phi(t+t_n,\theta_{-(t+t_n)}(\theta_t\omega),x_n),$$

and consequently, as $\phi$ is continuous,

$$\phi(t+t_n,\theta_{-(t+t_n)}(\theta_t\omega),x_n) \to \phi(t,\omega,y).$$

Since we have $t+t_n \to +\infty$ and $x_n \in D(\theta_{-t_n}\omega) = D(\theta_{-(t+t_n)}(\theta_t\omega))$, it follows that $\phi(t,\omega,y) \in \Lambda(\widehat{D},\theta_t\omega)$.

Now, we prove $\Lambda(\widehat{D},\theta_t\omega) \subset \phi(t,\omega,\Lambda(\widehat{D},\omega))$. For $y \in \Lambda(\widehat{D},\theta_t\omega)$ there exist a sequence $t_n \to +\infty$ and a sequence $x_n \in D(\theta_{-t_n}(\theta_t\omega))$ such that

$$\phi(t_n,\theta_{-t_n}(\theta_t\omega),x_n) \to y. \tag{11.25}$$

If $t_n \geq t$, we have

$$\begin{aligned}\phi(t_n,\theta_{-t_n}(\theta_t\omega),x_n) &= \phi(t+(t_n-t),\theta_{-(t_n-t)}\omega,x_n)\\ &= \phi(t,\omega,\phi(t_n-t,\theta_{-(t_n-t)}\omega,x_n)).\end{aligned} \tag{11.26}$$

But, as $\phi$ is pullback $\mathscr{D}$-a.c., $t_n - t \to +\infty$, and $x_n \in D(\theta_{-t_n}(\theta_t\omega)) = D(\theta_{-(t_n-t)}\omega)$, one can ensure that there exists a subsequence $\{(t_\mu,x_\mu)\} \subset \{(t_n,x_n)\}$ such that

$$\phi(t_\mu - t,\theta_{-(t_\mu-t)}\omega,x_\mu) \to z \in \Lambda(\widehat{D},\omega),$$

and then, by (11.26) and (11.25),

$$y = \phi(t,\omega,z) \in \phi(t,\omega,\Lambda(\widehat{D},\omega)). \quad \square$$

Now, we can prove the main Theorem.

*Proof (of Theorem 11.4).* The compactness of each $A(\omega)$ follows from Proposition 11.2. Also, by this proposition and the fact that $\bigcup_{\widehat{D}\in\mathscr{D}} \Lambda(\widehat{D},\omega) \subset A(\omega)$, we obtain

$$\lim_{t\to+\infty} \mathrm{dist}_X(\phi(t,\theta_{-t}\omega,D(\theta_{-t}\omega)),A(\omega)) = 0 \quad \text{for any} \quad \widehat{D} \in \mathscr{D}.$$

Now, observe that, by Proposition 11.3,

$$A(\theta_t\omega) = \overline{\bigcup_{\widehat{D}\in\mathscr{D}} \Lambda(\widehat{D},\theta_t\omega)} = \overline{\bigcup_{\widehat{D}\in\mathscr{D}} \phi(t,\omega,\Lambda(\widehat{D},\omega))} = \overline{\phi(t,\omega,\bigcup_{\widehat{D}\in\mathscr{D}} \Lambda(\widehat{D},\omega))},$$

and, as $\phi$ is continuous and $\overline{\bigcup_{\widehat{D}\in\mathscr{D}} \Lambda(\widehat{D},\omega)}$ is compact,

$$\overline{\phi(t,\omega,\bigcup_{\widehat{D}\in\mathscr{D}} \Lambda(\widehat{D},\omega))} = \phi(t,\omega,\overline{\bigcup_{\widehat{D}\in\mathscr{D}} \Lambda(\widehat{D},\omega)}).$$

Consequently, we have the invariance property $\phi(t,\omega,A(\omega)) = A(\theta_t\omega)$.

Now, let $\widehat{C} \in \mathscr{S}$ be a parameterized set such that $C(\omega)$ is closed and

$$\lim_{t\to+\infty} \operatorname{dist}_X(\phi(t,\theta_{-t}\omega,B(\theta_{-t}\omega)),C(\omega)) = 0.$$

Let $y$ be an element of $A(\omega)$, then

$$y = \lim_{n\to\infty} \phi(t_n,\theta_{-t_n}\omega,x_n),$$

for some sequences $t_n \to +\infty$, $x_n \in B(\theta_{-t_n}\omega)$, and consequently $y \in \overline{C(\omega)} = C(\omega)$. Thus, $A(\omega) \subset C(\omega)$. □

*Remark 11.4.* Observe that if in Theorem 11.4 we assume that $B(\omega)$ is closed for all $\omega \in \Omega$, and the attraction universe $\mathscr{D}$ is inclusion-closed (i.e., if $\widehat{D'} \in \mathscr{S}, \widehat{D} \in \mathscr{D}$ and $D'(\omega) \subset D(\omega)$ for all $\omega \in \Omega$, then $\widehat{D'} \in \mathscr{D}$), then it follows that the pullback $\mathscr{D}$-attractor $\widehat{A}$ belongs to $\mathscr{D}$, and hence it is the unique pullback $\mathscr{D}$-attractor which belongs to $\mathscr{D}$. This situation appears very often in applications, in particular, in our example in Sect. 11.4.

*Remark 11.5.* Note that we do not assume any structure (neither measurable nor topological) on the set of parameters $\Omega$. Consequently, this result can be applied to both non-autonomous and random dynamical systems.

Although the cocycle formalism is very useful, in particular, in the case of random dynamical systems or non-autonomous problems, we state below our main result (namely Theorem 11.4) in the language of evolutionary processes (see [225]) and then apply it in the theory of Navier–Stokes equations.

Let us consider an evolutionary process (also called a two-parameter semigroup or just a process) $U$ on $X$, i.e., a family $\{U(t,\tau)\}_{-\infty<\tau\le t<+\infty}$ of continuous mappings $U(t,\tau): X \to X$, such that $U(\tau,\tau)x = x$, and

$$U(t,\tau) = U(t,r)U(r,\tau) \quad \text{for all} \quad \tau \le r \le t. \tag{11.27}$$

The attraction universe $\mathscr{D}$ will now be a nonempty class of non-autonomous sets $\mathbb{D} = \{D(t)\}_{t\in\mathbb{R}}$ such that $D(t) \in \mathscr{P}(X)$ for all $t \in \mathbb{R}$.

**Definition 11.7.** The process $U$ is said to be pullback $\mathscr{D}$-asymptotically compact if for any $t \in \mathbb{R}$, any $\mathbb{D} \in \mathscr{D}$, any sequence $\tau_n \to -\infty$, and any sequence $x_n \in D(\tau_n)$, the sequence $\{U(t,\tau_n)x_n\}$ is relatively compact in $X$.

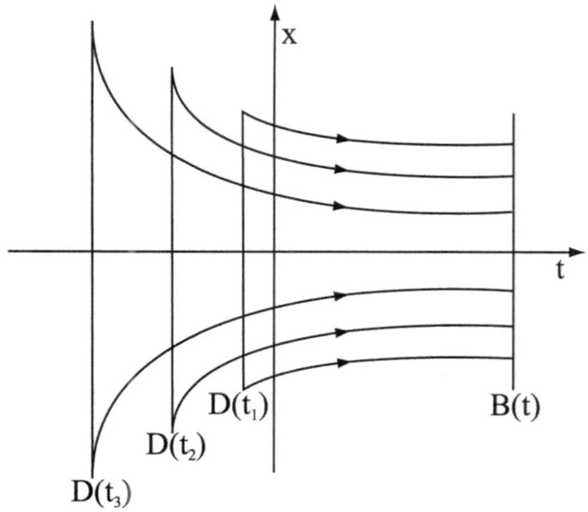


**Fig. 11.1** Illustration of a pullback $\mathscr{D}$-absorbing property. The sets $D(t_i)$ become larger as $t_i$ tend to $-\infty$ but still at time $t$ all trajectories originating from those sets are in $B(t)$

**Definition 11.8.** It is said that $\mathbb{B} = \{B(t)\}_{t\in\mathbb{R}} \in \mathscr{D}$ is pullback $\mathscr{D}$-absorbing for the process $U$ if for any $t \in \mathbb{R}$ and any $\mathbb{D} \in \mathscr{D}$, there exists a $\tau_0(t,\mathbb{D}) \le t$ such that

$$U(t,\tau)D(\tau) \subset B(t) \quad \text{for all} \quad \tau \le \tau_0(t,\mathbb{D}).$$

The notion of pullback $\mathscr{D}-$absorbing set is illustrated in Fig. 11.1.

We have the following result.

**Theorem 11.5.** *Suppose that the process $U$ is pullback $\mathscr{D}$-asymptotically compact and that $\mathbb{B} \in \mathscr{D}$ is a non-autonomous $\mathscr{D}$-absorbing set for $U$.*

*Then, the non-autonomous set $\mathbb{A} = \{A(t)\}_{t\in\mathbb{R}}$ with nonempty fibers $A(t) \in \mathscr{P}(X)$ for $t \in \mathbb{R}$ defined by*

$$A(t) = \Lambda(\mathbb{B},t) \quad \text{for} \quad t \in \mathbb{R},$$

*where for each $\mathbb{D} = \{D(t)\}_{t\in\mathbb{R}} \in \mathscr{D}$*

$$\Lambda(\mathbb{D},t) = \bigcap_{s\le t}\left(\overline{\bigcup_{\tau\le s} U(t,\tau)D(\tau)}\right),$$

*has the following properties:*

*(i) the sets $A(t)$ are compact for $t \in \mathbb{R}$,*
*(ii) the non-autonomous set $\mathbb{A}$ is pullback $\mathscr{D}$-attracting, i.e.,*

$$\lim_{\tau\to-\infty} \mathrm{dist}_X(U(t,\tau)D(\tau),A(t)) = 0 \quad \text{for all} \quad \mathbb{D} \in \mathscr{D},$$

*(iii) the non-autonomous set* $\mathbb{A}$ *is invariant, i.e.,*

$$U(t,\tau)A(\tau) = A(t) \quad \text{for} \quad -\infty < \tau \le t < +\infty,$$

*(iv) the sets* $A(t)$ *are given by*

$$A(t) = \overline{\bigcup_{\mathbb{D}\in\mathscr{D}} \Lambda(\mathbb{D},t)} \quad \text{for} \quad t \in \mathbb{R}.$$

*The non-autonomous set* $\mathbb{A}$*, termed the global pullback* $\mathscr{D}$*-attractor for the process* $U$*, is minimal in the sense that if* $\mathbb{C} = \{C(t)\}_{t\in\mathbb{R}}$ *with* $C(t) \in \mathscr{P}(X)$ *is a non-autonomous set such that* $C(t)$ *is closed and*

$$\lim_{\tau\to-\infty} \operatorname{dist}_X(U(t,\tau)B(\tau), C(t)) = 0,$$

*then* $A(t) \subset C(t)$*.*

*Proof.* It is enough to denote $\Omega = \mathbb{R}$, and $\theta_t\tau = \tau + t$, and to apply Theorem 11.4 to the non-autonomous dynamical system $\phi$ defined by

$$\phi(t,\tau,x) = U(t+\tau,\tau)x. \quad \square$$

*Remark 11.6.* The comments contained in Remark 11.4 also hold true for evolutionary processes.

## 11.4 Application to Two-Dimensional Navier–Stokes Equations in Unbounded Domains

Let $\Omega \subset \mathbb{R}^2$ be an open set, not necessarily bounded, with the boundary $\partial\Omega$, and suppose that $\Omega$ satisfies the Poincaré inequality, i.e., there exists a constant $\lambda_1 > 0$ such that

$$\lambda_1 \int_\Omega \phi^2\,dx \le \int_\Omega |\nabla\phi|^2\,dx \quad \text{for all} \quad \phi \in H_0^1(\Omega). \tag{11.28}$$

Consider the following two-dimensional Navier–Stokes problem:

$$\begin{cases} \dfrac{\partial u}{\partial t} - \nu\Delta u + \displaystyle\sum_{i=1}^{2} u_i \frac{\partial u}{\partial x_i} = f(t) - \nabla p \text{ in} & \Omega \times (\tau,+\infty), \\ \operatorname{div} u = 0 \qquad\qquad\qquad\qquad\qquad\quad \text{in} & \Omega \times (\tau,+\infty), \\ u = 0 \qquad\qquad\qquad\qquad\qquad\qquad\; \text{on} & \partial\Omega \times (\tau,+\infty), \\ u(\tau,x) = u_0(x) \qquad\qquad\qquad\qquad\; \text{for} & x \in \Omega. \end{cases}$$

To set our problem in the abstract framework, we consider the following usual abstract space

$$\tilde{V} = \left\{ u \in \left( C_0^\infty(\overline{\Omega}) \right)^2 \ : \ \operatorname{div} u \ = 0 \right\}.$$

We define $H$ as the closure of $\tilde{V}$ in $(L^2(\Omega))^2$ with the norm $|\cdot|$, and the inner product $(\cdot, \cdot)$ where for $u, v \in (L^2(\Omega))^2$ we set

$$(u, v) = \sum_{j=1}^{2} \int_\Omega u_j(x) v_j(x) \, dx.$$

Moreover we define $V$ as the closure of $\tilde{V}$ in $(H_0^1(\Omega))^2$ with the norm $\| \cdot \|$, and the associated scalar product $((\cdot, \cdot))$, where for $u, v \in (H_0^1(\Omega))^2$ we have

$$((u, v)) = \sum_{i,j=1}^{2} \int_\Omega \frac{\partial u_j}{\partial x_i} \frac{\partial v_j}{\partial x_i} \, dx.$$

It follows that $V \subset H \equiv H' \subset V'$, where the injections are continuous and dense.

Finally, we will use $\|\cdot\|_{V'}$ for the norm in $V'$ and $\langle \cdot, \cdot \rangle$ for the duality pairing between $V$ and $V'$.

Consider the trilinear form $b$ on $V \times V \times V$ given by

$$b(u, v, w) = \sum_{i,j=1}^{2} \int_\Omega u_i \frac{\partial v_j}{\partial x_i} w_j \, dx \quad \text{for} \quad u, v, w \in V.$$

Define $B : V \times V \to V'$ by

$$\langle B(u, v), w \rangle = b(u, v, w) \quad \text{for} \quad u, v, w \in V,$$

and denote

$$B(u) = B(u, u).$$

Assume now that $u_0 \in H, f \in L^2_{loc}(\mathbb{R}; V')$. For each $\tau \in \mathbb{R}$ we consider the problem

$$\begin{cases} u \in L^2(\tau, T; V) \cap L^\infty(\tau, T; H) \quad \text{for all} \quad T > \tau, \\ \dfrac{d}{dt}(u(t), v) + \nu((u(t), v)) + \langle B(u(t)), v \rangle = \langle f(t), v \rangle \quad \text{for all} \quad v \in V \\ \qquad \text{in the sense of scalar distributions on} \quad (\tau, +\infty), \\ u(\tau) = u_0. \end{cases} \tag{11.29}$$

It follows from the results of Chap. 7, cf. Theorem 7.2 and Remark 7.3, that problem (11.29) has a unique solution, $u(\cdot;\tau,u_0)$, that moreover belongs to $C([\tau,T];H)$ for all $T \geq \tau$.

Define

$$U(t,\tau)u_0 = u(t;\tau,u_0) \quad \text{for} \quad \tau \leq t \quad \text{and} \quad u_0 \in H. \tag{11.30}$$

From the uniqueness of solution to problem (11.29), it follows that

$$U(t,\tau)u_0 = U(t,r)U(r,\tau)u_0 \quad \text{for all} \quad \tau \leq r \leq t \quad \text{and} \quad u_0 \in H. \tag{11.31}$$

One can prove (cf. Chap. 7) that for all $\tau \leq t$, the mapping $U(\tau,t) : H \to H$ defined by (11.30) is continuous. Consequently, the family $\{U(t,\tau)\}_{\tau\leq t}$ defined by (11.30) is a process $U$ in $H$.

Moreover, it can be proved that $U$ is weakly continuous, and more exactly the following result holds true. Its proof is identical to the proof of Lemma 8.1 in [164], and so we omit it.

**Proposition 11.4.** *Let $\{u_{0n}\} \subset H$ be a sequence converging weakly in $H$ to an element $u_0 \in H$. Then*

$$U(t,\tau)u_{0n} \to U(t,\tau)u_0 \quad \text{weakly in} \quad H \quad \text{for all} \quad \tau \leq t, \tag{11.32}$$

$$U(\cdot,\tau)u_{0n} \to U(\cdot,\tau)u_0 \quad \text{weakly in} \quad L^2(\tau,T;V) \quad \text{for all} \quad \tau < T. \tag{11.33}$$

From now on, we denote

$$\sigma = \nu\lambda_1. \tag{11.34}$$

Let $\mathscr{R}_\sigma$ be the set of all functions $r : \mathbb{R} \to (0,+\infty)$ such that

$$\lim_{t\to-\infty} e^{\sigma t} r^2(t) = 0, \tag{11.35}$$

and denote by $\mathscr{D}_\sigma$ the class of all non-autonomous sets $\mathbb{D} = \{D(t)\}_{t\in\mathbb{R}}$ such that $D(t) \in \mathscr{P}(H)$ for $t \in \mathbb{R}$, for which there exists $r_{\mathbb{D}} \in \mathscr{R}_\sigma$ such that $D(t) \subset B(0,r_{\mathbb{D}}(t))$, where $B(0,r_{\mathbb{D}}(t))$ denotes the closed ball in $H$ centered at zero with radius $r_{\mathbb{D}}(t)$.

Now, we can prove the following result.

**Theorem 11.6.** *Suppose that $f \in L^2_{loc}(\mathbb{R};V')$ is such that*

$$\int_{-\infty}^{t} e^{\sigma\xi}\|f(\xi)\|^2_{V'}\,d\xi < +\infty \quad \text{for all} \quad t \in \mathbb{R}. \tag{11.36}$$

*Then, there exists a unique global pullback $\mathscr{D}_\sigma$-attractor for the process $U$ defined by (11.30).*

*Proof.* Let $\tau \in \mathbb{R}$ and $u_0 \in H$ be fixed, and denote

$$u(t) = u(t;\tau,u_0) = U(t,\tau)u_0 \quad \text{for all} \quad t \geq \tau.$$

Taking into account that $b(u,v,v) = 0$, it follows

$$\frac{d}{dt}\left(e^{\sigma t}|u(t)|^2\right) + 2\nu e^{\sigma t}\|u(t)\|^2 = \sigma e^{\sigma t}|u(t)|^2 + 2e^{\sigma t}\langle f(t), u(t)\rangle$$

in $\left(C_0^\infty(\tau,+\infty)\right)'$, and consequently, by (11.28),

$$e^{\sigma t}|u(t)|^2 \leq e^{\sigma \tau}|u_0|^2 + \frac{1}{\nu}\int_\tau^t e^{\sigma\xi}\|f(\xi)\|_{V'}^2\,d\xi \quad \text{for all} \quad \tau \leq t. \tag{11.37}$$

Let $\mathbb{D} \in \mathscr{D}_\sigma$ be given. From (11.37), we easily obtain

$$|U(t,\tau)u_0|^2 \leq e^{-\sigma(t-\tau)}r_{\mathbb{D}}^2(\tau) + \frac{e^{-\sigma t}}{\nu}\int_{-\infty}^t e^{\sigma\xi}\|f(\xi)\|_{V'}^2\,d\xi \tag{11.38}$$

for all $u_0 \in D(\tau)$, and all $t \geq \tau$.

Denote by $R_\sigma(t)$ the nonnegative number given for each $t \in \mathbb{R}$ by

$$(R_\sigma(t))^2 = \frac{2e^{-\sigma t}}{\nu}\int_{-\infty}^t e^{\sigma\xi}\|f(\xi)\|_{V'}^2\,d\xi,$$

and consider the non-autonomous set $\mathbb{B}_\sigma = \{B_\sigma(t)\}_{t\in\mathbb{R}}$ of closed balls in $H$ defined by

$$B_\sigma(t) = \{v \in H : \ |v| \leq R_\sigma(t)\}. \tag{11.39}$$

It is straightforward to check that $\mathbb{B}_\sigma \in \mathscr{D}_\sigma$, and moreover, by (11.35) and (11.38), the family $\mathbb{B}_\sigma$ is pullback $\mathscr{D}_\sigma$-absorbing for the process $U$.

According to Theorem 11.5 and Remark 11.6, to finish the proof of the theorem we only have to prove that $U$ is pullback $\mathscr{D}_\sigma$-asymptotically compact.

Let $\mathbb{D} \in \mathscr{D}_\sigma$, a sequence $\tau_n \to -\infty$, a sequence $u_{0n} \in D(\tau_n)$ and $t \in \mathbb{R}$, be fixed. We must prove that from the sequence $\{U(t,\tau_n)u_{0n}\}$ we can extract a subsequence that converges in $H$.

As the non-autonomous set $\mathbb{B}_\sigma$ is pullback $\mathscr{D}_\sigma$-absorbing, for each integer $k \geq 0$ there exists a $\tau_{\mathbb{D}}(k) \leq t-k$ such that

$$U(t-k,\tau)D(\tau) \subset B_\sigma(t-k) \quad \text{for all} \quad \tau \leq \tau_{\mathbb{D}}(k). \tag{11.40}$$

It is not difficult to conclude from (11.40), by a diagonal procedure, the existence of a subsequence $\{(\tau_{n'}, u_{0n'})\} \subset \{(\tau_n, u_{0n})\}$, and a sequence $\{w_k\}_{k=0}^{\infty} \subset H$, such that for all $k = 0, 1, 2, \ldots$ we have $w_k \in B_\sigma(t-k)$ and

$$U(t-k, \tau_{n'})u_{0n'} \to w_k \quad \text{weakly in} \quad H. \tag{11.41}$$

Observe that, by Proposition 11.4,

$$\begin{aligned} w_0 &= \text{weak} - \lim_{n'\to\infty} U(t, \tau_{n'})u_{0n'} \\ &= \text{weak} - \lim_{n'\to\infty} U(t, t-k)U(t-k, \tau_{n'})u_{0n'} \\ &= U(t, t-k)(\text{weak} - \lim_{n'\to\infty} U(t-k, \tau_{n'})u_{0n'}), \end{aligned}$$

i.e.,

$$U(t, t-k)w_k = w_0 \quad \text{for all} \quad k = 0, 1, 2, \ldots. \tag{11.42}$$

Then, by the weak lower semicontinuity of the norm,

$$|w_0| \le \liminf_{n'\to\infty} |U(t, \tau_{n'})u_{0n'}|. \tag{11.43}$$

If we now prove that also

$$\limsup_{n'\to\infty} |U(t, \tau_{n'})u_{0n'}| \le |w_0|, \tag{11.44}$$

then we will have

$$\lim_{n'\to\infty} |U(t, \tau_{n'})u_{0n'}| = |w_0|,$$

and this, together with the weak convergence, will imply the strong convergence in $H$ of $U(t, \tau_{n'})u_{0n'}$ to $w_0$.

In order to prove (11.44), we consider the Hilbert norm in $V$ given by

$$[u]^2 = \nu\|u\|^2 - \frac{\sigma}{2}|u|^2,$$

which by (11.28) is equivalent to the norm $\|u\|$ in $V$.

It is immediate that for all $\tau \le t$ and all $u_0 \in H$,

$$\begin{aligned} |U(t, \tau)u_0|^2 = {} & |u_0|^2 e^{\sigma(\tau - t)} \\ & + 2\int_\tau^t e^{\sigma(\xi - t)} \left(\langle f(\xi), U(\xi, \tau)u_0\rangle - [U(\xi, \tau)u_0]^2\right) d\xi, \end{aligned} \tag{11.45}$$

and, thus, for all $k \geq 0$ and all $\tau_{n'} \leq t-k$,

$$\begin{aligned} |U(t,\tau_{n'})u_{0n'}|^2 &= |U(t,t-k)U(t-k,\tau_{n'})u_{0n'}|^2 \\ &= e^{-\sigma k}|U(t-k,\tau_{n'})u_{0n'}|^2 \\ &\quad +2\int_{t-k}^{t} e^{\sigma(\xi-t)}\langle f(\xi), U(\xi,t-k)U(t-k,\tau_{n'})u_{0n'}\rangle\, d\xi \\ &\quad -2\int_{t-k}^{t} e^{\sigma(\xi-t)}[U(\xi,t-k)U(t-k,t-\tau_{n'})u_{0n'}]^2\, d\xi. \end{aligned} \tag{11.46}$$

As, by (11.40),

$$U(t-k,\tau_{n'})u_{0n'} \in B_\sigma(t-k) \quad \text{for all} \quad \tau_{n'} \leq \tau_{\mathbb{D}}(k) \quad \text{and} \quad k \geq 0,$$

we have

$$\limsup_{n'\to\infty}\left(e^{-\sigma k}|U(t-k,\tau_{n'})u_{0_{n'}}|^2\right) \leq e^{-\sigma k}R_\sigma^2(t-k) \quad \text{for all} \quad k \geq 0. \tag{11.47}$$

On the other hand, as $U(t-k,\tau_{n'})u_{0n'} \to w_k$ weakly in $H$, from Proposition 11.4 we have

$$U(\cdot,t-k)U(t-k,\tau_{n'})u_{0n'} \to U(\cdot,t-k)w_k \quad \text{weakly in} \quad L^2(t-k,t;V). \tag{11.48}$$

Taking into account that, in particular, $e^{\sigma(\xi-t)}f(\xi) \in L^2(t-k,t;V')$, we obtain from (11.48),

$$\begin{aligned} \lim_{n'\to\infty}\int_{t-k}^{t} e^{\sigma(\xi-t)}\langle f(\xi), U(\xi,t-k)U(t-k,\tau_{n'})u_{0n'}\rangle\, d\xi \\ = \int_{t-k}^{t} e^{\sigma(\xi-t)}\langle f(\xi), U(\xi,t-k)w_k\rangle\, d\xi. \end{aligned} \tag{11.49}$$

Moreover, as $\left(\int_{t-k}^{t} e^{\sigma(\xi-t)}[v(\xi)]^2\, d\xi\right)^{1/2}$ defines a norm in $L^2(t-k,t;V)$ which is equivalent to the usual one, we also obtain from (11.48),

$$\begin{aligned} &\int_{t-k}^{t} e^{\sigma(\xi-t)}[U(\xi,t-k)w_k)]^2\, d\xi \\ &\qquad \leq \liminf_{n'\to\infty}\int_{t-k}^{t} e^{\sigma(\xi-t)}[U(\xi,t-k)U(t-k,\tau_{n'})u_{0n'}]^2\, d\xi. \end{aligned} \tag{11.50}$$

Then, from (11.46), (11.47), (11.49), and (11.50), we easily obtain

$$\begin{aligned} &\limsup_{n'\to\infty}|U(t,\tau_{n'})u_{0n'}|^2 \\ &\leq e^{-\sigma k}R_\sigma^2(t-k) + 2\int_{t-k}^{t} e^{\sigma(\xi-t)}\left(\langle f(\xi), U(\xi,t-k)w_k\rangle - [U(\xi,t-k)w_k)]^2\right) d\xi. \end{aligned} \tag{11.51}$$

Now, from (11.42) and (11.45),

$$\begin{aligned}|w_0|^2 &= |U(t,t-k)w_k|^2 \\ &= |w_k|^2 e^{-\sigma k} + 2\int_{t-k}^{t} e^{\sigma(\xi-t)} \left(\langle f(\xi), U(\xi,t-k)w_k\rangle - [U(\xi,t-k)w_k]^2\right) d\xi.\end{aligned} \tag{11.52}$$

From (11.51) and (11.53) we have

$$\begin{aligned}\limsup_{n'\to\infty} |U(t,\tau_{n'})u_{0n'}|^2 &\le e^{-\sigma k} R_\sigma^2(t-k) + |w_0|^2 - |w_k|^2 e^{-\sigma k} \\ &\le e^{-\sigma k} R_\sigma^2(t-k) + |w_0|^2,\end{aligned}$$

and thus, taking into account that

$$e^{-\sigma k} R_\sigma^2(t-k) = \frac{2e^{-\sigma t}}{\nu} \int_{-\infty}^{t-k} e^{\sigma\xi} \|f(\xi)\|_{V'}^2 \, d\xi \to 0,$$

when $k \to +\infty$, we easily obtain (11.44) from the last inequality. □

In Chap. 13 we consider, in a similar framework, a problem from the theory of lubrication and prove the finite dimensionality of the pullback attractor.

## 11.5 Comments and Bibliographical Notes

The presentation in Sects. 11.1 and 11.2 follows [99]. Estimates on the number of determining modes and determining nodes can be found in [122]. We note here that another possibility of finite dimensional reduction of infinite dimensional dynamical systems is through the study of inertial manifolds [77] or of invariant manifolds [54]. For Navier–Stokes equations the approach by inertial manifolds leads to the so-called approximate inertial manifolds studied in [224], existence of inertial manifold for two-dimensional Navier–Stokes equations is still unresolved [103].

The results of Sects. 11.3 and 11.4 come from [45], where the notion of an asymptotically compact dynamical system as in Definition 11.2 was introduced.

The notion of pullback attractor is a direct generalization of that of global attractor to the non-autonomous case.

The first attempts to extend the notion of global attractor to the non-autonomous case led to the concept of the so-called uniform attractor (see [62]). The conditions ensuring the existence of the uniform attractor parallel those for autonomous systems. In this theory non-autonomous systems are lifted in [225] to autonomous ones by expanding the phase space. Then, the existence of uniform attractors relies on some compactness property of the solution operator associated to the resulting system. One disadvantage of uniform attractors is that they do not need to be invariant unlike the global attractor for autonomous systems. Moreover, they demand rather strong conditions on the time-dependent data.

At the same time, the theory of pullback (or cocycle) attractors has been developed for both the non-autonomous and random dynamical systems (see [80, 137, 152, 206, 232]), and has shown to be very useful in the understanding of the dynamics of non-autonomous dynamical systems. In this case, the concept of pullback (or cocycle or non-autonomous) attractor provides a time-dependent family of compact sets which attracts families of sets in a certain universe (e.g., the bounded sets in the phase space) and satisfying an invariance property, what seems to be a natural set of conditions to be satisfied for an appropriate extension of the autonomous concept of attractor. Moreover, this cocycle formulation allows to handle more general time-dependent terms in the models—not only the periodic, quasi-periodic, or almost periodic ones (see, for instance, [38], and [43] for non-autonomous models containing hereditary characteristics). Note that, however, the notion of pullback attractor has its limitations, namely it does not always capture the asymptotic behavior of the non-autonomous system as time advances to $+\infty$, see [140]. To counteract this limitation the synthesis of the approach by pullback attractors with the one by uniform attractors has been presented in [19, 53, 136]

In order to prove the existence of the attractor (in both the autonomous and non-autonomous cases) the simplest, and therefore the strongest, assumption is the compactness of the solution operator associated with the system, which is usually available for parabolic systems in bounded domains. However, this kind of compactness does not hold in general for parabolic equations in unbounded domains and second order in time equations on either bounded or unbounded domains. Instead we often have some kind of asymptotic compactness. In this situation, there are several approaches to prove the existence of the global (or non-autonomous) attractor. Roughly speaking, the first one ensures the existence of the global (resp. non-autonomous) attractor whenever a compact attracting set (resp. a family of compact attracting sets) exists. The second method consists in decomposing the solution operator (resp. the cocycle or two-parameter semigroup) into two parts: a compact part and another one which decays to zero as time goes to infinity. However, as it is not always easy to find such decomposition, one can use a third approach which is based on the use of the energy equations which are in direct connection with the concept of asymptotic compactness. This third method has been used in [164] (and also [181]) to extend to the non-autonomous situation the corresponding one in the autonomous framework (see [201] and also [11]), but related to uniform asymptotic compactness. The method used in this chapter is a generalization of the latter to the theory of pullback attractors.

More information on non-autonomous infinite dimensional dynamical systems can be found in monographs [53, 136] and the review article [10], cf. also [208]. For the random case, see, e.g., [30, 31, 39, 95, 135, 153]. Two-dimensional problems with delay have been studied in [37, 170].

# 12 Pullback Attractors and Statistical Solutions

This chapter is devoted to constructions of invariant measures and statistical solutions for non-autonomous Navier–Stokes equations in bounded and certain unbounded domains in $\mathbb{R}^2$.

After introducing some basic notions and results concerning attractors in the context of the Navier–Stokes equations, we construct the family of probability measures $\{\mu_t\}_{t\in\mathbb{R}}$ and prove the relations $\mu_t(E) = \mu_\tau(U(t,\tau)^{-1}E)$ for $t, \tau \in \mathbb{R}$, $t \geq \tau$ and Borel sets $E$ in $H$. Then we prove the Liouville and energy equations. Finally, we consider statistical solutions of the Navier–Stokes equations supported on the pullback attractor.

## 12.1 Pullback Attractors and Two-Dimensional Navier–Stokes Equations

We begin this section by recalling the notion of pullback attractor introduced in Chap. 11 in a more general context of cocycles.

Let us consider an evolutionary process $U$ (a process $U$—for short) on a metric space $X$, i.e., a family $\{U(t,\tau)\}_{-\infty<\tau\leq t<+\infty}$ of continuous mappings $U(t,\tau): X \to X$, such that $U(\tau,\tau)x = x$, and

$$U(t,\tau) = U(t,r)U(r,\tau) \quad \text{for all} \quad \tau \leq r \leq t.$$

Let $\mathscr{D}$ be an attraction universe, i.e., a nonempty family of non-autonomous sets $\mathbb{D} = \{D(t)\}_{t\in\mathbb{R}}$ such that $D(t) \in \mathscr{P}(X)$ for all $t \in \mathbb{R}$, where $\mathscr{P}(X)$ denotes the family of all nonempty subsets of $X$.



G. Łukaszewicz, P. Kalita, *Navier–Stokes Equations*, Advances in Mechanics and Mathematics 34, DOI 10.1007/978-3-319-27760-8_12

**Definition 12.1.** A process $U(\cdot,\cdot)$ is said to be pullback $\mathscr{D}$-asymptotically compact if for any $t \in \mathbb{R}$, any $\mathbb{D} \in \mathscr{D}$, any sequence $\tau_n \to -\infty$, and any sequence $x_n \in D(\tau_n)$, the sequence $\{U(t,\tau_n)x_n\}$ is relatively compact in $X$.

**Definition 12.2.** It is said that $\mathbb{B} \in \mathscr{D}$ is pullback $\mathscr{D}$-absorbing for the process $U(\cdot,\cdot)$ if for any $t \in \mathbb{R}$ and any $\mathbb{D} \in \mathscr{D}$, there exists a $\tau_0(t,\mathbb{D}) \le t$ such that

$$U(t,\tau)D(\tau) \subset B(t) \quad \text{for all} \quad \tau \le \tau_0(t,\mathbb{D}).$$

**Definition 12.3.** A non-autonomous set $\mathbb{A} = \{A(t)\}_{t\in\mathbb{R}}$ with $A(t) \in \mathscr{P}(X)$ for all $t \in \mathbb{R}$ is said to be a pullback $\mathscr{D}$-attractor for the process $U(\cdot,\cdot)$ if

(i) $A(t)$ is compact for all $t \in \mathbb{R}$,
(ii) $\mathbb{A}$ is pullback $\mathscr{D}$-attracting, i.e.,

$$\lim_{\tau\to-\infty} \operatorname{dist}_X(U(t,\tau)D(\tau), A(t)) = 0 \quad \text{for all} \quad \mathbb{D} \in \mathscr{D} \quad \text{and all} \quad t \in \mathbb{R},$$

(iii) $\mathbb{A}$ is invariant, i.e., $U(t,\tau)A(\tau) = A(t)$ for $-\infty < \tau \le t < +\infty$.

We consider the initial and boundary value problem for the Navier–Stokes equations studied already in Chap. 11. Namely, let $\Omega \subset \mathbb{R}^2$ be an open, bounded or unbounded, domain with smooth boundary $\partial\Omega$ such that there exists a constant $\lambda_1 > 0$ for which

$$\lambda_1 \int_\Omega \phi^2\,dx \le \int_\Omega |\nabla\phi|^2\,dx \quad \text{for all} \quad \phi \in H_0^1(\Omega).$$

Consider the following problem:

$$\begin{cases} \dfrac{\partial u}{\partial t} - \nu\Delta u + \displaystyle\sum_{i=1}^{2} u_i \frac{\partial u}{\partial x_i} = f(t) - \nabla p \text{ in} & \Omega \times (\tau,+\infty), \\ \operatorname{div} u = 0 \qquad\qquad\qquad\qquad\qquad\quad \text{in} & \Omega \times (\tau,+\infty), \\ u = 0 \qquad\qquad\qquad\qquad\qquad\qquad \text{on} & \partial\Omega \times (\tau,+\infty), \\ u(\tau,x) = u_0(x), \qquad\qquad\qquad\quad\;\; \text{for} & x \in \Omega. \end{cases}$$

This problem is set in a suitable abstract framework by considering the usual function space

$$\tilde{V} = \left\{ u \in \left(C_0^\infty(\overline{\Omega})\right)^2 \; : \; \operatorname{div} u \; = 0 \right\}.$$

The space $H$ is defined as the closure of $\tilde{V}$ in $(L^2(\Omega))^2$ with the norm $|\cdot|$, and the associated scalar product $(\cdot,\cdot)$ given by

$$(u,v) = \sum_{j=1}^{2} \int_\Omega u_j(x)v_j(x)\,dx \quad \text{for} \quad u,v \in (L^2(\Omega))^2.$$

The space $V$ is defined as the closure of $\tilde{V}$ in $(H_0^1(\Omega))^2$ with the norm $\|\cdot\|$, and the associated scalar product $((\cdot,\cdot))$, given by

$$(\nabla u, \nabla v) = ((u, v)) = \sum_{i,j=1}^{2} \int_\Omega \frac{\partial u_j}{\partial x_i} \frac{\partial v_j}{\partial x_i}\, dx \quad \text{for} \quad u, v \in (H_0^1(\Omega))^2.$$

It follows that $V \subset H \equiv H' \subset V'$, where the injections are continuous and dense.

Finally, we will use $\|\cdot\|_{V'}$ for the norm in $V'$ and $\langle\cdot,\cdot\rangle$ for the duality pairing between $V$ and $V'$.

Let $A : V \to V'$ be given by $\langle Au, v\rangle = ((u, v))$ for $u, v \in V$, and consider the trilinear form $b$ on $V \times V \times V$ given by

$$b(u, v, w) = \sum_{i,j=1}^{2} \int_\Omega u_i \frac{\partial v_j}{\partial x_i} w_j\, dx \quad \text{for} \quad u, v, w \in V.$$

Define $B : V \times V \to V'$ by $\langle B(u, v), w\rangle = b(u, v, w)$, for $u, v, w \in V$, and denote $B(u) = B(u, u)$. Assume now that $u_0 \in H, f \in L^2_{loc}(\mathbb{R}; V')$. For each $\tau \in \mathbb{R}$ we consider the problem

$$\begin{cases} u \in L^2(\tau, T; V) \cap L^\infty(\tau, T; H) \quad \text{for all} \quad T > \tau, \\ \dfrac{\mathrm{d}}{\mathrm{d}t}(u(t), v) + \nu((u(t), v)) + \langle B(u(t)), v\rangle = \langle f(t), v\rangle \quad \text{for all} \quad v \in V \\ \qquad \text{in the sense of scalar distributions on } (\tau, +\infty), \\ u(\tau) = u_0. \end{cases} \tag{12.1}$$

It follows from the results of Chap. 7, cf. Theorem 7.2 and Remark 7.3, that problem (12.1) has a unique solution, $u(\cdot; \tau, u_0)$, that moreover belongs to $C([\tau, T]; H)$ for all $T \geq \tau$.

Now we formulate a result about the existence of the pullback attractor for the above Navier–Stokes problem. We define the evolutionary process associated to (12.1) as follows. Let us consider the unique solution $u(\cdot; \tau, u_0)$ to (12.1). We set

$$U(t, \tau)u_0 = u(t; \tau, u_0) \quad \text{for} \quad \tau \leq t \quad \text{and} \quad u_0 \in H. \tag{12.2}$$

By the uniqueness of the solution we have

$$U(t, \tau)u_0 = U(t, r)U(r, \tau)u_0 \quad \text{for all} \quad \tau \leq r \leq t \quad \text{and} \quad u_0 \in H.$$

For all $\tau \leq t$, the mapping $U(\tau, t) : H \to H$ defined by (12.2) is continuous (cf. Theorem 7.2 and Remark 7.3). Thus, the family $\{U(t, \tau)\}_{\tau \leq t}$ defined by (12.2) is a process $U(\cdot,\cdot)$ in $H$.

Let $\sigma = \nu\lambda_1$ and $\mathscr{R}_\sigma$ be the set of all functions $r : \mathbb{R} \to (0, +\infty)$ such that

$$\lim_{t \to -\infty} e^{\sigma t} r^2(t) = 0,$$

and denote by $\mathcal{D}_\sigma$ the class of all non-autonomous sets $\mathbb{D} = \{D(t)\}_{t\in\mathbb{R}}$ with $D(t) \in \mathcal{P}(H)$ for all $t \in \mathbb{R}$ such that $D(t) \subset B(0, r_{\mathbb{D}}(t))$, for some $r_{\mathbb{D}} \in \mathcal{R}_\sigma$, where $B(0, r_{\mathbb{D}}(t))$ denotes the closed ball in $H$ centered at zero with radius $r_{\mathbb{D}}(t)$. Then we have (see Chap. 11).

**Theorem 12.1.** *Suppose that $f \in L^2_{loc}(\mathbb{R}; V')$ is such that*

$$\int_{-\infty}^{t} e^{\sigma\xi} \|f(\xi)\|^2_{V'} \, d\xi < +\infty \quad \textit{for all} \quad t \in \mathbb{R}. \tag{12.3}$$

*Then, the process $U(\cdot,\cdot)$ associated with problem (12.1) is $\mathcal{D}_\sigma$-asymptotically compact and there exists $\mathbb{B} \in \mathcal{D}_\sigma$ which is pullback $\mathcal{D}_\sigma$-absorbing for $U(t,\tau)$. In consequence, there exists a unique pullback $\mathcal{D}_\sigma$-attractor $\mathbb{A} = \{A(t)\}_{t\in\mathbb{R}}$ belonging to $\mathcal{D}_\sigma$ for the process $U(\cdot,\cdot)$ in $H$.*

The choice of the attraction universe $\mathcal{D}_\sigma$ comes here naturally from the first energy inequality (cf. Chap. 11)

$$|u(t)|^2 \le e^{-\sigma(t-\tau)}|u(\tau)|^2 + \frac{1}{\nu} e^{-\sigma t} \int_{\tau}^{t} e^{\sigma\xi} \|f(\xi)\|^2_{V'} \, d\xi \quad \text{for} \quad \tau \le t.$$

If $u(\tau) \in D(\tau) \subset B(0, r_{\mathbb{D}}(\tau))$, $r_{\mathbb{D}} \in \mathcal{R}_\sigma$, then the first term on the right-hand side converges to zero as $\tau \to -\infty$, and

$$|u(t)|^2 \le \frac{2}{\nu} e^{-\sigma t} \int_{-\infty}^{t} e^{\sigma\xi} \|f(\xi)\|^2_{V'} \, d\xi \equiv R(t) \quad \text{for all} \quad \tau \le \tau_0(t, \mathbb{D}). \tag{12.4}$$

Thus $U(t,\tau)D(\tau) \subset B(t)$, where $B(t)$ is the ball $B(0, R(t))$ in $H$. By (12.3), the non-autonomous set $\mathbb{B} = \{B(t)\}_{t\in\mathbb{R}}$ belongs to $\mathcal{D}_\sigma$ and is $\mathcal{D}_\sigma$-absorbing for the process $U(t,\tau)$. The uniqueness follows immediately from the fact that $\mathbb{A} \in \mathcal{D}_\sigma$ and from the attracting property.

Let us assume that the forcing $f$ is time-independent, with $f \in V'$. Then, from Theorem 12.1 there follows the existence of the global attractor $\mathcal{A} \subset H$ for the semigroup $\{S(t)\}_{t\ge 0}$ given by $S(t) = U(\tau + t, \tau)$, associated with the related autonomous Navier–Stokes problem (cf. Chap. 7).

**Theorem 12.2.** *There exists a unique global attractor for the semigroup $\{S(t)\}_{t\ge 0}$ in $H$ associated with the autonomous Navier–Stokes problem, i.e., problem (12.1) with $f \in V'$.*

Actually, the family $\mathcal{B}(H)$ of all bounded sets in $H$ constitutes, for this case, the attraction universe of non-autonomous (and, in this case, constant in time) sets $\mathcal{D}$ from Definition 12.2. From (12.4) we conclude that the ball $B(0, R)$ in $H$ with $R = 2/(\nu\sigma)\|f\|_{V'}$ absorbs all bounded sets in $H$, i.e., $S(t)B \subset B(0, R)$ for any bounded set $B \in \mathcal{B}(H)$ and all $t \ge t_0(B)$. The semigroup $\{S(t)\}_{t\ge 0}$ is $\mathcal{B}(H)$-asymptotically compact, that is, for any sequence $t_n \to \infty$ and any sequence $x_n$ bounded in $H$ the set $\{S(t_n)x_n\}_{n\in\mathbb{N}}$ is relatively compact in $H$. Moreover, $\mathcal{A} = \omega(B(0, R))$—the $\omega$-limit set of the ball $B(0, R)$.

## 12.2 Construction of the Family of Probability Measures

Our main result of this section (Theorem 12.3) concerns the construction of a family $\{\mu_t\}_{t\in\mathbb{R}}$ of time averaged probability measures in the phase space $H$ and their relation to the pullback attractor $\{A(t)\}_{t\in\mathbb{R}}$. Then (Lemma 12.3) we show how any two measures of the family are related to each other.

In Sects. 12.2–12.4 we work with somewhat stronger assumptions on the domain of the flow $\Omega$ and on the forcing $f$. Namely, we assume that $\Omega$ is an open bounded set in $\mathbb{R}^2$ of class $C^2$ and that $f \in L^2_{loc}(\mathbb{R}; H)$, with

$$\int_{-\infty}^{t} e^{\sigma\xi}|f(\xi)|^2 d\xi < \infty \tag{12.5}$$

for all $t \in \mathbb{R}$.

**Lemma 12.1.** *Let $\Omega$ be an open bounded set in $\mathbb{R}^2$ of class $C^2$, $f \in L^2_{loc}(\mathbb{R}; H)$ satisfy (12.5), and let $\{U(t,\tau)\}_{\tau\le t}$ be the evolutionary process associated with problem (12.1). Then, for every $t \in \mathbb{R}$ there exists a compact set $B(t)$ absorbing for $\{U(t,\tau)\}_{\tau\le t}$.*

*Proof.* The proof is based on the uniform Gronwall lemma (see Chap. 3) applied to the second energy inequality (see Chap. 7 and also Chap. III, Sect. 2 in [220])

$$\frac{d}{dt}\|u\|^2 + \sigma\|u\|^2 \le \frac{2}{\nu}|f|^2 + \frac{c}{\nu^3}|u|^2\|u\|^4.$$

As usual, we use the first energy inequality (see Chap. 7)

$$\frac{d}{dt}|u|^2 + \nu\|u\|^2 \le \frac{1}{\sigma}|f|^2 \tag{12.6}$$

and its consequence

$$|u(t)|^2 \le e^{-\sigma(t-\tau)}|u(\tau)|^2 + \frac{e^{-\sigma t}}{\sigma}\int_{-\infty}^{t} e^{\sigma\xi}|f(\xi)|^2 d\xi, \tag{12.7}$$

together with (12.5) to obtain a bound on the norm $\|u(t)\|$. However, the bound on the forcing (12.5) is not uniform in $t \in \mathbb{R}$, and this implies that our estimates are local in time. More precisely, we shall prove that for any $u(\tau) \in D(\tau)$ for $D(\tau)$ in any non-autonomous set $\mathbb{D} = \{D(t)\}_{t\in\mathbb{R}} \in \mathscr{D}$, any fixed $t \in \mathbb{R}$, and $T$, $r$, with $0 < r < T$, there exists $M(t,T,r)$ such that

$$\|U(\eta,\tau)u(\tau)\| \le M(t,T,r) \quad \text{for all} \quad \eta \in [t-T+r, t] \quad \text{and} \quad \tau \le \tau_0(t,T,\mathbb{D}). \tag{12.8}$$

Since $\Omega$ is a bounded set, $V$ is compactly embedded in $H$, and (12.8) implies that the closed ball $B(t) = B_V(0, M(t,T,r))$ in $V$ is a compact absorbing set for $U(t,\tau)$.

First, from (12.7) and (12.5) it follows that for all $\eta \in (t-T,t)$,

$$|U(\eta,\tau)u(\tau)|^2 \le 2\frac{e^{\sigma T}}{\sigma}\int_{-\infty}^{t} e^{\sigma\xi}|f(\xi)|^2 d\xi \equiv M_1(t,T) \quad \text{for all} \quad \tau \le \tau_0(t,T,\mathbb{D}). \tag{12.9}$$

Since $f \in L^2_{loc}(\mathbb{R};H)$, we have for all $s \in (t-T,t-r)$ and some $M_2(t,T,r) < \infty$,

$$\frac{1}{\sigma}\int_{s}^{s+r} |f(\xi)|^2 d\xi \le M_2(t,T,r). \tag{12.10}$$

Now, from (12.6) together with (12.9) and (12.10) we have for all $s \in (t-T,t-r)$ and some $M_3(t,T,r) < \infty$,

$$\sigma\int_{s}^{s+r} \|u(\xi)\|^2 d\xi \le M_3(t,T,r). \tag{12.11}$$

In the end, from (12.9) and (12.11) we obtain for all $s \in (t-T,t-r)$ and some $M_4(t,T,r) < \infty$,

$$\frac{c}{\nu^3}\int_{s}^{s+r} |u(\xi)|^2\|u(\xi)\|^2 d\xi \le M_4(t,T,r).$$

The last three estimates and the uniform Gronwall lemma give (12.8). □

Below we recall the notion of a Banach generalized limit introduced in Chap. 8. Since the existence of a generalized limit follows from the Hahn–Banach theorem which does not guarantee uniqueness, in the sequel we assume implicitly that we work with any given generalized limit.

**Definition 12.4.** By a Banach generalized limit we name any linear functional, denoted $\mathrm{LIM}_{T\to\infty}$, defined on the space of all bounded real-valued functions on $[0,\infty)$ and satisfying

(i) $\mathrm{LIM}_{T\to\infty} g(T) \ge 0$ for nonnegative functions $g$.
(ii) $\mathrm{LIM}_{T\to\infty} g(T) = \lim_{T\to\infty} g(T)$ if the usual limit $\lim_{T\to\infty} g(T)$ exists.

**Theorem 12.3.** *Let $\Omega$ be an open bounded set in $\mathbb{R}^2$ of class $C^2$ and $f \in L^2_{loc}(\mathbb{R};H)$ satisfy (12.5). Let $\{U(t,\tau)\}_{t\ge\tau}$ be the evolutionary process associated with the Navier–Stokes system (12.1) and $\{A(t)\}_{t\in\mathbb{R}}$ be the pullback attractor from Theorem 12.1. Let $u_0$ be an arbitrary element of the phase space $H$. Then there exists a family of Borel probability measures $\{\mu_t\}_{t\in\mathbb{R}}$ in the phase space $H$ such that the support of measure $\mu_t$ is contained in $A(t)$ and*

$$\mathrm{LIM}_{\tau\to-\infty}\frac{1}{t-\tau}\int_{\tau}^{t} \varphi(U(t,s)u_0)ds = \int_{A(t)} \varphi(u)d\mu_t(u) \tag{12.12}$$

*for all real-valued and continuous on $H$ functionals $\varphi$, ($\varphi \in C(H)$).*

*Proof.* First we prove a lemma.

**Lemma 12.2.** *For $u_0 \in H$ and $t \in \mathbb{R}$ the function*

$$s \to \varphi(U(t,s)u_0) \tag{12.13}$$

*is continuous and bounded on* $(-\infty, t]$.

*Proof.* We first prove that for any $\epsilon > 0$ there exists a $\delta > 0$ such that if $|r - s| < \delta$ then $|U(t,r)u_0 - U(t,s)u_0| < \epsilon$. We may assume that $r < s$.

It is easy to check (cf. Chap. 7) that if $u$ and $v$ are two solutions of the Navier–Stokes equations in a domain satisfying the Poincaré inequality then

$$|u(t) - v(t)|^2 \leq |u(0) - v(0)|^2 \exp\left\{C \int_0^t \|u(s)\|^2 ds\right\}.$$

We have thus,

$$\begin{aligned}|U(t,r)u_0 - U(t,s)u_0|^2 &= |U(t,s)U(s,r)u_0 - U(t,s)u_0|^2 \\ &\leq |U(s,r)u_0 - u_0|^2 \exp\left\{C \int_s^t \|U(\tau,s)u_0\|^2 d\tau\right\}.\end{aligned}$$

From the continuity of the function $r \leq \tau \to U(\tau, r)u_0$ with values in $H$ it follows that if $|r - s| < \delta$ is small enough then the right-hand side of the above inequality is as small as needed. Since $\varphi \in C(H)$ we conclude that the function in (12.13) is continuous.

As to its boundedness we know that there exists a compact absorbing set $B_1(t)$ for $\{U(t,\tau)\}$, in particular, there exists $\tau_0$ such that for $s \leq \tau_0$, $U(t,s)u_0 \in B_1(t)$. Hence, the function $s \to \varphi(U(t,s)u_0)$ is bounded on $(-\infty, \tau_0]$. It is also bounded on the compact interval $[\tau_0, t]$. □

We continue the proof of the theorem. Since the function in (12.13) is continuous and bounded, the left-hand side of (12.12) is well defined.

We have

$$\text{LIM}_{\tau\to-\infty} \frac{1}{t-\tau} \int_\tau^t \varphi(U(t,s)u_0)ds = \text{LIM}_{\tau\to-\infty} \frac{1}{t-\tau} \int_\tau^{\tau_0} \varphi(U(t,s)u_0)ds.$$

From this equality it follows that the left-hand side of (12.12) depends only on the values of $\varphi$ on the compact set $B_1(t)$. By the Tietze extension theorem, the left-hand side of (12.12) defines a linear positive functional on $C(B_1(t))$ and from the Kakutani–Riesz representation theorem it follows that there exists a measure $\mu_t^1$ on $B_1(t)$ defined on some $\sigma$-algebra $\mathscr{M}$ in $B_1(t)$ which contains all Borel sets in $B_1(t)$ and such that

$$\text{LIM}_{\tau\to-\infty} \frac{1}{t-\tau} \int_\tau^t \varphi(U(t,s)u_0)ds = \int_{B_1(t)} \varphi(u)d\mu_t^1(u).$$

We can extend this measure easily to all Borel sets in $H$ without enlarging its support.

We shall show that the support of the measure $\mu_t^1$ is contained in $A(t)$. Let $B_n(t)$, $n = 1, 2, \ldots$, be a sequence of compact absorbing sets such that $B_n(t) \supset B_{n+1}(t)$ and $\cap_{n=1}^{\infty} B_n(t) = A(t)$. From the above construction, for every $n$ there exists a measure $\mu_t^n$ of support contained in $B_n(t)$ and such that

$$\mathrm{LIM}_{\tau \to -\infty} \frac{1}{t-\tau} \int_{\tau}^{t} \varphi(U(t,s)u_0)ds = \int_{B_n(t)} \varphi(u)d\mu_t^n(u)$$

holds. As the left-hand side of this equation is independent of $n$, it follows that for $\varphi \in C(H)$,

$$\int_{B_1(t)} \varphi(u)d\mu_t^1(u) = \int_{B_n(t)} \varphi(u)d\mu_t^n(u). \tag{12.14}$$

By the Tietze extension theorem, every $\phi \in C(B_n(t))$ is of the form $\varphi|_{B_n(t)}$ for some $\varphi \in C(B_1(t))$. For any $\psi \in C_0(B_1(t) \setminus B_n(t))$ both $\varphi$ and $\varphi + \psi$ are continuous extensions of $\phi$. It then follows that

$$\int_{B_1(t) \setminus B_n(t)} \psi(u)d\mu_t^1(u) = 0.$$

For any compact $K$ in $B_1(t) \setminus B_n(t)$ there exist $\psi \in C_0(B_1(t) \setminus B_n(t))$ such that $\psi \geq \chi_K$ on $B_1(t) \setminus B_n(t)$ where $\chi_K$ is the characteristic function of the set $K$. We have

$$0 = \int_{B_1(t) \setminus B_n(t)} \psi(u)d\mu_t^1(u) \geq \int_{B_1(t) \setminus B_n(t)} \chi_K(u)d\mu_t^1(u) = \mu_t^1(K) \geq 0,$$

whence $\mu_t^1(K) = 0$ for every compact set $K$ in $B_1(t) \setminus B_n(t)$. From the regularity of $\mu_t^1$ it follows that $\mu_t^1(B_1(t) \setminus B_n(t)) = 0$. Therefore, the support of $\mu_t^1$ is contained in $B_n(t)$. From (12.14) and the Tietze extension theorem we conclude that

$$\int_{B_n(t)} \varphi(u)d\mu_t^1(u) = \int_{B_n(t)} \varphi(u)d\mu_t^n(u)$$

for all continuous functions $\varphi$ on $B_n(t)$. By the regularity of measures $\mu_t^1$ and $\mu_t^n$, the above equation holds for all $\mu_t^1$- and $\mu_t^n$-integrable functions. Thus $\mu_t^1 = \mu_t^n$ on $B_n(t)$ and on the whole phase space $H$. In the end, the support of $\mu_t^1$ is contained in every set $B_n(t)$ with $n \geq 1$ which implies that it is contained in $\cap_{n=1}^{\infty} B_n(t) = A(t)$. The resulting measure with support in $A(t)$ is denoted by $\mu_t$. Setting $\varphi = 1$ in (12.12) we obtain $\mu_t(A(t)) = 1$, and so $\mu_t$ is a probability measure. The proof of the theorem is complete. □

**Lemma 12.3.** *The measures* $\{\mu_t\}_{t\in\mathbb{R}}$ *satisfy the following relations:*

$$\mu_t(E) = \mu_\tau(U(t,\tau)^{-1}E) \quad \textit{for all} \quad t,\tau \in \mathbb{R} \quad \textit{such that} \quad t \geq \tau \tag{12.15}$$

*for every Borel set E in the phase space H.*

*Proof.* We have for $t \geq \tau$, and for all functionals $\varphi \in C(H)$,

$$\begin{aligned}
\int_H \varphi(u)d\mu_t(u) &= \mathrm{LIM}_{M\to-\infty}\frac{1}{t-M}\int_M^t \varphi(U(t,s)u_0)ds \\
&= \mathrm{LIM}_{M\to-\infty}\frac{1}{t-M}\int_M^\tau \varphi(U(t,s)u_0)ds \\
&\qquad +\mathrm{LIM}_{M\to-\infty}\frac{1}{t-M}\int_\tau^t \varphi(U(t,s)u_0)ds \\
&= \mathrm{LIM}_{M\to-\infty}\frac{1}{\tau-M}\int_M^\tau \varphi(U(t,\tau)U(\tau,s)u_0)ds \\
&= \int_H \varphi(U(t,\tau)u)d\mu_\tau(u),
\end{aligned}$$

whence

$$\int_H \varphi(u)d\mu_t(u) = \int_H \varphi(U(t,\tau)u)d\mu_\tau(u).$$

Using the regularity of $\mu_t$, the Urysohn lemma, and a suitable approximation argument we can extend this relation to all $\mu_t$-integrable functions $\varphi$ in $H$ and then it becomes equivalent to the generalized invariance property (12.15). □

## 12.3 Liouville and Energy Equations

In this section we prove that the family of probability measures constructed in Sect. 12.2 satisfies the Liouville equation for a suitable family of generalized moments and also the energy equation.

**Definition 12.5 (Cf. [96, 99]).** By $\mathscr{T}$ we denote the class of cylindrical test functionals consisting of the real-valued functionals $\Phi = \Phi(u)$ depending only on a finite number $k$ of components of $u$, namely,

$$\Phi(u) = \phi((u,g_1),\ldots,(u,g_k)),$$

where $\phi$ is a compactly supported $C^1$ scalar function on $\mathbb{R}^k$ and where $g_1,\ldots,g_k$ belong to $V$.

Let $\Phi'$ denote the differential of $\Phi$ in $H$, given as

$$\Phi'(u) = \sum_{j=1}^{k} \partial_j \phi((u, g_1), \ldots, (u, g_k)) g_j, \tag{12.16}$$

where $\partial_j \phi, j = 1, \ldots, k$, are partial derivatives of $\phi$ on $\mathbb{R}^k$. From (12.16) it follows that $\Phi'(u) \in V$.

**Theorem 12.4.** *The family of measures $\{\mu_t\}_{t\in\mathbb{R}}$ satisfies the following Liouville equation:*

$$\begin{aligned}\int_{A(t)} \Phi(u) d\mu_t(u) &- \int_{A(\tau)} \Phi(u) d\mu_\tau(u) \\ &= \int_\tau^t \int_{A(s)} (F(u,s), \Phi'(u)) d\mu_s(u) ds, \end{aligned}\tag{12.17}$$

*for all $t \geq \tau$, where $F(u,s) = f(s) - \nu Au - B(u)$, and $\Phi \in \mathscr{T}$.*

*Proof.* Let $u(t)$ be a solution of the Navier–Stokes system

$$\frac{du}{dt} + \nu Au + B(u) = f(t) \quad \text{and} \quad u(t)|_{t=s} = u_0 \quad \text{where} \quad s \in \mathbb{R}.$$

Since $u \in L^2(\tau, t; D(A))$ for $s < \tau < t$, it is not difficult to check that the function $t \to (f(t) - \nu Au(t) - B(u(t)), \Phi'(u(t)))$ is integrable on $(\tau, t)$. From the Navier–Stokes equation it follows that $u_t \in L^1(\tau, t, H)$. Thus for $\Phi \in \mathscr{T}$ we have

$$\begin{aligned}\frac{d}{dt}\Phi(u(t)) &= (u_t(t), \Phi'(u(t))) \\ &= (f(t) - \nu Au - B(u), \Phi'(u(t))) = (F(u(t), t), \Phi'(u(t)))\end{aligned}$$

almost everywhere in the considered interval. Integrating in $t$ we obtain

$$\Phi(u(t)) - \Phi(u(\tau)) = \int_\tau^t (F(u(\eta), \eta), \Phi'(u(\eta))) d\eta.$$

For an arbitrary $s < \tau$, let $u_0 \in H$ and $u(\eta) = U(\eta, s)u_0$ for $\eta \geq s$. Then, for the above solution we have

$$\Phi(U(t,s)u_0) - \Phi(U(\tau,s)u_0) = \int_\tau^t (F(U(\eta,s)u_0, \eta), \Phi'(U(\eta,s)u_0)) d\eta.$$

Applying the operator

$$\mathrm{LIM}_{M\to-\infty}\frac{1}{\sigma-M}\int_M^\sigma(\ldots)ds$$

for $\sigma<\tau$ we obtain

$$\begin{aligned}&\mathrm{LIM}_{M\to-\infty}\frac{1}{\sigma-M}\int_M^\sigma \Phi(U(t,s)u_0)ds-\mathrm{LIM}_{M\to-\infty}\frac{1}{\sigma-M}\int_M^\sigma \Phi(U(\tau,s)u_0)ds\\&=\mathrm{LIM}_{M\to-\infty}\frac{1}{\sigma-M}\int_M^\sigma\int_\tau^t(F(U(\eta,s)u_0,\eta),\Phi'(U(\eta,s)u_0))d\eta ds.\end{aligned}\tag{12.18}$$

Now, in view of (12.15), (cf. also Chap. II in [228]) the left-hand side equals

$$\begin{aligned}&\mathrm{LIM}_{M\to-\infty}\frac{1}{\sigma-M}\left\{\int_M^\sigma \Phi(U(t,\sigma)U(\sigma,s)u_0)ds-\int_M^\sigma \Phi(U(\tau,\sigma)U(\sigma,s)u_0)ds\right\}\\&=\int\Phi(U(t,\sigma)u)d\mu_\sigma(u)-\int\Phi(U(\tau,\sigma)u)d\mu_\sigma(u)\\&=\int\Phi(u)d\mu_t(u)-\int\Phi(u)d\mu_\tau(u).\end{aligned}$$

Similarly, the right-hand side of (12.18) equals

$$\begin{aligned}&\mathrm{LIM}_{M\to-\infty}\frac{1}{\sigma-M}\int_M^\sigma\int_\tau^t(F(U(\eta,s)u_0,\eta),\Phi'(U(\eta,s)u_0))d\eta ds\\&=\mathrm{LIM}_{M\to-\infty}\frac{1}{\sigma-M}\int_\tau^t\left\{\int_M^\sigma(F(U(\eta,s)u_0,\eta),\Phi'(U(\eta,s)u_0))ds\right\}d\eta\\&=\int_\tau^t\left\{\mathrm{LIM}_{M\to-\infty}\frac{1}{\sigma-M}\int_M^\sigma(F(U(\eta,s)u_0,\eta),\Phi'(U(\eta,s)u_0))ds\right\}d\eta.\end{aligned}$$

For $\eta\in(\tau,t)$ we have

$$\begin{aligned}&\mathrm{LIM}_{M\to-\infty}\frac{1}{\sigma-M}\int_M^\sigma(F(U(\eta,s)u_0,\eta),\Phi'(U(\eta,s)u_0))ds\\&=\mathrm{LIM}_{M\to-\infty}\frac{1}{\sigma-M}\int_M^\sigma(F(U(\eta,\sigma)U(\sigma,s)u_0,\eta),\Phi'(U(\eta,\sigma)U(\sigma,s)u_0))ds\\&=\int_H(F(U(\eta,\sigma)u,\eta),\Phi'(U(\eta,\sigma)u))d\mu_\sigma(u)\\&=\int_H(F(u,\eta),\Phi'(u))d\mu_\eta(u),\end{aligned}$$

whence

$$\mathrm{LIM}_{M\to-\infty}\frac{1}{\sigma-M}\int_M^\sigma\int_\tau^t (F(U(\eta,s)u_0,\eta),\Phi'(U(\eta,s)u_0))d\eta ds$$
$$=\int_\tau^t\int_H (F(u,\eta),\Phi'(u))d\mu_\eta(u)\,d\eta.$$

We have used the integrability of the function

$$(t,\tau)\times(M,\sigma)\ni(\eta,s)\to(F(U(\eta,s)u_0,\eta),\Phi'(U(\eta,s)u_0))$$

for $\sigma<\tau$, the Fubini theorem, and the linearity of the $\mathrm{LIM}_{M\to-\infty}$ operator.

In the end, since for every $t$ in $\mathbb{R}$ the support of the measure $\mu_t$ is contained in $A(t)$, we arrive at (12.17). □

**Theorem 12.5.** *The family of measures $\{\mu_t\}_{t\in\mathbb{R}}$ satisfies the following energy equation:*

$$\int_{A(t)}|u|^2d\mu_t(u)+2\nu\int_\tau^t\int_{A(s)}\|u\|^2d\mu_s ds$$
$$=2\int_\tau^t\int_{A(s)}(f(s),u)d\mu_s(u)ds+\int_{A(\tau)}|u|^2d\mu_\tau(u) \quad (12.19)$$

*for all $t,\tau\in\mathbb{R}$, $t\geq\tau$.*

*Proof.* The functionals $u\to\Phi(u)=|u|^2$ and $u\to\|u\|^2$ are bounded (even continuous in the topology of $H$) on every $A(t)$, $t\in\mathbb{R}$. Moreover, $\Phi'(u)=2u$. From (12.17) with this $\Phi$ and $F(u,s)=f(s)-\nu Au-B(u)$ we obtain (12.19) by an approximation argument. Let the functionals $\Phi_m\in\mathscr{T}$, $m=1,2,3,\ldots$, be such that $\Phi_m(u)=|P_mu|^2$ on some ball $B(0,R)$ in $H$ containing all $A(\eta)$ for $\tau\leq\eta\leq t$. Here, $P_m$ is the usual Galerkin projection operator in $H$ onto the space spanned by the first $m$ eigenmodes of the Stokes operator. The identity (12.19) thus holds for $P_mu$ in the place of $u$. Letting $m\to\infty$ and using the Lebesgue dominated convergence theorem we obtain (12.19). □

## 12.4 Time-Dependent and Stationary Statistical Solutions

A simple corollary from the above considerations is, that assuming any measure $\mu_\tau$ of the family $\{\mu_t\}_{t\in\mathbb{R}}$ to be the initial distribution, the family $\{\mu_t\}_{t\geq\tau}$ forms a statistical solutions of the two-dimensional Navier–Stokes system (Theorem 12.6, cf. Chap. 5 in [99]) for this particular initial data. This solution is unique and it reduces to the stationary time-average statistical solution in the case the forcing does not depend on time. The support of this stationary solution is contained in the corresponding global attractor (Theorem 12.7, cf. Chap. 8 and also Chap. 4 in [99]).

**Theorem 12.6.** *Let $\{U(t,\tau)\}_{t\geq\tau}$ be the evolutionary process associated with the Navier–Stokes system (12.1) and let $u_0$ be an arbitrary element of the phase space $H$. Denote by $\tilde{\mu}_0$ the time-average initial measure given by the relation*

$$\mathrm{LIM}_{\tau\to-\infty}\frac{1}{0-\tau}\int_{\tau}^{0}\varphi(U(0,s)u_0)ds=\int_H\varphi(u)d\tilde{\mu}_0(u)$$

*for all functionals $\varphi\in C(H)$. Then the support of $\tilde{\mu}_0$ is contained in the pullback attractor's section $A(0)$ and the kinetic energy of $\tilde{\mu}_0$ is finite*

$$\int_H|u|^2d\tilde{\mu}_0(u)<+\infty.$$

*Moreover, the family of measures $\{\mu_t\}_{t\geq 0}$ given by*

$$\mu_t(E)=\tilde{\mu}_0(U(t,0)^{-1}E)$$

*for all Borel sets in $H$, is the unique statistical solution of the two-dimensional Navier–Stokes system (12.1) with initial distribution $\tilde{\mu}_0$, namely, it satisfies the following conditions:*

*(S1) the Liouville equation*

$$\int_{A(t)}\Phi(u)d\mu_t(u)=\int_{A(0)}\Phi(u)d\tilde{\mu}_0(u)+\int_0^t\int_{A(s)}(F(u,s),\Phi'(u))d\mu_s(u)ds,$$

*holds for all $t\geq 0$, $F(u,s)=f(s)-\nu Au-B(u)$, and the family of moments $\Phi\in\mathscr{T}$,*

*(S2) the energy equation*

$$\int_{A(t)}|u|^2d\mu_t(u)+2\nu\int_0^t\int_{A(s)}\|u\|^2d\mu_s(u)ds$$
$$=\int_{A(0)}|u|^2d\tilde{\mu}_0(u)+2\int_0^t\int_{A(s)}(f(s),u)d\mu_s(u)ds$$

*holds for all $t\geq 0$,*

*(S3) the function*

$$t\to\int_{A(t)}\Phi(u)d\mu_t(u)$$

*is measurable on $\mathbb{R}^+$ for every bounded and continuous real-valued functional $\Phi$ on $H$. The function*

$$t \to \int_{A(t)} |u|^2 d\mu_t(u)$$

*is in $L^\infty(\mathbb{R}^+)$. The function*

$$t \to \int_{A(t)} \|u\|^2 d\mu_t(u)$$

*is in $L^1_{loc}(\mathbb{R}^+)$.*

If the system (12.1) is autonomous, that is, the forcing $f$ does not depend on time then Theorem 12.6 reduces to the following one.

**Theorem 12.7.** *Let $\{S(t)\}_{t\geq 0}$ be the semigroup associated with the autonomous Navier–Stokes system*

$$\frac{du}{dt} + \nu A u + B(u) = f \in H \quad \text{with} \quad u(t)|_{t=0} = u_0,$$

*and let $u_0$ be an arbitrary element of the phase space $H$. Then the time-average measure $\mu$ given by the relation*

$$\begin{aligned} \mathrm{LIM}_{\tau\to-\infty} \frac{1}{0-\tau} \int_\tau^0 \varphi(S(0-s)u_0)ds &= \mathrm{LIM}_{\tau\to+\infty} \frac{1}{\tau} \int_0^\tau \varphi(S(s)u_0)ds \\ &= \int_H \varphi(u) d\mu(u) \end{aligned}$$

*for all functionals $\varphi \in C(H)$ has support contained in the global attractor $\mathscr{A}$. Moreover, it is a stationary statistical solution of the considered Navier–Stokes system, namely, the following conditions hold:*

*(ST1)*

$$\int_H \|u\|^2 d\mu(u) < \infty,$$

*(ST2)*

$$\int_{\mathscr{A}} (F(u,s), \Phi'(u)) d\mu(u) = 0,$$

*where $F(u,s) = f(s) - \nu A u - B(u)$, and $\Phi$ belongs to* T,

*(ST3)*

$$\int_{E_1 \le |u|^2 < E_2} \left( \|u\|^2 - (f, u) \right) d\mu \le 0$$

*for all* $E_1, E_2$ *such that* $0 \le E_1 < E_2 \le \infty$.

## 12.5 The Case of an Unbounded Domain

In this section we do not assume that the domain of the flow is bounded. In consequence, we do not assume the existence of compact absorbing sets for the process $U(t, \tau)$, the assumption we needed in Sect. 12.2 to construct a family of invariant measures. Moreover, we show that we can obtain time-average invariant measures for $U(t, \tau)$ using as an initial condition any continuous mapping that is contained in the domain of attraction of the pullback attractor. Thus we obtain a twofold generalization of Theorem 12.3.

We need the following notion of continuity of the process $U(t, \tau)$.

**Definition 12.6.** A process $U(\cdot, \cdot)$ is said to be $\tau$-continuous if for every $u_0$ in $X$ and every $t \in \mathbb{R}$, the $X$-valued function

$$\tau \mapsto U(t, \tau)u_0 \tag{12.20}$$

is continuous and bounded on $(-\infty, t]$.

Our results for the Navier–Stokes equations in certain unbounded domains are direct consequences of the following abstract theorem:

**Theorem 12.8.** *Let* $U(\cdot, \cdot)$ *be a* $\tau$*-continuous evolutionary process in a complete metric space* $X$ *that has a pullback* $\mathscr{D}$*-attractor* $\mathbb{A} = \{A(t)\}_{t\in\mathbb{R}}$. *Fix a generalized Banach limit* $\mathrm{LIM}_{T\to\infty}$ *and let* $u : \mathbb{R} \to X$ *be a continuous map such that* $u(\cdot) \in \mathscr{D}$. *Then there exists a unique family of Borel probability measures* $\{\mu_t\}_{t\in\mathbb{R}}$ *in* $X$ *such that the support of the measure* $\mu_t$ *is contained in* $A(t)$ *and*

$$\mathrm{LIM}_{\tau\to-\infty} \frac{1}{t - \tau} \int_\tau^t \varphi(U(t, s)u(s))\, ds = \int_{A(t)} \varphi(v)\, d\mu_t(v) \tag{12.21}$$

*for any real-valued continuous functional* $\varphi$ *on* $X$. *In addition,* $\mu_t$ *is invariant in the sense that*

$$\int_{A(t)} \varphi(v)\, d\mu_t(v) = \int_{A(\tau)} \varphi(U(t, \tau)v)\, d\mu_\tau(v) \quad \text{for} \quad t \ge \tau. \tag{12.22}$$

*Remark 12.1.* Theorem 12.8 holds for any pullback attractor $\mathbb{A} = \{A(t)\}_{t\in\mathbb{R}}$. Assume that there exists a minimal pullback attractor $\mathbb{A}_m = \{A_m(t)\}_{t\in\mathbb{R}}$ for the process $U(\cdot,\cdot)$. Then the support of measure $\mu_t$ from Theorem 12.8 is contained in $A_m(t)$ and in formulas (12.21), (12.22) one can replace $A(t)$ by $A_m(t)$.

In the proof of the theorem we will use the following lemma, cf. [197]:

**Lemma 12.4.** *Let $(X,\rho)$ be a metric space, $f : X \to \mathbb{R}$ a continuous mapping, and let $K \subset X$ be compact. Then for every $\epsilon > 0$ there exists a $\delta = \delta_f(\epsilon) > 0$ such that if $u \in X$, $v \in K$, and $\rho(u,v) \le \delta$ then $|f(u) - f(v)| \le \epsilon$.*

*Proof (of Theorem 12.8).* Fix $u(\cdot) \in \mathscr{D}$ and $\varphi \in C(X)$. In view of the existence of the pullback attractor, the $\tau$-continuity of the process $U(t,\tau)$ and Lemma 12.4, the function $\tau \mapsto \frac{1}{t-\tau}\int_\tau^t \varphi(U(t,s)u(s))\,ds$ is bounded on the interval $(-\infty,t]$. It is continuous by an application of the dominated convergence theorem and the Lebesgue differential theorem.

It suffices to prove that the function $s \mapsto \varphi(U(t,s)u(s))$ is bounded on $(-\infty,t]$: it is continuous, since $U(\cdot,\cdot)$ is $\tau$-continuous and $u(\cdot)$ is continuous, and so it is bounded on every compact interval $[t_0,t]$. We now show that it is bounded on $(\infty,t_0]$ for $t_0$ sufficiently large and negative.

If this is not the case then there is a sequence $\{s_n\}$ with $s_n \to -\infty$ such that

$$|\varphi(U(t,s_n)u(s_n))| \to \infty. \tag{12.23}$$

However, from the existence of the pullback $\mathscr{D}$-attractor it follows that the process $U(\cdot,\cdot)$ is pullback $\mathscr{D}$-asymptotically compact, so there exists a subsequence of $\{s_n\}$ (which we relabel) such that $U(t,s_k)u(s_k) \to \omega$ for some $\omega \in X$ (since $u(\cdot) \in \mathscr{D}$). From the continuity of $\varphi$, we have $\varphi(U(t,s_k)u(s_k)) \to \varphi(\omega)$, contradicting (12.23).

We now define

$$L(\varphi) = \mathrm{LIM}_{\tau\to-\infty}\frac{1}{t-\tau}\int_\tau^t \varphi(U(t,s)u(s))\,ds, \tag{12.24}$$

and claim that $L(\varphi)$ depends only on the values of $\varphi$ on $A(t)$. Indeed, take functions $\varphi,\psi \in C(X)$ with $\varphi = \psi$ on $A(t)$, and given $\epsilon > 0$ use Lemma 12.4 to find a $\delta$ such that

$$\rho(u,v) \le \delta \quad \Rightarrow \quad |\varphi(u) - \varphi(v)| + |\psi(u) - \psi(v)| \le \epsilon \quad \text{for every} \quad v \in A(t). \tag{12.25}$$

Now let $\tau_0$ be such that

$$\mathrm{dist}_X(U(t,s)u(s),A(t)) \le \delta \qquad \text{for all} \quad s \le \tau_0.$$

Then for every $s \le \tau_0$ there exists a $v_s \in A(t)$ such that

$$\rho(U(t,s)u(s),v_s) \le \delta,$$

and so

$$\begin{aligned}|\varphi(U(t,s)u(s)) - \psi(U(t,s)u(s))| \\ \leq |\varphi(U(t,s)u(s)) - \varphi(v_s)| + |\varphi(v_s) - \psi(v_s)| \\ + |\psi(v_s) - \psi(U(t,s)u(s))| \leq \epsilon,\end{aligned}$$

using (12.25) and the fact that $\varphi(v_s) = \psi(v_s)$. Thus

$$\begin{aligned}|L(\varphi - \psi)| &= \left|\mathrm{LIM}_{\tau\to-\infty}\frac{1}{t-\tau}\int_\tau^t (\varphi(U(t,s)u(s)) - \psi(U(t,s)u(s)))\,ds\right| \\ &= \left|\mathrm{LIM}_{\tau\to-\infty}\frac{1}{t-\tau}\left(\int_\tau^{\tau_0} + \int_{\tau_0}^t\right)(\varphi(U(t,s)u(s)) - \psi(U(t,s)u(s)))\,ds\right| \\ &\leq \limsup_{\tau\to-\infty}\frac{(\tau_0-\tau)\epsilon}{t-\tau} \\ &\quad + \limsup_{\tau\to-\infty}\frac{(t-\tau_0)}{t-\tau}\sup_{s\in[\tau_0,t]}\{|\varphi(U(t,s)u(s))| + |\psi(U(t,s)u(s))|\} \leq \epsilon.\end{aligned}$$

Since $\epsilon > 0$ was arbitrary we conclude that $L(\varphi)$ depends only on the values of $\varphi$ on $A(t)$.

Define $G(\varphi) = L(l(\varphi))$ for $\varphi \in C(A(t))$, where $l(\varphi)$ is an extension of $\varphi$ given by the Tietze theorem. As $G$ is a linear and positive functional on $C(A(t))$, we have, by the Kakutani–Riesz representation theorem,

$$G(\varphi) = \int_{A(t)} \varphi(v)\,d\mu_t(v).$$

We extend $\mu_t$ (by zero) to a Borel measure on $X$, which we denote again by $\mu_t$, whence for every $\varphi \in C(X)$,

$$\int_{A(t)} \varphi(v)\,d\mu_t(v) = \int_X \varphi(v)\,d\mu_t(v),$$

and thus (12.21) holds.

To prove (12.22), observe that for $t \geq \tau$, and for all functionals $\varphi \in C(X)$,

$$\begin{aligned}\int_X \varphi(v)\,d\mu_t(v) &= \mathrm{LIM}_{M\to-\infty}\frac{1}{t-M}\int_M^t \varphi(U(t,s)u(s))\,ds \\ &= \mathrm{LIM}_{M\to-\infty}\frac{1}{t-M}\int_M^\tau \varphi(U(t,s)u(s))\,ds \\ &\quad + \mathrm{LIM}_{M\to-\infty}\frac{1}{t-M}\int_\tau^t \varphi(U(t,s)u(s))\,ds\end{aligned}$$

$$= \mathrm{LIM}_{M\to-\infty}\frac{1}{\tau - M}\int_M^\tau \varphi(U(t,\tau)U(\tau,s)u(s))\,ds$$

$$= \mathrm{LIM}_{M\to-\infty}\frac{1}{\tau - M}\int_M^\tau [\varphi\circ U(t,\tau)]\{U(\tau,s)u(s)\}\,ds$$

$$= \int_X [\varphi\circ U(t,\tau)](v)\,d\mu_\tau(v) = \int_X \varphi(U(t,\tau)v)\,d\mu_\tau(v),$$

where we have used the fact that $\varphi \circ U(t,\tau) \in C(X)$. □

Now, we can come back to the Navier–Stokes equations. We define the evolutionary process associated to (12.1) and the attraction universe $\mathscr{D}_\sigma$ exactly as in Sect. 12.1. We have already proved the existence of a unique pullback $\mathscr{D}_\sigma$-attractor $\mathbb{A} = \{A(t)\}_{t\in\mathbb{R}}$ belonging to $\mathscr{D}_\sigma$ for the process $U(t,\tau)$ in $H$ (cf. Theorem 12.1). By Lemma 12.2 the process $U(t,\tau)$ is $\tau$-continuous. Thus, the existence of a family of invariant measures $\mu_t$ for $U(t,\tau)$ follows from Theorem 12.8. Namely, we have the following result:

**Theorem 12.9.** *Let $\Omega \subset \mathbb{R}^2$ be an open, bounded or unbounded, set such that there exists a constant $\lambda_1 > 0$ for which*

$$\lambda_1 \int_\Omega \phi^2\,dx \le \int_\Omega |\nabla\phi|^2\,dx \quad \textit{for all} \quad \phi \in H_0^1(\Omega).$$

*Suppose that $f \in L^2_{loc}(\mathbb{R}; V')$ is such that*

$$\int_{-\infty}^t \varepsilon^{\sigma\xi}\|f(\xi)\|^2_{V'}\,d\xi < +\infty \quad \textit{for all} \quad t \in \mathbb{R}.$$

*Let $\{U(t,\tau)\}_{t\ge\tau}$ be the evolutionary process associated with the Navier–Stokes system (12.1). Fix a generalized Banach limit $\mathrm{LIM}_{T\to\infty}$ and let $u : \mathbb{R} \to X$ be a continuous map such that $u(\cdot) \in \mathscr{D}_\sigma$. Then there exists a unique family of Borel probability measures $\{\mu_t\}_{t\in\mathbb{R}}$ in the phase space $H$ such that the support of measure $\mu_t$ is contained in $A(t)$ and*

$$\mathrm{LIM}_{\tau\to-\infty}\frac{1}{t-\tau}\int_\tau^t \varphi(U(t,s)u(s))ds = \int_{A(t)} \varphi(u)d\mu_t(u)$$

*for any real-valued continuous functional $\varphi$ on $H$. In addition, $\mu_t$ is invariant in the sense that*

$$\int_{A(t)} \varphi(v)\,d\mu_t(v) = \int_{A(\tau)} \varphi(U(t,\tau)v)\,d\mu_\tau(v) \quad \textit{for} \quad t \ge \tau.$$

## 12.6 Comments and Bibliographical Notes

Theorem 12.1 was proved in [45]. Theorem 12.2 was proved first in [201], cf. also [220]. Theorem 12.8 generalizes Theorem 3.3 from [160] (for the non-autonomous two-dimensional Navier–Stokes equations, using compactness property of those equations) and Theorem 5 from [165] (for autonomous problems, but under weaker assumptions on the dynamical system). Lemma 12.4 comes from [197] stated there as Exercise 10.1, see also Lemma 2 in [165] and Lemma 3.1 in [56]. Theorem 12.9 was proved in [163].

# 13
# Pullback Attractors and Shear Flows

In this chapter we consider the problem of existence and finite dimensionality of the pullback attractor for a class of two-dimensional turbulent boundary driven flows which naturally appear in lubrication theory. We generalize here the results from Chap. 9 to the non-autonomous problem.

The new element in our study with respect to that in Chap. 9 is the allowance of the speed of rotation of the cylinder to depend on time. Our aim is first to prove the existence of the unique global in time solution of the problem, and then to study its time asymptotics in the frame of the dynamical systems theory. We use the theory of pullback attractors that allows us to pose quite general assumptions on the velocity of the boundary. Neither quasi-periodicity nor even boundedness is demanded to prove the existence and to estimate the fractal dimension of the corresponding pullback attractor. We shall apply the results from Chap. 11, reformulated here in the language of evolutionary processes.

## 13.1 Preliminaries

In this section we recall a theorem about the existence of pullback attractors for evolutionary processes.

Let us consider an evolutionary process $U$ on a metric space $X$, i.e., a family $\{U(t,\tau)\}_{-\infty<\tau\leq t<+\infty}$ of continuous mappings $U(t,\tau) : X \to X$, such that for $x \in X$ and $\tau \in \mathbb{R}$ $U(\tau,\tau)x = x$, and

$$U(t,\tau) = U(t,r)U(r,\tau) \quad \text{for all} \quad \tau \leq r \leq t.$$

By $\mathscr{P}(X)$ we denote the family of all nonempty subsets of $X$. Let $\mathscr{D}$ be an attraction universe, i.e., a nonempty class of non-autonomous sets $\mathbb{D} = \{D(t)\}_{t\in\mathbb{R}}$ such that $D(t) \in \mathscr{P}(X)$ for every $t \in \mathbb{R}$.

We have the following result, cf. Theorems 11.4 and 11.5 in Chap. 11.



G. Łukaszewicz, P. Kalita, *Navier–Stokes Equations*, Advances in Mechanics and Mathematics 34, DOI 10.1007/978-3-319-27760-8_13

**Theorem 13.1.** *Suppose that the process $U(t,\tau)$ is pullback $\mathscr{D}$-asymptotically compact, and that there exists a non-autonomous pullback $\mathscr{D}$-absorbing set for $U(\cdot,\cdot)$ belonging to the universe $\mathscr{D}$. Then there exists a minimal pullback $\mathscr{D}$-attractor $\mathbb{A} = \{A(t)\}_{t\in\mathbb{R}}$ for $U(\cdot,\cdot)$.*

*Remark 13.1.* Observe that there exists at most one pullback $\mathscr{D}$-attractor $\mathbb{A} \in \mathscr{D}$. In fact, if $\mathbb{A}_1$ and $\mathbb{A}_2$ are two global pullback $\mathscr{D}$-attractors belonging to $\mathscr{D}$, then by the invariance property,

$$\mathrm{dist}_X(A_1(t), A_2(t)) = \mathrm{dist}_X(U(t,\tau)A_1(\tau), A_2(t)).$$

But then, as $\mathbb{A}_1 \in \mathscr{D}$ and $\mathbb{A}_2$ is a global pullback $\mathscr{D}$-attractor,

$$\mathrm{dist}_X(A_1(t), A_2(t)) = \lim_{\tau\to-\infty} \mathrm{dist}_X(U(t,\tau)A_1(\tau), A_2(t)) = 0,$$

and therefore

$$A_1(t) \subset \overline{A_2(t)} = A_2(t).$$

Analogously, one obtains $A_2(t) \subset A_1(t)$.

*Remark 13.2.* It is said that the attraction universe $\mathscr{D}$ is inclusion-closed if, given the non-autonomous set $\mathbb{D} \in \mathscr{D}$, for any $\mathbb{D}' = \{D'(t)\}_{t\in\mathbb{R}}$ with $D'(t) \in \mathscr{P}(X)$ for $t \in \mathbb{R}$, such that $D'(t) \subset D(t)$ for all $t$, one also has $\mathbb{D}' \in \mathscr{D}$.

Taking into account that $\Lambda(\mathbb{B}, t) \subset \overline{B(t)}$, and Remark 13.1, we can assert that if the universe $\mathscr{D}$ is inclusion-closed and the sets $B(t)$ are closed, then the pullback $\mathscr{D}$-attractor $\mathbb{A}$ constructed in Theorem 13.1 belongs to the universe $\mathscr{D}$, and is the unique pullback $\mathscr{D}$-attractor for $U(\cdot,\cdot)$ belonging to this universe.

The rest of this chapter is organized as follows. In Sect. 13.2 we give a precise formulation of the considered problem and write it in a weak form, suitable for our further considerations. In Sect. 13.3 we derive the first energy inequality and establish the existence of unique global in time solution of the problem. In Sect. 13.4 we prove the existence of a pullback attractor for the corresponding evolutionary process by using the energy equation method. In Sect. 13.5 we obtain an upper bound of the pullback attractor dimension in terms of the data.

## 13.2 Formulation of the Problem

We consider two-dimensional Navier–Stokes equations,

$$u_t - \nu\Delta u + (u\cdot\nabla)u + \nabla p = 0, \tag{13.1}$$

$$\mathrm{div}\, u = 0, \tag{13.2}$$

in the channel

$$\Omega_\infty = \{x = (x_1, x_2) : \ -\infty < x_1 < \infty, \ \ 0 < x_2 < h(x_1)\},$$

where $h$ is a positive, smooth, and $L$-periodic function in $x_1$.

Let

$$\Omega = \{x = (x_1, x_2) : \ 0 < x_1 < L, \ \ 0 < x_2 < h(x_1)\},$$

and $\partial\Omega = \overline{\Gamma_C} \cup \overline{\Gamma_L} \cup \overline{\Gamma_D}$, where $\Gamma_C$ and $\Gamma_D$ are the bottom and the top, and $\Gamma_L$ is the lateral part of the boundary of $\Omega$.

We are interested in solutions of (13.1) and (13.2) in $\Omega$ which are $L$-periodic with respect to $x_1$, or, to be more precise, analogously to Chap. 10, both the velocity $u$ and the Cauchy stress vector $Tn$ on $\Gamma_L$ are $L$-periodic with respect to $x_1$. Moreover, we assume

$$u = 0 \quad \text{on} \quad \Gamma_D, \tag{13.3}$$

$$u = U_0(t)e_1 = (U_0(t), 0) \quad \text{on} \quad \Gamma_C, \tag{13.4}$$

where $U_0(t)$ is a locally Lipschitz continuous function of time $t$. Finally, the initial condition at some time $\tau \in \mathbb{R}$ is given as

$$u(x, \tau) = u_0(x) \quad \text{for} \quad x \in \Omega. \tag{13.5}$$

The problem is motivated by a flow in an infinite (rectified) journal bearing $\Omega \times (-\infty, +\infty)$, where $\Gamma_D \times (-\infty, +\infty)$ represents the outer cylinder, and the inner, rotating cylinder is represented by $\Gamma_C \times (-\infty, +\infty)$. We can assume that the rectification does not change the equations as the gap between cylinders is very small with respect to their radii.

In our considerations we use the background flow method and transform the boundary condition (13.4) by defining a smooth background flow, a simple version of the Hopf construction, described in detail in Sect. 13.3.

Let

$$u(x_1, x_2, t) = U(x_2, t)e_1 + v(x_1, x_2, t),$$

with

$$U(0, t) = U_0(t) \quad \text{and} \quad U(h(x_1), t) = 0 \quad \text{for} \quad x_1 \in (0, L) \quad \text{and} \quad t \in \mathbb{R}. \tag{13.6}$$

Then $v$ is $L$-periodic in $x_1$ and satisfies

$$v_t - \nu \Delta v + (v \cdot \nabla) v + \nabla p = G \quad \text{on} \quad \Omega \times (\tau, \infty), \tag{13.7}$$

$$\operatorname{div} v = 0 \quad \text{on} \quad \Omega \times (\tau, \infty), \tag{13.8}$$

$$v = 0 \quad \text{on} \quad (\Gamma_D \cup \Gamma_C) \times (\tau, \infty), \tag{13.9}$$

where $G = \nu U_{,x_2 x_2} e_1 - U_{,t} e_1 - U v_{,x_1} - (v)_2 U_{,x_2} e_1$, and the initial condition

$$v(x, \tau) = v_0(x) = u_0(x) - U(x_2, \tau) e_1 \quad \text{in} \quad \Omega. \tag{13.10}$$

By $(v)_2$ we denote the second component of $v$.

Now, we define a weak form of the above problem. To this end we need some notations. Let

$$\tilde{V} = \{v \in C^\infty(\overline{\Omega})^2 : \; v \text{ is } L\text{-periodic in } x_1, \; \operatorname{div} v = 0, \; v = 0 \text{ on } \Gamma_D \cup \Gamma_C\},$$

$$V = \text{ closure of } \tilde{V} \text{ in } H^1(\Omega)^2,$$

$$H = \text{ closure of } \tilde{V} \text{ in } L^2(\Omega)^2.$$

We define the scalar products

$$(u, v) = \int_\Omega u(x) \cdot v(x) dx,$$

and

$$((u, v)) = (\nabla u, \nabla v) = \int_\Omega \nabla u(x) \colon \nabla v(x) \, dx$$

in $H$ and $V$, respectively, and the norms

$$|v| = (v, v)^{\frac{1}{2}} \quad \text{and} \quad \|v\| = ((v, v))^{\frac{1}{2}}.$$

Let

$$b(u, v, w) = ((u \cdot \nabla) v, w),$$

and

$$a(u, v) = (\nabla u, \nabla v).$$

Then the natural weak formulation of the problem (13.7)–(13.10) is as follows.

**Problem 13.1.** Find the velocity $v$ such that for each $T > \tau$,

$$v \in C([\tau, T]; H) \cap L^2(\tau, T; V),$$

and for all $\Theta \in V$ and a.e. $t > \tau$,

$$\frac{d}{dt}(v(t), \Theta) + \nu a(v(t), \Theta) + b(v(t), v(t), \Theta) = F(v(t), \Theta), \tag{13.11}$$

with

$$v(x, \tau) = v_0(x) \quad \text{on} \quad \Omega,$$

where

$$F(v, \Theta) = -\nu a(\xi, \Theta) - b(\xi, v, \Theta) - b(v, \xi, \Theta) - (\xi_{,t}, \Theta), \tag{13.12}$$

and $\xi = Ue_1$ is a suitable background flow.

In the next section we shall derive the first energy inequality and establish the existence result for the above problem.

## 13.3 Existence and Uniqueness of Global in Time Solutions

To study the pullback attractor associated with a particular problem we first need to know that the latter has a unique global in time solution. In our particular case we have the following existence theorem.

**Theorem 13.2.** *Let $U_0$ be a locally Lipschitz continuous function on the real line. Then there exists a unique weak solution of Problem 13.1 such that for all $\eta$, $T$, $\tau < \eta < T$ we have $v \in L^2(\eta, T; H^2(\Omega))$ and for each $t > \tau$ the map $v_0 \mapsto v(t)$ is continuous as a map in $H$.*

*Proof.* We shall start the proof from the derivation of the first energy estimate for the solutions, namely, estimate (13.25) below.

Taking $\Theta = v$ in (13.11), we obtain

$$\frac{1}{2}\frac{d}{dt}|v|^2 + \nu a(v, v) + b(v, v, v) = F(v, v).$$

Since $v = 0$ on $\Gamma_D \cup \Gamma_C$, $v$ is $L$-periodic in $x_1$, and $\operatorname{div} v = 0$, we have $b(v, v, v) = 0$, and

$$\frac{1}{2}\frac{d}{dt}|v|^2 + \nu \|v\|^2 = F(v, v). \tag{13.13}$$

Consider the term $F(v, v)$ on the right-hand side of (13.13). By (13.12) we have

$$F(v, v) = -\nu a(\xi, v) - b(v, \xi, v) - (\xi_{,t}, v). \tag{13.14}$$

We have

$$|\nu a(\xi, v)| \leq \nu \|\xi\| \|v\| \leq 2\nu \|\xi\|^2 + \frac{\nu}{8}\|v\|^2. \tag{13.15}$$

To estimate the second term in (13.14) we need the following lemma.

**Lemma 13.1.** *For any $t \in \mathbb{R}$ there exists a smooth extension*

$$\xi = \xi(x_2, t) = U(x_2, t)e_1 = (U(x_2, t), 0)$$

*of the boundary condition for u, such that*

$$|b(v, \xi(t), v)| \leq \frac{\nu}{4}\|v\|^2 \quad \text{for all} \quad v \in V. \tag{13.16}$$

*Proof.* Let $\rho : [0, \infty) \to [0, 1]$ be a smooth function such that

$$\rho(0) = 1 \quad \text{and} \quad \operatorname{supp} \rho \subset [0, 1/2], \quad \text{and} \quad \max |\rho'(s)| \leq \sqrt{8}.$$

Let $h_0 = \min_{0 \leq x_1 \leq L} h(x_1)$, and

$$\varepsilon = \varepsilon(t) = \begin{cases} 2 & \text{if} \quad |U_0(t)| \leq \nu/(8h_0), \\ \nu/(4h_0|U_0(t)|) & \text{if} \quad |U_0(t)| \geq \nu/(8h_0). \end{cases} \tag{13.17}$$

Thus, $0 < \varepsilon(t) \leq 2$. Define

$$U(x_2, t) = U_0(t)\rho(x_2/(h_0\varepsilon(t))). \tag{13.18}$$

Then $U$ matches the boundary conditions (13.6), is $L$-periodic in $x_1$, and $\operatorname{div} Ue_1 = 0$. Moreover, for $Q_\varepsilon = (0, L) \times (0, h_0\varepsilon/2)$,

$$|b(v, Ue_1, v)| = |b(v, v, Ue_1)| \leq \left|\frac{v}{x_2}\right|_{L^2(Q_\varepsilon)} |\nabla v|_{L^2(Q_\varepsilon)} |x_2 U(x_2, t)|_{L^\infty(Q_\varepsilon)}.$$

Now,

$$|x_2 U(x_2, t)|_{L^\infty(Q_\varepsilon)} \leq \frac{h_0\varepsilon}{2}|U_0|,$$

and

$$\left|\frac{v}{x_2}\right|^2_{L^2(Q_\varepsilon)} \le \int_0^L \int_0^{h(x_1)} \left|\frac{v(x_1,x_2)}{x_2}\right|^2 dx_2dx_1 \le 4\int_0^L \int_0^{h(x_1)} \left|\frac{\partial v(x_1,x_2)}{\partial x_2}\right|^2 dx_2dx_1$$

by the Hardy inequality. Thus, we obtain (13.16). □

We estimate the last term on the right-hand side of (13.14) using the Poincaré inequality

$$|v| \le \frac{1}{\sqrt{\lambda_1}}\|v\| \quad \text{for} \quad v \in V, \tag{13.19}$$

where we can take $1/\sqrt{\lambda_1} = h_M/\sqrt{2}$ with $h_M = \max_{0\le x_1\le L} h(x_1)$. We get

$$|(\xi_{,t}, v)| \le |\xi_{,t}||v| \le |\xi_{,t}|\frac{1}{\sqrt{\lambda_1}}\|v\| \le \frac{2}{\nu\lambda_1}|\xi_{,t}|^2 + \frac{\nu}{8}\|v\|^2. \tag{13.20}$$

In view of estimates (13.15), (13.16), and (13.20),

$$|F(v,v)| \le 2\nu\|\xi\|^2 + \frac{2}{\nu\lambda_1}|\xi_{,t}|^2 + \frac{\nu}{2}\|v\|^2,$$

and by (13.13),

$$\frac{1}{2}\frac{d}{dt}|v|^2 + \frac{\nu}{2}\|v\|^2 \le 2\nu\|\xi\|^2 + \frac{2}{\nu\lambda_1}|\xi_{,t}|^2. \tag{13.21}$$

To estimate the right-hand side in terms of the data, we use the following lemma.

**Lemma 13.2.** *Let U be as in (13.18). Then for almost all t,*

$$\int_\Omega |U(x_2,t)|^2 dx_1dx_2 \le \frac{1}{2}Lh_0U_0^2(t)\varepsilon(t), \tag{13.22}$$

$$\int_\Omega |U_{,x_2}(x_2,t)|^2 dx_1dx_2 \le \frac{4LU_0^2(t)}{h_0}\frac{1}{\varepsilon(t)}, \tag{13.23}$$

*and*

$$\int_\Omega |U_{,t}(x_2,t)|^2 dx_1dx_2 \le \left(|U_0'(t)| + \sqrt{2}|U_0(t)|\frac{|\varepsilon'(t)|}{\varepsilon(t)}\right)^2 \frac{Lh_0\varepsilon(t)}{2}. \tag{13.24}$$

*Proof.* We have $|U(x_2, t)| \le |U_0(t)|$ and supp $U \subset Q_\varepsilon$, which gives (13.22). Moreover,

$$\int_\Omega |U_{,x_2}(x_2, t)|^2 \, dx = \frac{U_0^2}{h_0^2 \varepsilon^2} \int_\Omega \left| \rho' \left( \frac{x_2}{h_0 \varepsilon} \right) \right|^2 dx$$

$$= \frac{U_0^2}{h_0^2 \varepsilon^2} L \int_0^{\frac{h_0 \varepsilon}{2}} \left| \rho' \left( \frac{x_2}{h_0 \varepsilon} \right) \right|^2 dx_2 \le \frac{4 L U_0^2}{h_0} \frac{1}{\varepsilon},$$

as $|\rho'| \le \sqrt{8}$, which gives (13.23). In order to obtain (13.24) we use (13.18) and the definition of $\rho$ to get

$$|U_{,t}(x_2, t)| \le |U_0'(t)| \rho \left( \frac{x_2}{h_0 \varepsilon(t)} \right) + |U_0(t)| \frac{1}{\varepsilon^2(t)} \left| \frac{x_2 \varepsilon'(t)}{h_0} \right| \left| \rho' \left( \frac{x_2}{h_0 \varepsilon(t)} \right) \right|$$

$$\le |U_0'(t)| + \sqrt{2} |U_0(t)| \frac{|\varepsilon'(t)|}{\varepsilon(t)}$$

for $0 \le x_2 \le h_0 \varepsilon / 2$ and almost all $t$, from which the assertion follows easily. □

Applying now Lemma 13.2 to inequality (13.21) with $\varepsilon$ as in (13.17) we obtain the first energy estimate,

$$\frac{1}{2} \frac{d}{dt} |v(t)|^2 + \frac{\nu}{2} \|v(t)\|^2 \le f^2(t) \quad \text{a.e. in} \quad t, \tag{13.25}$$

where

$$f^2(t) = c_1(\nu, \Omega)(1 + |U_0(t)|^3 + |U_0'(t)|^2). \tag{13.26}$$

From (13.25) and (13.19) we conclude further that

$$\frac{1}{2} \frac{d}{dt} |v(t)|^2 + \frac{\sigma}{2} |v(t)|^2 \le f^2(t),$$

with $\sigma = \nu \lambda_1$, and that

$$|v(t)|^2 \le e^{-\sigma(t-\tau)} |v(\tau)|^2 + 2 \int_\tau^t e^{\sigma(s-t)} f^2(s) ds. \tag{13.27}$$

To prove the uniqueness of solutions and their continuous dependence on the initial data we assume that $v$ and $\bar{v}$ are two solutions of Problem 13.1, write (13.11) for their difference, $w = v - \bar{v}$, and set $\Theta = w$, to get

$$\frac{1}{2} \frac{d}{dt} |w(t)|^2 + \frac{\nu}{2} \|w(t)\|^2 + b(w, v, w) = -b(w, \xi, w).$$

Now we use the Ladyzhenskaya inequality (cf. Lemma 10.2),

$$\|v\|_{L^4(\Omega)} \le c_2(\Omega)^{1/4}|v|^{1/2}\|v\|^{1/2}$$

for all $v \in V$ (we can take $c_2(\Omega) = 18\max\{2, 1 + (h_M/L)^2\}$), to get

$$\begin{aligned}|b(w, v, w)| &\le \|w\|_{L^4(\Omega)}\|v\|\,\|w\|_{L^4(\Omega)} \\ &\le c_2(\Omega)^{1/2}|w|\,\|w\|\,\|v\| \le \frac{1}{8}\nu\|w\|^2 + \frac{2}{\nu}c_2(\Omega)|w|^2\|v\|^2.\end{aligned}$$

From this, together with (13.16), we get

$$\frac{d}{dt}|w(t)|^2 + \frac{\nu}{4}\|w(t)\|^2 \le \frac{4}{\nu}c_2(\Omega)\|v\|^2|w|^2,$$

and

$$\frac{d}{dt}|w(t)|^2 + \frac{\sigma}{4}|w(t)|^2 \le \frac{4}{\nu}c_2(\Omega)\|v\|^2|w|^2.$$

By (13.25) and (13.27) the function $s \to \|v(s)\|^2$ is locally integrable on the real line, whence the Gronwall lemma gives

$$|w(t)|^2 \le |w(\tau)|^2 \exp\left\{-\int_\tau^t \left(\frac{\sigma}{4} - \frac{4}{\nu}c_2(\Omega)\|v(s)\|^2\right) ds\right\}. \tag{13.28}$$

This means that the map $v(\tau) \to v(t)$ for $t > \tau$ is continuous in $H$, and, in particular, that the considered problem is uniquely solvable.

The existence part of the proof is based on inequality (13.25), the Galerkin approximations, and the compactness method, and is very similar to the proof of Theorem 9.6. □

## 13.4 Existence of the Pullback Attractor

In this section we prove the existence of the pullback attractor for the evolutionary process associated with the considered problem.

For $t \ge \tau$ let us define the map $U(t, \tau)$ in $H$ by

$$U(t, \tau)v_0 = v(t; \tau, v_0) \quad \text{for} \quad t \ge \tau \quad \text{and} \quad v_0 \in H, \tag{13.29}$$

where $v(t; \tau, v_0)$ is the solution of Problem 13.1. From the uniqueness of solution to this problem, one immediately obtains that

$$U(t, \tau)v_0 = U(t, r)U(r, \tau)v_0, \quad \text{for all} \quad \tau \le r \le t \quad \text{and} \quad v_0 \in H.$$

From Theorem 13.2 it follows that for all $t \geq \tau$, the mapping $U(t,\tau) : H \to H$ defined by (13.29) is continuous. Consequently, the family $\{U(t,\tau)\}_{\tau \leq t}$ defined by (13.29) is a process in $H$.

Moreover, it can be proved that $U(\tau, t)$ is sequentially weakly continuous, more exactly, we have the following result.

**Proposition 13.1.** *Let $\{v_{0n}\} \subset H$ be a sequence converging weakly in $H$ to an element $v_0 \in H$. Then*

$$U(t,\tau)v_{0n} \to U(t,\tau)v_0 \quad \text{weakly in} \quad H \quad \text{for all} \quad t \geq \tau,$$

$$U(\cdot,\tau)v_{0n} \to U(\cdot,\tau)v_0 \quad \text{weakly in} \quad L^2(\tau,T;V) \quad \text{for all} \quad T > \tau.$$

*Proof.* The proof is almost identical to that of Lemma 2.1 in [201] and we omit it. □

Now, we define the universe of the non-autonomous sets. For $\sigma = \nu\lambda_1$, let

$$\mathscr{D}_\sigma = \{D : \mathbb{R} \to \mathscr{P}(H) : \lim_{t \to -\infty} e^{\sigma t}|D(t)|^2 = 0\}, \tag{13.30}$$

where $|D(t)| = \sup\{|y| : y \in D(t)\}$.

We can prove the following result:

**Theorem 13.3.** *Let $U_0$ be a locally Lipschitz continuous function on the real line such that*

$$\int_{-\infty}^{t} e^{\sigma s}(|U_0(s)|^3 + |U_0'(s)|^2)\,ds < +\infty \quad \text{for all} \quad t \in \mathbb{R}. \tag{13.31}$$

*Then, there exists a unique pullback $\mathscr{D}_\sigma$-attractor $\mathbb{A} \in \mathscr{D}_\sigma$ for the process $U(t,\tau)$ defined by (13.29).*

*Proof.* Let $D \in \mathscr{D}_\sigma$ be given. From the first energy estimate (13.27) and from (13.31) we obtain

$$|U(t,\tau)v_0|^2 \leq e^{-\sigma(t-\tau)}|D(\tau)|^2 + 2e^{-\sigma t}\int_{-\infty}^{t} e^{\sigma s} f(s)^2 ds < \infty \tag{13.32}$$

for all $v_0 \in D(\tau)$, and all $t \geq \tau$.

Denote by $R_\sigma(t)$ the positive number given for each $t \in \mathbb{R}$ by

$$(R_\sigma(t))^2 = 3e^{-\sigma t}\int_{-\infty}^{t} e^{\sigma s} f(s)^2 ds,$$

and consider the non-autonomous set $\mathbb{B}_\sigma = \{B_\sigma(t)\}_{t\in\mathbb{R}}$ where $B_\sigma(t)$ are closed balls in $H$ defined by

$$B_\sigma(t) = \{w \in H : \ |w| \le R_\sigma(t)\}.$$

In view of (13.30) it is immediate to see that $B_\sigma \in \mathscr{D}_\sigma$. Moreover, by (13.30) and (13.32), the family $B_\sigma$ is pullback $\mathscr{D}_\sigma$-absorbing for the process $U(t, \tau)$.

Clearly, the attraction universe $\mathscr{D}_\sigma$ is inclusion-closed and the sets $B_\sigma(t)$ are closed, and thus, according to Theorem 13.1 and Remark 13.2, to complete the proof of the theorem we have to prove that $U(t, \tau)$ is pullback $\mathscr{D}_\sigma$-asymptotically compact.

Let $\mathbb{D} = \{D(t)\}_{t\in\mathbb{R}} \in \mathscr{D}_\sigma$, a sequence $\tau_n \to -\infty$, a sequence $v_{0n} \in D(\tau_n)$, and $t \in \mathbb{R}$ be fixed. We must prove that we can extract a subsequence that converges in $H$ from the sequence $\{U(t, \tau_n)v_{0n}\}$.

As the non-autonomous set $\mathbb{B}_\sigma = \{B_\sigma(t)\}_{t\in\mathbb{R}}$ is pullback $\mathscr{D}_\sigma$-absorbing, for each integer $k \ge 0$ there exists a $\tau_{\mathbb{D}}(k) \le t - k$ such that

$$U(t-k, \tau)D(\tau) \subset B_\sigma(t-k) \quad \text{for all} \quad \tau \le \tau_{\mathbb{D}}(k). \tag{13.33}$$

It is not difficult to deduce from (13.33), by a diagonal procedure, the existence of a subsequence $\{(\tau_{n'}, v_{0n'})\} \subset \{(\tau_n, v_{0n})\}$, and a sequence $\{w_k\}_{k=0}^{\infty} \subset H$, such that for all $k = 0, 1, 2, \ldots$ we have $w_k \in B_\sigma(t-k)$ and

$$U(t-k, \tau_{n'})v_{0n'} \rightharpoonup w_k \quad \text{weakly in} \quad H.$$

Now, observe that, by Proposition 13.1,

$$\begin{aligned} w_0 &= \text{weak-} \lim_{n'\to\infty} U(t, \tau_{n'})v_{0'_n} \\ &= \text{weak-} \lim_{n'\to\infty} U(t, t-k)U(t-k, \tau_{n'})v_{0n'} \\ &= U(t, t-k)(\text{weak-} \lim_{n'\to\infty} U(t-k, \tau_{n'})v_{0n'}), \end{aligned}$$

i.e.,

$$U(t, t-k)w_k = w_0 \quad \text{for all} \quad k = 0, 1, 2, \ldots . \tag{13.34}$$

Then, by the weak lower semicontinuity of the norm,

$$|w_0| \le \liminf_{n'\to\infty} |U(t, \tau_{n'})v_{0_{n'}}|.$$

If we now prove that also

$$\limsup_{n'\to\infty} |U(t, \tau_{n'})v_{0_{n'}}| \le |w_0|, \tag{13.35}$$

then we will have

$$\lim_{n'\to\infty} |U(t,\tau_{n'})v_{0_{n'}}| = |w_0|,$$

and this, together with the weak convergence, will imply the strong convergence in $H$ of $U(t,\tau_{n'})v_{0_{n'}}$ to $w_0$.

Our aim is to prove (13.35). We shall use the notation,

$$\langle g(t), w\rangle = -\nu a(\xi(t), w) - (\xi_{,t}, w) \quad \text{for} \quad w \in V,$$

and

$$Q(t,w) = \nu\|w\|^2 - \frac{\sigma}{2}|w|^2 + b(w,\xi(t),w) \quad \text{for} \quad (t,w) \in \mathbb{R}\times V.$$

Then, the energy equation (13.13) can be written as

$$\frac{d}{dt}|v|^2 + \sigma|v|^2 = 2(\langle g(t), v\rangle - Q(t,v)),$$

whence for all $t \geq \tau$, and all $v_0 \in H$,

$$\begin{aligned} |U(t,\tau)v_0|^2 &= |v_0|^2 e^{\sigma(\tau-t)} \\ &\quad + 2\int_\tau^t e^{\sigma(s-t)}\left(\langle g(s), U(s,\tau)v_0\rangle - Q(s,U(s,\tau)v_0)\right) ds, \end{aligned} \tag{13.36}$$

and, thus, for all $k = 0,1,2,\dots$ and all $\tau_{n'} \leq t-k$,

$$\begin{aligned} |U(t,\tau_{n'})v_{0n'}|^2 &= |U(t,t-k)U(t-k,\tau_{n'})v_{0n'}|^2 \\ &= e^{-\sigma k}|U(t-k,\tau_{n'})v_{0n'}|^2 \\ &\quad + 2\int_{t-k}^t e^{\sigma(s-t)}\langle g(s), U(s,t-k)U(t-k,\tau_{n'})v_{0n'}\rangle\, ds \\ &\quad - 2\int_{t-k}^t e^{\sigma(s-t)}Q(s,U(s,t-k)U(t-k,t-\tau_{n'})v_{0n'})\, ds. \end{aligned} \tag{13.37}$$

As, by (13.33),

$$U(t-k,\tau_{n'})v_{0_{n'}} \in B_\sigma(t-k) \quad \text{for all} \quad \tau_{n'} \leq \tau_{\mathbb{D}}(k) \quad \text{and} \quad k = 0,1,2,\dots,$$

we have

$$\limsup_{n'\to\infty}\left(e^{-\sigma k}[U(t-k,\tau_{n'})v_{0n'}]^2\right) \leq e^{-\sigma k}R_\sigma^2(t-k) \quad \text{for} \quad k = 0,1,2,\dots. \tag{13.38}$$

On the other hand, as $U(t-k,\tau_{n'})v_{0n'} \to w_k$ weakly in $H$, from Proposition 13.1 we have

$$U(\cdot,t-k)U(t-k,\tau_{n'})v_{0n'} \to U(\cdot,t-k)w_k \tag{13.39}$$

weakly in $L^2(t-k,t;V)$.

Taking into account that, in particular, $e^{\sigma(\cdot-t)}g(\cdot) \in L^2(t-k,t;V')$, we obtain from (13.39),

$$\lim_{n'\to\infty}\int_{t-k}^{t} e^{\sigma(s-t)}\langle g(s), U(s,t-k)U(t-k,\tau_{n'})v_{0n'}\rangle\, ds$$
$$= \int_{t-k}^{t} e^{\sigma(s-t)}\langle g(s), U(s,t-k)w_k\rangle\, ds. \tag{13.40}$$

Moreover, it is not difficult to see, using (13.16), the continuous injection of $H^1(\Omega)$ into $L^4(\Omega)$, and the fact that by (13.23), $\nabla\xi \in L^\infty(t-k,t;L^2(\Omega)^{2\times 2})$, that the functional

$$\Psi(w) = \int_{t-k}^{t} e^{\sigma(s-t)}Q(s,w(s))\, ds$$

is convex and continuous, and consequently weakly lower semicontinuous on the space $L^2(t-k,t;V)$. Thus, we also obtain from (13.39)

$$\int_{t-k}^{t} e^{\sigma(s-t)}Q(s,U(s,t-k)w_k)\, ds \tag{13.41}$$
$$\le \liminf_{n'\to\infty}\int_{t-k}^{t} e^{\sigma(s-t)}Q(s,U(s,t-k)U(t-k,\tau_{n'})v_{0n'})\, ds.$$

Then, from (13.37), (13.38), (13.40), and (13.41), we obtain

$$\limsup_{n'\to\infty}|U(t,\tau_{n'})v_{0n'}|^2 \le e^{-\sigma k}R_\sigma^2(t-k) \tag{13.42}$$
$$+ 2\int_{t-k}^{t} e^{\sigma(s-t)}\left(\langle g(s), U(s,t-k)w_k\rangle - Q(s,U(s,t-k)w_k)\right) ds.$$

Now, from (13.34) and (13.36),

$$|w_0|^2 = |U(t,t-k)w_k|^2 = |w_k|^2e^{-\sigma k} \tag{13.43}$$
$$2\int_{t-k}^{t} e^{\sigma(s-t)}\left(\langle g(s), U(s,t-k)w_k\rangle - Q(s,U(s,t-k)w_k)\right) ds.$$

From (13.42) and (13.43), we have

$$\begin{aligned}\limsup_{n'\to\infty} |U(t,\tau_{n'})v_{0n'}|^2 &\leq e^{-\sigma k}R_\sigma^2(t-k) + |w_0|^2 - |w_k|^2 e^{-\sigma k}\\ &\leq e^{-\sigma k}R_\sigma^2(t-k) + |w_0|^2,\end{aligned}$$

and thus, taking into account that

$$e^{-\sigma k}R_\sigma^2(t-k) = \frac{3e^{-\sigma t}}{\nu}\int_{-\infty}^{t-k} e^{\sigma s}\|g(s)\|_{V'}^2\,ds \to 0,$$

when $k \to +\infty$, we easily obtain (13.35) from the last inequality. □

## 13.5 Fractal Dimension of the Pullback Attractor

In this section we prove

**Theorem 13.4.** *Let $U_0$ be a locally Lipschitz continuous function on the real line such that for some $t^\star \in \mathbb{R}$, $r > 0$, $M_b > 0$, $M > 0$, all $t \leq t^\star$, and all $s \leq t^\star - r$,*

$$|U_0(t)| \leq M_b \quad \text{and} \quad \int_s^{s+r} |U_0'(\eta)|^2 d\eta \leq M. \tag{13.44}$$

*Then the attractor $\mathbb{A} = \{A(t)\}_{t\in\mathbb{R}}$ from Theorem 13.3 has finite fractal dimension, namely,*

$$d_f^H(A(t)) \leq d \tag{13.45}$$

*for all $t \in \mathbb{R}$ and some constant $d$.*

To this end we shall use a result from [42] which in our notation can be expressed as follows.

**Theorem 13.5.** *Suppose there exist constants $K_0, K_1, \theta > 0$ such that*

$$|A(t)| = \sup\{|y| : \ y \in A(t)\} \leq K_0|t|^\theta + K_1 \tag{13.46}$$

*for all $t \in \mathbb{R}$. Assume that for any $t \in \mathbb{R}$ there exist $T = T(t)$, $l = l(t,T) \in [1,+\infty)$, $\delta = \delta(t,T) \in (0, 1/\sqrt{2})$, and $N = N(t)$ such that for any $u, v \in A(\tau)$ and $\tau \leq t - T$,*

$$|U(\tau+T,\tau)u - U(\tau+T,\tau)v| \leq l|u-v|, \tag{13.47}$$

$$|Q_N(U(\tau+T,\tau)u - U(\tau+T,\tau)v)| \leq \delta|u-v|, \tag{13.48}$$

*where $Q_N$ is the projector mapping from $H$ onto some subspace $H_N^{\perp}$ of codimension $N \in \mathbb{N}$. Then, for any $\eta = \eta(t) > 0$ such that $\sigma = \sigma(t) = (6\sqrt{2}l)^N(\sqrt{2}\sigma)^{\eta} < 1$, the fractal dimension of $A(t)$ is bounded, with*

$$d_f^H(A(t)) \leq N + \eta. \tag{13.49}$$

*Proof (of Theorem 13.4).* First we show that for some $M_0 > 0$ and all $t \leq t^{\star}$,

$$\|A(t)\| = \sup\{\|y\| : \ y \in A(t)\} \leq M_0. \tag{13.50}$$

Here we need the second energy estimate. With $\Theta = Av$ in (13.11) we get

$$\frac{1}{2}\frac{d}{dt}\|v(t)\|^2 + \nu|Av(t)|^2 + b(v(t), v(t), Av) = F(v(t), Av), \tag{13.51}$$

where

$$F(v, Av) = \nu(\Delta\xi, Av) - b(\xi, v, Av) - b(v, \xi, Av) - (\xi_{,t}, Av).$$

To estimate the nonlinear term on the left-hand side of (13.51) we use the Agmon inequality

$$\|v\|_{L^{\infty}(\Omega)} \leq c|v|^{1/2}|Av|^{1/2},$$

that holds for $v \in D(A)$. This estimate follows from the usual Agmon inequality

$$\|w\|_{L^{\infty}(\Omega_3)} \leq c\|w\|_{L^2(\Omega_3)}^{1/2}\|w\|_{H^2(\Omega_3)}^{1/2},$$

where $\Omega_3$ is a smooth set such that $\Omega \subset \Omega_3 \subset \Omega_{\infty}$ and $w = \varphi v$, $\varphi$ being the suitable cut-off function, and the inequality $\|v\|_{H^2(\Omega)} \leq c|Av|$ for $v \in D(A)$, cf. the proof of Theorem 9.6.

We have thus

$$|b(v, v, Av)| \leq \|v\|_{L^{\infty}(\Omega)}\|v\||Av| \leq c|v|^{1/2}\|v\||Av|^{3/2} \leq \frac{\nu}{8}|Av|^2 + \frac{c'}{\nu^3}|v|^2\|v\|^4.$$

Moreover,

$$|(\xi_{,t}, Av)| \leq \frac{\nu}{8}|Av|^2 + \frac{2}{\nu}|\xi_{,t}|^2,$$

and

$$|b(\xi, v, Av)| \leq \|\xi\|_{L^{\infty}(\Omega)}\|v\||Av| \leq \frac{\nu}{8}|Av|^2 + \frac{2}{\nu}\|\xi\|_{L^{\infty}(\Omega)}^2\|v\|^2.$$

Similarly,

$$|b(v,\xi,Av)| \leq |v|\|\nabla\xi\|_{L^\infty(\Omega)}|Av| \leq \frac{\nu}{8}|Av|^2 + \frac{2}{\nu}\|\nabla\xi\|^2_{L^\infty(\Omega)}|v|^2.$$

Finally,

$$|\nu(\Delta\xi, Av)| \leq \frac{\nu}{8}|Av|^2 + 2\nu|\Delta\xi|^2.$$

Therefore, from (13.51) we obtain

$$\frac{1}{2}\frac{d}{dt}\|v(t)\|^2 + \frac{1}{2}\nu|Av(t)|^2 \leq a(t)\|v(t)\|^2 + b(t), \tag{13.52}$$

where

$$a(t) = \frac{c'}{\nu^3}|v(t)|^2\|v(t)\|^2 + \frac{2}{\nu}\|\xi(t)\|^2_{L^\infty(\Omega)}, \tag{13.53}$$

and

$$b(t) = \frac{2}{\nu}|\xi_{,t}(t)|^2 + \frac{2}{\nu}\|\nabla\xi(t)\|^2_{L^\infty(\Omega)}|v(t)|^2 + 2\nu[\Delta\xi(t)]^2. \tag{13.54}$$

Now, recall that for all $v_0 \in A(\tau)$ and all $t \geq \tau$, we have

$$|U(t,\tau)v_0|^2 \leq e^{-\sigma(t-\tau)}|A(\tau)|^2 + 2\int_{-\infty}^{t} e^{\sigma(s-t)}f^2(s)ds. \tag{13.55}$$

Let us fix some $T > r$ and an arbitrary $t \leq t^*$. Let $v_0 \in A(\tau)$ for $\tau$ near $-\infty$. Then, from (13.55) and (13.30) we deduce that for all $\eta \in (t-T, t)$,

$$\begin{aligned} |v(\eta)|^2 &\leq 3e^{\sigma T}\int_{-\infty}^{t} e^{\sigma(s-t)}f^2(s)ds \\ &= 3e^{\sigma T}C_1(\nu,\Omega)\left(\frac{1}{\sigma} + e^{-\sigma t}\int_{-\infty}^{t} e^{\sigma s}[|U_0(s)|^3 + |U_0'(s)|^2]ds\right). \end{aligned} \tag{13.56}$$

From (13.54) we have

$$\begin{aligned} \int_s^{s+r} b(\eta)d\eta \leq & \frac{2Lh_0}{\nu}(3+2\sqrt{2})\int_s^{s+r} |U_0'(\eta)]^2 d\eta + \frac{8}{\nu^3}\int_s^{s+r} |U_0(\eta)|^4)|v(\eta)|^2 d\eta \\ & + \frac{L(4h_0)^3}{\nu^2}\left(\sup_{\eta\in\mathbb{R}}\rho''(\eta)\right)^2\int_s^{s+r} |U_0(\eta)|^5 d\eta \quad \text{if} \quad U_0(\eta) \geq \frac{\nu}{8h_0}, \end{aligned}$$

and

$$\int_s^{s+r} b(\eta)d\eta \le Lh_0 \int_s^{s+r} |U_0'(\eta)|^2 d\eta + \frac{\nu}{8^3 h_0^4} \int_s^{s+r} |v(\eta)|^2 d\eta$$
$$+ \frac{L\nu^3}{8^3 h_0^2} \left( \sup_{\eta \in \mathbb{R}} \rho''(\eta) \right)^2 \quad \text{if} \quad U_0(\eta) \le \frac{\nu}{8h_0}.$$

Thus, by (13.44) and (13.56) we conclude that there exists $M_2(r)$ such that

$$\int_s^{s+r} b(\eta)d\eta \le M_2(r)$$

for all $s \in (t-T, t-r)$.

From (13.25), (13.26), (13.31), (13.44), and (13.56) we deduce further that there exists $M_3(r)$ such that

$$\int_s^{s+r} \|v(\eta)\|^2 d\eta \le M_3(r) \tag{13.57}$$

for all $s \in (t-T, t-r)$.

Finally, by (13.53) together with (13.56), (13.57), and (13.44), there exists $M_1(r)$ such that

$$\int_s^{s+r} a(\eta)d\eta \le M_1(r) \tag{13.58}$$

for all $s \in (t-T, t-r)$.

With these estimates we apply now the uniform Gronwall lemma to inequality (13.52) to get, in particular,

$$\|v(t)\|^2 \le \left\{ \frac{M_3(r)}{r} + 2M_2(r) \right\} \exp 2M_1(r) \equiv \tilde{M}(r),$$

which gives (13.50).

We shall prove now that in our case of the Navier–Stokes flow inequality (13.46) holds true for $t \le t^\star$, and inequalities (13.47) and (13.48) hold true for $\tau + T \le t \le t^\star$. This restriction for the time variable $t$ in the hypotheses of Theorem 13.5 produces a similar change in its assertion, namely, inequality (13.49) holds then true for $t \le t^\star$. Such restricted result is, however, sufficient in our case of the Navier–Stokes flow in order to obtain (13.45) for all $t$ as it will be shown in the end of the proof.

Observe that in view of (13.50) inequality (13.46) is satisfied for $t \le t^\star$, namely,

$$|A(t)| \le \frac{1}{\sqrt{\lambda_1}} M_0 \quad \text{for} \quad t \le t^\star.$$

Inequality (13.47) for $\tau + T \le t \le t^\star$ follows from inequality (13.28). In fact, let $v(\tau)$ and $\bar{v}(\tau)$ be in $A(\tau)$, and denote $\omega = v - \bar{v}$. Then from (13.28) and (13.50) it follows that

$$|\omega(\tau + T)|^2 \le l(T)|\omega(\tau)|^2, \tag{13.59}$$

with $l(T) = \exp\{(-\frac{\sigma}{4} + \frac{4}{\nu}c_2(\Omega)M_0^2)T\}$ (we may always assume that the constant $l(T)$ in (13.59) is greater or equal than one).

Now, we shall prove that inequality (13.48) holds true. Multiplying

$$\frac{d\omega}{dt} + \nu A\omega + B(\omega, \bar{v}) + B(v, \omega) = B(w, \xi)$$

by $Q_m\omega$ defined by $Q_m\omega(x,t) = \sum_{j=m+1}^{\infty} \tilde{\omega}(t)w_j(x)$ ($Q_m = I - P_m$, $P_m$ being the projection on the space spanned by the first $m$ eigenfuntions of the Stokes operator with the periodic boundary conditions on $\Gamma_L$ and Dirichlet boundary conditions on $\Gamma_C \cup \Gamma_D$), we obtain

$$\begin{aligned} &\frac{1}{2}\frac{d}{dt}|Q_m\omega|^2 + \nu\|Q_m\omega\|^2 + b(\omega, v, Q_m\omega) + b(\bar{v}, \omega, Q_m\omega) \\ &\quad = b(w, \xi, Q_m\omega). \end{aligned} \tag{13.60}$$

Our first step is to obtain a differential inequality for $|Q_m\omega|^2$ of the form (13.65) below. To this end we have to estimate the nonlinear terms in (13.60).

We have,

$$b(\omega, v, Q_m\omega) = b(P_m\omega, v, Q_m\omega) + b(Q_m\omega, v, Q_m\omega).$$

Now,

$$\begin{aligned} |b(P_m\omega, v, Q_m\omega)| &= |b(P_m\omega, Q_m\omega, v)| \\ &\le c_2(\Omega)^{\frac{1}{2}}|P_m\omega|^{\frac{1}{2}}\|P_m\omega\|^{\frac{1}{2}}\|Q_m\omega\||v|^{\frac{1}{2}}\|v\|^{\frac{1}{2}}, \end{aligned}$$

and $\|P_m\omega\|^2 \le \lambda_m|P_m\omega|^2$, whence, from (13.50) we conclude

$$|b(P_m\omega, v, Q_m\omega)| \le \frac{2c_2(\Omega)M_0^2}{\nu\sqrt{\lambda_1}}\lambda_m^{\frac{1}{2}}|\omega|^2 + \frac{\nu}{8}\|Q_m\omega\|^2. \tag{13.61}$$

We have also

$$|b(Q_m\omega, v, Q_m\omega)| \le \frac{2c_2(\Omega)M_0^2}{\nu}|Q_m\omega|^2 + \frac{\nu}{8}\|Q_m\omega\|^2. \tag{13.62}$$

We estimate the last term on the left-hand side of (13.60) as follows,

$$\begin{aligned} |b(\bar{v},\omega,Q_m\omega)| &= |b(\bar{v},P_m\omega,Q_m\omega)| \\ &\le c_2(\Omega)^{\frac{1}{2}}|v|^{\frac{1}{2}}\|\bar{v}\|^{\frac{1}{2}}|P_m\omega|^{\frac{1}{2}}\|P_m\omega\|^{\frac{1}{2}}\|Q_m\omega\| \\ &\le \frac{2c_2(\Omega)M_0^2}{\nu\sqrt{\lambda_1}}\lambda_m^{\frac{1}{2}}|\omega|^2+\frac{\nu}{8}\|Q_m\omega\|^2. \end{aligned} \tag{13.63}$$

Using (13.44) we treat the term on the right-hand side of (13.60) as follows,

$$\begin{aligned} |b(\omega,\xi,Q_m\omega)| &= |b(\omega,Q_m\omega,\xi)| \le \|\xi\|_{L^\infty(\Omega)}|\omega|\|Q_m\omega\| \\ &\le \frac{2}{\nu}M_b^2|\omega|^2+\frac{\nu}{8}\|Q_m\omega\|^2. \end{aligned} \tag{13.64}$$

From (13.60) together with (13.61)–(13.64), and in view of

$$\lambda_{m+1}|Q_m\omega|^2 \le \|Q_m\omega\|^2,$$

we obtain, in particular, for $\tau \le s \le \tau+T \le t \le t^\star$,

$$\frac{d}{ds}|Q_m\omega(s)|^2+\sigma_m|Q_m\omega(s)|^2 \le \gamma_m|\omega(s)|^2, \tag{13.65}$$

where

$$\sigma_m=\left(\nu\lambda_{m+1}-\frac{4c_2(\Omega)M_0^2}{\nu}\right) \quad \text{and} \quad \gamma_m=\frac{8c_2(\Omega)M_0^2}{\nu\sqrt{\lambda_1}}\lambda_m^{\frac{1}{2}}+\frac{4M_b^2}{\nu}.$$

We can see that for $m$ large enough we have $\sigma_m>0$.

Now, using the bound $|\omega(s)|^2 \le l(T)|\omega(\tau)|^2$ for $\tau \le s \le \tau+T$, we obtain

$$\frac{d}{ds}|Q_m\omega(s)|^2+\sigma_m|Q_m\omega(s)|^2 \le \gamma_m l(T)|\omega(\tau)|^2.$$

Multiplying both sides by $\exp\{\sigma_m(s-\tau)\}$ we get

$$\frac{d}{ds}\{|Q_m\omega(s)|^2\exp\{\sigma_m(s-\tau)\}\} \le \gamma_m l(T)|\omega(\tau)|^2\exp\{\sigma_m(s-\tau)\}.$$

Integrating with respect to $s$ in the interval $(\tau,\tau+T)$ we obtain,

$$|Q_m\omega(\tau+T)|^2 \le \delta^2(T)|\omega(\tau)|^2,$$

with

$$\delta^2=\exp\{-\sigma_m T\}+\frac{\gamma_m l(T)}{\sigma_m}<\frac{1}{2}$$

for $m$ large enough, as $\sigma_m$ is of the order of $\lambda_{m+1}$. This gives inequality (13.48) and thus completes the proof of a finite dimensionality of the sets $A(t)$, for $t \leq t^\star$, with

$$d_f^H(A(t)) \leq m + \eta.$$

The same estimate holds also for $t > t^\star$ in view of the property of the fractal dimension under Lipschitz continuous mappings, cf. Theorem 9.1. □

## 13.6 Comments and Bibliographical Notes

To prove the existence of the pullback attractor we used, in Sect. 13.4, the energy equation method, developed in [44, 45] for the pullback attractor theory. This method works also in the case of certain unbounded domains of the flow as it bypasses the usual compactness argument. In turn, to estimate the pullback attractor dimension we used, in Sect. 13.5, the method proposed in [42], alternative to the usual one [149, 220]. It is also possible to study the finite dimensionality of pullback attractors through so-called *pullback exponential attractors*, see [51, 52, 151].

When $h = const$ and $U_0 = const$, problem (13.1)–(13.5) was intensively studied in several contexts. Most results concern the autonomous Navier–Stokes equations (cf. Sect. 9.4). In [89], the domain of the flow is an elongated rectangle $\Omega = (0, L) \times (0, h)$, $L \gg h$. In this case the attractor dimension can be estimated from above by $c\frac{L}{h}Re^{3/2}$, where $c$ is a universal constant and $Re = \frac{Uh}{\nu}$ is the Reynolds number. In [241] optimal bounds of the attractor dimension are given for a flow in a rectangle $(0, 2\pi L) \times (0, 2\pi L/\alpha)$, with periodic boundary conditions and given external forcing. The estimates are of the form $c_0/\alpha \leq d_f^H(\mathscr{A}) \leq c_1/\alpha$, see also [180]. Some free boundary conditions are considered in [242], see also [222], and an upper bound on the attractor dimension established with the use of a suitable anisotropic version of the Lieb–Thirring inequality, similarly to [89]. Dirichlet-periodic and free-periodic boundary conditions and domains with more general geometry were considered in [22, 24] (cf. also [25]) where still other forms of the Lieb–Thirring inequality were established to study the dependence of the attractor dimension on the shape of the domain of the flow. Slip boundary condition and unbounded domain case were considered in [183].

The autonomous case with $h \neq const$ was considered in [22, 24].

Boundary driven flows in smooth and bounded two-dimensional domains for a non-autonomous Navier–Stokes system were considered in [178], by using an approach of Chepyzhov and Vishik, cf. [62]. An extension to some unbounded domains can be found in [181], cf. also [164].

# 14
# Trajectory Attractors and Feedback Boundary Control in Contact Problems

*I hail a semigroup when I see one and I seem to see them everywhere. Friends have observed, however, that there are mathematical objects which are not semigroups.*

– Einar Hille

In this chapter we consider two-dimensional nonstationary incompressible Navier–Stokes shear flows with nonmonotone and multivalued leak boundary conditions on a part of the boundary of the flow domain. Our considerations are motivated by feedback control problems for fluid flows in domains with semipermeable walls and membranes and by the theory of lubrication. Our aim is to prove the existence of global in time solutions of the considered problem which is governed by a partial differential inclusion, and then to prove the existence of a trajectory attractor and a weak global attractor for the associated multivalued semiflow.

## 14.1 Setting of the Problem

The problem we consider is as follows. The flow of an incompressible fluid in a two-dimensional domain $\Omega$ is described by the equation of motion

$$u_t - \nu \Delta u + (u \cdot \nabla) u + \nabla p = 0 \quad \text{for} \quad (x,t) \in \Omega \times \mathbb{R}^+, \tag{14.1}$$

where the viscosity coefficient $\nu > 0$, and the incompressibility condition

$$\operatorname{div} u = 0 \quad \text{for} \quad (x,t) \in \Omega \times \mathbb{R}^+. \tag{14.2}$$

To define the domain $\Omega$ of the flow let us consider the channel

$$\Omega_\infty = \{x = (x_1, x_2) : \ -\infty < x_1 < \infty, \ 0 < x_2 < h(x_1)\},$$



G. Łukaszewicz, P. Kalita, *Navier–Stokes Equations*, Advances in Mechanics and Mathematics 34, DOI 10.1007/978-3-319-27760-8_14

where the function $h : \mathbb{R} \to \mathbb{R}$ is positive, smooth, and $L$-periodic. Then we set

$$\Omega = \{x = (x_1, x_2) :\ 0 < x_1 < L,\ 0 < x_2 < h(x_1)\},$$

and $\partial\Omega = \overline{\Gamma_C} \cup \overline{\Gamma_L} \cup \overline{\Gamma_D}$, where $\Gamma_C$ and $\Gamma_D$ are the bottom and the top, and $\Gamma_L$ is the lateral part of the boundary of $\Omega$. The domain $\Omega$ is schematically presented in Fig. 14.1. By $n$ we will always denote the unit normal vector on $\partial\Omega$.

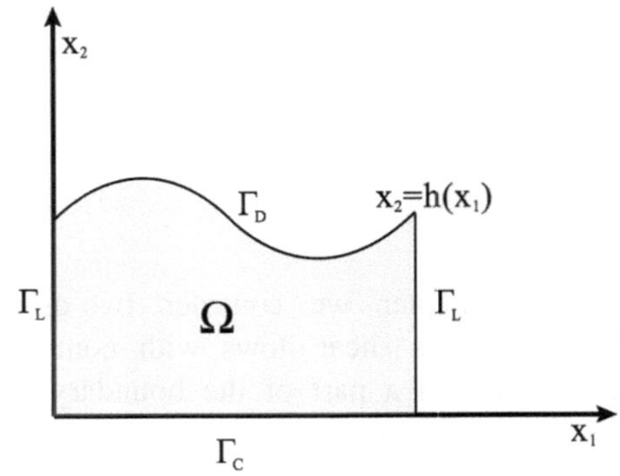


**Fig. 14.1** Schematic overview of the domain $\Omega$ and parts $\Gamma_C, \Gamma_D, \Gamma_L$ of its boundary

We are interested in solutions of (14.1)–(14.2) such that the velocity $u$ and the Cauchy stress vector $Tn$, where $T$ is the stress tensor given by the constitutive law $T_{ij} = -p\delta_{ij} + \nu(u_{i,j} + u_{j,i})$, are $L$-periodic with respect to $x_1$ on $\Gamma_L$. We assume that

$$u = 0 \quad \text{on} \quad \Gamma_D \times \mathbb{R}^+. \tag{14.3}$$

On the bottom $\Gamma_C$ we impose the following conditions. The tangential component $u_\tau$ of the velocity vector on $\Gamma_C$ is given, namely, for some $s \in \mathbb{R}$, we have

$$u_\tau = u - u_n n = (s, 0) \quad \text{on} \quad \Gamma_C \times \mathbb{R}^+, \quad \text{where} \quad u_n = u \cdot n. \tag{14.4}$$

Furthermore, we assume the following subdifferential boundary condition

$$\tilde{p}(x,t) \in \partial j(u_n(x,t)) \quad \text{on} \quad \Gamma_C \times \mathbb{R}^+, \tag{14.5}$$

where $\tilde{p} = p + \frac{1}{2}|u|^2$ is the total pressure (called the Bernoulli pressure), $j : \mathbb{R} \to \mathbb{R}$ is a given locally Lipschitz potential, and $\partial j$ is a Clarke subdifferential of $j$ (see Sect. 3.7 and [72, 84] for the definition and properties of Clarke subdifferential).

Let, moreover,

$$u(0) = u_0 \quad \text{in} \quad \Omega. \tag{14.6}$$

The considered problem is motivated by the examination of a certain two-dimensional flow in an infinite (rectified) journal bearing $\Omega \times (-\infty, +\infty)$, where the outer cylinder is represented by $\Gamma_D \times (-\infty, +\infty)$, and $\Gamma_C \times (-\infty, +\infty)$ represents the inner, rotating cylinder. In the lubrication problems the gap $h$ between cylinders is never constant. We can assume that the rectification does not change the equations as the gap between cylinders is very small with respect to their radii.

A physical interpretation of the boundary condition (14.5) can be as follows. The potential $j$ in our control problem is not convex as its subdifferential corresponds to the *nonmonotone* relation between the normal velocity $u_n$ and the total pressure $\tilde{p}$ at $\Gamma_C$. Assuming that, left uncontrolled, the total pressure at $\Gamma_C$ would increase with the increase of the normal velocity of the fluid at $\Gamma_C$, we control $\tilde{p}$ by a hydraulic device which opens wider the boundary orifices at $\Gamma_C$ when $u_n$ attains a certain value and thus $\tilde{p}$ *drops* at this value of $u_n$. Particular examples of such relations are provided in [175, 176].

## 14.2 Weak Formulation of the Problem

In this section we introduce the basic notations and define a notion of a weak solution $u$ of the initial boundary value problem (14.1)–(14.6). For convenience, we shall work with a homogeneous problem whose solution $v$ has the tangential component $v_\tau$ at $\Gamma_C$ equal to zero, and then $u = v + w$ for a suitable extension $w$ of the boundary data.

In order to define a weak formulation of the homogeneous problem (14.1)–(14.6) we need to introduce some function spaces and operators. Let

$$\begin{aligned} W = \{w \in C^\infty(\overline{\Omega})^2 :\ \operatorname{div} w = 0 \text{ in } \Omega,\ \ w \text{ is } L\text{-periodic in } x_1, \\ w = 0 \text{ on } \Gamma_D,\ \ w_\tau = 0 \text{ on } \Gamma_C\}, \end{aligned}$$

and let $V$ and $H$ be the closures of $W$ in the norms of $H^1(\Omega)^2$ and $L^2(\Omega)^2$, respectively. In the sequel we will use the notation $\|\cdot\|, |\cdot|$ to denote, respectively, the norms in $V$ and $H$. We denote the trace operator $V \to L^2(\Gamma_C)^2$ by $\gamma$. By the trace theorem, $\gamma$ is linear and bounded; we will denote its norm by $\|\gamma\| := \|\gamma\|_{\mathscr{L}(V;L^2(\Gamma_C)^2)}$. In the sequel we will write $u$ instead of $\gamma u$ for the sake of notation simplicity.

Let the operators $A : H^1(\Omega)^2 \to V'$ and $B[\cdot] : H^1(\Omega)^2 \to V'$ be defined by

$$\langle Au, v\rangle = \nu \int_\Omega \operatorname{rot} u \cdot \operatorname{rot} v \, dx \tag{14.7}$$

for all $u \in H^1(\Omega)^2$, $v \in V$, and $B[u] = B(u, u)$, where

$$\langle B(u, w), z\rangle = \int_\Omega (\operatorname{rot} u \times w) \cdot z \, dx \tag{14.8}$$

for $u, w \in H^1(\Omega)^2$, $z \in V$.

According to the hydrodynamical interpretation of the considered problem given in Sect. 14.1, we can understand the rot operator in the following way. For $u(x_1, x_2) = (u_1(x_1, x_2), u_2(x_1, x_2))$ and $\bar{u}(\bar{x}) = (u_1(x_1, x_2), u_2(x_1, x_2), 0)$, where $\bar{x} = (x_1, x_2, x_3)$, we set $\operatorname{rot} u(x_1, x_2) = \operatorname{rot} \bar{u}(\bar{x})$.

Let $G = \Omega \times (0, 1)$ and $f(\bar{x}) = f(x_1, x_2)$ be a scalar function. Then

$$\int_\Omega f(x_1, x_2)\, dx = \int_G f(\bar{x})\, d\bar{x}. \tag{14.9}$$

In particular, for $u, v$ in $V$,

$$\begin{aligned} \langle Au, v\rangle &= \nu \int_\Omega \operatorname{rot} u(x) \cdot \operatorname{rot} v(x)\, dx = \nu \int_G \operatorname{rot} \bar{u}(\bar{x}) \cdot \operatorname{rot} \bar{v}(\bar{x})\, d\bar{x} \\ &= \nu \int_G \nabla \bar{u}(\bar{x}) \cdot \nabla \bar{v}(\bar{x})\, d\bar{x} = \nu \int_\Omega \nabla u(x) \cdot \nabla v(x)\, dx. \end{aligned} \tag{14.10}$$

To work with the boundary condition (14.5) we rewrite equation of motion (14.1) in the Lamb form,

$$u_t + \nu \operatorname{rot}\operatorname{rot} u + \operatorname{rot} u \times u + \nabla \tilde{p} = 0 \quad \text{in} \quad \Omega \times \mathbb{R}^+. \tag{14.11}$$

Further, to transform the problem into the homogeneous one, for $u \in H^1(\Omega)^2$ let $w \in H^1(\Omega)^2$ be such that $\operatorname{div} w = 0$, $w = 0$ at $\Gamma_D$, $w$ is $L$-periodic on $\Gamma_L$, and $w_\tau = u_\tau = s$ and $w_n = 0$ on $\Gamma_C$, and let $u = v + w$. Then $v \in V$ as $v_\tau = 0$ at $\Gamma_C$. Moreover, $v_n = u_n$ on $\Gamma_C$.

Multiplying (14.11) by $z \in V$ and using the Green formula we obtain

$$\langle v'(t) + Av(t) + B[v(t)], z\rangle + \int_{\Gamma_C} \tilde{p} z_n\, dS = \langle F, z\rangle + \langle G(v), z\rangle, \tag{14.12}$$

where

$$\langle F, z\rangle = \nu \int_\Omega \operatorname{rot} w \cdot \operatorname{rot} z\, dx - \langle B(w, w), z\rangle, \tag{14.13}$$

and

$$\langle G(v), z\rangle = -\langle B(v, w) + B(w, v), z\rangle. \tag{14.14}$$

Above we have used the formula

$$\int_\Omega \operatorname{rot} R \cdot a\, dx = \int_\Omega R \cdot \operatorname{rot} a\, dx + \int_{\partial\Omega} (R \times a) \cdot n\, dS, \tag{14.15}$$

with $R = \operatorname{rot} v$ or $R = \operatorname{rot} w$. Formula (14.15) is easy to get using the three-dimensional vector calculus and (14.9). Observe that if $a_\tau = 0$ on $\partial\Omega$, then we have

$$(R\times a)\cdot n = (R\times a_n n)\cdot n = 0.$$

We need the following assumptions on the potential $j$:

(j1) $j : \mathbb{R} \to \mathbb{R}$ is locally Lipschitz,
(j2) $\partial j$ satisfies the growth condition $|\xi| \le c_1 + c_2|u|$ for all $u \in \mathbb{R}$ and all $\xi \in \partial j(u)$, with $c_1 > 0$ and $c_2 > 0$,
(j3) $\partial j$ satisfies the dissipativity condition $\inf_{\xi\in\partial j(u)} \xi u \ge d_1 - d_2|u|^2$, for all $u \in \mathbb{R}$, where $d_1 \in \mathbb{R}$ and $d_2 \in \left(0, \frac{\nu}{\|\gamma\|^2}\right)$.

From (14.12) we obtain the following weak formulation of the homogeneous problem.

**Problem 14.1.** Let $v_0 \in H$. Find

$$v \in L^2_{loc}(\mathbb{R}^+; V) \cap L^\infty_{loc}(\mathbb{R}^+; H),$$

with $v' \in L^{\frac{4}{3}}_{loc}(\mathbb{R}^+; V')$, $v(0) = v_0$, and such that

$$\langle v'(t) + Av(t) + B[v(t)], z\rangle + (\xi(t), z_n)_{L^2(\Gamma_C)} = \langle F, z\rangle + \langle G(v(t)), z\rangle, \quad (14.16)$$

$$\xi(t) \in S^2_{\partial j(v_n(\cdot,t))},$$

for a.e. $t \in \mathbb{R}^+$ and for all $z \in V$.

In the above definition we use the notation

$$S^2_U = \{u \in L^2(\Gamma_C) \, : \, u(x) \in U(x) \quad \text{for a.e.} \quad x \in \Gamma_C\},$$

valid for a multifunction $U : \Gamma_C \to 2^{\mathbb{R}}$. Note that from the fact that $v \in L^2_{loc}(\mathbb{R}^+; V)$ and $v' \in L^{\frac{4}{3}}_{loc}(\mathbb{R}^+; V')$ it follows that $v \in C(\mathbb{R}^+; V')$, and hence the initial condition makes sense. We define $C_w([0,T]; H)$ as the space of all functions $u : [0,T] \to H$ which are weakly continuous, i.e., such that for any $\phi \in H$ the scalar product $(u(t), \phi)$ in $H$ is a continuous function of $t$ for $t \in [0,T]$. Since $v \in L^\infty_{loc}(\mathbb{R}^+; H)$ then (see Lemma 3.18 or [227], Theorem II.1.7) $v \in C_w(\mathbb{R}^+; H)$, and thus the initial condition makes sense in the phase space $H$.

One can see (cf. [176]) that if $v \in L^2_{loc}(\mathbb{R}^+; V)$ is a sufficiently smooth solution of the partial differential inclusion (14.16) then there exists a distribution $\tilde{p}$ such that the conditions (14.11) and (14.5) hold for $u = v + w$. In conclusion, the function $u = v + w$ can be regarded as a weak solution of the initial boundary value problem (14.1)–(14.6), provided $v$ is a solution of Problem 14.1 with $v(0) = u_0 - w$, $u_0 \in H$.

The trajectory space $\mathscr{K}^+$ of Problem 14.1 is defined as the set of those of its solutions with some $v_0 \in H$ that satisfy the following inequality,

$$\frac{1}{2}\frac{d}{dt}|v(t)|^2 + C_1\|v(t)\|^2 \leq C_2(1 + \|F\|_{V'}^2), \tag{14.17}$$

where $C_1, C_2 > 0$ and inequality (14.17) is understood in the sense that for all $0 \leq t_1 < t_2$ and for all $\psi \in C_0^\infty((t_1, t_2))$, $\psi \geq 0$ we have

$$-\frac{1}{2}\int_{t_1}^{t_2} |v(t)|^2\psi'(t)\, dt + C_1\int_{t_1}^{t_2} \|v(t)\|^2\psi(t)\, dt \leq C_2(1 + \|F\|_{V'}^2)\int_{t_1}^{t_2} \psi(t)\, dt. \tag{14.18}$$

Note that since we cannot guarantee that for every solution of Problem 14.1 $\langle v'(t), v(t)\rangle = \frac{1}{2}\frac{d}{dt}|v(t)|^2$ holds, we cannot derive (14.17) for every solution of Problem 14.1.

In the next section we prove that the trajectory space is not empty.

## 14.3 Existence of Global in Time Solutions

In this section we give a proof of the existence of solutions of Problem 14.1 that satisfy inequality (14.17). The proof will be based on the standard technique that uses the regularization of the multivalued term and in main points will follow the existence proofs in Chap. 10 (see also [162, 176]).

The operators $A$ and $B$ defined in (14.7) and (14.8) and restricted to $V$ have the following properties:

(1) $A : V \to V'$ is a linear, continuous, symmetric operator such that

$$\langle Av, v\rangle = \nu\|v\|^2 \quad \text{for} \quad v \in V, \tag{14.19}$$

(2) $B : V \times V \to V'$ is a bilinear, continuous operator such that

$$\langle B(u, v), v\rangle = 0 \quad \text{for} \quad v \in V. \tag{14.20}$$

**Lemma 14.1.** *Given $\lambda > 0$ and $s \in \mathbb{R}$ there exists a smooth function $w \in H^1(\Omega)^2$ such that* $\operatorname{div} w = 0$ *in $\Omega$, $w = 0$ on $\Gamma_D$, $w$ is L-periodic in $x_1$, $w_\tau = (s, 0)$, $w_n = 0$ on $\Gamma_C$, and for all $v \in V$*

$$|\langle B(v, w), v\rangle| \leq \lambda\|v\|^2. \tag{14.21}$$

*Proof.* Let $w$ be of the form $w(x_2) = (s\rho(x_2/h_0), 0)$, where $\rho : [0, \infty) \to [0, 1]$ is a smooth function such that $\rho(0) = 1$, $\rho'(0) = 0$, $\operatorname{supp}\rho \subset [0, \min\{\frac{\lambda}{2|s|}, h_0, 1)\}]$, and

$h_0$ is the minimum value of $h(x_1)$ on $[0, L]$. It is clear that all the stated properties of $w$ other than (14.21) hold. To prove (14.21) observe that under our assumptions

$$\int_\Omega |\text{rot } w|^2 dx = \int_\Omega |\nabla w|^2 \, dx, \tag{14.22}$$

and then

$$\begin{aligned} |\langle B(v, w), v\rangle| &\le |\nabla v| \|x_2 w(x_2)\|_{L^\infty([0,h_0])} \left| \frac{v}{x_2} \right| \\ &\le \|v\| \frac{\lambda}{2} 2 \|v\| = \lambda \|v\|^2 \end{aligned} \tag{14.23}$$

in view of the Hardy inequality. □

**Lemma 14.2.** *Let $j : \mathbb{R} \to \mathbb{R}$ satisfy* (*j*1)–(*j*3). *For any solution $v$ of Problem 14.1,*

$$\|v'(t)\|_{V'} \le C_3 (1 + \|v(t)\| + |v(t)|^{1/2} \|v(t)\|^{\frac{3}{2}}) \quad \textit{for a.e.} \quad t \in \mathbb{R}^+, \tag{14.24}$$

*with a constant $C_3 > 0$ independent of $v$.*

*Proof.* Let $v$ be a solution of Problem 14.1. For any test function $z \in V$ we have, for a.e. $t \in \mathbb{R}^+$,

$$|\langle v'(t), z\rangle| \le \|F\|_{V'} \|z\| + \|\xi(t)\|_{L^2(\Gamma_C)} \|\gamma\| \|z\| + |\langle G(v(t)) - Av(t) - B[v(t)], z\rangle|.$$

From the growth condition (*j*2) it follows that $\|\xi(t)\|_{L^2(\Gamma_C)} \le C(1 + \|v(t)\|)$ with a constant $C > 0$. Moreover, $|\langle Av(t), z\rangle| + |\langle G(v(t)), z\rangle| \le C\|v(t)\| \|z\|$. It remains to estimate the nonlinear term. For all $w \in V$ we have (14.22). Now, from the Hölder and Ladyzhenskaya inequalities we obtain

$$\begin{aligned} |\langle B[v(t)], z\rangle| &\le \int_\Omega |\text{rot } v(t)| |v(t)| |z| \, dx \le \|v(t)\| \, \|v(t)\|_{L^4(\Omega)^2} \|z\|_{L^4(\Omega)^2} \\ &\le C |v(t)|^{1/2} \|v(t)\|^{3/2} \|z\|. \end{aligned}$$

In this way we obtain (14.24). □

**Theorem 14.1.** *Let the potential $j$ satisfy* (*j*1)–(*j*3), *$F \in V'$. Then for every $v_0 \in H$ there exists $v \in \mathscr{K}^+$ such that $v(0) = v_0$.*

*Proof.* Let $\varrho \in C_0^\infty((-1, 1))$ be a mollifier such that $\int_{-1}^1 \varrho(s) \, ds = 1$ and $\varrho(s) \ge 0$. We define $\varrho_k : \mathbb{R} \to \mathbb{R}$ by $\varrho_k(s) = k\varrho(ks)$ for $k \in \mathbb{N}$ and $s \in \mathbb{R}$. Then supp $\varrho_k \subset \left(-\frac{1}{k}, \frac{1}{k}\right)$. We consider $j_k : \mathbb{R} \to \mathbb{R}$ defined by the convolution

$$j_k(r) = \int_{\mathbb{R}} \varrho_k(s) j(r - s) \, ds \quad \text{for} \quad r \in \mathbb{R}.$$

By assertion $(A)$ of Theorem 3.21 we have $j_k \in C^\infty(\mathbb{R})$. Let $s \in \mathbb{R}$. We calculate

$$j_k'(s)s = \lim_{h\to 0+} \frac{j_k(s+hs)-j_k(s)}{h} = \lim_{h\to 0+} \int_{\operatorname{supp}\varrho_k} \varrho_k(r)\frac{j(s+hs-r)-j(s-r)}{h}\,dr.$$

Estimating the integrand from below, by the Lebourg mean value theorem (cf. Theorem 3.20) we get, for $\theta \in (0,1)$ dependent on $s, r, h$, and $\xi \in \partial j(s-r+\theta hs)$,

$$\begin{aligned}
\varrho_k(r)\frac{j(s+hs-r)-j(s-r)}{h} &= \varrho_k(r)\xi(s-r+\theta hs+r-\theta hs)\\
&\geq \varrho_k(r)\left(d_1 - d_2|s-r+\theta hs|^2 - (c_1+c_2|s-r+\theta hs|)(|r|+h|s|)\right)\\
&\geq \varrho_k(r)\left(d_1 - d_2\left(|s|+1+h|s|\right)^2 - \left(c_1+c_2\left(|s|+1+h|s|\right)\right)\left(1+h|s|\right)\right),
\end{aligned}$$

where we used the fact that $|r| \leq \frac{1}{k} \leq 1$. After integrating over $r$, passing with $h$ to zero and noting that $k \geq 1$ we get

$$j_k'(s)s \geq d_1 - d_2(|s|+1)^2 - c_1 - c_2(|s|+1) \geq \widetilde{d_1} - \widetilde{d_2}|s|^2,$$

whence regularized functions $j_k$ still satisfy $(j3)$, where the constants $d_1, d_2$ are different than the ones for $j$, but independent on $k$, and still we have $\widetilde{d_2} \in \left(0, \frac{\nu}{\|\gamma\|^2}\right)$.

Let us furthermore take the sequence $V_k$ of finite dimensional spaces such that $V_k$ is spanned by the first $k$ eigenfunctions of the Stokes operator with the Dirichlet and periodic boundary conditions given in the definition of the space $V$. Then $\{V_k\}_{k=1}^\infty$ approximate $V$ from inside, i.e., $\overline{\bigcup_{k=1}^\infty V_k} = V$. Moreover we take the sequence $v_{0k} \to v_0$ strongly in $H$ such that $v_{0k} \in V_k$. We formulate the regularized Galerkin problems for $k \in \mathbb{N}$:

Find a continuous function $v_k : \mathbb{R}^+ \to V_k$ such that for a.e. $t \in \mathbb{R}^+$ $v_k$ is differentiable and

$$\begin{aligned}
\langle v_k'(t) &+ Av_k(t) + B[v_k(t)], z\rangle + (j_k'((v_k)_n(\cdot,t)), z_n)_{L^2(\Gamma_C)} \qquad (14.25)\\
&= \langle F + G(v_k(t)), z\rangle,\\
v_k(0) &= v_{0k} \qquad (14.26)
\end{aligned}$$

for a.e. $t \in \mathbb{R}^+$ and for all $z \in V_k$.

We first show that if $v_k$ solves (14.25) then an estimate analogous to (14.17) holds.

We take $z = v_k(t)$ in (14.25) and, using (14.19) and (14.20) as well as the fact that $\langle B(v,w),v\rangle \le \lambda\|v\|^2$ for all $v \in V$, where $\lambda$ can be made arbitrarily small (Lemma 14.1), we obtain

$$\frac{1}{2}\frac{d}{dt}|v_k(t)|^2 + \nu\|v_k(t)\|^2 + (j_k'((v_k)_n(\cdot,t)),(v_k)_n(t))_{L^2(\Gamma_C)} \le \|F\|_{V'}\|v_k(t)\| + \lambda\|v_k(t)\|^2$$

for a.e. $t \in \mathbb{R}^+$. Using ($j3$) and the Cauchy inequality with some $\varepsilon > 0$ we obtain

$$\frac{1}{2}\frac{d}{dt}|v_k(t)|^2 + \nu\|v_k(t)\|^2 + \widetilde{d_1}m_1(\Gamma_C) - \widetilde{d_2}\|(v_k)_n(t)\|^2_{L^2(\Gamma_C)} \le C(\varepsilon)\|F\|^2_{V'} + (\lambda+\varepsilon)\|v_k(t)\|^2$$

for a.e. $t \in \mathbb{R}^+$, where $\varepsilon > 0$ is arbitrary and the constant $C(\varepsilon)$ is positive. Note that by the trace theorem $\|(v_k)_n(t)\|^2_{L^2(\Gamma_C)} \le \|v_k(t)\|^2_{L^2(\Gamma_C)^2} \le \|\gamma\|^2\|v_k(t)\|^2$. It follows that

$$\frac{1}{2}\frac{d}{dt}|v_k(t)|^2 + (\nu - \widetilde{d_2}\|\gamma\|^2)\|v_k(t)\|^2 \le C(\varepsilon)\|F\|^2_{V'} + (\lambda+\varepsilon)\|v_k(t)\|^2 + |\widetilde{d_1}|L$$

for a.e. $t \in \mathbb{R}^+$. It is enough to take $\lambda = \varepsilon = \frac{\nu - \widetilde{d_2}\|\gamma\|^2}{4}$. We get, with $C_1, C_2 > 0$ independent of $t$,

$$\frac{1}{2}\frac{d}{dt}|v_k(t)|^2 + C_1\|v_k(t)\|^2 \le C_2(1 + \|F\|^2_{V'}). \tag{14.27}$$

Note that, after integration,

$$\frac{1}{2}|v_k(t)|^2 + C_1\int_0^t \|v_k(s)\|^2\,ds \le \frac{1}{2}|v_{0k}|^2 + C_2(1+\|F\|^2_{V'})t \quad \text{for all} \quad t \ge 0. \tag{14.28}$$

Existence of the solution to the Galerkin problem (14.25)–(14.26) is standard and follows by the Carathéodory theorem and estimate (14.28). Note that for all $k \in \mathbb{N}$ solutions of (14.25) satisfy the estimate of Lemma 14.2, where the constants do not depend on initial conditions and the dimension $k$. We deduce from (14.27) that for all $T > 0$

$$v_k \quad \text{is bounded in} \quad L^2(0,T;V) \cap L^\infty(0,T;H). \tag{14.29}$$

We denote by $\iota : V \to L^2(\Gamma_C)$ the linear and bounded map given as $\iota v = (\gamma v)_n$ and by $\iota^* : L^2(\Gamma_C) \to V'$ its adjoint operator. Now if $P_k$ is the projection operator onto the space $V_k$, (14.25) is equivalent to the following equation in $V'$,

$$v_k'(t) + Av_k(t) + P_kB[v_k(t)] + P_k\iota^*j_k'((v_k)_k(\cdot,t)) = P_kF + P_kG(v_k(t)).$$

From the choice of the space $V_k$ it follows by Theorem 3.17 that $\|P_k u\|_{V'} \le \|u\|_{V'}$ and by the argument of Lemma 14.2 it follows that for all $T > 0$

$$v_k' \quad \text{is bounded in} \quad L^{\frac{4}{3}}(0, T; V'). \tag{14.30}$$

In view of (14.29) and (14.30), by diagonalization, we can construct a subsequence such that

$$v_k \to v \quad \text{weakly in} \quad L^2_{loc}(\mathbb{R}^+; V) \tag{14.31}$$

and

$$v_k' \to v' \quad \text{weakly in} \quad L^{\frac{4}{3}}_{loc}(\mathbb{R}^+; V'). \tag{14.32}$$

First we show that $v(0) = v_0$ in $H$. To this end choose $T > 0$. We have, for $t \in (0, T)$,

$$v_k(t) = v_{k0} + \int_0^t v_k'(s)\, ds, \qquad v(t) = v(0) + \int_0^t v'(s)\, ds,$$

where the equalities hold in $V'$. Take $\phi \in V$. Then

$$\begin{aligned}
&\int_0^T \langle v_k(t) - v(t), \phi\rangle\, dt \\
&\quad = \int_0^T \langle v_{k0} - v(0), \phi\rangle\, dt + \int_0^T \left\langle \int_0^t (v_k'(s) - v'(s))\, ds, \phi \right\rangle dt \\
&\quad = T\langle v_{k0} - v(0), \phi\rangle + \int_0^T \langle v_k'(t) - v'(t), (T-t)\phi\rangle\, dt.
\end{aligned}$$

The left-hand side of this equality goes to zero with $k \to \infty$ as $v_k \to v$ weakly in $L^2(0, T; V)$. The last integral on the right-hand side also goes to zero with $k \to \infty$ as $v_k' \to v'$ weakly in $L^{\frac{4}{3}}(0, T; V')$ and $(T - t)\phi \in L^4(0, T; V)$. Hence, $v_{k0} \to v_0$ weakly in $V'$ and $v(0) = v_0$ in $H$.

Now we pass with $k \to \infty$ in Eq. (14.25). To this end let us choose $T > 0$. We multiply (14.25) by $\phi \in C([0, T])$ and integrate with respect to $t$ in $[0, T]$. Passing to the limit in linear terms $A$ and $G$ is standard. We focus on the multivalued term and the convective term.

Let $Z = H^{1-\delta}(\Omega)^2 \cap V$ for $\delta \in \left(0, \frac{1}{2}\right)$ be endowed with $H^{1-\delta}$ norm. Since $V \subset Z$ and $Z \subset V'$, with compact and continuous embeddings, respectively, from the Aubin–Lions compactness theorem it follows that, for a subsequence, $v_k \to v$ strongly in $L^2(0, T; Z)$, and, by the continuity of the trace operator and also of the corresponding Nemytskii operator, we have for all $T > 0$,

$$(v_k)_n \to v_n \quad \text{strongly in} \quad L^2(0, T; L^2(\Gamma_C)). \tag{14.33}$$

By assertion $(B)$ of Theorem 3.21, the functions $j_k'$ satisfy the growth condition $(j2)$ with the constants independent of $k$. It follows that the sequence $j_k'((v_k)_n(\cdot,\cdot))$ is bounded in $L^2(0,T;L^2(\Gamma_C))$, and we can extract a subsequence, not renumbered, such that for all $T>0$

$$j_k'((v_k)_n(\cdot,\cdot)) \to \xi \quad \text{weakly in} \quad L^2(0,T;L^2(\Gamma_C)). \tag{14.34}$$

We need to show that $\xi(t) \in S^2_{\partial j(v_n(\cdot,t))}$ for a.e. $t \in \mathbb{R}^+$. We use assertion $(C)$ of Theorem 3.21, whence we have, for all $T>0$,

$$\xi(x,t) \in \partial j(v_n(x,t)) \quad \text{for a.e.} \quad (x,t) \in \Gamma_C \times (0,T),$$

and the desired result follows. Hence we can pass to the limit in the multivalued term.

Now we show the convergence in the nonlinear term, namely, that (for a subsequence)

$$\int_0^T \int_\Omega (\operatorname{rot} v_k(t) \times v_k(t)) \cdot z(x)\phi(t)\,dx\,dt \to \int_0^T \int_\Omega (\operatorname{rot} v(t) \times v(t)) \cdot z(x)\phi(t)\,dx\,dt. \tag{14.35}$$

First we prove that

$$\begin{aligned}\int_0^T \int_\Omega (\operatorname{rot} v_k(t) \times v_k(t)) \cdot z(x)\phi(t)\,dx\,dt &= \frac{1}{2}\int_0^T \int_{\Gamma_C} (v_k)_n^2(x,t) z_n(x)\phi(t)\,dS\,dt \\ &- \int_0^T \int_\Omega (v_k(x,t)\cdot\nabla) z(x) v_k(x,t)\phi(t)\,dx\,dt. \end{aligned} \tag{14.36}$$

From (14.15), as well as the formulas $(a\times b)\cdot c = (b \times c)\cdot a$, and

$$\nabla\cdot F = 0, \nabla\cdot G = 0 \quad \Longrightarrow \quad \operatorname{rot}(F\times G) = (G\cdot\nabla)F - (F\cdot\nabla)G \tag{14.37}$$

we have

$$\begin{aligned}\langle B[v_k(t)], z\rangle &= \int_\Omega (v_k(t)\times z)\cdot \operatorname{rot} v_k(t)\,dx \\ &= \int_\Omega \operatorname{rot}(v_k(t)\times z)\cdot v_k(t)\,dx + \int_{\partial\Omega} (v_k(t)\times(v_k(t)\times z))\cdot n\,dS.\end{aligned} \tag{14.38}$$

The surface integral is equal to zero and hence, using (14.37) in the right-hand side, we obtain

$$\langle B[v_k(t)], z\rangle = \int_\Omega (z\cdot\nabla)v_k(t)\cdot v_k(t)dx - \int_\Omega (v_k(t)\cdot\nabla)z\cdot v_k(t)\,dx. \tag{14.39}$$

Integration by parts in the first integral on the right-hand side and then integration in the time variable give the result.

Since we can deal with the second term on the right-hand side of (14.36) as in the usual Navier–Stokes theory [219], we consider only the surface integral. Taking the difference of the corresponding terms and setting $z_n(x)\phi(t) = \psi(x,t)$ we obtain

$$\left| \int_0^T \int_{\Gamma_C} ((v_k)_n(x,t) - v_n(x,t))((v_k)_n(x,t) + v_n(x,t))\psi(x,t)\,dS\,dt \right|$$
$$\leq \|(v_k)_n - v_n\|_{L^2(0,T;L^2(\Gamma_C))} \|v_k + v\|_{L^2(0,T;L^4(\Gamma_C)^2)} \max_{t\in[0,T]} \|\psi(t)\|_{L^4(\Gamma_C)}.$$

This proves (14.35) since the embedding $H^{1/2}(\Gamma_C)^2 \subset L^r(\Gamma_C)^2$ is continuous for $r \geq 1$, $v_k$ is bounded in $L^2(0,T;L^4(\Gamma_C)^2)$, and $\|(v_k)_n - v_n\|_{L^2(0,T;L^2(\Gamma_C))} \to 0$ by (14.33).

Hence, the limit $v$ solves Problem 14.1. It remains to show (14.18). The proof follows the lines of that of Theorem 8.1 in [61]. Let us fix $0 \leq t_1 < t_2$ and choose $\psi \in C_0^\infty((t_1,t_2))$ with $\psi \geq 0$. We multiply (14.27) by $\psi$ and integrate by parts. We have

$$-\frac{1}{2}\int_{t_1}^{t_2} |v_k(t)|^2\psi'(t)\,dt + C_1\int_{t_1}^{t_2} \|v_k(t)\|^2\psi(t)\,dt \leq C_2(1+\|F\|^2_{V'})\int_{t_1}^{t_2}\psi(t)\,dt. \tag{14.40}$$

We need to pass to the limit ($k\to\infty$) in the last inequality. From (14.31) and (14.32) by the Aubin–Lions compactness theorem we conclude that $v_k \to v$ strongly in $L^2(t_1,t_2;H)$. From inequality

$$\int_{t_1}^{t_2} (|v_k(t)| - |v(t)|)^2\,dt \leq \int_{t_1}^{t_2} |v_k(t) - v(t)|^2\,ds$$

it follows that $|v_k(t)| \to |v(t)|$ strongly in $L^2(t_1,t_2)$, and hence, for a subsequence, $|v_k(t)|^2 \to |v(t)|^2$ for almost every $t \in (t_1,t_2)$. Since from (14.28) it follows that functions $|v_k(t)|^2\psi'(t)$ have an integrable majorant on $(t_1,t_2)$, we have, by the Lebesgue dominated convergence theorem that

$$\int_{t_1}^{t_2} |v_k(t)|^2\psi'(t)\,dt \to \int_{t_1}^{t_2} |v(t)|^2\psi'(t)\,dt. \tag{14.41}$$

Now, from (14.31) it follows that $v_k(t)(\psi(t))^{\frac{1}{2}} \to v(t)(\psi(t))^{\frac{1}{2}}$ weakly in $L^2(t_1,t_2;V)$ and hence, by the sequential weak lower semicontinuity of the norm we obtain

$$\int_{t_1}^{t_2} \|v(t)\|^2\psi(t)\,dt \leq \liminf_{k\to\infty}\int_{t_1}^{t_2} \|v_k(t)\|^2\psi(t)\,dt. \tag{14.42}$$

Thereby, we can pass to the limit in (14.40) which gives us (14.18) and the proof is complete. □

## 14.4 Existence of Attractors

In this section we prove the existence of a trajectory attractor and a weak global attractor for the considered problem.

We have shown that for any initial condition $v_0 \in H$ Problem 14.1 has at least one solution $v \in \mathscr{K}_+$. The key idea behind the trajectory attractor (see [61, 62, 226, 227]) is, in contrast to the direct study of the solutions asymptotic behavior, the investigation of the family of shift operators $\{T(t)\}_{t\geq 0}$ defined on $\mathscr{K}^+$ by the formula

$$(T(t)v)(s) = v(s+t).$$

Before we pass to a theorem on the existence of the trajectory attractor, which is the main result of this section, we recall some definitions and results (see [61, 62, 226, 227]).

Let

$$\mathscr{F}(0,T) = \{u \in L^2(0,T;V) \cap L^\infty(0,T;H) : \ u' \in L^{\frac{4}{3}}(0,T;V')\},$$

and

$$\mathscr{F}_+^{loc} = \{u \in L^2_{loc}(\mathbb{R}^+;V) \cap L^\infty_{loc}(\mathbb{R}^+;H) : \ u' \in L^{\frac{4}{3}}_{loc}(\mathbb{R}^+;V')\}.$$

Note that $\mathscr{F}(0,T) \subset C_w([0,T];H)$ and from the Lions–Magenes lemma (cf. Lemma 1.4, Chap. 3 in [219], Theorem II,1.7 in [62] or Lemma 2.1 in [227]) it follows that for all $u \in \mathscr{F}(0,T)$ we have

$$|u(t)| \leq \|u\|_{L^\infty(0,T;H)} \quad \text{for all} \quad t \in [0,T]. \tag{14.43}$$

Furthermore, let us define the Banach space

$$\mathscr{F}_+^b = \{u \in \mathscr{F}_+^{loc} : \ \|u\|_{\mathscr{F}_+^b} < \infty\},$$

where the norm in $\mathscr{F}_+^b$ is given by

$$\|u\|_{\mathscr{F}_+^b} = \sup_{h\geq 0} \|\Pi_{0,1}u(\cdot + h)\|_{\mathscr{F}(0,1)}.$$

In the above definition $\Pi_{0,1}u(\cdot)$ is a restriction of $u(\cdot)$ to the interval $(0,1)$, and

$$||v||_{\mathscr{F}(0,1)} = \|v\|_{L^2(0,1;V)} + ||v||_{L^\infty(0,1;H)} + \|v'\|_{L^{\frac{4}{3}}(0,1;V')}.$$

Finally, we define the topology by $\Theta_+^{loc}$ in the space $\mathscr{F}_+^{loc}$ in the following way: the sequence $\{u_k\}_{k=1}^\infty \subset \mathscr{F}_+^{loc}$ is said to converge to $u \in \mathscr{F}_+^{loc}$ in the sense of $\Theta_+^{loc}$ if

$$u_k \to u \quad \text{weakly in} \quad L^2_{loc}(\mathbb{R}^+; V),$$
$$u_k \to u \quad \text{weakly-}* \text{ in} \quad L^\infty_{loc}(\mathbb{R}^+; H),$$
$$u_k' \to u' \quad \text{weakly in} \quad L^{\frac{4}{3}}_{loc}(\mathbb{R}^+; V').$$

Note (see, for example, [227]) that the topology $\Theta_+^{loc}$ is stronger than the topology of $C_w([0,T]; H)$ for all $T \geq 0$ and hence from the fact that $v_k \to v$ in $\Theta_+^{loc}$ it follows that $v_k(t) \to v(t)$ weakly in $H$ for all $t \geq 0$.

**Definition 14.1.** A set $\mathscr{P} \subset \mathscr{K}^+$ is said to be *absorbing* for the shift semigroup $\{T(t)\}_{t\geq 0}$ if for any set $\mathfrak{B} \subset \mathscr{K}^+$ bounded in $\mathscr{F}_+^b$ there exists $\tau = \tau(\mathfrak{B}) > 0$ such that $T(t)\mathfrak{B} \subset \mathscr{P}$ for all $t \geq \tau$.

**Definition 14.2.** A set $\mathscr{P} \subset \mathscr{K}^+$ is said to be *attracting* for the shift semigroup $\{T(t)\}_{t\geq 0}$ in the topology $\Theta_+^{loc}$ if for any set $\mathfrak{B} \subset \mathscr{K}^+$ bounded in $\mathscr{F}_+^b$ and for any neighborhood $\mathscr{O}(\mathscr{P})$ of $\mathscr{P}$ in the topology $\Theta_+^{loc}$ there exists $\tau = \tau(\mathfrak{B}, \mathscr{O}(\mathscr{P})) > 0$ such that $T(t)\mathfrak{B} \subset \mathscr{O}(\mathscr{P})$ for all $t \geq \tau$.

**Definition 14.3.** A set $\mathfrak{U} \subset \mathscr{K}^+$ is called a *trajectory attractor* of the shift semigroup $\{T(t)\}_{t\geq 0}$ on $\mathscr{K}^+$ in the topology $\Theta_+^{loc}$ if

(a) $\mathfrak{U}$ is bounded in $\mathscr{F}_+^b$ and compact in the topology $\Theta_+^{loc}$,
(b) $\mathfrak{U}$ is an attracting set in the topology $\Theta_+^{loc}$,
(c) $\mathfrak{U}$ is invariant, i.e., $T(t)\mathfrak{U} = \mathfrak{U}$ for any $t \geq 0$.

In order to show the existence of a trajectory attractor for Problem 14.1 we will use the following theorem (see, for instance, Theorem 4.1 in [227]).

**Theorem 14.2.** *Assume that the trajectory set $\mathscr{K}^+$ is contained in the space $\mathscr{F}_+^b$ and that*

$$T(t)\mathscr{K}^+ \subset \mathscr{K}^+. \tag{14.44}$$

*Suppose that the semigroup $\{T(t)\}_{t\geq 0}$ has an attracting set $\mathscr{P}$ that is bounded in the norm of $\mathscr{F}_+^b$ and compact in the topology $\Theta_+^{loc}$. Then the shift semigroup $\{T(t)\}_{t\geq 0}$ has a trajectory attractor $\mathfrak{U} \subset \mathscr{P}$.*

We are in position to formulate the main theorem of this section.

**Theorem 14.3.** *The shift semigroup $\{T(t)\}_{t\geq 0}$ defined on the set $\mathscr{K}^+$ of solutions of Problem 14.1 has a trajectory attractor $\mathfrak{U}$ which is bounded in $\mathscr{F}_+^b$ and compact in the topology $\Theta_+^{loc}$.*

Before we pass to the proof of this theorem we will need three auxiliary lemmas.

**Lemma 14.3.** *Let $v \in \mathscr{K}^+$. For all $h \geq 0$ the following estimates hold:*

$$\|v\|^2_{L^2(h,h+1;V)} + \|v\|^2_{L^\infty(h,h+1;H)} \leq C_4 + C_5\|v\|^2_{L^\infty(0,1;H)}e^{-C_6 h}, \tag{14.45}$$

$$\|v'\|^{\frac{4}{3}}_{L^{\frac{4}{3}}(h,h+1;V')} \le C_7 + C_8 \|v\|^{\frac{8}{3}}_{L^\infty(0,1;H)} e^{-C_9 h}, \tag{14.46}$$

*where $C_4, C_5, C_6, C_7, C_8$, and $C_9$ are positive constants independent of $h$, $v$.*

*Proof.* Let us fix $h \ge 0$ and $v \in \mathscr{K}^+$. Since for all $u \in V$, $\lambda_1 |u|^2 \le \|u\|^2$ with the constant $\lambda_1 > 0$, it follows from (14.18), that for all $0 \le t_1 < t_2$ and $\psi \in C_0^\infty((t_1, t_2))$ with $\psi \ge 0$,

$$-\frac{1}{2}\int_{t_1}^{t_2} |v(t)|^2 \psi'(t)\, dt + C_1\lambda_1 \int_{t_1}^{t_2} |v(t)|^2 \psi(t)\, dt \le C_2(1 + \|F\|^2_{V'}) \int_{t_1}^{t_2} \psi(t)\, dt. \tag{14.47}$$

We are in position to use Lemma 9.2 from [61] to deduce that there exists a Lebesgue null set $Q \subset \mathbb{R}^+$ such that for all $t, \tau \in \mathbb{R}^+ \setminus Q$ we have

$$|v(t)|^2 e^{C_1\lambda_1(t-\tau)} - |v(\tau)|^2 \le \frac{e^{C_1\lambda_1(t-\tau)} - 1}{C_1\lambda_1} C_2(1 + \|F\|^2_{V'}). \tag{14.48}$$

Now, since the function $t \to v(t)$ is weakly continuous, it follows that $t \to |v(t)|^2$ is lower semicontinuous and (14.48) holds for all $t \ge \tau$ (but only for a.e. $\tau \in \mathbb{R}^+$). Hence, for all $t > 0$ we can find $\tau \in (0, \min\{1, t\})$ such that

$$|v(t)|^2 e^{C_1\lambda_1(t-\tau)} \le \|v\|^2_{L^\infty(0,1;H)} + \frac{e^{C_1\lambda_1(t-\tau)}}{C_1\lambda_1} C_2(1 + \|F\|^2_{V'}), \tag{14.49}$$

and moreover, for all $t > 0$, we have

$$|v(t)|^2 \le e^{C_1\lambda_1(1-t)} \|v\|^2_{L^\infty(0,1;H)} + \frac{C_2(1 + \|F\|^2_{V'})}{C_1\lambda_1}. \tag{14.50}$$

By Corollary 9.2 in [61] it follows from (14.18) that for almost all $h > 0$,

$$\frac{1}{2}|v(h+1)|^2 + C_1 \int_h^{h+1} \|v(t)\|^2\, dt \le C_2(1 + \|F\|^2_{V'}) + \frac{1}{2}|v(h)|^2.$$

Using (14.50) we obtain, for a.e. $h > 0$,

$$\int_h^{h+1} \|v(t)\|^2\, dt \le \frac{C_2}{C_1}(1 + \|F\|^2_{V'}) + \frac{C_2(1 + \|F\|^2_{V'})}{2C_1^2\lambda_1} + \frac{\|v\|^2_{L^\infty(0,1;H)} e^{C_1\lambda_1(1-h)}}{2C_1},$$

and the estimate (14.45) follows for all $h \ge 0$ since both left- and right-hand side of the above inequality are continuous functions of $h$. Now, using (14.24) we get

$$\|v'(t)\|^{\frac{4}{3}}_{V'} \le C_{10}\left(1 + \left(1 + |v(t)|^{\frac{2}{3}}\right)\|v(t)\|^2\right) \quad \text{for a.e.} \quad t \in \mathbb{R}^+,$$

where $C_{10} > 0$. Integrating this inequality from $h$ to $h+1$ we obtain

$$\int_h^{h+1} \|v'(t)\|_{V'}^{4/3}\, dt \le C_{10} + C_{10}\left(1 + ||v(t)||^{\frac{2}{3}}_{L^\infty(h,h+1;H)}\right) \int_h^{h+1} \|v(t)\|^2\, dt.$$

Inequality (14.46) follows from an application of (14.45). □

**Lemma 14.4.** *The set $\mathscr{K}^+$ is closed in the topology $\Theta_+^{loc}$.*

*Proof.* Assume that for a sequence $\{v_k\}_{k=1}^\infty \subset \mathscr{K}^+$ we have

$$\begin{aligned} v_k &\to v \quad \text{weakly in} \quad L^2_{loc}(\mathbb{R}^+; V), \\ v_k &\to v \quad \text{weakly-}* \text{ in} \quad L^\infty_{loc}(\mathbb{R}^+; H), \\ v_k' &\to v' \quad \text{weakly in} \quad L^{\frac{4}{3}}_{loc}(\mathbb{R}^+; V^*). \end{aligned}$$

We need to show that $v$ satisfies (14.16) and (14.18). Since $\{v_k\}_{k=1}^\infty \subset \mathscr{K}^+$, we have, for all $k \in \mathbb{N}$ and $z \in V$,

$$\begin{aligned} \langle v_k'(t) + Av_k(t) + B[v_k(t)], z\rangle + (\xi_k(t), z_n)_{L^2(\Gamma_C)} &= \langle F, z\rangle + \langle G(v_k(t)), z\rangle, \\ \xi_k(t) &\in S^2_{\partial j((v_k)_n(\cdot,t))} \end{aligned}$$

for a.e. $t \in \mathbb{R}^+$. Passing to the limit in terms with $A, B, G$ is analogous to that in the proof of Theorem 14.1 (see also [62, 176, 219, 220]).

In order to pass to the limit in the multivalued term observe that from ($j2$) it follows that for a.e. $t \in \mathbb{R}^+$

$$\|\xi_k(t)\|^2_{L^2(\Gamma_C)} \le 2c_1^2 m_1(\Gamma_C) + 2c_2\|\gamma\|^2\|v_k(t)\|^2.$$

Hence, after integration in the time variable we get, for all $T \in \mathbb{R}^+$,

$$\|\xi_k\|^2_{L^2(0,T;L^2(\Gamma_C))} \le 2Tc_1^2 L + 2c_2\|\gamma\|^2\|v_k\|^2_{L^2(0,T;V)}.$$

By a diagonal argument it follows that there exists $\xi \in L^2_{loc}(\mathbb{R}^+; L^2(\Gamma_C))$ such that, for a subsequence,

$$\xi_k \to \xi \quad \text{weakly in} \quad L^2_{loc}(\mathbb{R}^+; L^2(\Gamma_C)).$$

As in the proof of Theorem 14.1,

$$(v_k)_n \to v_n \quad \text{strongly in} \quad L^2_{loc}(\mathbb{R}^+; L^2(\Gamma_C)).$$

Using assertion ($C$) of Theorem 3.21 we deduce that

$$\xi(x,t) \in \partial j(v_n(x,t)) \quad \text{a.e. in} \quad \Gamma_C \times \mathbb{R}^+,$$

and moreover,

$$\xi(t) \in S^2_{\partial j(v_n(\cdot,t))} \quad \text{a.e. in} \quad \mathbb{R}^+.$$

Hence we can pass to the limit in the term $(\xi_k(t), z_n)_{L^2(\Gamma_C)}$. It remains to show that $v$ satisfies (14.18). The argument is similar to that in the proof of Theorem 14.1. We choose $0 \le t_1 < t_2$ and $\psi \in C_0^\infty((t_1,t_2))$. We have, after possible refining to a subsequence, $|v_k(t)|^2\psi'(t) \to |v(t)|^2\psi'(t)$ for a.e. $t \in (t_1,t_2)$. Moreover, since weakly-* convergent sequences in $L^\infty(t_1,t_2;H)$ are bounded in this space it follows that there exists a majorant for $|v_k(t)|^2\psi'(t)$. Hence, by the Lebesgue dominated convergence theorem and since $v_k(t)(\psi(t))^{\frac{1}{2}} \to v(t)(\psi(t))^{\frac{1}{2}}$ weakly in $L^2(t_1,t_2;V)$, we can pass to the limit in (14.18) written for $v_k$, which completes the proof. □

*Proof (of Theorem 14.3).* Observe that from Lemma 14.3 it follows that for every $v \in \mathscr{K}^+$, $\|v\|_{\mathscr{F}^b_+} < \infty$. Note that since the problem is autonomous, for any $v \in \mathscr{K}^+$ and any $h \in \mathbb{R}^+$, the function $v_h$, defined as $v_h(t) = v(t+h)$ for all $t \ge 0$, belongs to $\mathscr{K}^+$ and hence (14.44) holds.

Now let us estimate $\|T(s)v\|_{\mathscr{F}^b_+}$ for $v \in \mathscr{K}^+$ and $s \ge 0$. We have

$$\|T(s)v\|_{\mathscr{F}^b_+} = \sup_{h\ge s}\left\{\|v\|_{L^2(h,h+1;V)} + ||v(t)||_{L^\infty(h,h+1;H)} + \|v'\|_{L^{\frac{4}{3}}(h,h+1;V')}\right\}.$$

Using Lemma 14.3 we get

$$\|T(s)v\|_{\mathscr{F}^b_+} \le \sup_{h\ge s} C\left\{2\left(1+\|v\|^2_{L^\infty(0,1;H)}e^{-C_6h}\right)^{\frac{1}{2}} + \left(1+\|v\|^{\frac{8}{3}}_{L^\infty(0,1;H)}e^{-C_9h}\right)^{\frac{3}{4}}\right\},$$

with a constant $C > 0$ dependent only on $C_4, C_5, C_7, C_8$ of Lemma 14.3. By a simple calculation we obtain, for all $s \in \mathbb{R}^+$,

$$\|T(s)v\|_{\mathscr{F}^b_+} \le R_0 + C\|v\|^\beta_{L^\infty(0,1;H)}e^{-\delta s} \le R_0 + C\|v\|^\beta_{\mathscr{F}(0,1)}e^{-\delta s}, \tag{14.51}$$

where $C, R_0, \beta, \delta > 0$ do not depend on $s, v$.

Now we define the set $\mathscr{P}$ as

$$\mathscr{P} = \{v \in \mathscr{K}^+ : \ \|v\|_{\mathscr{F}^b_+} \le 2R_0\}.$$

We will show that $\mathscr{P}$ is absorbing (and hence also attracting) for $\{T(t)\}_{t\ge 0}$. Let $\mathfrak{B}$ be bounded in $\mathscr{F}^b_+$. Hence $\mathfrak{B}$ is bounded in $\mathscr{F}(0,1)$. Let $\|v\|_{\mathscr{F}(0,1)} \le R$ for $v \in \mathfrak{B}$. Now we choose $s_0 > 0$ such that $CR^\beta e^{-\delta s_0} \le R_0$. From (14.51) we have for $s \ge s_0$ that

$$\|T(s)v\|_{\mathscr{F}^b_+} \le 2R_0,$$

and we deduce that $\mathscr{P}$ is absorbing.

It suffices to show that $\mathscr{P}$ is compact in the topology $\Theta_+^{loc}$. The fact that it is relatively compact follows from its boundedness and basic properties of weak compactness in reflexive Banach spaces. It remains to show that it is closed. Let $v_k \overset{\Theta_+^{loc}}{\rightarrow} v$ and $\{v_k\} \subset \mathscr{P}$. From the weak lower semicontinuity of the norm it follows that

$$\|v\|_{\mathscr{F}_+^b} \leq \liminf_{k\to\infty} \|v_k\|_{\mathscr{F}_+^b} \leq 2R_0.$$

Moreover from Lemma 14.4 it follows that $v \in \mathscr{K}^+$. Thus $v \in \mathscr{P}$ and the proof is complete. □

Now we show that any section of the trajectory attractor is a weak global attractor. We will start from the definition of a weak global attractor (cf., e.g., [227]).

**Definition 14.4.** A set $\mathscr{A} \subset H$ is called a *weak global attractor* of the set $\mathscr{K}^+$ if it is weakly compact in $H$ and the following properties hold:

(*i*) for any set $\mathfrak{B} \subset \mathscr{K}^+$ bounded in the norm of $\mathscr{F}_+^b$ its section $\mathfrak{B}(t)$ is attracted to $\mathscr{A}$ in the weak topology of $H$ as $t \to \infty$, that is, for any neighborhood $\mathscr{O}_w(\mathscr{A})$ of $\mathscr{A}$ in the weak topology of $H$ there exists a time $\tau = \tau(\mathfrak{B}, \mathscr{O}_w(\mathscr{A}))$ such that

$$\mathfrak{B}(t) \subset \mathscr{O}_w(\mathscr{A}) \quad \text{for all} \quad t \geq \tau,$$

(*ii*) $\mathscr{A}$ is the minimal weakly closed set in $H$ that attracts the sections of all bounded sets in $\mathscr{K}^+$ in the weak topology of $H$ as $t \to \infty$.

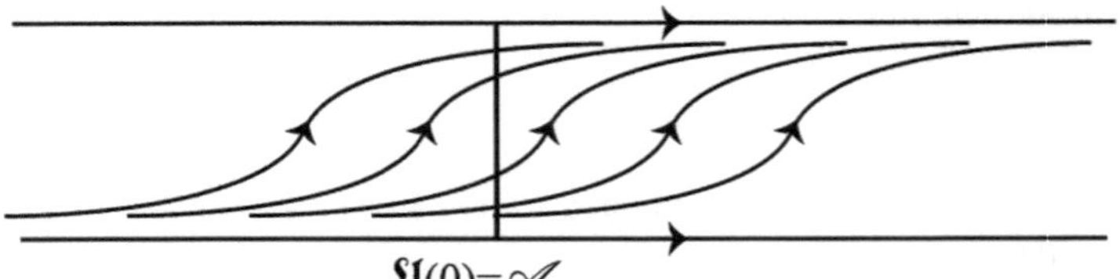


**Fig. 14.2** Schematic illustration of trajectory attractor for the semiflow governed by the solution map of the one-dimensional ODE $\dot{x} = |x|(1-x)$. The trajectory attractor $\mathfrak{U}$ consists of two constant trajectories located at the stationary points 0 and 1, and of the family of bounded complete trajectories that connect those points. The global attractor $\mathscr{A}$ is the section of the trajectory attractor $\mathscr{A} = \mathfrak{U}(0) = \mathfrak{U}(t)$ for all $t \in \mathbb{R}$

We prove the following Theorem.

**Theorem 14.4.** *The set* $\mathscr{A} = \mathfrak{U}(0) \subset H$ *is a weak global attractor of* $\mathscr{K}^+$.

*Proof.* The argument is standard in the theory of trajectory attractors and follows the lines of the proof of Assertion 5.3 in [227]. Since $\mathfrak{U}$ is bounded in $\mathscr{F}_+^b$ and thus in $\mathscr{F}(0,1)$, by (14.43) we deduce that $\mathfrak{U}(0)$ is bounded in $H$. Moreover, since the topology $\Theta_+^{loc}$ is stronger than the topology of $C_w([0,T];H)$ it follows that if $u_k \to u$ in the topology $\Theta_+^{loc}$ then $u_k(0) \to u(0)$ weakly in $H$. From the fact that $\mathfrak{U}$ is compact it follows that $\mathfrak{U}(0)$ is weakly compact in $H$.

To show $(i)$ let us take $\mathfrak{B} \subset \mathscr{K}^+$ bounded in the norm of $\mathscr{F}_+^b$. This set is attracted to $\mathfrak{U}$ in the topology $\Theta_+^{loc}$. Since for every sequence $u_k \to u$ in the topology $\Theta_+^{loc}$ we have $u_k(0) \to u(0)$ weakly in $H$, it follows that $(T(t)\mathfrak{B})(0)$ is attracted to $\mathfrak{U}(0)$ in the weak topology of $H$, or, in other words, $\mathfrak{B}(t)$ is attracted to $\mathfrak{U}(0)$ in the weak topology of $H$.

To show $(ii)$ let $\mathscr{E}$ be an arbitrary weakly closed subset of $H$ that weakly attracts the sections of bounded sets $\mathfrak{B} \subset \mathscr{K}^+$, that is, for any weak neighborhood $\mathscr{O}_w(\mathscr{E})$, $\mathfrak{B}(t) \subset \mathscr{O}_w(\mathscr{E})$ for all $t \geq \tau = \tau(\mathfrak{B}, \mathscr{O}_w(\mathscr{E}))$. Let us take $\mathfrak{B} = \mathfrak{U}$. We have $\mathfrak{U}(t) \subset \mathscr{O}_w(\mathscr{E})$ for all $t \geq \tau = \tau(\mathfrak{U}, \mathscr{O}_w(\mathscr{E}))$. From the invariance of the trajectory attractor we obtain that $\mathscr{A} = \mathfrak{U}(0) = \mathfrak{U}(t) \subset \mathscr{O}_w(\mathscr{E})$. Since $\mathscr{E}$ is weakly closed in $H$, it follows that $\mathscr{A} \subset \mathscr{E}$ and the proof is complete. □

The example which illustrates the fact that the global attractor is the section of the trajectory attractor is presented in Fig. 14.2.

## 14.5 Comments and Bibliographical Notes

In this chapter we followed [124].

The system of Eqs. (14.1)–(14.2) with *no-slip* boundary conditions: (14.3) at $\Gamma_D$ for $h = const$ and $u = const$ on $\Gamma_C$ (instead of (14.4)–(14.5) on $\Gamma_C$) was intensively studied in several contexts, some of them mentioned in the introduction of Boukrouche and Łukaszewicz [26]. The autonomous case with $h \neq const$ and with $u = const$ on $\Gamma_C$ was considered in [22, 24] (also see Chap. 9) and the non-autonomous case $h \neq const$, $u = U(t)e_1$ on $\Gamma_C$ was considered in [28] (also see Chap. 13). Existence of exponential attractors for the Navier–Stokes fluids and global attractors for Bingham fluids, with the Tresca boundary condition on $\Gamma_C$ was proved in [162] (also see Chap. 10) and [27], respectively. Recently, attractors for two-dimensional Navier–Stokes flows with Dirichlet boundary conditions were studied in [79], where the time continuous problem has a unique solution but a solution of the semidiscrete problem constructed by the implicit Euler scheme can be nonunique, and hence the theory of multivalued semiflows is needed to study the time discretized systems.

Asymptotic behavior of solutions for the problems governed by partial differential inclusions where the multivalued term has the form of Clarke subdifferential was studied in [132, 133], where the reaction-diffusion problem with multivalued semilinear term was considered, and in [123], where the strongly damped wave equation with multivalued boundary conditions was analyzed.

For the problem considered in this chapter, existence of weak solutions for the case $u_\tau = 0$ in place of (14.4) was shown in [176]. Our assumptions ($j1$)–($j3$) are more general than the corresponding assumptions of Theorem 1 in [176], save for the fact that $j$ is assumed there to depend on space and time variables directly.

To prove the existence of attractors we used the theory of trajectory attractors, which, instead of the direct analysis of the multivalued semiflow (i.e., map that assigns to initial condition the set of states obtainable after some time $t$, see [171, 239] or the review paper [10]), focuses on the shift operator defined on the space of trajectories for the studied problem. This approach was introduced in papers [60, 167, 210] as a method to avoid the nonuniqueness of solutions, indeed, the shift operator is *uniquely* defined even if the dynamics of the problem is governed by the *multivalued* semiflow. Recent results and open problems in the theory of trajectory attractors are discussed in the survey papers [10, 227].

# 15 Evolutionary Systems and the Navier–Stokes Equations

This chapter is devoted to the study of three-dimensional nonstationary Navier–Stokes equations with the multivalued frictional boundary condition. We use the formalism of evolutionary systems to prove the existence of weak global attractor for the studied problem.

## 15.1 Evolutionary Systems and Their Attractors

We remind the framework of evolutionary systems [65, 66]. Let $(X, d_s(\cdot,\cdot))$ be a metric space. The metric $d_s$ will be referred to as the strong metric. We denote by $d_w$ another metric on $X$, referred to as the weak metric, satisfying the following conditions:

1. the space $X$ is $d_w$ compact,
2. if $\lim_{k\to\infty} d_s(u_k, v_k) = 0$ for some sequences $u_k, v_k \in X$, then also $\lim_{k\to\infty} d_w(u_k, v_k) = 0$.

Any strongly compact (i.e., $d_s$ compact) subset of $X$ is also weakly compact (i.e., $d_w$ compact). Any strongly closed (i.e., $d_s$ closed) subset of $X$ is also weakly closed (i.e., $d_w$ closed) and, as a subset of the $d_w$ compact set $X$, also $d_w$ compact. To define the evolutionary system, let

$$\mathscr{T} = \{I : \; I = [T,\infty) \subset \mathbb{R} \quad \text{or} \quad I = (-\infty,\infty)\}.$$

By $\mathscr{F}(I)$ we denote the set of all $X$-valued functions having the domain $I \in \mathscr{T}$.



G. Łukaszewicz, P. Kalita, *Navier–Stokes Equations*, Advances in Mechanics and Mathematics 34, DOI 10.1007/978-3-319-27760-8_15

**Definition 15.1.** A map $\mathscr{E}$ that associates to each $I \in \mathscr{T}$ a subset $\mathscr{E}(I) \subset \mathscr{F}(I)$ is called an evolutionary system if the following conditions hold:

$(i)$ $\mathscr{E}([0,\infty)) \neq \emptyset$,
$(ii)$ $\mathscr{E}(I+s) = \{u(\cdot): \ u(\cdot - s) \in \mathscr{E}(I)\}$ for all $s \in \mathbb{R}$,
$(iii)$ $\{u(\cdot)|_{I_2}: \ u(\cdot) \in \mathscr{E}(I_1)\} \subset \mathscr{E}(I_2)$ for all $I_1, I_2 \in \mathscr{T}$ such that $I_2 \subset I_1$,
$(iv)$ $\mathscr{E}((-\infty,\infty)) = \{u(\cdot): \ u(\cdot)|_{[T,\infty)} \in \mathscr{E}([T,\infty))$ for all $T \in \mathbb{R}\}$.

For $I = [T,\infty)$ the set $\mathscr{E}(I)$ is referred to as the set of trajectories on the interval $I$, and $\mathscr{E}((-\infty,\infty))$ is called the set of complete trajectories. Having the evolutionary system we can define the map $\mathscr{G}: [0,\infty) \times X \to 2^X$ by

$$\mathscr{G}(t,x) = \{u(t): \ u(0) = x \quad \text{for some} \quad u \in \mathscr{E}([0,\infty))\}.$$

We can naturally extend $\mathscr{G}$ to $[0,\infty) \times 2^X$ by defining $\mathscr{G}(t,A) = \bigcup_{x\in A} \mathscr{G}(t,x)$.

*Remark 15.1.* If we additionally assume the existence, i.e., that for any $x_0 \in X$ there exists $u \in \mathscr{E}([0,\infty))$ such that $u(0) = x_0$, then the map $\mathscr{G}$ becomes an $m$-semiflow in the sense of Melnik and Valero [171], that is:

$(i)$ $\mathscr{G}(0,x) = \{x\}$ for all $x \in X$,
$(ii)$ $\mathscr{G}(t+s,X) \subset \mathscr{G}(t,\mathscr{G}(s,x))$ for all $x \in X$ and $s,t \geq 0$.

The illustration of an evolutionary system is presented in Fig. 15.1.

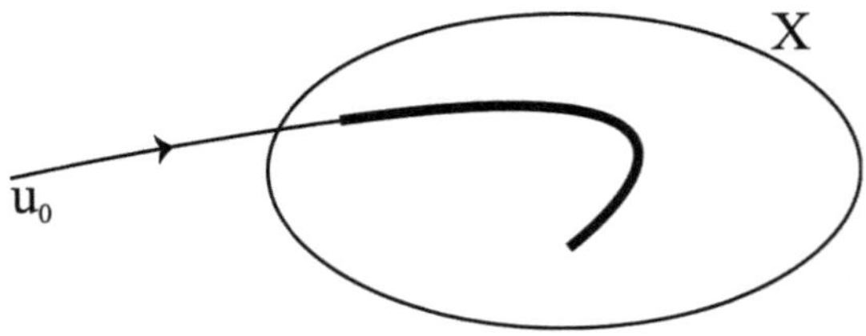


**Fig. 15.1** Illustration of an evolutionary system. If a semiflow (or $m$-semiflow) is dissipative, i.e., has a bounded absorbing set $X$, then this set, endowed with appropriate metrics becomes the metric space for the evolutionary system. Moreover, the endings of all trajectories, which must be contained in the absorbing set (*thickened line* in the figure), constitute the family $\mathscr{E}(I)$

Let $A \subset X$ and $r \geq 0$. We denote $B_w(A,r) = \{x \in X: \operatorname{dist}_w(x,A) < r\}$, where $\operatorname{dist}_w(x,A) = \inf_{y\in A} d_w(x,y)$. Moreover, for sets $A, B \subset X$ we define the weak Hausdorff semi-distance $\operatorname{dist}_w(A,B) = \sup_{x\in A} \operatorname{dist}_w(x,B)$. We pass to the definition of uniformly weakly attracting sets and weak attractors.

**Definition 15.2.** A set $A \subset X$ is uniformly weakly attracting (uniformly $d_w$ attracting) if for any $\varepsilon > 0$ there exists $t_0 > 0$ such that $\mathscr{G}(t,X) \subset B_w(A,\varepsilon)$ for all $t \geq t_0$, or, equivalently

$$\lim_{t\to\infty} \operatorname{dist}_w(\mathscr{G}(t,X),A) = 0.$$

**Definition 15.3.** A set $\mathscr{A} \subset X$ is a weak global attractor ($d_w$ global attractor) if it is a minimal weakly closed uniformly weakly attracting set.

Of course, the weak global attractor must be weakly compact. The following theorem gives the existence of the weak global attractor.

**Theorem 15.1 (cf. [65, Theorem 3.9]).** *Every evolutionary system possesses a weak global attractor $\mathscr{A} \subset X$.*

Let $a, b \in \mathbb{R}$ be such that $a < b$. Consider the Banach space $C_w([a,b];X)$ of weakly continuous functions from the interval $[a,b]$ to the space $X$. This space is a metric one, equipped with the metric

$$d_{C_w([a,b];X)}(u,v) = \sup_{t\in[a,b]} d_w(u(t),v(t)).$$

This metric can be naturally extended to $C_w([a,\infty);X)$ by the formula

$$d_{C_w([a,\infty);X)}(u,v) = \sum_{k\in\mathbb{N}} \frac{1}{2^k} \frac{\sup_{t\in[a,a+k]} d_w(u(t),v(t))}{1+\sup_{t\in[a,a+k]} d_w(u(t),v(t))}.$$

We make the following assumption on the evolution system:

$(A1)$ the set $\mathscr{E}([0,\infty))$ is compact in $C_w([0,\infty);X)$.

Assumption $(A1)$ will yield the invariance of the global attractor. To define the notion of invariance we first set

$$\tilde{\mathscr{G}}(t,A) = \{u(t): \ u(0) \in A \quad \text{for} \quad u \in \mathscr{E}((-\infty,\infty))\}, \quad \text{where} \quad A \subset X, \ t \in \mathbb{R}.$$

**Definition 15.4.** The set $A \subset X$ is said to be invariant (in the sense of Foiaş–Cheskidov) if $\tilde{\mathscr{G}}(t,A) = A$ for all $t \geq 0$.

*Remark 15.2.* Since $\tilde{\mathscr{G}}(t,A) \subset \mathscr{G}(t,A)$, then the fact that $A$ is invariant implies that $A \subset \mathscr{G}(t,A)$, i.e., $A$ is negatively semi-invariant in the sense of Melnik and Valero (see [171]).

**Theorem 15.2 (cf. [65, Theorem 5.8]).** *Let $\mathscr{E}$ be an evolutionary system that satisfies $(A1)$. Then the weak global attractor $\mathscr{A}$ is the maximal invariant set. Moreover,*

$$\mathscr{A} = \{x \in X: \ x = u(0) \quad \textit{for some} \quad u \in \mathscr{E}((-\infty,\infty))\},$$

*where, by invariance, we can also take values at any $t \geq 0$ instead of zero. Moreover, for any $\varepsilon > 0$ there exists $t_0 > 0$ such that for any $t^* > t_0$ and for any trajectory $u \in \mathscr{E}([0,\infty))$ we have*

$$d_{C_w([t^*,\infty);X)}(u,v) < \varepsilon$$

*for a certain complete trajectory $v \in \mathscr{E}((-\infty,\infty))$.*

The last assertion of the above theorem is known as the *tracking property*.

## 15.2 Three-Dimensional Navier–Stokes Problem with Multivalued Friction

We will consider a flow between two surfaces, the flat one at the bottom, and the periodic one at the top. Define $\Omega_\infty = \{(x_1, x_2, x_3) \in \mathbb{R}^3 : 0 < x_3 < h(x_1, x_2)\}$, where the function $h(\cdot, x_2)$ is assumed to be $L_1$-periodic, and $h(x_1, \cdot)$ is assumed to be $L_2$-periodic. Moreover we assume that $h$ is Lipschitz and that

$$\min_{(x_1,x_2)\in[0,L_1]\times[0,L_2]} h(x_1, x_2) > 0.$$

The periodicity of $h$ implies that it suffices to consider the flow inside the domain

$$\Omega = \{(x_1, x_2, x_3) \in \mathbb{R}^3 : \ (x_1, x_2) \in (0, L_1) \times (0, L_2), \ 0 < x_3 < h(x, y)\}.$$

Let $t_0 \in \mathbb{R}$ be the initial time. The momentum equation and the incompressibility condition are, respectively, given by

$$u_t - \nu\Delta u + (u \cdot \nabla)u + \nabla p = f \quad \text{on} \quad \Omega \times (t_0, \infty), \tag{15.1}$$

$$\operatorname{div} u = 0 \quad \text{on} \quad \Omega \times (t_0, \infty), \tag{15.2}$$

where we look for the velocity $u : \Omega\times(t_0, \infty) \to \mathbb{R}^3$ and pressure $p : \Omega\times(t_0, \infty) \to \mathbb{R}$. The function $f : \Omega \to \mathbb{R}^3$ is the mass force density, and $\nu > 0$ is the viscosity coefficient. The boundary of $\Omega$ is divided into the following parts:

$$\Gamma_D = (0, L_1) \times (0, L_2) \times \{0\},$$

$$\Gamma_C = \{(x_1, x_2, x_3) \in \mathbb{R}^3 : \ (x_1, x_2) \in (0, L_1) \times (0, L_2), x_3 = h(x_1, x_2)\},$$

$$\Gamma_L = \Gamma_{L1} \cup \Gamma_{L2},$$

$$\Gamma_{L1} = \{0, L_1\} \times \{(x_2, x_3) \in \mathbb{R}^2 : \ 0 < x_2 < L_2, 0 < x_3 < h(0, x_2)\},$$

$$\Gamma_{L2} = \{(x_1, x_2, x_3) \in \mathbb{R}^3 : \ 0 < x_1 < L_1, x_2 \in \{0, L_2\}, 0 < x_3 < h(x_1, 0)\}.$$

The domain $\Omega$ and three parts of its boundary are schematically presented in Fig. 15.2. The stress tensor $T$ is associated with the symmetrized displacement gradient $D(u) = \frac{1}{2}(\nabla u + \nabla u^{\mathsf{T}})$ and the pressure by the linear constitutive law $T = -pI + 2\nu D(u)$. On the boundary of $\Omega$ the stress can be decomposed into normal and tangential components: $T_n = (Tn) \cdot n$ and $T_\tau = Tn - T_n n$.

On $\Gamma_D$ we assume the homogeneous Dirichlet boundary condition

$$u = 0 \quad \text{on} \quad \Gamma_D \times (t_0, \infty). \tag{15.3}$$

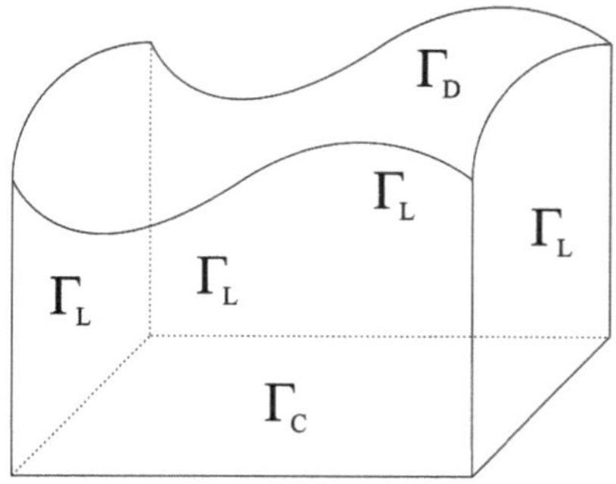


**Fig. 15.2** The example setup of the studied problem. We assume the frictional contact between the flowing liquid and the foundation on $\Gamma_C$, periodic boundary conditions on $\Gamma_L$, and the homogeneous Dirichlet boundary conditions on $\Gamma_D$

We assume the periodicity of normal stresses and velocities on $\Gamma_L$, i.e.,

$$(x_1, x_2, x_3, t) \in \Gamma_{L1} \times [t_0, \infty) \quad \Rightarrow \quad u(0, x_2, x_3, t) = u(L_1, x_2, x_3, t) \qquad (15.4)$$
$$\wedge\ (Tn)(0, x_2, x_3, t) = (Tn)(L_1, x_2, x_3, t),$$
$$(x_1, x_2, x_3, t) \in \Gamma_{L2} \times [t_0, \infty) \quad \Rightarrow \quad u(x_1, 0, x_3, t) = u(x_1, L_2, x_3, t) \qquad (15.5)$$
$$\wedge\ (Tn)(x_1, 0, x_3, t) = (Tn)(x_1, L_2, x_3, t).$$

Note that, due to the periodicity of $u$, the periodicity of normal stresses on $\Gamma_{L1}$ means that, for $(x_1, x_2, x_3) \in \Gamma_{L1}$ we have

$$p(0, x_2, x_3, t) = p(L_1, x_2, x_3, t) \text{ and } u_{i,1}(0, x_2, x_3, t) = u_{i,1}(L_1, x_2, x_3, t) \text{ for } i \neq 1,$$

and on $\Gamma_{L2}$ we have

$$p(x_1, 0, x_3, t) = p(x_1, L_2, x_3, t) \text{ and } u_{i,2}(x_1, 0, x_3, t) = u_{i,2}(x_1, L_2, x_3, t) \text{ for } i \neq 2.$$

Assume that the velocity $u$ is sufficiently smooth. The fact that on $\Gamma_{L1}$ the function $u(\cdot, x_2, x_3)$ is $L_1$-periodic implies the same periodicity of partial derivatives $u_{i,2}$ and $u_{i,3}$ for $i = 1, 2, 3$, and, by the incompressibility condition, also of $u_{1,1}$. Similarly, from $L_2$-periodicity of $u(x_1, \cdot, x_3)$ on $\Gamma_{L2}$ we obtain the same $L_2$-periodicity of the partial derivatives $u_{i,1}$ and $u_{i,3}$ for $i = 1, 2, 3$, and, by the incompressibility condition, also of $u_{2,2}$.

On $\Gamma_C$ we assume that

$$u_n = 0 \quad \text{on} \quad \Gamma_C \times (t_0, \infty), \qquad (15.6)$$
$$-T_\tau \in g(u_\tau) \quad \text{on} \quad \Gamma_C \times (t_0, \infty), \qquad (15.7)$$

where $g : \Gamma_C \times \mathbb{R} \to 2^{\mathbb{R}}$ is a *friction multifunction*, $u_n = u \cdot n$ is the normal velocity, and $u_\tau = u - u_n n$ is the tangential velocity. Finally, we need the initial condition

$$u = u_0 \quad \text{on} \quad \Omega \quad \text{for} \quad t = t_0. \qquad (15.8)$$

We pass to the weak formulation for the studied problem. First, we define the spaces:

$$\begin{aligned}\tilde{V} =&\{v \in C^\infty(\overline{\Omega})^3 : \ \operatorname{div} v = 0 \text{ in } \Omega, \quad v = 0 \text{ on } \Gamma_D, \quad v_n = 0 \text{ on } \Gamma_C,\\ &v(\cdot, x_2, x_3) \text{ is } L_1\text{-periodic on } \Gamma_{L1}, \quad v(x_1, \cdot, x_3) \text{ is } L_2\text{-periodic on } \Gamma_{L2}\},\end{aligned}$$

$$V = \text{ closure of } \tilde{V} \text{ in } H^1(\Omega)^3 \quad \text{and} \quad H = \text{ closure of } \tilde{V} \text{ in } L^2(\Omega)^3.$$

The spaces $V \subset H \subset V'$ constitute the evolution triple with dense and continuous embeddings. Moreover, the embedding $V \subset H$ is compact. As the Dirichlet boundary is nontrivial, the norm in the space $V$ equivalent with the standard $H^1$ norm is given by $\|v\|^2 = \int_\Omega \nabla v(x) \colon \nabla v(x)\, dx$. We will also use the space $W = L^2(\Gamma_C)^3$ as well as the linear and continuous trace operator $\gamma : V \to W$, whose norm we will denote by $\|\gamma\|_{\mathscr{L}(V;W)} = \|\gamma\|$. The calligraphic symbols will be used for time-dependent spaces: $\mathscr{V} = L^2(t_0, T; V)$, $\mathscr{V}' = L^2(t_0, T; V')$.

To motivate the definition of the weak solution assume that smooth functions $u, p$ satisfy (15.1)–(15.8). Multiplying the momentum equation (15.1) by $v \in \tilde{V}$ and integrating over $\Omega$, after application of the incompressibility condition (15.2), the boundary conditions (15.3)–(15.7), and integration by parts, we arrive at the following weak form of the problem,

$$\int_\Omega u_t \cdot v\, dx + \nu \int_\Omega \nabla u \cdot \nabla v\, dx + \int_\Omega (u \cdot \nabla) u \cdot v\, dx + \int_{\Gamma_C} \xi \cdot v_\tau\, dS = \int_\Omega f \cdot v\, dx,$$

where $\xi(x,t) \in g(x, u_\tau(x,t))$ for $(x,t) \in \Gamma_C \times [t_0, \infty)$.

**Exercise 15.1.** Derive the weak formulation of the considered problem.

*Hint.* The derivation of the above weak form is similar to that in Chap. 5. Note that due to the fact that the contact boundary is a flat surface we can consider the bilinear form $\int_\Omega \nabla u(x) \colon \nabla v(x)\, dx$ instead of more general $\int_\Omega D(u)(x) \colon D(v)(x)\, dx$ (see Remark 5.1). Moreover, the periodic conditions on $u, Tn$, and the definition of the space $\tilde{V}$ imply that all boundary integrals on $\Gamma_L$ cancel.

## 15.3 Existence of Leray–Hopf Weak Solution

We assume that $(t_0, T)$, the time interval, is given. Using the Galerkin method we will establish the existence of the weak solution for considered problem, understood in the Leray–Hopf sense, on the interval $(t_0, T)$. First, we define the operator $A : V \to V'$ by the formula

$$\langle Au, v \rangle = \int_\Omega \nabla u(x) \colon \nabla v(x)\, dx \quad \text{for} \quad u, v \in V.$$

Obviously, the operator $A$ is linear and bounded, moreover $\langle Au, u\rangle = \|u\|^2$ for $u \in V$. We define the Nemytskii operator $\mathscr{A} : \mathscr{V} \to \mathscr{V}'$ by the formula $(\mathscr{A}u)(t) = A(u(t))$ for a.e. $t \in (t_0, T)$. This Nemytskii operator is also linear and bounded, and hence it is continuous and moreover also weakly sequentially continuous, i.e., $u_k \to u$ weakly in $\mathscr{V}$ implies that $\mathscr{A}u_k \to \mathscr{A}u$ weakly in $\mathscr{V}'$. Next, we define the bilinear operator $B : V \times V \to V'$ as

$$\langle B(u, w), v\rangle = \int_\Omega (u \cdot \nabla)w \cdot v \, dx.$$

Similar as in Lemmata 5.4 and 5.5 we have for $u, v, w \in V$,

$$\langle B(u, w), v\rangle \le c_B \|u\| \|w\| \|v\|, \tag{15.9}$$

$$\langle B(u, v), v\rangle = 0, \tag{15.10}$$

$$\langle B(u, w), v\rangle = -\langle B(u, v), w\rangle, \tag{15.11}$$

and $B$ is sequentially weakly continuous, i.e.,

$$w_k \to w \quad \text{and} \quad u_k \to u \quad \text{weakly in} \quad V \tag{15.12}$$
$$\Rightarrow \quad \langle B(w_k, u_k), v\rangle \to \langle B(w, u), v\rangle \quad \text{for all} \quad v \in V.$$

The Nemytskii operator for $B$ has values only in $L^{\frac{4}{3}}(t_0, T; V')$ and not in $\mathscr{V}'$. Indeed, let us consider the function $(t_0, T) \ni t \to B(u(t), v(t)) \in V'$, where $u, v \in \mathscr{V} \cap L^\infty(t_0, T; H)$. We have

$$\|B(u(t), v(t))\|_{V'} = \sup_{\|w\|=1} \int_\Omega (u(t) \cdot \nabla)v(t) \cdot w \, dt = \sup_{\|w\|=1} \int_\Omega (u(t) \cdot \nabla)w \cdot v(t) \, dt$$
$$\le \|u(t)\|_{L^4(\Omega)^3} \|v(t)\|_{L^4(\Omega)^3}. \tag{15.13}$$

We use the following Ladyzhenskaya interpolation inequality valid for all $u \in H^1(\Omega)^3$ (see Theorem 3.11)

$$\|u\|_{L^4(\Omega)^3} \le C \|u\|^{\frac{1}{4}}_{L^2(\Omega)^3} \|u\|^{\frac{3}{4}}_{H^1(\Omega)^3}.$$

In our case we have

$$\|u\|_{L^4(\Omega)^3} \le C |u|^{\frac{1}{4}} \|u\|^{\frac{3}{4}}, \tag{15.14}$$

and hence

$$\|B(u(t), v(t))\|^{\frac{4}{3}}_{V'} \le C^2 |u(t)|^{\frac{1}{3}} \|u(t)\| |v(t)|^{\frac{1}{3}} \|v(t)\|,$$

whereas we can define

$$\mathscr{B} : (\mathscr{V} \cap L^\infty(t_0, T; H)) \times (\mathscr{V} \cap L^\infty(t_0, T; H)) \to L^{\frac{4}{3}}(t_0, T; V')$$

by $(\mathscr{B}(u, v))(t) = B(u(t), v(t))$. We have the following result.

**Lemma 15.1.** *Let the sequences $u_k \to u$ and $v_k \to v$ both converge weakly in $L^2(t_0, T; V)$ and strongly in $L^2(t_0, T; H)$. Then we have $\mathscr{B}(u_k, v_k) \to \mathscr{B}(u, v)$ weakly in $L^{\frac{4}{3}}(t_0, T; V')$.*

*Proof.* The proof follows the lines of the proof of Lemma III.3.2 in [219]. It is enough to prove that for smooth $w : (t_0, T) \times \Omega \to \mathbb{R}^3$ we have

$$\int_{t_0}^T \langle B(u_k(t), v_k(t)), w(t)\rangle \, dt \to \int_{t_0}^T \langle B(u(t), v(t)), w(t)\rangle \, dt.$$

The result follows by the density of smooth functions in $L^4(t_0, T; V)$, a predual space of $L^{\frac{4}{3}}(t_0, T; V')$. We have

$$\int_{t_0}^T \langle B(u_k(t), v_k(t)), w(t)\rangle \, dt = -\sum_{i,j=1}^3 \int_{t_0}^T \int_\Omega (u_k(t))_i \frac{\partial w_j(t)}{\partial x_i} (v_k(t))_j \, dx \, dt.$$

The last integrals converge to

$$-\sum_{i,j=1}^3 \int_{t_0}^T \int_\Omega (u(t))_i \frac{\partial w_j(t)}{\partial x_i} (v(t))_j \, dx \, dt = \int_{t_0}^T \langle B(u(t), v(t)), w(t)\rangle \, dt,$$

and the proof is complete. □

We pick the initial condition $u_0 \in H$ and the source term $f \in V'$.

We assume that the friction multifunction $g : \Gamma_C \times \mathbb{R}^3 \to 2^{\mathbb{R}^3}$ satisfies:

($g1$) for every $s \in \mathbb{R}^3$ the multifunction $x \to g(x, s)$ has a measurable selection, i.e., there exists $g_s : \Gamma_C \to \mathbb{R}^3$, a measurable function, such that $g_s(x) \in g(x, s)$ for all $x \in \Gamma_C$,

($g2$) the set $g(x, s)$ is nonempty, compact, and convex in $\mathbb{R}^3$, for all $s \in \mathbb{R}^3$ and for a.e. $x \in \Gamma_C$,

($g3$) graph of $s \to g(x, s)$ is closed in $\mathbb{R}^6$ for a.e. $x \in \Gamma_C$,

($g4$) we have $|\xi| \le c_1 + c_2|s|$ for all $s \in \mathbb{R}^3$, a.e. $x \in \Gamma_C$ and all $\xi \in g(x, s)$, with $c_2 \ge 0$ and $c_1 \in \mathbb{R}$,

($g5$) we have the following dissipativity condition $\xi \cdot s \ge -d_1 - d_2|s|^2$ for all $s \in \mathbb{R}^3$, a.e. $x \in \Gamma_C$ and all $\xi \in g(x, s)$ with $d_1 \ge 0$ and $d_2 \in \left(0, \frac{\nu}{\|\gamma\|^2}\right)$.

Finally, we define the multivalued operators $N : W \to 2^W$ and $\mathscr{N} : L^2(t_0, T; W) \to 2^{L^2(t_0,T;W)}$ as

$$N(v) = \{\eta \in W : \ \eta(x) \in g(x, v(x)) \quad \text{a.e. on} \quad \Gamma_C\},$$
$$\mathcal{N}(v) = \{\eta \in L^2(t_0, T; W) : \ \eta(x,t) \in g(x, v(x,t)) \quad \text{a.e. on} \quad \Gamma_C \times (t_0, T)\}.$$

If $v, \eta \in L^2(t_0, T; W)$, then we have

$$\begin{aligned} \eta \in \mathcal{N}(v) \quad &\Leftrightarrow \quad \eta(t) \in N(v(t)) \quad \text{for a.e.} \quad t \in (t_0, T) \\ &\Leftrightarrow \quad \eta(x,t) \in g(x, v(x,t)) \quad \text{for a.e.} \quad (x,t) \in \Gamma_C \times (t_0, T). \end{aligned}$$

Let $0 < \lambda_1 \leq \lambda_2 \leq \ldots \leq \lambda_k \leq \ldots$ be the sequence of the eigenvalues of the operator $A$, with the corresponding eigenvectors $\{z_k\}_{k=1}^{\infty}$ being orthonormal in $H$ and orthogonal in $V$. Define the finite dimensional spaces $V_k = \text{span}\{z_1, \ldots, z_k\}$. These spaces approximate $V$ from the inside, i.e., $\bigcup_{k=1}^{\infty} V_k$ is dense in $V$. By $P_k v$ we denote the projection onto $V_k$ of the function $v \in H$. Then $|P_k v| \leq |v|$ and $P_k v \to v$ strongly in $H$ as $k \to \infty$. We start with the following Galerkin problem.

**Problem 15.1.** Find $u \in AC([t_0, T]; V_k)$ such that $u(t_0) = P_k u_0$ and for all $v \in V_k$ and a.e. $t \in (t_0, T)$ we have

$$\int_\Omega u' \cdot v \, dx + \nu \int_\Omega \nabla u \colon \nabla v \, dx + \int_\Omega (u \cdot \nabla) u \cdot v \, dx + \int_{\Gamma_C} \xi \cdot v_\tau \, dS = \langle f, v \rangle, \qquad (15.15)$$

with $\xi \in \mathcal{N}(u_\tau)$.

**Theorem 15.3.** *Let* $(g1)$–$(g4)$ *hold. Problem 15.1 has at least one solution.*

*Proof.* We will prove the existence of solution to this ordinary differential inclusion by means of the Kakutani–Fan–Glicksberg fixed point theorem (see Theorem 3.8). Fix $\eta \in L^2(t_0, T; W)$ and $w \in L^2(t_0, T; V)$, and consider the following auxiliary linear problem.

**Problem 15.2.** Find $u \in AC([t_0, T]; V_k)$ such that $u(t_0) = P_k u_0$ and for all $v \in V_k$ and a.e. $t \in (t_0, T)$ we have

$$\int_\Omega u'(t) \cdot v \, dx + \nu \int_\Omega \nabla u(t) \colon \nabla v \, dx + \int_\Omega (w(t) \cdot \nabla) u(t) \cdot v \, dx + \int_{\Gamma_C} \eta(t) \cdot v_\tau \, dS = \langle f, v \rangle. \qquad (15.16)$$

This linear system of ODEs has a unique solution $u$. We define a multifunction $\Lambda : \mathcal{V} \times L^2(t_0, T; W) \to \mathcal{V} \times 2^{L^2(t_0, T; W)}$ which assigns to the pair $(w, \eta)$ the set $\{u\} \times \mathcal{N}(u_\tau)$. First, we note that by assertion $(H1)$ of Theorem 3.23, the set $\mathcal{N}(u_\tau)$ is nonempty and convex, so the multifunction $\Lambda$ assumes nonempty and convex values. We will prove, by the Kakutani–Fan–Glicksberg fixed point theorem (see Theorem 3.8), that this multifunction has a fixed point. This fixed point is a solution of Problem 15.1. Taking $v = u(t)$ in (15.16), we get

$$\frac{1}{2} \frac{d}{dt} |u(t)|^2 + \nu \|u(t)\|^2 \leq \|\eta(t)\|_W \|\gamma\| \|u(t)\| + \|f\|_{V'} \|u(t)\|.$$

It follows that

$$\frac{d}{dt}|u(t)|^2 + \nu \|u(t)\|^2 \le \frac{2\|\gamma\|^2}{\nu}\|\eta(t)\|_W^2 + \frac{2\|f\|_{V'}^2}{\nu}.$$

After integration on $(t_0, t)$, for any $t \in (t_0, T)$ we get

$$|u(t)|^2 + \nu \int_{t_0}^t \|u(s)\|^2\, ds \le |u_0|^2 + \frac{2\|\gamma\|^2}{\nu}\|\eta\|_{L^2(t_0,t;W)}^2 + \frac{2\|f\|_{V'}^2(t-t_0)}{\nu}.$$

In particular,

$$\|u\|_{\mathscr{V}}^2 \le \frac{1}{\nu}|u_0|^2 + \frac{2\|\gamma\|^2}{\nu^2}\|\eta\|_{L^2(t_0,T;W)}^2 + \frac{2\|f\|_{V'}^2(T-t_0)}{\nu^2}, \tag{15.17}$$

and

$$\|u(t)\|^2 \le C_k^2|u(t)|^2 \le C_k^2|u_0|^2 + \frac{C_k^2 2\|\gamma\|^2}{\nu}\|\eta\|_{L^2(t_0,T;W)}^2 + \frac{C_k^2 2\|f\|_{V'}^2(T-t_0)}{\nu}, \tag{15.18}$$

where the constant $C_k$ depends only on the dimension of the space $V_k$. Let $\xi \in \mathscr{N}(u)$. We have

$$\begin{aligned}\|\xi(t)\|_W = &\sup_{\|z\|_W=1}\int_{\Gamma_C}\xi(x,t)\cdot z(x)\,dx \le \sup_{\|z\|_W=1}\int_{\Gamma_C}(c_1 + c_2|u(x,t)|)|z(x)|\,dx\\ \le &c_1\sqrt{m_2(\Gamma_C)} + c_2\|\gamma u(t)\|_W \le c_1\sqrt{L_1L_2} + c_2\|\gamma\|\,\|u(t)\|.\end{aligned}$$

Hence

$$\begin{aligned}\|\xi(t)\|_W^2 &\le 2c_1^2L_1L_2 + 2c_2^2\|\gamma\|^2\|u(t)\|^2 \qquad (15.19)\\ &\le 2c_1^2L_1L_2+2c_2^2\|\gamma\|^2\left(C_k^2|u_0|^2+\frac{C_k^2 2\|\gamma\|^2}{\nu}\|\eta\|_{L^2(t_0,T;W)}^2+\frac{C_k^2 2\|f\|_{V'}^2(T-t_0)}{\nu}\right),\end{aligned}$$

and, after integration on $(t_0, T)$,

$$\begin{aligned}\|\xi\|_{L^2(t_0,T;W)}^2 \le &2(T-t_0)(c_1^2L_1L_2 + c_2^2\|\gamma\|^2C_k^2|u_0|^2) \qquad (15.20)\\ &+\frac{4(T-t_0)c_2^2C_k^2\|\gamma\|^4}{\nu}\|\eta\|_{L^2(t_0,T;W)}^2 + \frac{4c_2^2\|\gamma\|^2C_k^2\|f\|_{V'}^2(T-t_0)^2}{\nu}.\end{aligned}$$

Clearly, from (15.17) and (15.20), if only $(T-t_0) < \frac{\nu}{4c_2^2C_k^2\|\gamma\|^4}$, we can find $R_1 > 0$ and $R_2 > 0$ such that $\Lambda$ is a map from $S$ to $2^S$, where

$$S = \{(u,\xi) \in \mathscr{V} \times L^2(t_0,T;W) : \ \|u\|_{\mathscr{V}} \le R_1, \|\xi\|_{L^2(0,T;W)} \le R_2\}.$$

Consequently, the fixed point argument will give us the existence of solution on the interval $[t_0, T]$ for $(T - t_0) < \frac{\nu}{4c_2^2 C_k^2 \|\gamma\|^4}$, and by subsequent continuation this solution can be elongated to arbitrarily long time interval. It remains to prove that the graph of $\Lambda$ is weakly closed in $S \times S$. In view of the fact that $\mathscr{V} \times L^2(t_0, T; W)$ is a reflexive Banach space, and $S$ is a bounded, closed, and convex set, it is enough to show the sequential weak closedness. Let $w_k \to w$ weakly in $\mathscr{V}$ and $\eta_k \to \eta$ weakly in $L^2(0, T; W)$, with $(w_k, \eta_k) \in S$. Let moreover $(u_k, \xi_k) \in \Lambda(w_k, \eta_k)$ be such that $u_k \to u$ weakly in $\mathscr{V}$ and $\eta_k \to \eta$ weakly in $L^2(t_0, T; W)$. We know that $(u_k, \xi_k) \in S$. Let us introduce an equivalent norm on the finite dimensional space $V_k$ by the formula

$$\|z\|_{V_k'} = \sup_{v \in V_k, \|v\|=1} \int_\Omega z \cdot v \, dx.$$

We have

$$\begin{aligned} \|u_k'(t)\|_{V_k'} &= \sup_{\substack{v \in V_k \\ \|v\|=1}} \int_\Omega u_k'(t) \cdot v \, dx = \sup_{\substack{v \in V_k \\ \|v\|=1}} \Bigg( \langle f, v \rangle - \nu \int_\Omega \nabla u_k(t) : \nabla v \, dx \\ &\quad - \int_\Omega (w_k(t) \cdot \nabla) u_k(t) \cdot v \, dx - \int_{\Gamma_C} \eta_k(t) \cdot v_\tau \, dS \Bigg) \\ &\le \|f\|_{V'} + \nu \|u_k(t)\| + \|w_k(t)\| \|u_k(t)\| + \|\eta_k(t)\|_W \|\gamma\|. \end{aligned}$$

Using (15.18) and (15.19) it follows from the fact that $w_k$ is uniformly bounded in $L^2(t_0, T; W)$ that $u_k'$ is uniformly bounded in $L^2(t_0, T; V_k')$ and moreover in $\mathscr{V}$, as the space $V_k$ is finite dimensional. Hence, for a subsequence, not renumbered, $u_k' \to u'$ weakly in $\mathscr{V}$ and $u_k \to u$ strongly in $\mathscr{V}$. Moreover, $(u_k)_\tau \to u_\tau$ strongly in $L^2(t_0, T; W)$. As $\xi_k \to \xi$ weakly in $L^2(t_0, T; W)$ and $\xi_k \in \mathscr{N}((u_k)_\tau)$, by assertion $(H2)$ of Theorem 3.23 it follows that $\xi \in \mathscr{N}(u_\tau)$. It remains to prove that $u$ solves Problem 15.2 with $w$ and $\eta$, using the fact the $u_k$ solves this problem with $w_k$ and $\eta_k$. To this end, it suffices to multiply Eq. (15.16) written for $u_k, w_k, \eta_k$ by the function $\phi \in L^2(t_0, T)$ and integrate the resultant equation over the interval $(t_0, T)$. The assertion follows by passing to the limit in the obtained equation. As the details are technical, their verification is left to the reader as an exercise. □

We are in position to define the Leray–Hopf weak solution of the considered problem. Since we need two types of weak solutions, we call this class a type I Leray–Hopf weak solution.

**Definition 15.5.** By the type I Leray–Hopf weak solution of the three-dimensional Navier–Stokes problem with multivalued friction we mean a function $u \in L^2_{loc}([t_0, \infty); V) \cap L^\infty_{loc}([t_0, \infty); H)$ with $u' \in L^{\frac{4}{3}}_{loc}([t_0, \infty); V')$ such that:

- $u(t_0) = u_0$,
- for all $v \in V$ and a.e. $t > t_0$ we have

$$\langle u'(t) + \nu Au(t) + B(u(t), u(t)) - f, v\rangle + \int_{\Gamma_C} \xi(t) \cdot v_\tau \, dS = 0, \tag{15.21}$$

with $\xi \in L^2_{loc}([t_0, \infty); W)$ such that $\xi(t) \in N(u_\tau(t))$ for a.e. $t > t_0$,

- for a.e. $t \geq t_0$ (including $t_0$) and for all $s > t$ we have

$$\frac{1}{2}|u(s)|^2 + \nu \int_t^s \|u(r)\|^2 \, dr + \int_t^s (\xi(r), u_\tau(r))_W - \langle f, u(r)\rangle \, dr \leq \frac{1}{2}|u(t)|^2. \tag{15.22}$$

*Remark 15.3.* Every type I Leray–Hopf weak solution defined above belongs to the spaces $C([t_0, \infty); V')$ and $C_w([t_0, \infty); H)$. Moreover, (15.22) implies that $|u(t)|^2$ is continuous from the right at $t_0$.

**Theorem 15.4.** *Assume* $(g1)$–$(g5)$. *For any* $u_0 \in H$ *there exists at least one type I Leray–Hopf weak solution given by Definition 15.5.*

*Proof.* The proof is standard in the theory of three-dimensional Navier–Stokes equations and it is based on the Galerkin method. It is enough to prove, for certain given $T > t_0$, the existence of a function $u \in L^2(t_0, T; V) \cap L^\infty(t_0, T; H)$ with $u' \in L^{\frac{4}{3}}(t_0, T; V')$ and corresponding $\xi \in L^2(t_0, T; W)$ which satisfies the initial condition, (15.21) for a.e. $t \in (t_0, T)$, and (15.22) for a.e. $t \in (t_0, T)$ and for $t = t_0$. The solution on the interval $(t_0, \infty)$ can then be obtained by concatenating the weak solutions obtained on the intervals of length $T - t_0$ by taking the value of the previous solution at the endpoint of the time interval as the initial condition for the subsequent problem. We will proceed in the standard way obtaining the Leray–Hopf weak solution by passing to the limit in the Galerkin problems. Let $u_k \in AC([t_0, T]; V_k)$ be the solutions of Problem 15.1 with the initial condition $P_k u_0$. We denote by $\xi_k$ the corresponding selections of $\mathcal{N}((u_k)_\tau)$. Taking $v = u_k(t)$ in (15.15) we get for a.e. $t \in (t_0, T)$,

$$\frac{1}{2}\frac{d}{dt}|u_k(t)|^2 + \nu\|u_k(t)\|^2 + \int_{\Gamma_C} \xi_k(t) \cdot (u_k)_\tau(t) \, dS = \langle f, u_k(t)\rangle. \tag{15.23}$$

Using $(g4)$ we get

$$\begin{aligned}\int_{\Gamma_C} |\xi_k(t) \cdot (u_k)_\tau(t)| \, dS &\leq \int_{\Gamma_C} (c_1 + c_2|(u_k)_\tau(x,t)|)|(u_k)_\tau(x,t)| \, dS \\ &\leq (c_2 + 1)\|u_k(t)\|_W^2 + \frac{c_1^2 m_2(\Gamma_C)}{4}. \end{aligned} \tag{15.24}$$

Let $\delta \in (0, \frac{1}{2})$ and let $Z = V \cap H^{1-\delta}(\Omega; \mathbb{R}^3)$ be endowed with $H^{1-\delta}(\Omega; \mathbb{R}^3)$ topology. The embedding $V \subset Z$ is compact and $Z \subset H$ is continuous. Moreover, the trace operator from $Z$ to $W$ is linear and continuous. Thus, we can use the Ehrling lemma (cf. Lemma 3.17) to deduce that for any $\varepsilon > 0$ there exists $C(\varepsilon) > 0$ such that $\|v\|_W \leq \varepsilon\|v\| + C(\varepsilon)|v|$ for all $v \in V$. Thus, from (15.24) we get

$$\int_{\Gamma_C} |\xi_k(t) \cdot (u_k)_\tau(t)|\, dS \le \varepsilon \|u_k(t)\|^2 + C_1(\varepsilon)|u_k(t)|^2 + C_2,$$

with constants $C_1(\varepsilon), C_2 > 0$. Coming back to (15.23) we get

$$\frac{1}{2}\frac{d}{dt}|u_k(t)|^2 + \nu\|u_k(t)\|^2 \le 2\varepsilon\|u_k(t)\|^2 + \frac{1}{4\varepsilon}\|f\|_{V'} + C_1(\varepsilon)|u_k(t)|^2 + C_2. \qquad (15.25)$$

It suffices to take $\varepsilon = \frac{\nu}{4}$. Now, we use the Gronwall lemma (cf. Lemma 3.20) to deduce that $u_k$ is uniformly bounded in $\mathscr{V}$ and $L^\infty(t_0, T; H)$. To prove that $\xi_k$ is bounded in $L^2(t_0, T; W)$ we observe that

$$\|\xi_k(t)\|_W^2 \le \int_{\Gamma_C} (c_1 + c_2|(u_k)_\tau(x,t)|)^2\, dS \le 2c_1^2 m_2(\Gamma_C) + 2c_2^2\|\gamma\|^2\|u_k(t)\|^2,$$

and the bound of $u_k$ in $\mathscr{V}$ implies that, indeed, $\xi_k$ is bounded in $L^2(t_0; T; W)$. To obtain the estimates of $u_k'$ we will use the fact that we chose the eigenfunctions of the operator $A$ as a basis of the space $V_k$. The projection operator $P_k : V' \to V_k$ is given by $P_k v = \sum_{i=1}^k \langle v, z_i\rangle z_i$, where $z_i$ is the $i$-th eigenfunction of the operator $A$. Define $\iota : V \to W$ by the tangential component of the trace, i.e., $\iota v = (\gamma v)_\tau$ and $\iota^* : W \to V'$ its adjoint operator given by $\langle \iota^* v, w\rangle = (v, w_\tau)_W$. Then we have

$$\int_{\Gamma_C} \xi_k(t) \cdot v_\tau\, dS = \langle \iota^* \xi_k(t), v\rangle.$$

The Galerkin equation (15.15) can be rewritten in the following way:

$$u_k'(t) + \nu A u_k(t) + P_k B(u_k(t), u_k(t)) + P_k \iota^* \xi_k(t) = P_k f.$$

As, by Theorem 3.17, we have $\|P_k v\|_{V'} \le \|v\|_{V'}$, it follows that

$$\begin{aligned}\|u_k'(t)\|_{V'} \le & \nu\|A\|_{\mathscr{L}(V;V')}\|u_k(t)\| + \|f\|_{V'} \\ & + \|\iota^*\|_{\mathscr{L}(W;V')}\|\xi_k(t)\|_W + \|B(u_k(t), u_k(t))\|_{V'}.\end{aligned} \qquad (15.26)$$

Using (15.13) and (15.14) we get

$$\|B(u_k(t), u_k(t))\|_{V'} \le C|u_k(t)|^{\frac{1}{2}}\|u_k(t)\|^{\frac{3}{2}}.$$

The last two bounds imply that $u_k'$ is bounded in $L^{\frac{4}{3}}(t_0, T; V')$. Thus, for a subsequence, not renumbered, we have

$$u_k \to u \quad \text{weakly in} \quad \mathscr{V}, \qquad (15.27)$$

$$u_k \to u \quad \text{weakly-}* \text{ in} \quad L^\infty(t_0, T; H), \qquad (15.28)$$

$$u_k' \to u' \quad \text{weakly in} \quad L^{\frac{4}{3}}(t_0, T; H), \tag{15.29}$$

$$u_k \to u \quad \text{strongly in} \quad L^2(t_0, T; H), \tag{15.30}$$

$$u_k(t) \to u(t) \quad \text{strongly in} \quad H \quad \text{for almost all} \quad t \in (t_0, T), \tag{15.31}$$

$$u_k \to u \quad \text{strongly in} \quad C([t_0, T]; V'), \tag{15.32}$$

$$u_k(t) \to u(t) \quad \text{weakly in} \quad H \quad \text{for all} \quad t \in [t_0, T], \tag{15.33}$$

$$\xi_k \to \xi \quad \text{weakly in} \quad L^2(t_0, T; W), \tag{15.34}$$

$$(u_k)_\tau \to u_\tau \quad \text{strongly in} \quad L^2(t_0, T; W). \tag{15.35}$$

The convergence (15.30) follows from the compact embedding

$$\{u \in \mathscr{V} : \ u' \in L^{\frac{4}{3}}(t_0, T; V')\} \quad \subset \quad L^2(t_0, T; H),$$

the convergence (15.32) follows from the compact embedding

$$\{u \in L^\infty(t_0, T; H) : \ u' \in L^{\frac{4}{3}}(t_0, T; V')\} \quad \subset \quad C([t_0, T]; V'),$$

while (15.34) follows from the compact embedding

$$\{u \in L^2(t_0, T; V) : \ u' \in L^{\frac{4}{3}}(t_0, T; V')\} \quad \subset \quad L^2(0, T; Z),$$

and the continuity of the Nemytskii trace operator from $L^2(0, T; Z)$ to $L^2(0, T; W)$. The convergence (15.33) follows by the computation

$$(u_k(t), v) = (P_k u_0, v) + \int_{t_0}^t \langle u_k'(s), v\rangle \, ds,$$

valid for $v \in H$, the strong convergence $P_k u_0 \to u_0$ in $H$, and (15.29). Multiplying (15.15) by $\varphi \in C^1((t_0, T))$ and integrating over $(t_0, T)$ we have, for $v \in V_k$,

$$\int_{t_0}^T \langle u_k'(t) + \nu A u_k(t) + B(u_k(t), u_k(t)) - f, v\varphi(t)\rangle \, dt + \int_{t_0}^T (\xi_k(t), v_\tau \varphi(t))_W \, dt = 0.$$

Using (15.27), (15.29), (15.30), (15.33), and Lemma 15.1 we can pass to the limit in above equation, getting

$$\int_{t_0}^T \langle u'(t) + \nu A u(t) + B(u(t), u(t)) - f, v\varphi(t)\rangle \, dt + \int_{t_0}^T (\xi(t), v_\tau \varphi(t))_W \, dt = 0,$$

valid for $v \in \bigcup_{k=1}^{\infty} V_k$ and $\varphi \in C^1((t_0, T))$, whence (15.21) follows. Integrating (15.23) on $(t, s)$ we get, for all $t, s$ with $t_0 \le t < s \le T$,

$$\frac{1}{2}|u_k(s)|^2 + \nu \int_t^s \|u_k(r)\|^2 \, dr + \int_t^s (\xi_k(r), (u_k)_\tau(r))_W - \langle f, u_k(r)\rangle \, dr \le \frac{1}{2}|u_k(t)|^2. \tag{15.36}$$

Using (15.27), (15.31), (15.34), and (15.35) as well as the sequential weak lower semicontinuity of the norm of the space $L^2(t, s; V)$ we get, for a.e. $t \in (t_0, T)$ and a.e. $s \in (t, T)$,

$$\frac{1}{2}|u(s)|^2 + \nu \int_t^s \|u(r)\|^2 \, dr + \int_t^s (\xi(r), u_\tau(r))_W - \langle f, u(r)\rangle \, dr \le \frac{1}{2}|u(t)|^2.$$

The convergence (15.33) and the sequential weak lower semicontinuity of the norm $|\cdot|$ imply that the last inequality holds for all $s \in [t, T]$. Integrating (15.23) on $(t_0, s)$ we get, for all $s \in (t_0, T]$,

$$\frac{1}{2}|u_k(s)|^2 + \nu \int_{t_0}^s \|u_k(r)\|^2 \, dr + \int_{t_0}^s (\xi_k(r), (u_k)_\tau(r))_W - \langle f, u_k(r)\rangle \, dr \le \frac{1}{2}|P_k u_0|^2.$$

As we did in (15.36), we can pass to the limit in the above inequality, to get

$$\frac{1}{2}|u(s)|^2 + \nu \int_{t_0}^s \|u(r)\|^2 \, dr + \int_{t_0}^s (\xi(r), u_\tau(r))_W - \langle f, u(r)\rangle \, dr \le \frac{1}{2}|u_0|^2.$$

We have proved (15.22). The fact that $u(t_0) = u_0$ follows from (15.33) for $t = t_0$ and the fact that $u_k(0) = P_k u_0 \to u_0$ strongly in $H$ as $k \to \infty$. Finally, (15.34), (15.35), and assertion $(H2)$ of Theorem 3.23 imply that $\xi \in \mathcal{N}(u_\tau)$. The proof is complete. □

*Remark 15.4.* There are two "sources" of possible nonuniqueness of Leray–Hopf weak solution of the studied problem. First, the nonlinear term $B$ which, in three-dimensional case, makes the weak solution uniqueness unknown (cf., e.g., [99, 219]), and, moreover, the nonmonotone friction multifunction $g$ which also can contribute to the lack of solution uniqueness.

## 15.4 Existence and Invariance of Weak Global Attractor, and Weak Tracking Property

First we discuss the dissipativity of the type I Leray–Hopf weak solutions. We recall a useful lemma.

**Lemma 15.2 (cf. [11, Lemma 7.2]).** *Let $\theta : [t_0, \infty) \to \mathbb{R}$ be lower semicontinuous, continuous at $t_0$, $\theta \in L^1(0,T)$ for all $T > 0$ and let, for some constant $c \geq 0$*

$$\theta(s) + c\int_t^s \theta(\tau)\, d\tau \leq \theta(t),$$

*for all $s \geq t$, for a.e. $t > t_0$ and for $t = t_0$. Then*

$$\theta(t) \leq \theta(t_0)e^{-c(t-t_0)} \quad \textit{for all} \quad t \geq t_0.$$

We use the above lemma to get a dissipativity result for type I Leray–Hopf weak solutions.

**Lemma 15.3.** *If $u$ is a type I Leray–Hopf weak solution given by Definition 15.5, then for all $t \geq t_0$ we have*

$$|u(t)|^2 \leq e^{-\lambda_1(\nu - d_2\|\gamma\|^2)(t-t_0)}|u_0|^2 + \frac{\|f\|_{V'}^2 + 2d_1L_1L_2(\nu - d_2\|\gamma\|^2)}{\lambda_1(\nu - d_2\|\gamma\|^2)^2}. \tag{15.37}$$

*Proof.* Using (g5) in (15.22) we get

$$\frac{1}{2}|u(s)|^2 + \nu\int_t^s \|u(r)\|^2\, dr + \int_t^s\int_{\Gamma_C} -d_1 - d_2|u_\tau(x,r)|^2\, dS\, dr$$
$$\leq \frac{1}{2}|u(t)|^2 + \varepsilon\int_t^s \|u(r)\|^2\, dr + \frac{\|f\|_{V'}^2(s-t)}{4\varepsilon},$$

for all $s \geq t$ and for a.e. $t > t_0$ and for $t = t_0$, where $\varepsilon > 0$ is arbitrary. Using the trace inequality, we get

$$\frac{1}{2}|u(s)|^2 + (\nu - d_2\|\gamma\|^2 - \varepsilon)\int_t^s \|u(r)\|^2 dr \leq \frac{1}{2}|u(t)|^2 + \left(\frac{\|f\|_{V'}^2}{4\varepsilon} + d_1L_1L_2\right)(s-t).$$

We take $\varepsilon = \frac{\nu - d_2\|\gamma\|^2}{2}$. Using the Poincaré inequality $\lambda_1|v|^2 \leq \|v\|^2$ valid for $v \in V$, we get

$$|u(s)|^2 + \lambda_1(\nu - d_2\|\gamma\|^2)\int_t^s |u(r)|^2 dr$$
$$\leq |u(t)|^2 + \left(\frac{\|f\|_{V'}^2}{\nu - d_2\|\gamma\|^2} + 2d_1L_1L_2\right)(s-t).$$

Denote $A = \lambda_1(\nu - d_2\|\gamma\|^2)$ and $B = \frac{\|f\|_{V'}^2}{\nu - d_2\|\gamma\|^2} + 2d_1L_1L_2$. We have

$$|u(s)|^2 - \frac{B}{A} + A\int_t^s |u(r)|^2 - \frac{B}{A}\, dr \leq |u(t)|^2 - \frac{B}{A}.$$

As $u \in C_w([t_0,\infty);H)$ and $|u(t)|^2$ is continuous from the right at $t_0$ (cf. Remark 15.3), it follows that the function $\theta(r) = |u(r)|^2 - \frac{B}{A}$ satisfies all assumptions of Lemma 15.2. It follows that

$$\left(|u(t)|^2 - \frac{B}{A}\right) \leq \left(|u_0|^2 - \frac{B}{A}\right) e^{-A(t-t_0)} \quad \text{for all} \quad t \geq t_0,$$

whence we get the assertion of the lemma. The proof is complete. □

Define

$$X = \{u \in H : \ |u|^2 \leq R\},$$

where $R$ is any number strictly greater than $\frac{\|f\|_{V'}^2 + 2d_1 L_1 L_2 (\nu - d_2\|\gamma\|^2)}{\lambda_1 (\nu - d_2\|\gamma\|^2)^2}$. Directly from (15.37) we get the following result.

**Lemma 15.4.** *Let $K \subset H$ be a bounded set, and let $t_0 \in \mathbb{R}$. For every type I Leray–Hopf weak solution $u$ with the initial condition $u(t_0) = u_0 \in K$ there exists a time $t_1 = t_1(t_0, K)$ such that for all $t \geq t_1$ $u(t) \in X$.*

We endow $X$ with two metrics. The strong metric $d_s : X \times X \to [0,\infty)$ is given by $d_s(u,v) = |u - v|$. As the weak topology of $H$ is metrizable on the closed ball $X$ we can define the metric $d_w : X \times X \to [0,\infty)$ such that $d_w(u_k, u) \to 0$ if and only if $u_k \to u$ weakly in $H$. Note that the set $X$ is compact in $d_w$.

We need to define a class of type II Leray–Hopf weak solutions (see [65, 66]).

**Definition 15.6.** By the type II Leray–Hopf solution of the three-dimensional Navier–Stokes problem with multivalued friction we mean a function $u \in L^2_{loc}([t_0,\infty);V) \cap L^\infty_{loc}([t_0,\infty);H)$ with $u' \in L^{\frac{4}{3}}_{loc}([t_0,\infty);V')$ such that:

- for all $v \in V$ and a.e. $t > t_0$ we have

$$\langle u'(t) + \nu A u(t) + B(u(t),u(t)) - f, v\rangle + \int_{\Gamma_C} \xi(t)\cdot v_\tau \, dS = 0, \tag{15.38}$$

  with $\xi \in L^2_{loc}(t_0,\infty;W)$ such that $\xi(t) \in N(u_\tau(t))$ for a.e. $t > t_0$,
- for a.e. $t \geq t_0$ (**not necessarily for** $t_0$) and for all $s > t$ we have

$$\frac{1}{2}|u(s)|^2 + \nu \int_t^s \|u(r)\|^2\, dr + \int_t^s (\xi(r), u_\tau(r))_W - \langle f, u(r)\rangle \, dr \leq \frac{1}{2}|u(t)|^2. \tag{15.39}$$

Note that every type I Leray–Hopf weak solution is a type II Leray–Hopf weak solution. Moreover, a restriction of either type I or type II Leray–Hopf weak solution given on interval $[t_0,\infty)$ to a smaller interval $[T,\infty) \subset [t_0,\infty)$ is always a type II Leray–Hopf weak solution on $[T,\infty)$. It is, however, impossible to concatenate type II Leray–Hopf weak solutions, as the energy inequality (15.39) does not have to hold at the initial time $t = t_0$. We can define

$$\begin{aligned}\mathscr{E}([T,\infty)) = \{u: \ u \quad &\text{is a type II Leray–Hopf weak solution on} \quad [T,\infty),\\ &\text{and} \quad u(t) \in X \quad \text{for all} \quad t \geq T\}. \end{aligned} \tag{15.40}$$

Obviously, $\mathscr{E}([T,\infty))$ satisfies the first three axioms of Definition 15.1, and hence it defines the evolutionary system with $\mathscr{E}((-\infty,\infty))$ given by axiom $(iv)$ of Definition 15.1. We can use Theorem 15.1 to conclude the following result on the existence of a weak global attractor.

**Theorem 15.5.** *The evolutionary system defined by (15.40) has a weak global attractor* $\mathscr{A}$.

To study the attractor invariance we need the next result.

**Theorem 15.6.** *Let $u_k$ be the sequence of type II Leray–Hopf weak solutions on $[t_0,\infty)$ such that $u_k(t) \in X$ for all $t \geq t_0$. Let $T > t_0$ be given arbitrarily. Then there exists a subsequence, not renumbered, which converges in $C_w([t_0,T];H)$ to some type II Leray–Hopf weak solution u, i.e.,*

$$(u_k(t), v) \to (u(t), v) \quad \textit{uniformly on} \quad [t_0,T] \quad \textit{for all} \quad v \in H \quad \textit{as} \quad k \to \infty.$$

*Proof.* Let the sequence $t_m \to t_0$ be such that (15.39) holds for $t = t_m$. Proceeding in the same way as in the proof of Lemma 15.3 we get

$$\int_{t_m}^{T} \|u_k(r)\|^2 \, dr \leq C,$$

from (15.39). Passing with $m \to \infty$ we obtain that $u_k$ is bounded in $L^2(t_0;T;V)$. Proceeding as in (15.26) we find that from (15.38) it follows that $u_k'$ is bounded in $L^{\frac{4}{3}}(t_0,T;V')$. In the similar way as we obtained (15.27)–(15.35) in the proof of existence, we have

$$\begin{aligned}
u_k &\to u \quad \text{weakly in} \quad \mathscr{V},\\
u_k &\to u \quad \text{weakly-}* \text{ in} \quad L^\infty(t_0,T;H),\\
u_k' &\to u' \quad \text{weakly in} \quad L^{\frac{4}{3}}(t_0,T;H),\\
u_k &\to u \quad \text{strongly in} \quad L^2(t_0,T;H),\\
u_k(t) &\to u(t) \quad \text{strongly in} \quad H \quad \text{for almost all} \quad t \in (t_0,T),\\
u_k &\to u \quad \text{strongly in} \quad C([t_0,T];V'),\\
u_k(t) &\to u(t) \quad \text{weakly in} \quad H \quad \text{for all} \quad t \in [t_0,T],
\end{aligned} \tag{15.41}$$

$$\xi_k \to \xi \quad \text{weakly in} \quad L^2(t_0, T; W),$$
$$(u_k)_\tau \to u_\tau \quad \text{strongly in} \quad L^2(t_0, T; W).$$

Now, exactly the same as in the proof of Theorem 15.4, it follows that $u$ satisfies (15.38) and (15.39) for a.e. $t \in (t_0, T)$, whence $u$ is a certain type II Leray–Hopf weak solution. To show the uniform convergence in $C_w([t_0, T]; H)$, let $v \in H$ and let $\varepsilon > 0$. We can find $v_\varepsilon \in V$ with $|v_\varepsilon - v| \leq \varepsilon$. We have

$$\begin{aligned}\sup_{t\in[t_0,T]} |(u_k(t) - u(t), v)| &= \sup_{t\in[t_0,T]} |\langle u_k(t) - u(t), v_\varepsilon\rangle + (u_k(t) - u(t), v - v_\varepsilon)| \\ &\leq \sup_{t\in[t_0,T]} \|u_k(t) - u(t)\|_{V'} \|v_\varepsilon\| + 2\sqrt{R}\varepsilon,\end{aligned}$$

where we have used the fact that $u_k(t), u(t) \in X$. We pass with $k \to \infty$ and use (15.41), to get

$$\limsup_{k\to\infty} \sup_{t\in[t_0,T]} |(u_k(t) - u(t), v)| \leq 2\sqrt{R}\varepsilon.$$

As $\varepsilon$ was arbitrary, it follows that $\lim_{k\to\infty} \sup_{t\in[t_0,T]} |(u_k(t) - u(t), v)| = 0$, and the proof is complete. □

**Theorem 15.7.** *$\mathscr{E}([0, \infty))$ is compact in $C_w([0, \infty); H)$.*

*Proof.* The proof follows the lines of the proof of Lemma 3.12 in [66]. Take any sequence $u_k \in \mathscr{E}([0, \infty))$. By Theorem 15.6 there exists a subsequence, still denoted by $u_k$, which converges in $C_w([0, 1]; H)$ to some $u^1$, a type II Leray–Hopf solution. Passing again to subsequence, not renumbered, $u_k$ converges in $C_w([0, 2]; H)$ to some $u^2$, a type II Leray–Hopf solution such that $u^1 = u^2$ on $[0, 1]$. Continuing this diagonalization process, $u_k$ converges in $C_w([0, \infty); H)$ to some $u$, a type II Leray–Hopf solution. As $|u(t)| \leq \liminf_{k\to\infty} |u_k(t)|$ for all $t \geq 0$, it follows that $u(t) \in X$, and in consequence $u \in \mathscr{E}([0, \infty))$. The proof is complete. □

As we have shown that $(A1)$ holds, we can use Theorem 15.2 to conclude the last result of this section.

**Theorem 15.8.** *The weak global attractor $\mathscr{A}$ is a maximal invariant set. Moreover,*

$$\mathscr{A} = \{x \in X : \ x = u(0) \quad \textit{for some} \quad u \in \mathscr{E}((-\infty, \infty))\},$$

*and the tracking property holds, i.e., for every $\varepsilon > 0$ there exists $t_0 > 0$ such that for any $t^* > t_0$ and for any trajectory $u \in \mathscr{E}([0, \infty))$ we have*

$$d_{C_w([t^*,\infty);X)}(u, v) < \varepsilon$$

*for a certain complete trajectory $v \in \mathscr{E}((-\infty, \infty))$.*

The tracking property established in Theorem15.8 is illustrated in Fig. 15.3.

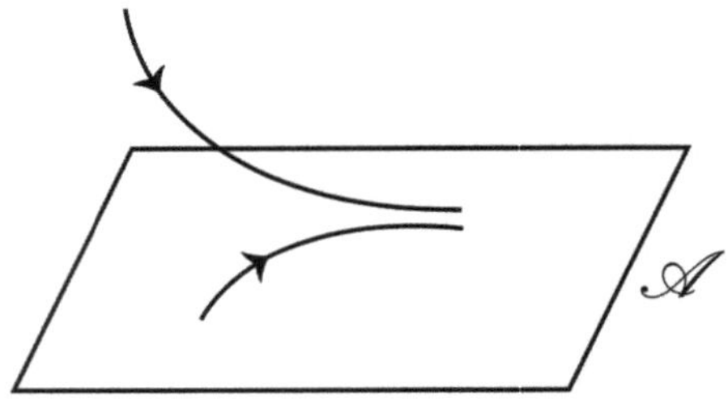


**Fig. 15.3** Illustration of the tracking property. Every trajectory has a corresponding complete trajectory on the attractor to which it converges as the time goes to infinity

## 15.5 Comments and Bibliographical Notes

There are several mathematical formalisms useful to study the global attractors for autonomous evolution problems without uniqueness of solutions. Among them we can list:

- generalized semiflows introduced and studied by Ball [11],
- multivalued semiflows ($m$-semiflows) anticipated by Babin and Vishik [9] and introduced and studied by Melnik and Valero in [171] and [172],
- trajectory attractors introduced independently by Chepyzhov and Vishik [59], Málek and Nečas [167], and Sell [210]. The application of this approach to two-dimensional incompressible Navier–Stokes problem with multivalued feedback control boundary condition is presented in Chap. 14,
- evolutionary systems introduced by Cheskidov and Foiaş and [66] and later extended by Cheskidov in [65].

The formalism of evolutionary systems chosen by us in this chapter appears to be especially well suited to the problems governed by the three-dimensional Navier–Stokes equations. Still, such problems can be and have been dealt with by the other methods: trajectory attractors [59, 167, 210], multivalued semiflows [128, 141], and generalized semiflows [11]. The approach based on generalized semiflows has been compared with the one based on multivalued semiflows in [40], and both these approached were related to the one based on the trajectory attractors in [129]. More information on the structure of weak attractor for three-dimensional case with periodic boundary conditions can be found in [102] and on the structure of the attainability sets of weak solutions in [138].

If we assume the periodic or homogeneous Dirichlet boundary conditions on whole $\partial\Omega$ it is known that a weak solution for three-dimensional Navier–Stokes equations, so-called Leray–Hopf solution (see [115, 154]), exists for arbitrarily long time but its uniqueness is still an open problem. On the other hand, the regular solution is known to exist and be unique only locally in time (see [98] for local regularity results in three-dimensional case and global regularity results in two-dimensional case for periodic boundary conditions). There are many additional conditions under which the weak solution becomes strong and the associated weak global attractor is the strong one, too, but so far it remains an open problem whether any of such conditions is in fact satisfied [34, 66–68, 73, 82, 143, 144, 189]. Only such conditional results are available on the existence of strong global attractor in the three-dimensional case [65, 66, 128, 202]. An interesting recent idea is to use the results of numerical simulations to verify the existence of strong solutions [64]. A lot of work has been put in recent years to study the *partial regularity* of solutions. This work includes the analysis of the so-called *singular set* of the weak solutions, i.e., the set of space-time points such that the solution is unbounded in the neighborhood of such points, this work started from the works of Scheffer [205] and Caffarelli et al. [32]. For the recent progress in such studies see the forthcoming book [200].

There are some approaches relying on the modifications or simplifications of three-dimensional equations to improve their properties, for example, one can study globally modified Navier–Stokes equations [47, 48, 139], Navier–Stokes equations with fractional operators [86], or $\alpha$-model [45, 63, 100].

# 16
# Attractors for Multivalued Processes in Contact Problems

In this chapter we consider further non-autonomous and multivalued evolution problems, this time in the frame of the theory of pullback attractors for multivalued processes.

First we prove an abstract theorem on the existence of a pullback $\mathscr{D}$-attractor and then apply it to study a two-dimensional incompressible Navier–Stokes flow with a general form of multivalued frictional contact conditions. Such conditions represent the frictional contact between the fluid and the wall, where the friction force depends in a nonmonotone and even discontinuous way on the slip rate, and are a generalization of the conditions considered in Chap. 10.

In contrast to Chap. 10 we are able to obtain only the attractor existence, while, for example, the question of its fractal dimension for the case without uniqueness remains an open problem.

## 16.1 Abstract Theory of Pullback $\mathscr{D}$-Attractors for Multivalued Processes

In this section we recall basic definitions of the theory of pullback $\mathscr{D}$-attractors for multivalued processes, prove a theorem which gives some useful criteria of existence of pullback $\mathscr{D}$-attractors for such processes, and introduce the notion of "condition (NW)."

Let $(H, \varrho)$ be a complete metric space, and $\mathscr{P}(H)$ be the family of all nonempty subsets of $H$. By $\mathrm{dist}_H(A, B)$ we denote the *Hausdorff semi-distance* between the sets $A, B \subset H$ defined as

$$\mathrm{dist}_H(A, B) = \sup_{x\in A} \inf_{y\in B} \varrho(x, y).$$



G. Łukaszewicz, P. Kalita, *Navier–Stokes Equations*, Advances in Mechanics and Mathematics 34, DOI 10.1007/978-3-319-27760-8_16

If the set $A \subset H$ is bounded, than we define its *Kuratowski measure of noncompactness* $\kappa(A)$ (cf. [144]) as

$$\kappa(A) = \{\inf \ \delta > 0 : \ A \text{ has finite open cover of sets of diameter less than } \delta \}.$$

For the properties of $\kappa$, see, for example, [111, 125]. Subsets of $\mathbb{R} \times H$ will be called *non-autonomous sets*. For a non-autonomous set $\mathbb{D} \subset \mathbb{R} \times H$ we identify $\mathbb{D}$ with the family of its fibers, i.e.,

$$\mathbb{D} = \{D(t)\}_{t \in \mathbb{R}}, \quad \text{where} \quad x \in D(t) \quad \Leftrightarrow \quad (x,t) \in \mathbb{D}.$$

A family $\mathscr{D}$ of non-autonomous sets such that for every $\mathbb{D} \in \mathscr{D}$ and all $t \in \mathbb{R}$ the fiber $D(t)$ is nonempty (i.e., $D(t) \in \mathscr{P}(H)$) is called an *attraction universe* over $H$. We denote $\mathbb{R}_d = \{(t,\tau) \in \mathbb{R}^2 : t \geq \tau\}$.

**Definition 16.1.** The map $U : \mathbb{R}_d \times H \to \mathscr{P}(H)$ is called a *multivalued process (m-process)* if:

(1) $U(\tau,\tau,z) = z$ for all $z \in H$,
(2) $U(t,\tau,z) \subset U(t,s,U(s,\tau,z))$ for all $z \in H$ and all $t \geq s \geq \tau$.

If in (2), in place of inclusion we have the equality $U(t,\tau,z) = U(t,s,U(s,\tau,z))$, then the $m$-process is said to be *strict*.

**Definition 16.2.** Let $\mathscr{D}$ be an attraction universe over $H$. The $m$-process $U$ in $H$ is *pullback $\omega$-$\mathscr{D}$-limit compact* if for every $\mathbb{D} = \{D(\tau)\}_{\tau \in \mathbb{R}} \in \mathscr{D}$ and $t \in \mathbb{R}$ we have

$$\kappa\left(\bigcup_{\tau \leq s} U(t,\tau,D(\tau))\right) \to 0 \quad \text{as} \quad s \to -\infty.$$

**Definition 16.3.** Let $\mathscr{D}$ be an attraction universe over $H$. The $m$-process $U$ in $H$ is *pullback $\mathscr{D}$-asymptotically compact* if for any $t \in \mathbb{R}$, every $\mathbb{D} \in \mathscr{D}$, and all sequences $\tau_k \to -\infty$ and $\xi_k \in U(t,\tau_k,D(\tau_k))$, there exists a subsequence $\{\xi_{k_m}\}$ such that for some $\xi \in H$, $\xi_{k_m} \to \xi$ in $H$.

**Lemma 16.1.** *Let $\mathscr{D}$ be an attraction universe over $H$. The m-process $U$ in $H$ is pullback $\omega$-$\mathscr{D}$-limit compact if and only if it is pullback $\mathscr{D}$-asymptotically compact.*

*Proof.* We first prove that pullback $\omega$-$\mathscr{D}$-limit compactness implies pullback $\mathscr{D}$-asymptotic compactness. Fix $t \in \mathbb{R}$. Let $\mathbb{D} \in \mathscr{D}$ and let $\tau_k$ be such that

$$\kappa\left(\bigcup_{\tau \leq \tau_k} U(t,\tau,D(\tau))\right) \leq \frac{1}{k} \quad \text{and} \quad \tau_k \to -\infty. \tag{16.1}$$

Let $t_i \to -\infty$ and $\xi_i \in U(t,t_i,D(t_i))$. We shall prove that $\kappa(\{\xi_i\}_{i=1}^{\infty}) = 0$. For every $m \in \mathbb{N}$ we have

$$\kappa(\{\xi_i\}_{i=1}^{\infty}) = \kappa(\{\xi_i\}_{i=1}^{m} \cup \{\xi_i\}_{i=m+1}^{\infty}) \leq \kappa(\{\xi_i\}_{i=m+1}^{\infty}).$$

For every $k \in \mathbb{N}$ and $m$ such that $t_{m+1} \leq \tau_k$ we have $\kappa(\{\xi_i\}_{i=m+1}^{\infty}) \leq \frac{1}{k}$, whence $\kappa(\{\xi_i\}_{i=1}^{\infty}) = 0$. This proves, in turn, the precompactness of $\{\xi_i\}_{i=1}^{\infty}$, and, in consequence, the pullback $\mathscr{D}$-asymptotic compactness of $U$.

Now let $U$ be pullback $\mathscr{D}$-asymptotically compact. We shall prove first that for every $\mathbb{D} \in \mathscr{D}$ and $t \in \mathbb{R}$ the set

$$\omega(\mathbb{D}, t) = \bigcap_{s \leq t} \overline{\bigcup_{\tau \leq s} U(t, \tau, D(\tau))} \tag{16.2}$$

is nonempty.

Indeed, let $t_k \to -\infty$ and $\xi_k \in U(t, t_k, D(t_k))$ and let, for a subsequence, still denoted by $k$, $\xi_k \to \xi$. For every $s \leq t$ and every index $k$ such that $t_k \leq s$, we have $\xi_k \in \bigcup_{\tau \leq s} U(t, \tau, D(\tau))$. Since $\xi_k \to \xi$, then $\xi \in \overline{\bigcup_{\tau \leq s} U(t, \tau, D(\tau))}$ for every $s \leq t$. Thus $\xi \in \omega(\mathbb{D}, t)$.

Now we prove that for every $\mathbb{D} \in \mathscr{D}$ and every $t \in \mathbb{R}$,

$$\text{dist}_H(U(t, \tau, D(\tau)), \omega(D, t)) \to 0 \quad \text{as} \quad \tau \to -\infty.$$

Assume, to the contrary, that there exists the non-autonomous set $\mathbb{D} \in \mathscr{D}$, the number $\bar{t} \in \mathbb{R}$, and the sequences $t_k \to -\infty$ and $\xi_k \in U(\bar{t}, t_k, D(t_k))$ such that $\text{dist}_H(\xi_k, \omega(\mathbb{D}, t)) \geq \epsilon > 0$. By the asymptotic compactness property, for a subsequence, still denoted by $k$, we have $\xi_k \to \xi$ in $H$. But $\xi \in \omega(\mathbb{D}, t)$, which gives a contradiction.

Let $\mathbb{D} \in \mathscr{D}$, $t \in \mathbb{R}$ and let $x_n$ be a sequence in $\omega(\mathbb{D}, t)$. We prove that this sequence has a subsequence that converges to some element in $\omega(\mathbb{D}, t)$ and thus the set $\omega(\mathbb{D}, t)$ is compact. As

$$x_n \in \overline{\bigcup_{\tau \leq s} U(t, \tau, D(\tau))} \quad \text{for every} \quad s \leq t,$$

then, for any sequence $t_k \to -\infty$ there exists $\xi_{m_k} \in U(t, t_{m_k}, D(t_{m_k}))$ such that $\rho(x_k, \xi_{m_k}) \leq \frac{1}{k}$. By pullback $\mathscr{D}$-asymptotic compactness, there exists a subsequence $\xi_\nu$ of $\xi_{m_k}$ converging to some $\xi \in \omega(\mathbb{D}, t)$. Thus, also $x_\nu \to \xi$.

Now let us fix $\epsilon > 0$. We need to show that there exists $s < t$ such that the set $\bigcup_{\tau \leq s} U(t, \tau, D(\tau))$ can be covered by a finite number of sets with diameter $\epsilon$. From compactness of $\omega(\mathbb{D}, t)$ it follows that there exists a finite number of points $\{x_i\}_{i=1}^{k}$ such that $\omega(\mathbb{D}, t) \subset \bigcup_{i=1}^{k} B(x_i, \frac{\epsilon}{2})$. Now choose $s < t$ such that

$$\text{dist}_H(U(t, \tau, D(\tau)), \omega(\mathbb{D}, t)) < \frac{\varepsilon}{2},$$

whenever $\tau \leq s$. It follows that for all $\tau \leq s$ we have $G(t, \tau, D(\tau)) \subset \bigcup_{i=1}^{k} B(x_i, \epsilon)$ and the proof is complete. □

**Definition 16.4.** Let $\mathcal{D}$ be an attraction universe over $H$. We say that an $m$-process $U$ has a *pullback $\mathcal{D}$-absorbing non-autonomous set* $\mathbb{B} = \{B(t)\}_{t\in\mathbb{R}} \in \mathcal{D}$ if for every $t \in \mathbb{R}$ and $\mathbb{D} \in \mathcal{D}$ there exists $\tau_0 \leq t$ depending on $t$ and $\mathbb{D}$ such that for $\tau \leq \tau_0$,

$$U(t, \tau, D(\tau)) \subset B(t).$$

**Definition 16.5.** Let $\mathcal{D}$ be an attraction universe over $H$ and let $U$ be an $m$-process on $H$. The non-autonomous set $\mathbb{A} = \{A(t)\}_{t\in\mathbb{R}} \subset \mathbb{R} \times H$ is called a *pullback $\mathcal{D}$-attractor* for $U$ if:

(1) $A(t)$ is compact in $H$ and nonempty for every $t \in \mathbb{R}$,
(2) $A(t) \subset U(t, \tau, A(\tau))$ for every $t \in \mathbb{R}$ and $\tau \leq t$ ($\mathbb{A}$ is negatively semi-invariant),
(3) $\mathbb{A}$ is pullback $\mathcal{D}$-attracting, that is, for every $t \in \mathbb{R}$ and $\mathbb{D} \in \mathcal{D}$,

$$\mathrm{dist}_H(U(t, \tau, D(\tau)), A(t)) \to 0 \quad \text{as} \quad \tau \to -\infty,$$

(4) if $\mathbb{C}$ is a non-autonomous set such that $C(t)$ are closed for $t \in \mathbb{R}$ and $\mathbb{C}$ is pullback $\mathcal{D}$-attracting then $A(t) \subset C(t)$ for every $t \in \mathbb{R}$ (minimality property).

*Remark 16.1.* Condition (4) in the above definition implies that pullback $\mathcal{D}$-attractor is always defined uniquely.

*Remark 16.2.* If we assume that, by definition, $\mathbb{A} \in \mathcal{D}$ then condition (4) (minimality) follows from conditions (1)–(3). In such a case, by Remark 16.1, conditions (1)–(3) suffice to guarantee the uniqueness of $\mathbb{A}$.

The minimality property can be proved as follows. Suppose that there exists a non-autonomous set $\mathbb{C}$ such that $C(t)$ are closed for all $t \in \mathbb{R}$, $\mathbb{C}$ is attracting, and for some $t \in \mathbb{R}$, $C(t) \subset A(t)$ with proper inclusion. As $\mathbb{A} \in \mathcal{D}$, we have $\mathrm{dist}_H(U(t, \tau, A(\tau)), C(t)) \to 0$ for $\tau \to -\infty$. But $A(t) \subset U(t, \tau, A(\tau))$, and we have $\mathrm{dist}_H(A(t), C(t)) \leq \mathrm{dist}_H(U(t, \tau, A(t)), C(t)) \to 0$ as $\tau \to -\infty$, whence $A(t) \subset C(t)$, a contradiction.

**Definition 16.6.** The $m$-process $U$ on $H$ is *closed* if for all $\tau \leq t$ the graph of the multivalued mapping $x \to U(t, \tau, x)$ is a closed set in the product topology on $H \times H$.

Now we formulate a theorem about the existence of pullback $\mathcal{D}$-attractors in complete metric spaces. A similar result is proved in [4, 35, 169] (see also [171, 172] for the case of $m$-semiflows and [41, 239] for the case of the universe of bounded and constant in time sets).

**Theorem 16.1.** *Let $H$ be a complete metric space and let the $m$-process $U$ on $H$ be closed. Assume that*

*(i) $U$ has a pullback $\mathcal{D}$-absorbing non-autonomous set $\mathbb{B} \in \mathcal{D}$.*
*(ii) $U$ is pullback $\omega$-$\mathcal{D}$-limit compact (equivalently, $U$ is pullback $\mathcal{D}$-asymptotically compact).*

*Then there exists a pullback $\mathcal{D}$-attractor for $U$.*

*Proof.* The proof is similar to that of Theorem 2.1 in [125].

**Step 1.** We define a candidate set for a pullback $\mathscr{D}$-attractor by setting

$$A(t) = \bigcap_{s \le t} \overline{\bigcup_{\tau \le s} U(t, \tau, B(\tau))}, \tag{16.3}$$

where $\mathbb{B} = \{B(\tau)\}_{t \in \mathbb{R}}$ is a non-autonomous pullback $\mathscr{D}$-absorbing set.

**Step 2.** $A(t)$ is nonempty and compact. We have, by $\omega$-$\mathscr{D}$-limit compactness of $U$,

$$\kappa \left( \overline{\bigcup_{\tau \le s} U(t, \tau, B(\tau))} \right) \to 0 \quad \text{as} \quad s \to -\infty. \tag{16.4}$$

Since the sets $X_s = \overline{\bigcup_{\tau \le s} U(t, \tau, B(\tau))}$ are nonempty, bounded, and closed in $H$, and the family of sets $\{X_s\}_{s \le t}$ is nonincreasing as $s \to -\infty$, we can apply a well-known property of $\kappa$ (cf., e.g., [111]) to get the claims.

**Step 3.** Attraction property. Since for every $t \in \mathbb{R}$, every non-autonomous set $\mathbb{D} \in \mathscr{D}$, and every $r \le t$ it is $U(t, \tau, D(\tau)) \subset U(t, r, U(r, \tau, D(\tau))) \subset U(t, r, B(r))$ for all $\tau \le \tau_0$, for some $\tau_0(\mathbb{D}, r) \le r$, it suffices to prove that $\text{dist}_H(U(t, \tau, B(\tau)), A(t)) \to 0$ as $\tau \to -\infty$. We argue by contradiction. Assume, to the contrary, that there exist sequences $\tau_k \to -\infty$ and $\xi_k \in U(t, \tau_k, B(\tau_k))$ such that $\inf_{y \in A(t)} \varrho(\xi_k, y) \ge \epsilon > 0$. By the pullback $\mathscr{D}$-asymptotic compactness property, there exists a convergent subsequence, $\xi_\mu \to \xi$ in $H$, and by the definition of the pullback $\mathscr{D}$-attractor in (16.3), $\xi \in A(t)$, which gives a contradiction.

**Step 4.** We prove that $A(t) \subset U(t, \tau, A(\tau))$ for all $\tau \le t$. Let $x \in A(t)$ and $\tau \le t$. We shall prove that $x \in U(t, \tau, p)$ for some $p \in A(\tau)$. Since $x \in A(t)$ defined in (16.3), there exist $t_k \to -\infty$ and $\xi_k \in U(t, t_k, B(t_k))$ such that $\xi_k \to x$ in $H$. We have

$$\xi_k \in U(t, t_k, B(t_k)) \subset U(t, \tau, U(\tau, t_k, B(t_k)))$$

for large $k$, whence there exist $z_k \in B(t_k)$ such that $\xi_k \in U(t, \tau, U(\tau, t_k, z_k))$, and $p_k \in U(\tau, t_k, z_k)$ such that $\xi_k \in U(t, \tau, p_k)$. By the pullback $\mathscr{D}$-asymptotic compactness it follows that for a subsequence of $\{p_k\}$ we have $p_\mu \to p$ in $H$. From (16.3) it follows that $p \in A(\tau)$. Since $\xi_\mu \to x$ and $p_\mu \to p$ with $\xi_\mu \in U(t, \tau, p_\mu)$, from the fact that $U$ is closed it follows that $x \in U(t, \tau, p)$. The proof is complete.

**Step 5.** To prove the minimality property assume that there exists a pullback $\mathscr{D}$-attracting non-autonomous set $\mathbb{C}$ such that $C(t)$ are closed. Then due to the fact that $\mathbb{B} \in \mathscr{D}$ we have $\text{dist}_H(U(t, \tau, B(\tau)), C(t)) \to 0$ as $\tau \to -\infty$ . Let $x \in A(t)$. By (16.3) there exist sequences $\tau_k \to -\infty$ and $\xi_k \in U(t, \tau_k, B(\tau_k))$ such that $\xi_k \to x$, whereas, as $C(t)$ is closed, we have $x \in C(t)$ and the proof is complete.

□

*Remark 16.3.* The above theorem still holds if we assume that, in definition of the pullback $\mathscr{D}$-asymptotic compactness, any set $\mathbb{D} \in \mathscr{D}$ is replaced with the pullback $\mathscr{D}$-absorbing set $\mathbb{B}$. In such a case this set does not have to belong to the universe $\mathscr{D}$ (see [4, 35, 169]).

*Remark 16.4.* The existence of the pullback $\mathscr{D}$-attractor implies condition (ii) in the above theorem, that is, the $\omega$-$\mathscr{D}$-limit compactness of $U$. To prove it we follow the argument provided, e.g., in [240]. Denote by $\mathscr{O}_\varepsilon(A(t))$ the closed $\varepsilon$-neighborhood of $A(t)$. Since the set $A(t)$ is compact, then $\kappa(A(t)) = 0$, and $\kappa(\mathscr{O}_\varepsilon(A(t))) \leq 2\varepsilon$. From the attraction property of $\mathbb{A}$, for any $\mathbb{D} \in \mathscr{D}$ we have $\mathrm{dist}_H(U(t,\tau,D(\tau)), A(t)) \leq 2\varepsilon$ for all $\tau \leq \tau_0$ for some $\tau_0(\mathbb{D}, t)$. Thus

$$\kappa\left(\bigcup_{\tau\leq\tau_0} U(t,\tau,D(\tau))\right) \leq 2\varepsilon,$$

which implies (ii).

*Remark 16.5.* In general, the existence of a pullback $\mathscr{D}$-absorbing non-autonomous set in $\mathscr{D}$ does not follow from the existence of a pullback $\mathscr{D}$-attractor.

Indeed, consider the equation $u'(t) = 0$ in $\mathbb{R}$. Then $U(t,\tau) = S(t-\tau) = id$. Let the universe $\mathscr{D}$ consist of only one non-autonomous set $\mathbb{D}$, given by $D(\tau) = \{\sin\tau\}$. Then the attractor is given as $A(t) = A = [-1, 1]$ for $t \in \mathbb{R}$. There is no choice for a family of $\mathscr{D}$-absorbing set but $B(\tau) = D(\tau)$. But $U(t,\tau,D(\tau)) = D(\tau) \neq B(t)$ as $\sin\tau \neq \sin t$. Thus, in this case there does not exist a pullback $\mathscr{D}$-absorbing non-autonomous set in $\mathscr{D}$. The presented example is illustrated in Fig. 16.1.

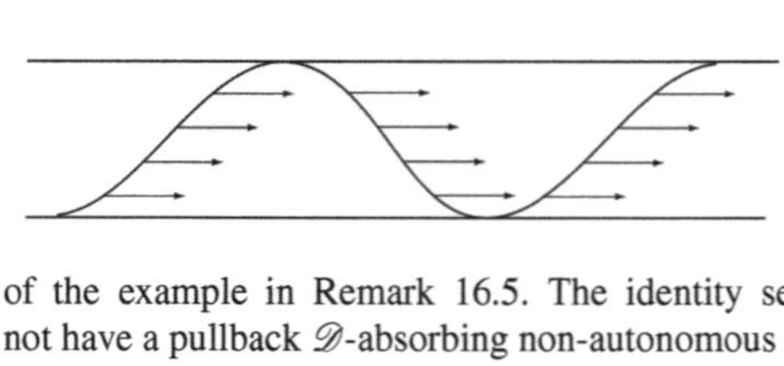

**Fig. 16.1** Illustration of the example in Remark 16.5. The identity semiflow has a pullback $\mathscr{D}$-attractor but it does not have a pullback $\mathscr{D}$-absorbing non-autonomous set in $\mathscr{D}$

It is natural to ask what assumptions on the universe $\mathscr{D}$ are required to guarantee that existence of pullback $\mathscr{D}$-attractor implies (i). We make the following assumptions:

($H\mathscr{D}1$) The universe $\mathscr{D}$ is *inclusion-closed*, that is if $\mathbb{D} = \{D(t)\}_{t\in\mathbb{R}} \in \mathscr{D}$ and $\mathbb{D}' = \{D'(t)\}_{t\in\mathbb{R}}$ is such that $D'(t) \subset \overline{D(t)}$ and $D'(t) \in \mathscr{P}(H)$ for all $t \in \mathbb{R}$, then $\mathbb{D}' \in \mathscr{D}$.

($H\mathscr{D}2$) If $\mathbb{D} = \{D(t)\}_{t\in\mathbb{R}} \in \mathscr{D}$, then there exists $\varepsilon > 0$ such that the non-autonomous set $\{\mathscr{O}_\varepsilon(D(t))\}_{t\in\mathbb{R}}$ also belongs to $\mathscr{D}$.

**Corollary 16.1.** *Let the universe $\mathscr{D}$ satisfy ($H\mathscr{D}1$) and let the m-process $U$ satisfy the assumptions of Theorem 16.1. Then we have $\mathbb{A} \in \mathscr{D}$. If we moreover assume that $U$ is strict then $\mathbb{A}$ is invariant, i.e., $U(t,\tau,A(\tau)) = A(t)$ for all $t \in \mathbb{R}$ and $\tau \leq t$.*

*Proof.* The proof follows the lines of the proofs of item (6) in Theorem 3 in [169] and items (5) (6) in Theorem 3.10 in [4]. Since $\mathbb{B}$ is pullback $\mathscr{D}$-absorbing we have $\mathrm{dist}_H(U(t,\tau,D(\tau)),\overline{B(t)}) = 0$ for all $\mathbb{D} \in \mathscr{D}$ and $\tau \le t_0(\mathbb{D},t)$. Hence as $\tau \to -\infty$ we have $\mathrm{dist}_H(U(t,\tau,D(\tau)),\overline{B(t)}) \to 0$ and $\{\overline{B(t)}\}_{t\in\mathbb{R}}$ is pullback $\mathscr{D}$-attracting. Then from the attractor minimality we have $A(t) \subset \overline{B(t)}$ and the assertion follows from $(H\mathscr{D}1)$.

For the proof of the positive invariance, let $t \ge \tau$ be fixed and let $r \le 0$. We have

$$U(t,\tau,A(\tau)) \subset U(t,\tau,U(\tau,\tau+r,A(\tau+r))) = U(t,\tau+r,A(\tau+r)).$$

Since $\mathbb{A} \in \mathscr{D}$, it pullback attracts itself, and

$$\lim_{r\to-\infty} \mathrm{dist}_H(U(t,\tau+r,A(\tau+r)),A(t)) = 0.$$

Hence

$$\lim_{r\to-\infty} \mathrm{dist}_H(U(t,\tau,A(\tau)),A(t)) = 0$$

and $\mathrm{dist}_H(U(t,\tau,A(\tau)),A(t)) = 0$, whence, by the closedness of $A(t)$, it follows that $U(t,\tau,A(\tau)) \subset A(t)$ and the assertion is proved. □

**Corollary 16.2.** *Let the universe $\mathscr{D}$ satisfy $(H\mathscr{D}2)$ and let the m-process $U$ have the pullback $\mathscr{D}$-attractor $\mathbb{A} \in \mathscr{D}$. Then $U$ has a pullback $\mathscr{D}$-absorbing non-autonomous set $\mathbb{B} \in \mathscr{D}$.*

*Proof.* Define $B(t) = \mathscr{O}_\varepsilon(A(t))$, where $\varepsilon$ is given in $(H\mathscr{D}2)$. Then $\mathbb{B} = \{B(t)\}_{t\in\mathbb{R}}$ belongs to $\mathscr{D}$. Suppose, on the contrary, that $B(t)$ is not pullback $\mathscr{D}$-absorbing. This means that there exist sequences $t_k \to -\infty$ and $\xi_k \in U(t,t_k,D(t_k))$ such that for large $k$ we have $\xi_k \notin B(t)$. From Remark 16.4 it follows that $U$ is pullback $\mathscr{D}$-asymptotically compact, and, for a subsequence, $\xi_\mu \to \xi$. As $\mathbb{A}$ is pullback $\mathscr{D}$-attracting and $A(t)$ is closed, we have $\xi \in A(t)$, a contradiction with $\xi_k \notin \mathscr{O}_\varepsilon(A(t))$. □

Summarizing, we have a following theorem which is a consequence of Theorem 16.1, Remark 16.4, and Corollaries 16.1 and 16.2.

**Theorem 16.2.** *Let $\mathscr{D}$ be a universe of non-autonomous sets satisfying $(H\mathscr{D}1)$ and $(H\mathscr{D}2)$ and let $U$ be a closed m-process. Then $U$ has a pullback $\mathscr{D}$-attractor $\mathbb{A}$ belonging to $\mathscr{D}$ if and only if $U$ is pullback $\mathscr{D}$-asymptotically compact and has a pullback $\mathscr{D}$-absorbing non-autonomous set $\mathbb{B} \in \mathscr{D}$.*

In the sequel of this section we confine ourselves to Banach spaces where we will consider both strong and weak topologies. We introduce condition (*NW*), "norm-to-weak," that generalizes to the non-autonomous multivalued case the norm-to-weak continuity assumed in [240] for semigroups (see Definition 3.4 in [240]) and in [125] for multivalued semiflows (see Definition 2.11 in [125]). A similar condition is introduced for the non-autonomous multivalued case in [231] (see condition (3) in

Definition 2.6 in [231]), where only the strict case is considered and, instead of a subsequence, whole sequence is assumed to converge weakly.

**Definition 16.7.** The $m$-process $U$ on a Banach space $H$ satisfies *condition (NW)* if for every $t \in \mathbb{R}$ and $\tau \leq t$, from $x_k \to x$ in $H$ and $\xi_k \in U(t, \tau, x_k)$ it follows that there exists a subsequence $\{\xi_{m_k}\}$, such that $\xi_{m_k} \to \xi$ weakly in $H$ with $\xi \in U(t, \tau, x)$.

**Lemma 16.2.** *If a multivalued process $U$ on a Banach space $H$ satisfies condition (NW) then it is closed.*

*Proof.* An elementary proof follows directly from the definitions. □

From Theorem 16.1 together with Lemma 16.2 we have the following

**Corollary 16.3.** *If the m-process $U$ on a Banach space $H$ has a pullback $\mathscr{D}$-absorbing non-autonomous set $\mathbb{B} \in \mathscr{D}$, is pullback $\mathscr{D}$-asymptotically compact (equivalently, $\omega$-$\mathscr{D}$-limit compact), and satisfies condition (NW) then there exists a pullback $\mathscr{D}$-attractor for $U$.*

The above corollary provides useful criteria of existence of pullback $\mathscr{D}$-attractors for multivalued processes. In the next section we apply them to study the time asymptotic behavior of solutions of a problem originating from contact mechanics.

## 16.2 Application to a Contact Problem

**Problem Formulation** The flow of an incompressible fluid in a two-dimensional domain $\Omega$ is described by the equation of motion

$$u'(t) - \nu \Delta u(t) + (u(t) \cdot \nabla) u(t) + \nabla p(t) = f(t) \quad \text{in} \quad \Omega \times (t_0, \infty), \tag{16.5}$$

and the incompressibility condition

$$\operatorname{div} u(t) = 0 \quad \text{in} \quad \Omega \times (t_0, \infty), \tag{16.6}$$

where the unknowns are the velocity $u : \Omega \times (t_0, \infty) \to \mathbb{R}^2$ and pressure $p : \Omega \times (t_0, \infty) \to \mathbb{R}$, $\nu > 0$ is the viscosity coefficient and $f : \Omega \times (t_0, \infty) \to \mathbb{R}^2$ is the volume mass force density. To define the domain $\Omega$ of the flow, let $\Omega_\infty$ be the infinite channel

$$\Omega_\infty = \{x = (x_1, x_2) \in \mathbb{R}^2 : \ x_2 \in (0, h(x_1))\},$$

where $h : \mathbb{R} \to \mathbb{R}$ is a positive function (we assume that $h(x_1) \geq h_0 > 0$ for all $x_1 \in \mathbb{R}$), smooth, and $L$-periodic in $x_1$. Then we set

$$\Omega = \{x = (x_1, x_2) \in \mathbb{R}^2 : \ x_1 \in (0, L), \ \ x_2 \in (0, h(x_1))\},$$

with the boundary $\partial\Omega = \overline{\Gamma_D} \cup \overline{\Gamma_C} \cup \overline{\Gamma_L}$, where $\Gamma_D = \{(x_1, h(x_1)) : x_1 \in (0, L)\}$, $\Gamma_C = (0, L) \times \{0\}$, and $\Gamma_L = \{0, L\} \times (0, h(0))$ are the top, bottom, and lateral parts of $\partial\Omega$, respectively. We will use the notation $e_1 = (1, 0)$ and $e_2 = (0, 1)$ for the canonical basis of $\mathbb{R}^2$. Note that on $\Gamma_C$ the outer normal unit vector is given by $n = -e_2$.

We are interested in the solutions of (16.5)–(16.6) periodic in the sense that for $x_2 \in [0, h(0)]$ and $t \in \mathbb{R}^+$ we have $u(0, x_2, t) = u(L, x_2, t)$, $\frac{\partial u_2(0,x_2,t)}{\partial x_1} = \frac{\partial u_2(L,x_2,t)}{\partial x_1}$, and $p(0, x_2, t) = p(L, x_2, t)$. The first condition represents the $L$-periodicity of velocities, while the latter two ones the $L$-periodicity of normal stresses in the space of divergence free functions. Moreover, we assume that

$$u(t) = 0 \quad \text{on} \quad \Gamma_D \times (t_0, \infty). \tag{16.7}$$

On the contact boundary $\Gamma_C$ we decompose the velocity into the normal component $u_n = u \cdot n$, where $n$ is the unit outward normal vector, and the tangential one $u_\tau = u \cdot e_1$. Note that since the domain $\Omega$ is two-dimensional it is possible to consider the tangential components as scalars, for the sake of the ease of notation. Likewise, we decompose the stress on the boundary $\Gamma_C$ into its normal component $T_n = Tn \cdot n$ and the tangential one $T_\tau = Tn \cdot e_1$. The stress tensor is related to the velocity and pressure through the linear constitutive law $T_{ij} = -p\delta_{ij} + \nu(u_{i,j} + u_{j,i})$.

We assume that there is no flux across $\Gamma_C$,

$$u_n(t) = 0 \quad \text{on} \quad \Gamma_C \times (t_0, \infty), \tag{16.8}$$

and that the tangential component of the velocity $u_\tau$ on $\Gamma_C$ is in the following relation with the tangential stresses $T_\tau$,

$$-T_\tau(x, t) \in \partial j(x, t, u_\tau(x, t)) \quad \text{on} \quad \Gamma_C \times (t_0, \infty). \tag{16.9}$$

In the above formula, $j : \Gamma_C \times \mathbb{R} \times \mathbb{R} \to \mathbb{R}$ is a potential which is locally Lipschitz and not necessarily convex with respect to the last variable, and $\partial$ is the subdifferential in the sense of Clarke (see Sect. 3.7) taken with respect to the last variable. Note that the function $j$ is assumed to depend on both space and time variables, however, sometimes the dependence on the space variable $x \in \Gamma_C$ is skipped for the ease of notation. In the end, we have the initial condition

$$u(x, t_0) = u_0(x) \quad \text{in} \quad \Omega. \tag{16.10}$$

Relation (16.9) is a generalized version of the Tresca friction law. We assume that the friction force is related by the multivalued and nonmonotone law with the tangential velocity. Our setup is more general than in the previous works [126, 162]. In contract to [126] we do not assume any monotonicity type conditions on the potential $j$ and hence the solutions to the formulated problem can be nonunique here. Similar laws were used, for example, in [12, 213] for static problems in elasticity, see Fig. 4.6 in [12] and Fig. 2(a) in [213] for examples of particular nonmonotone friction laws that satisfy the assumptions ($j1$)–($j4$) listed below. Complex nonmonotone behavior of the friction force density is related to the presence of asperities on the surface [191]. Similar friction law is also used, for example, in the study of the motion of tectonic plates, see [119, 190, 207] and the references therein.

**Weak Formulation and Existence of Solutions** We introduce the variational formulation of the problem and, for the convenience of the readers, we describe the relations between the classical and the weak formulations.

We begin with some basic definitions. Let

$$\tilde{V} = \{u \in C^\infty(\overline{\Omega})^2 \ : \ \operatorname{div} u = 0 \ \text{ in } \ \Omega, \quad u \text{ is } L\text{-periodic in } x_1, \\ u = 0 \text{ at } \Gamma_D, \quad u \cdot n = 0 \text{ at } \Gamma_C\}$$

and

$$V = \ \text{closure of } \ \tilde{V} \ \text{ in } \ H^1(\Omega)^2, \qquad H = \ \text{closure of } \ \tilde{V} \ \text{ in } \ L^2(\Omega)^2.$$

We define scalar products in $H$ and $V$, respectively, by

$$(u\,,\,w) = \int_\Omega u(x)\cdot w(x)\,dx \quad \text{and} \quad (\nabla u\,,\,\nabla w) = \int_\Omega \nabla u(x)\colon \nabla w(x)\,dx,$$

and their associated norms by

$$|u| = (u,u)^{\frac{1}{2}} \quad \text{and} \quad \|u\| = (\nabla u\,,\,\nabla u)^{\frac{1}{2}}.$$

The duality in $V' \times V$ is denoted as $\langle \cdot, \cdot \rangle$. Moreover, let, for $u$, $w$, and $z$ in $V$,

$$a(u\,,\,w) = (\nabla u\,,\,\nabla w) \quad \text{and} \quad b(u\,,\,z\,,\,w) = ((u\cdot\nabla)z\,,\,w).$$

We define a linear operator $A : V \to V'$ as $\langle Au, w\rangle = a(u,w)$, and a nonlinear operator $\langle Bu, w\rangle = b(u,u,w)$. For $w \in V$ we have the Poincaré inequality $\lambda_1|w|^2 \leq \|w\|^2$, where $\lambda_1$ is the first eigenvalue of the Stokes operator $A$ in $V$.

The linear and continuous trace mapping leading from $V$ to $L^2(\Gamma_C)^2$ is denoted by $\gamma$, we will use the same symbol to denote the Nemytskii trace operator on the spaces of time-dependent functions. The norm of the trace operator will be denoted by $\|\gamma\| \equiv \|\gamma\|_{\mathcal{L}(V;L^2(\Gamma_C)^2)}$.

If $u \in L^2(\Gamma_C)$ we will denote by $S^2_{\partial j(t,u)}$ the set of all $L^2$ selections of $\partial j(t,u)$, i.e., all functions $\xi \in L^2(\Gamma_C)$ such that $\xi(x) \in \partial j(x,t,u(x))$ for a.e. $x \in \Gamma_C$ (note that dependence on $x \in \Gamma_C$ is not written explicitly for the sake of notation brevity).

The variational formulation of the problem is as follows.

**Problem 16.1.** Given $u_0 \in H$, find $u : (t_0, \infty) \to H$ such that for all $T > t_0$,

$$u \in C([t_0, T]; H) \cap L^2(t_0, T; V) \quad \text{with} \quad u' \in L^2(t_0, T; V'),$$

and

$$\langle u'(t) + \nu Au(t) + Bu(t), z\rangle + (\xi(t), z_\tau)_{L^2(\Gamma_C)} = \langle f(t), z\rangle, \tag{16.11}$$

$$-\xi(t) \in S^2_{\partial j(t, u_\tau(t))} \tag{16.12}$$

for a.e. $t \geq t_0$ and for all $z \in V$.

The assumptions on the problem data are the following. The functions $f : \mathbb{R} \to H$ and $j : \Gamma_C \times \mathbb{R} \times \mathbb{R} \to \mathbb{R}$ satisfy:

$(f1)$ $f \in L^2_{loc}(\mathbb{R}; H)$,
$(f2)$ for certain $t \in \mathbb{R}$ (and thus for all $t \in \mathbb{R}$) we have

$$\int_{-\infty}^{t} e^{(\nu - d\|\gamma\|^2)\lambda_1 s} |f(s)|^2\, ds < \infty,$$

$(j1)$ $j(x, t, \cdot)$ is locally Lipschitz for a.e. $(x, t) \in \Gamma_C \times \mathbb{R}$,
$(j2)$ $j(\cdot, \cdot, s)$ is measurable for all $s \in \mathbb{R}$,
$(j3)$ $\partial j$ satisfies the growth condition $|\xi| \leq a + b|s|$ for all $s \in \mathbb{R}$ and $\xi \in \partial j(x, t, s)$ with $a, b > 0$ a.e. $(x, t) \in \Gamma_C \times \mathbb{R}$,
$(j4)$ $\partial j$ satisfies the condition $\xi s \geq c - d|s|^2$ for all $s \in \mathbb{R}$ and $\xi \in \partial j(x, t, s)$ with $c \in \mathbb{R}$ and $d \in \left(0, \frac{\nu}{\|\gamma\|^2}\right)$ a.e. $(x, t) \in \Gamma_C \times \mathbb{R}$.

Note that by $(j3)$ it follows that for $u \in L^2(\Gamma_C)$ the set $S^2_{\partial j(t,u)}$ is in fact the set of all measurable selections of $\partial j(t, u)$. Condition $(j4)$ is known as *dissipativity condition.*

We have the following relations between classical and weak formulations.

**Proposition 16.1.** *Every classical solution of* (16.5)–(16.10) *is also a solution of Problem 16.1. On the other hand, every solution of Problem 16.1 which is smooth enough is also a classical solution of* (16.5)–(16.10).

*Proof.* The proof is only sketched here, the details are left to the reader. Let $u$ be a classical solution of (16.5)–(16.10). As it is (by assumption) sufficiently regular, we have to check only (16.11). Remark first that (16.5) can be written as

$$u'(t) - \operatorname{div} T(u(t), p(t)) + (u(t) \cdot \nabla)u(t) = f(t). \tag{16.13}$$

Let $z \in V$. Multiplying (16.13) by $z$, integrating by parts and using the Green formula we obtain

$$\int_\Omega u'(t) \cdot z\,dx + \int_\Omega T_{ij}(u(t), p(t)) z_{i,j}\,dx + b(u(t), u(t), z)$$
$$= \int_{\partial\Omega} T_{ij}(u(t), p(t)) n_j z_i\,dS + \int_\Omega f(t) \cdot z\,dx \qquad (16.14)$$

for $t \in (t_0, \infty)$. As $u(t)$ and $z$ are in $V$, after some calculations we obtain

$$\int_\Omega T_{ij}(u(t), p(t)) z_{i,j}\,dx = \nu a(u(t), z). \qquad (16.15)$$

Taking into account the boundary conditions we get

$$\int_{\partial\Omega} T_{ij}(u(t), p(t)) n_j z_i\,dS = \int_{\Gamma_C} T_\tau(u(t), p(t)) z_\tau\,dS + \int_{\Gamma_C} T_n(u(t), p(t)) z_n\,dS,$$

and the last integral equals zero as $z$ satisfies (16.8) a.e. on $\Gamma_C$. Thus, (16.11) holds.

Conversely, suppose that $u$ is a sufficiently smooth solution to Problem 16.1. We have immediately (16.6)–(16.8) and (16.10). Now, let $z$ be in the space

$$(H^1_0(div, \Omega))^2 = \{z \in V : z = 0 \quad \text{on} \quad \partial\Omega\}.$$

We take such $z$ in (16.11) to get

$$\langle u'(t) - \nu\Delta u(t) + (u(t) \cdot \nabla) u(t) - f(t)\,,\, z\rangle = 0 \quad \text{for every} \quad z \in (H^1_0(div, \Omega))^2.$$

Thus, there exists a distribution $p(t)$ on $\Omega$ such that

$$u'(t) - \nu\Delta u(t) + (u(t) \cdot \nabla) u(t) - f(t) = \nabla p(t) \quad \text{in} \quad \Omega \times (t_0, \infty), \qquad (16.16)$$

so that (16.5) holds.

Now, we shall derive the subdifferential boundary condition (16.9) from the weak formulation. Take $z \in V$. We have

$$\int_\Omega T_{ij}(u(t), p(t)) z_{i,j}\,dx = -\int_\Omega \operatorname{div} T(u(t), p(t)) \cdot z\,dx + \int_{\partial\Omega} Tn \cdot z\,dS. \qquad (16.17)$$

Since $z \in V$, we have $z_n = 0$ a.e. on $\Gamma_C$ and hence $Tn \cdot z = T_\tau z_\tau$ on $\Gamma_C$. Applying (16.15) and (16.17) to (16.11) we get

$$\int_\Omega (u'(t) - \operatorname{div} T(u(t), p(t)) + (u(t) \cdot \nabla) u(t) - f(t)) \cdot z\,dx$$
$$+ \int_{\Gamma_L} T(u(t), p(t)) n \cdot z\,dS + (\xi(t), z_\tau)_{L^2(\Gamma_C)} + \int_{\Gamma_C} T_\tau(t) z_\tau\,dS = 0.$$

By (16.16) we have (16.13) and so the first integral on the left-hand side vanishes. In the same way as in the proof of Proposition 10.1 we assume enough smoothness of the classical solution, such that the Cauchy stress vector $Tn$ on $\Gamma_L$ is $L$-periodic with respect to $x_1$. This means that the second integral on the left-hand side of the last equality vanishes for any $z \in V$. Thus we get $\int_{\Gamma_C}(T_\tau - \xi)z_\tau \, dS$. Analogously to the proof of Proposition 10.1, we use the results of Chapter 1.2 in [146] on the extension of smooth functions on the boundary to divergence free vector fields to deduce that in the last inequality $z_\tau$ can be replaced by any sufficiently smooth function. It follows that $-T_\tau = \xi$ for a.e. $x \in \Gamma_C$ and the proof is complete. □

**Theorem 16.3.** *Assuming* $(f1)$, *and* $(j1)$–$(j3)$, *Problem 16.1 has a solution.*

The proof uses the standard approach by the mollification of the nonsmooth term and the Galerkin method. Since it is, on one hand, technical and quite involved, and, on the other hand the methodology is the same as in the proof of Theorem 10.8 in Chap. 10 and Theorem 14.1 in Chap. 14, with the necessary modifications for the non-autonomous terms, we skip the proof here.

**Multivalued Process and Its Attractor** We associate with Problem 16.1 the multifunction $U : \mathbb{R}_d \times H \to \mathscr{P}(H)$, where $U(t, t_0, u_0)$ is the set of states attainable at time $t$ from the initial condition $u_0$ taken at time $t_0$. Observe that $U$ is a strict $m$-process.

**Lemma 16.3.** *Assume* $(f1)$, $(j1)$–$(j4)$. *If the function* $u \in L^2_{loc}(t_0, +\infty; V)$ *with* $u' \in L^2_{loc}(t_0, +\infty; V')$ *solves Problem 16.1, then*

$$\frac{d}{dt}|u(t)|^2 + (\nu - d\|\gamma\|^2)\|u(t)\|^2 \le C_1(1 + |f(t)|^2), \tag{16.18}$$

$$\|u'(t)\|_{V'} \le C_2(1 + |f(t)|) + C_3(1 + |u(t)|)\|u(t)\| \tag{16.19}$$

*for a.e.* $t \in (t_0, +\infty)$ *with* $C_1, C_2, C_3 > 0$ *independent on* $t_0, u_0, t$.

*Proof.* We take $z = u(t)$ in (16.11) and make use of

$$b(w, w, w) = 0 \quad \text{for} \quad w \in V \tag{16.20}$$

to get

$$\frac{1}{2}\frac{d}{dt}|u(t)|^2 + \nu\|u(t)\|^2 + (\xi(t), u_\tau(t))_{L^2(\Gamma_C)} = (f(t), u(t)),$$

where $\xi(t) \in S^2_{\partial j(t, u_\tau(t))}$ for a.e. $t \ge t_0$. From $(j4)$ we obtain, with arbitrary $\varepsilon > 0$,

$$\frac{1}{2}\frac{d}{dt}|u(t)|^2 + \nu\|u(t)\|^2 \le \frac{\varepsilon}{2}\|u(t)\|^2 + \frac{1}{2\varepsilon}\|f(t)\|^2_{V'} + d\|u_\tau(t)\|^2_{L^2(\Gamma_C)} - cL.$$

Taking $\varepsilon = \nu - d\|\gamma\|^2$ we obtain, with a constant $C > 0$,

$$\frac{1}{2}\frac{d}{dt}|u(t)|^2 + \frac{\nu - d\|\gamma\|^2}{2}\|u(t)\|^2 \leq C(1 + \|f(t)\|_{V'}^2),$$

which proves (16.18), as $H \subset V'$ is a continuous embedding.

To show (16.19) let us observe that for $z \in V$ and a.e. $t \in \mathbb{R}^+$ we have

$$\langle u'(t), z\rangle \leq \|A\|_{\mathscr{L}(V;V^*)}\|u(t)\|\,\|z\| + \|f(t)\|_{V'}\|z\| \\ + \|\xi(t)\|_{L^2(\Gamma_C)}\|\gamma\|\,\|z\| + c|u(t)|\,\|u(t)\|\,\|z\|,$$

with $\xi(t) \in S^2_{\partial j(u_\tau(t))}$, where we used the fact that, in view of the Ladyzhenskaya inequality (cf. Lemma 10.2), for $w, z \in V$ we have

$$|b(w, w, z)| \leq c|w|\,\|w\|\,\|z\|. \tag{16.21}$$

Now (16.19) follows directly from the growth condition $(j3)$ and the trace inequality. □

Denote $\sigma = (\nu - d\|\gamma\|^2)\lambda_1$ and define $R(t)$ by

$$R(t)^2 = \frac{C_1}{\sigma} + C_1\int_{-\infty}^{t} e^{\sigma s}|f(s)|^2\,ds + 1 \quad \text{for} \quad t \in \mathbb{R}. \tag{16.22}$$

By $(f2)$ the quantity $R(t)$ is well defined and finite. We denote by $\mathscr{R}_\sigma$ the family of all functions $r : \mathbb{R} \to (0, \infty)$ such that

$$\lim_{t\to-\infty} e^{\sigma t} r^2(t) = 0.$$

By $\mathscr{D}_\sigma$ we denote the family of all non-autonomous sets $\mathbb{D} = \{D(t)\}_{t\in\mathbb{R}}$ such that $D(t) \in \mathscr{P}(H)$ for all $t \in \mathbb{R}$ and there exists $r_\mathbb{D} \in \mathscr{R}_\sigma$ such that for all $t \in \mathbb{R}$ and for all $w \in D(t)$, $|w| \leq r_\mathbb{D}(t)$. The family $\mathscr{D}_\sigma$ will be our attraction universe.

**Lemma 16.4.** *Let assumptions $(f1)$–$(f2)$ and $(j1)$–$(j4)$ hold. The non-autonomous set $\mathbb{B} = \{B(t)\}_{t\in\mathbb{R}}$ defined as $B(t) = \{v \in H : |v| \leq R(t)\}$ is $\mathscr{D}_\sigma$-pullback absorbing and $\mathbb{B} \in \mathscr{R}_\delta$.*

*Proof.* Obviously $R \in \mathscr{R}_\sigma$ and hence $\mathbb{B} \in \mathscr{D}_\sigma$. Take $t \in \mathbb{R}$ and $\mathbb{D} \in \mathscr{D}_\sigma$. There exists $r_\mathbb{D} \in \mathscr{R}_\sigma$ with $|w| \leq r_\mathbb{D}(s)$ for all $w \in D(s)$ and $s \in \mathbb{R}$. We can choose $t_0 = t_0(\mathbb{D}, t)$ small enough such that $r_\mathbb{D}(\tau)^2 e^{\sigma\tau} \leq e^{\sigma t}$ for all $\tau \leq t_0$. We will consider a solution $u$ of Problem 16.1 with the initial time $\tau \leq t_0$ and the initial condition $u_0 \in D(\tau)$. From (16.18) by the Poincaré inequality we obtain

$$\frac{d}{dt}|u(s)|^2 + (\nu - d\|\gamma\|^2)\lambda_1|u(s)|^2 \leq C_1(1 + |f(s)|^2) \quad \text{for a.e.} \quad s \in (\tau, t).$$

From the Gronwall inequality we get

$$|u(t)|^2 \le |u(\tau)|^2 e^{-\sigma(t-\tau)} + \frac{C_1}{\sigma} + C_1 e^{-\sigma t} \int_\tau^t e^{\sigma s} |f(s)|^2 \, ds. \tag{16.23}$$

Hence

$$|u(t)|^2 \le |u(\tau)|^2 e^{-\sigma(t-\tau)} + R^2(t) - 1.$$

Since $u(\tau) \in D(\tau)$, then

$$|u(t)|^2 \le r_{\mathbb{D}}(\tau)^2 e^{\sigma\tau} e^{-\sigma t} - 1 + R^2(t).$$

Due to the bound $r_{\mathbb{D}}(\tau)^2 e^{\sigma\tau} \le e^{\sigma t}$, we have

$$|u(t)|^2 \le R^2(t),$$

whence the assertion follows. □

We pass to the next lemma which states that the $m$-process $U$ is $\mathscr{D}_\sigma$-pullback asymptotically compact.

**Lemma 16.5.** *Assume* (*f*1)–(*f*2) *and* (*j*1)–(*j*4)*. The m-process U is* $\mathscr{D}_\sigma$*-asymptotically compact.*

*Proof.* The proof uses the technique of Proposition 7.4 in [11] (alternatively it is also possible to obtain the same assertion by the method of [201]) and is based on the energy equation. Let $\{D(t)\}_{t\in\mathbb{R}} = \mathbb{D} \in \mathscr{D}_\sigma$ and let $u_0^k \in D(t_0^k)$ with $t_0^k \to -\infty$. Let moreover $z_k \in U(t, t_0^k, u_0^k)$. We show that the sequence $\{z_k\}$ is relatively compact in $H$. There exists a sequence of functions $u_k \in L^2(t_0^k, t+1; V) \cap C([t_0^k, t+1]; H)$ with $u_n' \in L^2(t_0^k, t+1; V')$, solutions of Problem 16.1, such that $u_k(t_0^k) = u_0^k$ and $u_k(t) = z_k$. From Lemma 16.4 it follows that there exists $N_0$ such that for all natural $k \ge N_0$ we have $|u_k(t-1)| \le R(t-1)$. From now on we will consider the sequence $\{u_k\}_{k=N_0}^\infty$. The selections of $S^2_{\partial j(t,u_{k\tau}(t))}$ given in (16.12) will be denoted by $\xi_k$. Note that the restrictions of $u_k, \xi_k$ to the interval $[t-1, t+1]$ solve Problem 16.1 on this interval, with the initial conditions $u_k(t-1)$ taken at $t-1$. Using (16.18) it follows that the sequence $u_k$ is uniformly bounded in $L^2(t-1, t+1; V) \cap C([t-1, t+1]; H)$, and, by (16.19) it follows that $u_k'$ is uniformly bounded in $L^2(t-1, t+1; V')$. The growth condition (*j*3) implies that $\xi_k$ are bounded in $L^2(t-1, t+1; L^2(\Gamma_C))$. These bounds are sufficient to extract a subsequence, not renumbered, such that for certain $u \in L^2(t-1, t+1; V)$ with $u' \in (t-1, t+1; V')$ and $\xi \in L^2(t-1, t+1; L^2(\Gamma_C))$ the following convergences hold:

$$
\begin{aligned}
u_k &\to u \quad \text{weakly in} \quad L^2(t-1,t+1;V),\\
u_k' &\to u' \quad \text{weakly in} \quad L^2(t-1,t+1;V'),\\
u_k &\to u \quad \text{strongly in} \quad L^2(t-1,t+1;H), && (16.24)\\
u_k(s) &\to u(s) \quad \text{weakly in} \quad H \quad \text{for all} \quad s \in [t-1,t+1], && (16.25)\\
\gamma u_{k\tau} &\to \gamma u_\tau \quad \text{strongly in} \quad L^2(t-1,t+1;L^2(\Gamma_C)), && (16.26)\\
\xi_k &\to \xi \quad \text{weakly in} \quad L^2(t-1,t+1;L^2(\Gamma_C)). && (16.27)
\end{aligned}
$$

In view of (16.25) it is sufficient to show that $|u_k(t)| \to |u(t)|$, whence, as $H$ is a Hilbert space, it follows that $u_n(t) \to u(t)$ strongly in $H$ and the assertion will be proved. Note that by (16.24), for another subsequence, also not renumbered, the strong convergence $u_k(s) \to u(s)$ holds in $H$ for a.e. $s \in (t-1,t+1)$. Taking the test function $u_k(t)$ in (16.11) written for $u_k$, where the corresponding $\xi$ is denoted by $\xi_k$, and integrating from $t-1$ to $s \in [t-1,t+1]$, we get

$$
\begin{aligned}
&\frac{1}{2}|u_k(s)|^2 + \nu \int_{t-1}^{s} a(u_k(r),u_k(r))\,dr + \int_{t-1}^{s} (\xi_k(r),u_{k\tau}(r))_{L^2(\Gamma_C)}\,dr\\
&\qquad = \frac{1}{2}|u_k(t-1)|^2 + \int_{t-1}^{s} (f(r),u_k(r))\,dr. \qquad (16.28)
\end{aligned}
$$

We define the functions $V_k : [t-1,t+1] \to \mathbb{R}$ as

$$
V_k(s) = \frac{1}{2}|u_k(t-1)|^2 - \nu \int_{t-1}^{s} a(u_k(r),u_k(r))\,dr.
$$

Note that the coercivity of $a$ implies that the functions $V_k$ are nonincreasing. By the energy equation (16.28) we have

$$
V_k(s) = \frac{1}{2}|u_k(s)|^2 + \int_{t-1}^{s} (\xi_k(r),u_{k\tau}(r))_{L^2(\Gamma_C)}\,dr - \int_{t-1}^{s} (f(r),u_k(r))\,dr.
$$

From (16.24), (16.26), and (16.27) it follows that

$$
\lim_{k\to\infty} V_k(s) = \frac{1}{2}|u(s)|^2 + \int_{t-1}^{s} (\xi(r),u_\tau(r))_{L^2(\Gamma_C)}\,dr - \int_{t-1}^{s} (f(r),u(r))\,dr
$$

for a.e. $s \in (t-1,t+1)$. We denote the right-hand side of the above equation by $V(s)$. We can choose sequences $p_m \nearrow t$ and $q_m \searrow t$ such that $V_k(p_m) \to V(p_m)$ and $V_k(q_m) \to V(q_m)$ as $k \to \infty$ for all $m \in \mathbb{N}$. We have

$$
V_k(p_m) \geq V_k(t) \geq V_k(q_m),
$$

and hence

$$V(p_m) = \lim_{k\to\infty} V_k(p_m) \geq \limsup_{k\to\infty} V_k(t) \geq \liminf_{k\to\infty} V_k(t) \geq \lim_{k\to\infty} V_k(q_m) = V(q_m).$$

We pass with $m$ to infinity and use the fact that $V$, by definition, is continuous. We get

$$V(t) \geq \limsup_{k\to\infty} V_k(t) \geq \liminf_{k\to\infty} V_k(t) \geq V(t),$$

whence $V_k(t) \to V(t)$ as $k \to \infty$. But we have

$$\int_{t-1}^{t} (\xi_k(r), \gamma u_{k\tau}(r))_{L^2(\Gamma_C)}\, dr - \int_{t-1}^{t} (f(r), u_k(r))\, dr$$
$$\longrightarrow \int_{t-1}^{t} (\xi(r), \gamma u_\tau(r))_{L^2(\Gamma_C)}\, dr - \int_{t-1}^{t} (f(r), u(r))\, dr,$$

which gives us the convergence $|u_n(t)| \to |u(t)|$, and the proof is complete. □

**Lemma 16.6.** *If for every integer $k$, $u_k$ solves Problem 16.1 with the initial condition $u_k^0$ taken at $t_0$, and $u_k^0 \to u_0$ strongly in $H$ then for any $t > t_0$, for a subsequence, $u_k(t) \to u(t)$ weakly in $H$, where $u$ is a solution of Problem 16.1 with the initial condition $u_0$ taken at $t_0$. In consequence, the m-process $U$ on $H$ satisfies condition (NW).*

*Proof.* The proof is standard since the a priori estimates of Lemma 16.3 provide enough convergence to pass to the limit. Indeed, the sequence $u_k$ is bounded in $L^2(t_0, t; V)$ with $u_k'$ bounded in $L^2(t_0, t; V')$. Hence, for a subsequence, $u_k \to u$ weakly in $L^2(t_0, t; V)$ with $u_k' \to u'$ weakly in $L^2(t_0, t; V')$. In consequence, for all $s \in [0, t]$, $u_k(s) \to u(s)$ weakly in $H$, which means that $u(t_0) = u_0$ and $u_k(t) \to u(t)$ weakly in $H$. It also follows that $u_{k\tau} \to u_\tau$ strongly in $L^2(t_0, t; L^2(\Gamma_C))$ and in $L^2(\Gamma_C \times (t_0, t))$. We must show that $u$ solves Problem 16.1 on $(t_0, t)$. We only discuss passing to the limit in the multivalued term since for other terms it is standard. Let $\xi_k \in L^2(t_0, t; L^2(\Gamma_C))$ be such that $\xi_k(s) \in S^2_{\partial j(s, u_{k\tau}(s))}$ a.e. $s \in (t_0, t)$ and (16.11) holds. From the growth condition ($j3$) it follows that $\xi_k$ is bounded in $L^2(t_0, t; L^2(\Gamma_C))$ and thus, for a subsequence, we have $\xi_k \to \xi$ weakly in $L^2(t_0, t; L^2(\Gamma_C))$ and weakly in $L^2(\Gamma_C \times (t_0, t))$. Using assertion ($H2$) in Theorem 3.24 it follows that for a.e. $(x, s) \in \Gamma_C \times (t_0, t)$ $\xi(x, s) \in \partial j(x, s, u_\tau(x, s))$, which, in turn, implies that $\xi(s) \in S^2_{\partial j(s, u_\tau(s))}$ for a.e. $s \in (t_0, t)$, and the proof is complete. □

From Lemmas 16.4–16.6 it follows that all assumptions of Corollary 16.3 hold and hence we have shown the following Theorem.

**Theorem 16.4.** *The m-process $U$ on $H$ associated with Problem 16.1 has a $\mathscr{D}_\sigma$-pullback attractor. This attractor is moreover invariant and it belongs to $\mathscr{D}_\sigma$.*

In the above theorem the attractor invariance and the fact that it belongs to the universe $\mathscr{D}_\sigma$ follow from Corollary 16.1 as the $m$-process is strict and the attraction universe $\mathscr{D}_\sigma$ satisfies $(H\mathscr{D}1)$.

## 16.3 Comments and Bibliographical Notes

In this chapter we were interested in the evolution problems for which the solutions for a given initial conditions are not unique. To deal with such problems, the theory of $m$-semiflows and their global attractors was introduced in [171, 172] and later extended to non-autonomous case [41]. Results on the existence of pullback $\mathscr{D}$-attractors for the problems without uniqueness of solutions were obtained in [4, 35, 169]. The abstract result on the existence of the pullback $\mathscr{D}$-attractor shown in this chapter was based on these works, however, we showed some extension to the results presented there: our alternative proof of the pullback $\mathscr{D}$-attractor existence was based on the notion of $\omega$-$\mathscr{D}$-limit-compactness and seems particularly transparent. We weakened the upper semicontinuity assumption from [4, 35, 169] replacing it by the graph closedness, and we additionally studied not only necessary but also the necessary and sufficient conditions for the attractor existence.

We introduced the so-called condition (*NW*), earlier studied for autonomous case in [240] and recently in [125], and for non-autonomous case in [231] in context of uniform attractors. This condition is convenient to verify for particular problems of mathematical physics governed by PDEs.

For the notions of uniform attractors, pullback attractors, and skew-product flows we refer to [10, 53, 55, 130, 131, 239]. For similar problems in contact mechanics, cf. [126, 162] and, in particular, [12, 213]. The results of this chapter come from [127].

# References


1. D.J. Acheson, *Elementary Fluid Dynamics* (Oxford University Press, Oxford, 1990)
2. R. Adams, *Sobolev Spaces* (Academic Press, New York, San Francisco, London, 1975)
3. C.D. Aliprantis, K.C. Border, *Infinite Dimensional Analysis, A Hitchhiker's Guide*, 3rd edn. (Springer, Berlin, Heidelberg, 2006)
4. M. Anguiano, Attractors for nonlinear and non-autonomous parabolic PDEs in unbounded domains. Ph.D. thesis, Universidad de Sevilla, 2011
5. R. Aris, *Vectors, Tensors and the Basic Equations of Fluid Mechanics* (Prentice-Hall, Englewood Cliffs, NJ, 1962)
6. J.-P. Aubin, Un théorème de compacité. C. R. Acad. Sci. Paris **256**, 5042–5044 (1963)
7. J.-P. Aubin, A. Cellina, *Differential Inclusions* (Springer, Berlin, 1984)
8. J.-P. Aubin, H. Frankowska, *Set-Valued Analysis* (Birkhäuser, Basel, 1990)
9. A.V. Babin, M.I. Vishik, Maximal attractors of semigroups corresponding to evolution differential equations. Math. USSR Sb. **54**, 387–408 (1986)
10. F. Balibrea, T. Caraballo, P.E. Kloeden, J. Valero, Recent developments in dynamical systems: three perspectives. Int. J. Bifurcation Chaos **20**, 2591–2636 (2010)
11. J.M. Ball, Continuity properties and global attractors of generalized semiflows and the Navier-Stokes equations. Nonlinear Sci. **7**, 475–502 (1997). Erratum, ibid 8 (1998) 233. Corrected version appears in "Mechanics: from Theory to Computation" (Springer, Berlin, 2000), pp. 447–474
12. C.C. Baniotopoulos, J. Haslinger, Z. Morávková, Mathematical modeling of delamination and nonmonotone friction problems by hemivariational inequalities. Appl. Math. **50**, 1–25 (2005)
13. V. Barbu, I. Lasiecka, R. Triggiani, *Tangential Boundary Stabilization of Navier-Stokes Equations*. Memoirs of the American Mathematical Society, vol. 181 (American Mathematical Society, Providence, RI, 2006)
14. G.K. Batchelor, *An Introduction to Fluid Dynamics* (Cambridge University Press, New York, 2007)
15. G. Bayada, M. Boukrouche, On a free boundary problem for Reynolds equation derived from the Stokes system with Tresca boundary conditions. J. Math. Anal. Appl. **282**, 212–231 (2003)
16. G. Birkhoff, *Hydrodynamics. A Study in Logic, Fact and Similitude* (Princeton University Press, Princeton, NJ, 1960)
17. G.W. Bluman, S.C. Anco, *Symmetry and Integration Methods for Differential Equations* (Springer, New York, 2002)
18. J. Bolte, A. Daniliidis, O. Ley, L. Mazet, Characterizations of Łojasiewicz inequalities: subgradient flows, talweg, convexity. Trans. Am. Math. Soc. **362**, 3319–3363 (2010)




G. Łukaszewicz, P. Kalita, *Navier–Stokes Equations*, Advances in Mechanics and Mathematics 34, DOI 10.1007/978-3-319-27760-8

19. M.C. Bortolan, A.N. Carvalho, J.A. Langa, Structure of attractors for skew product semiflows, J. Differ. Equ. **257**, 490–522 (2014)
20. M. Boukrouche, R. El-Mir, On a non-isothermal, non-Newtonian lubrication problem with Tresca law. Existence and the behavior of weak solutions. Nonlinear Anal. Real **9**, 674–692 (2008)
21. M. Boukrouche, G. Łukaszewicz, Asymptotic analysis of solutions of a thin film lubrication problem with Coulomb fluid solid interface law. Int. J. Eng. Sci. **41**, 521–537 (2003)
22. M. Boukrouche, G. Łukaszewicz, An upper bound on the attractor dimension of a 2D turbulent shear flow in lubrication theory. Nonlinear Anal. Theory **59**, 1077–1089 (2004)
23. M. Boukrouche, G. Łukaszewicz, On a lubrication problem with Fourier and Tresca boundary conditions. Math. Models Methods Appl. Sci. **14**, 913–941 (2004)
24. M. Boukrouche, G. Łukaszewicz, An upper bound on the attractor dimension of a 2D turbulent shear flow with a free boundary condition, in *Regularity and Other Aspects of the Navier–Stokes Equations*, vol. 70 (Banach Center Publications, Warsaw, 2005), pp. 61–72
25. M. Boukrouche, G. Łukaszewicz, Attractor dimension estimate for plane shear flow of micropolar fluid with free boundary. Math. Methods Appl. Sci. **28**, 1673–1694 (2005)
26. M. Boukrouche, G. Łukaszewicz, Shear flows and their attractors, in *Partial Differential Equations and Fluid Mechanics*, ed. by J.L. Rodrigo, J.C. Robinson. London Mathematical Society Lecture Note Series, vol. 364 (Cambridge University Press, Cambridge, 2009), pp. 1–27
27. M. Boukrouche, G. Łukaszewicz, On global in time dynamics of a planar Bingham flow subject to a subdifferential boundary condition. Discrete Cont. Dyn. Syst. **34**, 3969–3983 (2014)
28. M. Boukrouche, G. Łukaszewicz, J. Real, On pullback attractors for a class of two-dimensional turbulent shear flows. Int. J. Eng. Sci. **44**, 830–844 (2006)
29. A.C. Bronzi, C.F. Mondaini, R. Rosa, Trajectory statistical solutions for three-dimensional Navier–Stokes-like systems. SIAM J. Math. Anal. **46**, 1893–1921 (2014)
30. Z. Brzeźniak, Y. Li, Asymptotic compactness of 2D stochastic Navier-Stokes equations on some unbounded domains. Mathematics Research Reports, Department of Mathematics, The University of Hull, vol. 15, 2002
31. Z. Brzeźniak, T. Caraballo, J.A. Langa, Y. Li, G. Łukaszewicz, J. Real, Random attractors for stochastic 2D-Navier–Stokes equations in some unbounded domains. J. Differ. Equ. **255**, 3897–3919 (2013)
32. L. Caffarelli, R. Kohn, L. Nirenberg, Partial regularity of suitable weak solutions of the Navier–Stokes equations. Commun. Pure Appl. Math. **35**, 771–831 (1982)
33. B.J. Cantwell, *Introduction to Symmetry Analysis* (Cambridge University Press, Cambridge, 2002)
34. C.S. Cao, E.S. Titi, Regularity criteria for the three-dimensional Navier–Stokes equations. Indiana Univ. Math. J. **57**, 2643–2661 (2008)
35. T. Caraballo, P.E. Kloeden, Non-autonomous attractor for integro-differential evolution equations. Discrete Cont. Dyn. Syst. Ser. S **2**, 17–36 (2009)
36. T. Caraballo, J.A. Langa, On the upper semicontinuity of cocycle attractors for nonautonomous and random dynamical systems. Dynam. Cont. Dis. Ser. A **10**, 491–514 (2003)
37. T. Caraballo, J. Real, Asymptotic behaviour of two–dimensional Navier–Stokes equations with delays. Proc. R. Soc. Lond. Ser. A **459**, 3181–3194 (2003)
38. T. Caraballo, J. Real, Attractors for 2D-Navier-Stokes models with delays. J. Differ. Equ. **205**, 271–297 (2004)
39. T. Caraballo, J.A. Langa, T. Taniguchi, The exponential behaviour and stabilizability of stochastic 2D-Navier–Stokes equations. J. Differ. Equ. **179**, 714–737 (2002)
40. T. Caraballo, P. Marín-Rubio, J.C. Robinson, A comparison between two theories for multi-valued semiflows and their asymptotic behaviour. Set-Valued Anal. **11**, 297–322 (2003)
41. T. Caraballo, J.A. Langa, V.S. Melnik, J. Valero, Pullback attractors of nonautonomous and stochastic multivalued dynamical systems. Set-Valued Anal. **11**, 153–201 (2003)

42. T. Caraballo, J.A. Langa, J. Valero, The dimension of attractors of nonautonomous partial differential equations. ANZIAM J. **45**, 207–222 (2003)
43. T. Caraballo, P.E. Kloeden, J. Real, Pullback and forward attractors for a damped wave equation with delays. Stoch. Dynam. **4**(3), 405–423 (2004)
44. T. Caraballo, G. Łukaszewicz, J. Real, Pullback attractors for non-autonomous 2D-Navier-Stokes equations in some unbounded domains. C. R. Acad. Sci. I Math. **342**, 263–268 (2006)
45. T. Caraballo, G. Łukaszewicz, J. Real, Pullback attractors for asymptotically compact non-autonomous dynamical systems. Nonlinear Anal. Theory **64**, 484–498 (2006)
46. T. Caraballo, A.M. Márquez-Durán, J. Real, Pullback and forward attractors for a 3D LANS-$\alpha$ model with delay. Discrete Cont. Dyn. Syst. **15**(2), 559–578 (2006)
47. T. Caraballo, P.E. Kloeden, J. Real, Invariant measures and statistical solutions of the globally modified Navier–Stokes equations. Discrete Cont. Dyn. B **10**, 761–781 (2008)
48. T. Caraballo, J. Real, A.M. Márquez, Three-dimensional system of globally modified Navier–Stokes equations with delay. Int. J. Bifurcation Chaos **20**, 2869–2883 (2010)
49. S. Carl, Existence and extremal solutions of parabolic variational-hemivariational inequalities. Monatsh. Math. **172**, 29–54 (2013)
50. S. Carl, V.K. Le, D. Motreanu, *Nonsmooth Variational Problems and Their Inequalities* (Springer, New York, 2007)
51. A.N. Carvalho, S. Sonner, Pullback exponential attractors for evolution processes in Banach spaces: theoretical results. Commun. Pure Appl. Anal. **12**, 3047–3071 (2013)
52. A.N. Carvalho, S. Sonner, Pullback exponential attractors for evolution processes in Banach spaces: properties and applications. Commun. Pure Appl. Anal. **13**, 1141–1165 (2014)
53. A.N. Carvalho, J.A. Langa, J.C. Robinson, *Attractors for Infinite-Dimensional Nonautonomous Dynamical Systems* (Springer, New York, 2012)
54. A.N. Carvalho, J. Cholewa, G. Lozada-Cruz, M.R.T. Primo, Reduction of infinite dimensional systems to finite dimensions: compact convergence approach. SIAM J. Math. Anal. **45**, 600–638 (2013)
55. D.N. Cheban, *Global Attractors of Non-autonomous Dissipative Dynamical Systems*. Interdisciplinary Mathematical Sciences (World Scientific, Singapore, 2004)
56. M.D. Chekroun, N.E. Glatt-Holtz, Invariant measures for dissipative dynamical systems: abstract results and applications. Commun. Math. Phys. **316**, 723–761 (2012)
57. V.V. Chepyzhov, A.A. Ilyin, A note on the fractal dimension of attractors of dissipative dynamical systems. Nonlinear Anal. **44**, 811–819 (2001)
58. V.V. Chepyzhov, A.A. Ilyin, On the fractal dimension of invariant sets; applications to Navier–Stokes equations. Discrete Cont. Dyn. Syst. **10**, 117–135 (2004)
59. V.V. Chepyzhov, M.I. Vishik, Trajectory attractors for evolution equations. C. R. Acad. Sci. I Math. **321**, 1309–1314 (1995)
60. V.V. Chepyzhov, M.I. Vishik, Trajectory attractors for reaction-diffusion systems. Topol. Methods Nonlinear Anal. **7**, 49–76 (1996)
61. V.V. Chepyzhov, M.I. Vishik, Evolution equations and their trajectory attractors. J. Math. Pure Appl. **76**, 913–964 (1997)
62. V.V. Chepyzhov, M.I. Vishik, *Attractors for Equations of Mathematical Physics* (American Mathematical Society, Providence, RI, 2002)
63. V.V. Chepyzhov, E.S. Titi, M.I. Vishik, On the convergence of solutions of the Leray-alpha model to the trajectory attractor of the 3D Navier-Stokes system. Discrete Cont. Dyn. Syst. **17**, 481–500 (2007)
64. S.I. Chernyshenko, P. Constantin, J.C. Robinson, E.S. Titi, A posteriori regularity of the three-dimensional Navier–Stokes equations from numerical computations. J. Math. Phys. **48** (2007). Article ID: 065204
65. A. Cheskidov, Global attractors of evolution systems. J. Dyn. Diff. Equ. **21**, 249–268 (2009)
66. A. Cheskidov, C. Foiaş, On global attractors of the 3D Navier–Stokes equations. J. Differ. Equ. **231**, 714–754 (2006)

67. A. Cheskidov, R. Shvydkoy, The regularity of weak solutions of the 3D Navier–Stokes equations in $B^{-1}_{\infty,\infty}$. Arch. Ration. Mech. Anal. **195**, 159–169 (2010)
68. A. Cheskidov, S. Friedlander, R. Shvydkoy, On the energy equality for weak solutions of the 3D Navier–Stokes equations, in *Advances in Mathematical Fluid Mechanics*, ed. by R. Rannacher, A. Sequeira, Springer Berlin, Heidelberg (2010), pp. 171–175
69. J.W. Cholewa, T. Dłotko, *Global Attractors in Abstract Parabolic Problems*. London Mathematical Society Lecture Note Series, vol. 278 (Cambridge University Press, Cambridge, 2000)
70. I.D. Chueshov, *Introduction to the Theory of Infinite-Dimensional Dissipative Systems* (Acta, Kharkov, 1999) (Russian). English transl.: Acta, Kharkov, 2002
71. F.H. Clarke, *Methods of Dynamic and Nonsmooth Optimization* (SIAM, Philadelphia, 1989)
72. F.H. Clarke, *Optimization and Nonsmooth Analysis* (SIAM, Philadelphia, 1990)
73. P. Constantin, C. Fefferman, Direction of vorticity and the problem of global regularity for the Navier–Stokes equations. Indiana Univ. Math. J. **42**, 775–789 (1993)
74. P. Constantin, C. Foiaş, Global Lyapunov exponents, Kaplan-Yorke formulas and the dimension of the attractors for 2D Navier-Stokes equations. Commun. Pure Appl. Math. **38**, 1–27 (1985)
75. P. Constantin, C. Foiaş, *Navier-Stokes Equations* (University of Chicago Press, Chicago, IL, 1989)
76. P. Constantin, C. Foiaş, R. Temam, *Attractors Representing Turbulent Flows*. Memoirs of the American Mathematical Society (American Mathematical Society, Providence, RI, 1985)
77. P. Constantin, C. Foiaş, B. Nicolaenko, R. Temam, *Integral Manifolds and Inertial Manifolds for Dissipative Partial Differential Equations*. Applied Mathematical Sciences, vol. 70 (Springer, New York, 1989)
78. N. Costea, V. Rădulescu, Inequality problems of quasi-hemivariational type involving set-valued operators and a nonlinear term. J. Glob. Optim. **52**, 743–756 (2012)
79. M. Coti Zelati, F. Tone, Multivalued attractors and their approximation: applications to the Navier-Stokes equations. Numer. Math. **122**, 421–441 (2012)
80. H. Crauel, A. Debussche, F. Flandoli, Random attractors. J. Dyn. Differ. Equ. **9**(2), 307–341 (1997)
81. O. Darrigol, *Worlds of Flows: A History of Hydrodynamics from Bernoullis to Prandtl* (Oxford University Press, Oxford, 2005)
82. M. Dashti, J.C. Robinson, A simple proof of uniqueness of the particle trajectories for solutions of the Navier–Stokes equations. Nonlinearity **22**, 735–746 (2009)
83. P.A. Davidson, *Turbulence. An Introduction to Scientists and Engineers* (Oxford University Press, Oxford, 2004)
84. Z. Denkowski, S. Migórski, N.S. Papageorgiou, *An Introduction to Nonlinear Analysis: Theory* (Kluwer Academic/Plenum Publishers, Dordrecht/New York, Boston, London, 2003)
85. J. Diestel, J.J. Uhl Jr., *Vector Measures*. Mathematical Surveys (American Mathematical Society, Providence, RI, 1977)
86. T. Dłotko, New look at the Navier–Stokes equation (2015). arXiv:1501.02085
87. Ch.R. Doering, P. Constantin, Energy dissipation in shear driven turbulence. Phys. Rev. Lett. **69**, 1648–1651 (1991)
88. Ch.R. Doering, J.D. Gibbon, *Applied Analysis of the Navier-Stokes Equations* (Cambridge University Press, Cambridge, 1995)
89. Ch.R. Doering, X. Wang, Attractor dimension estimates for two-dimensional shear flows. Phys. D **123**, 206–222 (1998)
90. G. Duvaut, Équilibre d'un solide élastique avec contact unilatéral et frottement de Coulomb. C. R. Acad. Sci. Paris **290**, 263–265 (1980)
91. G. Duvaut, J.L. Lions, *Les inéquations en mécanique et en physique* (Dunod, Paris, 1972)
92. A.C. Eringen, Theory of micropolar fluids. J. Math. Mech. **16**(1), 1–18 (1966)
93. L.C. Evans, *Partial Differential Equations*, 2nd edn. (American Mathematical Society, Providence, RI, 2010)
94. E. Feireisl, D. Pražák, *Asymptotic Behavior of Dynamical Systems in Fluid Mechanics*. American Institute of Mathematical Sciences (American Mathematical Society, Providence, RI, 2010)

95. F. Flandoli, B. Schmalfuss, Random attractors for the 3D stochastic Navier-Stokes equation with multiplicative white noise. Stoch. Stoch. Rep. **59**, 21–45 (1996)
96. C. Foiaş, Statistical study of Navier-Stokes equations, I. Rend. Sem. Mat. Univ. Padova **48**, 219–348 (1972)
97. C. Foiaş, Statistical study of Navier-Stokes equations, II. Rend. Semin. Mat. Univ. Padova **49**, 9–123 (1973)
98. C. Foiaş, R. Temam, Gevrey class regularity for the solutions of the Navier-Stokes equations. J. Funct. Anal. **87**, 359–369 (1989)
99. C. Foiaş, O.P. Manley, R. Rosa, R. Temam, *Navier-Stokes Equations and Turbulence* (Cambridge University Press, Cambridge, 2001)
100. C. Foiaş, D.D. Holm, E.S. Titi, The Navier–Stokes-alpha model of fluid turbulence. Phys. D **152**, 505–519 (2001)
101. C. Foiaş, M.S. Jolly, O.P. Manley, R. Rosa, Statistical estimates for the Navier-Stokes equations and the Kraichnan theory of 2-D fully developed turbulence. J. Stat. Phys. **108**(3/4), 591–645 (2002)
102. C. Foiaş, R. Rosa, R. Temam, Topological properties of the weak global attractor of the three-dimensional Navier–Stokes equations. Discrete Cont. Dyn. Syst. **27**, 1611–1631 (2010)
103. C. Foiaş, M.S. Jolly, R. Kravchenko, E.S. Titi, A unified approach to determining forms for the 2D Navier–Stokes equations - the general interpolants case. Russ. Math. Surv. **69**, 359–381 (2014)
104. C. Foiaş, R. Rosa, R. Temam, Convergence of time averages of weak solutions of the three-dimensional Navier–Stokes equations. J. Stat. Phys. **160**, 519–531 (2015)
105. A. Friedman, *Partial Differential Equations* (Holt, Rinehart and Winston, Austin, TX, 1969)
106. A. Friedman, *Foundations of Modern Analysis* (Holt, Rinehart and Winston, New York, 1970)
107. H. Gajewski, K. Gröger, K. Zacharias, *Nichtlineare Operatorgleichungen und Operatordifferentialgleichungen* (Academie, Berlin, 1974)
108. G.P. Galdi, *An Introduction to the Mathematical Theory of the Navier-Stokes Equations* (Springer, Berlin, 1994)
109. G. Gallavotti, *Foundations of Fluid Dynamics* (Springer, Berlin, 2005)
110. D. Gilbarg, N.S. Trudinger, *Elliptic Partial Differential Equations of Second Order* (Springer, Berlin, 1983)
111. J.K. Hale, *Asymptotic Behavior of Dissipative Systems* (American Mathematical Society, Providence, RI, 1988)
112. G.H. Hardy, J.E. Littlewood, G. Polya, *Inequalities* (Cambridge University Press, Cambridge, 1934)
113. J. Haslinger, I. Hlaváček, J. Nečas, Numerical methods for unilateral problems in solid mechanics, in *Handbook of Numerical Analysis*, ed. by P.G. Ciarlet, J.L. Lions, vol. IV (1996), Elsevier, pp. 313–485
114. J. Haslinger, M. Miettinen, P.D. Panagiotopoulos, *Finite Element Method for Hemivariational Inequalities: Theory, Methods and Applications* (Kluwer, Dordrecht, 1999)
115. E. Hopf, Uber die Anfangswertaufgabe für die hydrodynamischen Grundgleichungen. Math. Nachr. **4**, 213–231 (1950)
116. E. Hopf, Statistical hydrodynamics and functional calculus. J. Ration. Mech. Anal. **1**, 87–123 (1952)
117. B.R. Hunt, V.Y. Kaloshin, Regularity of embeddings of infinite-dimensional fractal sets into finite-dimensional spaces. Nonlinearity **12**, 1263–1275 (1999)
118. V.C.L. Hutson, J.S. Pym, *Applications of Functional Analysis and Operator Theory* (Academic, London, New York, Toronto, Sydney, San Francisco, 1980)
119. I.R. Ionescu, Q.L. Nguyen, Dynamic contact problems with slip dependent friction in viscoelasticity. Int. J. Appl. Math. Comput. **12**, 71–80 (2002)
120. B.O. Jacobson, At the boundary between lubrication and wear, in *New Directions in Tribology*, ed. by I. Hutchings (Mechanical Engineering Publications, London, 1997), pp. 291–298

121. B.O. Jacobson, B.J. Hamrock, Non-Newtonian fluid model incorporated into elastohydrodynamic lubrication of rectangular contacts. J. Tribol. **106**, 275–284 (1984)
122. D.A. Jones, E.S. Titi, Upper bounds on the number of determining modes, nodes, and volume elements for the Navier–Stokes equations. Indiana Univ. Math. J. **42**, 875–887 (1993)
123. P. Kalita, Decay of energy for second-order boundary hemivariational inequalities with coercive damping. Nonlinear Anal. Theory **74**, 1164–1181 (2011)
124. P. Kalita, G. Łukaszewicz, Attractors for Navier-Stokes flows with multivalued and nonmonotone subdifferential boundary conditions. Nonlinear Anal. Real **19**, 75–88 (2014)
125. P. Kalita, G. Łukaszewicz, Global attractors for multivalued semiflows with weak continuity properties. Nonlinear Anal. Theory **101**, 124–143 (2014)
126. P. Kalita, G. Łukaszewicz, On large time asymptotics for two classes of contact problems, in *Advances in Variational and Hemivariational Inequalities*, ed. by W. Han, S. Migórski, M. Sofonea. Theory Numerical Analysis, and Applications (Springer, Berlin, 2015), pp. 299–332
127. P. Kalita, G. Łukaszewicz, Attractors for multivalued processes with weak continuity properties, in *Continuous and Distributed Systems*, ed. by V.A. Sadovnichiy, M.Z. Zgurovsky (Springer, Berlin, 2015), pp. 149–166
128. O.V. Kapustyan, J. Valero, Weak and strong attractors for the 3D Navier–Stokes system. J. Differ. Equ. **240**, 249–278 (2007)
129. O.V. Kapustyan, J. Valero, Comparison between trajectory and global attractors for evolution systems without uniqueness of solutions. Int. J. Bifurcation Chaos **20**, 2723–2734 (2010)
130. O.V. Kapustyan, P.O. Kasyanov, J. Valero, Pullback attractors for some class of extremal solutions of 3D Navier–Stokes system. J. Math. Anal. Appl. **373**, 535–547 (2011)
131. O.V. Kapustyan, P.O. Kasyanov, J. Valero, M.Z. Zgurovsky, Structure of uniform global attractor for general non-autonomous reaction-diffusion system, in *Continuous and Distributed Systems*, ed. by M.Z. Zgurovsky, V.A. Sadovnichiy (Springer, Berlin, 2014), pp. 163–180
132. P.O. Kasyanov, L. Toscano, N.V. Zadoyanchuk, Long-time behaviour of solutions for autonomous evolution hemivariational inequality with multidimensional "reaction-displacement" law. Abstr. Appl. Anal. **2012** (2012). Article ID: 450984
133. P.O. Kasyanov, L. Toscano, N.V. Zadoyanchuk, Regularity of weak solutions and their attractors for a parabolic feedback control problem. Set-Valued Var. Anal. **21**, 271–282 (2013)
134. R.R. Kerswell, Unification of variational principles for turbulent shear flows: the background method of Doering-Constantin and the mean-fluctuation formulation of Howard-Busse. Phys. D **121**, 175–192 (1998)
135. P.E. Kloeden, J.A. Langa, Flattening, squeezing and the existence of random attractors. Proc. R. Soc. Lond. Ser. A **463**, 163–181 (2007)
136. P.E. Kloeden, M. Rasmussen, *Nonautonomous Dynamical Systems*. Mathematical Surveys and Monographs, vol. 176 (American Mathematical Society, Providence, RI, 2011)
137. P.E. Kloeden, B. Schmalfuss, Asymptotic behaviour of nonautonomous difference inclusions. Syst. Control Lett. **33**(4), 275–280 (1998)
138. P.E. Kloeden, J. Valero, The weak connectedness of the attainability set of weak solutions of the three-dimensional Navier–Stokes equations. Proc. R. Soc. Lond. Ser. A **463**, 1491–1508 (2007)
139. P.E. Kloeden, J.A. Langa, J. Real, Pullback V-attractors of the 3-dimensional globally modified Navier–Stokes equations. Commun. Pure Appl. Anal. **6**, 937–955 (2007)
140. P.E. Kloeden, C. Pötzsche, M. Rasmussen, Limitations of pullback attractors for processes. J. Differ. Equ. Appl. **18**, 693–701 (2012)
141. P.E. Kloeden, P. Marín-Rubio, J. Valero, The envelope attractor of non-strict multivalued dynamical systems with application to the 3D Navier–Stokes and reaction-diffusion equations. Set-Valued Var. Anal. **21**, 517–540 (2013)
142. A. Kufner, O. John, S. Fučik, *Function Spaces* (Academia, Prague, 1977)
143. I. Kukavica, M. Ziane, One component regularity for the Navier–Stokes equations. Nonlinearity **19**, 453–469 (2006)

144. I. Kukavica, M. Ziane, Navier–Stokes equations with regularity in one direction. J. Math. Phys. **48** (2007). Article ID: 065203
145. K. Kuratowski, Sur les espaces complets. Fundam. Math. **15**, 301–309 (1930)
146. O.A. Ladyzhenskaya, *The Mathematical Theory of Viscous Incompressible Flow*, 2nd edn, (Gordon and Breach, New York, 1969)
147. O.A. Ladyzhenskaya, Finite-dimensionality of bounded invariant sets for Navier–Stokes systems and other dissipative systems. J. Soviet Math. **28**, 714–726 (1985)
148. O.A. Ladyzhenskaya, G. Seregin, On semigroups generated by initial-boundary problems describing two dimensional visco-plastic flows, in *Nonlinear Evolution Equations*. American Mathematical Society Translation Series 2, vol. 164 (American Mathematical Society, Providence, RI, 1995), pp. 99–123
149. L.D. Landau, E.M. Lifshitz, *Fluid Mechanics*, 2nd edn. Course of Theoretical Physics Series (Butterworth-Heinemann, Oxford, 1987)
150. J.A. Langa, G. Łukaszewicz, J. Real, Finite fractal dimension of pullback attractor for non-autonomous 2-D Navier-Stokes equations in some unbounded domains. Nonlinear Anal. Theory **66**, 735–749 (2007)
151. J.A. Langa, A. Miranville, J. Real, Pullback exponential attractors. Discrete Cont. Dyn. Syst. **26**, 1329–1357 (2010)
152. J.A. Langa, B. Schmalfuss, Finite dimensionality of attractors for non-autonomous dynamical systems given by partial differential equations. Stochastics Dyn. **4**(3), 385–404 (2004)
153. J.A. Langa, J. Real, J. Simon, Existence and regularity of the pressure for the stochastic Navier–Stokes equations. Appl. Math. Opt. **48**, 95–110 (2003)
154. J. Leray, Sur le mouvement d'un liquide visquex emplissent l'espace. Acta Math. J. **63**, 193–248 (1934)
155. J.L. Lions, *Quelques Méthodes de Résolution des Problèmes aux Limites Non Linéaires* (Dunod, Paris, 1969)
156. P.L. Lions, *Mathematical Topics in Fluid Mechanics. Volume 1: Incompressible Models* (Oxford University Press, Oxford, 1996)
157. P.L. Lions, *Mathematical Topics in Fluid Mechanics. Volume 2: Compressible Models* (Oxford University Press, Oxford, 1998)
158. L.A. Ljusternik, V.I. Sobolev, *Elements of Functional Analysis* (Hindustan Publishing Corporation, Delhi; Halstadt Press, New York, 1974)
159. G. Łukaszewicz, *Micropolar Fluids. Theory and Applications* (Birkhauser, Boston, 1999)
160. G. Łukaszewicz, Pullback attractors and statistical solutions for 2-D Navier-Stokes equations. Discrete Cont. Dyn. B **9**, 643–659 (2008)
161. G. Łukaszewicz, On pullback attractors in $H_0^1$ for nonautonomous reaction-diffusion equations. Int. J. Bifurcation Chaos **20**, 2637–2644 (2010)
162. G. Łukaszewicz On the existence of an exponential attractor for a planar shear flow with the Tresca friction condition. Nonlinear Anal. Real **14**, 1585–1600 (2013)
163. G. Łukaszewicz, J.C. Robinson, Invariant measures for non-autonomous dissipative dynamical systems. Discrete Cont. Dyn. Syst. **34**, 4211–4222 (2014)
164. G. Łukaszewicz, W. Sadowski, Uniform attractor for 2D magneto-micropolar fluid flow in some unbounded domains. Z. Angew. Math. Phys. **55**, 247–257 (2004)
165. G. Łukaszewicz, J. Real, J.C. Robinson, Invariant measures for dissipative systems and generalized Banach limits. J. Dyn. Differ Equ. **23**, 225–250 (2011)
166. A.J. Majda, A.L. Bertozzi, *Vorticity and Incompressible Flow* (Cambridge University Press, Cambridge, 2002)
167. J. Málek, J. Nečas, A finite-dimensional attractor for three-dimensional flow of incompressible fluids. J. Differ Equ. **127**, 498–518 (1996)
168. J. Málek, D. Pražák, Large time behavior via the method of l-trajectories. J. Differ. Equ. **181**, 243–279 (2002)
169. P. Marín-Rubio, J. Real, Pullback attractors for 2d-Navier–Stokes equations with delays in continuous and sub-linear operators. Discrete Cont. Dyn. Syst. **26**, 989–1006 (2010)

170. P. Marín-Rubio, J. Real, J. Valero, Pullback attractors for a two-dimensional Navier–Stokes model in an infinite delay case. Nonlinear Anal. Theory **74**, 2012–2030 (2011)
171. V.S. Melnik, J. Valero, On attractors of multivalued semiflows and differential inclusions. Set-Valued Anal. **6**, 83–111 (1998)
172. V.S. Melnik, J. Valero, Addendum to "On Attractors of Multivalued Semiflows and Differential Inclusions" [Set-Valued Anal. 6, 83–111 (1998)]. Set-Valued Anal. **16**, 507–509 (2008)
173. M. Miettinen, P.D. Panagiotopoulos, On parabolic hemivariational inequalities and applications. Nonlinear Anal. **35**, 885–915 (1999)
174. S. Migórski, A. Ochal, Boundary hemivariational inequality of parabolic type. Nonlinear Anal. Theory **57**, 579–596 (2004)
175. S. Migórski, A. Ochal, Hemivariational inequalities for stationary Navier-Stokes equations. J. Math. Anal. Appl. **306**, 197–217 (2005)
176. S. Migórski, A. Ochal, Navier-Stokes models modeled by evolution hemivariational inequalities. Discrete Cont. Dyn. Syst. Supplement, 731–740 (2007)
177. S. Migórski, A. Ochal, M. Sofonea, *Nonlinear Inclusions and Hemivariational Inequalities. Models and Analysis of Contact Problems*. Advances in Mechanics and Mathematics, vol. 26 (Springer, New York, 2013)
178. A. Miranville, X. Wang, Attractor for nonautonomous nonhomogeneous Navier-Stokes equations. Nonlinearity **10**, 1047–1061 (1997)
179. A. Miranville, S. Zelik, Attractors for dissipative partial differential equations in bounded and unbounded domains, in *Evolutionary Equations*. Handbook of Differential Equations, vol. IV (Elsevier, North-Holland, Amsterdam, 2008), pp. 103–200
180. A. Miranville, M. Ziane, On the dimension of the attractor for the Bénard problem with free surfaces. Russ. J. Math. Phys. **5**, 489–502 (1997)
181. I. Moise, R. Rosa, X. Wang, Attractors for noncompact nonautonomous systems via energy equations. Discrete Cont. Dyn. Syst. **10**(1–2), 473–496 (2004)
182. Ch.B. Morrey, *Multiple Integrals in the Calculus of Variations* (Springer, Berlin, Heidelberg, New York, 1966)
183. P. Mucha, W. Sadowski, Long time behaviour of a flow in infinite pipe conforming to slip boundary conditions. Math. Methods Appl. Sci. **28**, 1867–1880 (2005)
184. Z. Naniewicz, P.D. Panagiotopoulos, *Mathematical Theory of Hemivariational Inequalities and Applications* (Dekker, New York, 1995)
185. J. Nečas, *Les Méthodes Directes en Théorie des Equations Elliptiques* (Mason, Paris, 1967)
186. V.V. Nemytskii, V.V. Stepanov, *Qualitative Theory of Differential Equations* (Dover, New York, 1989)
187. A. Novotný, I. Straškraba, *Introduction to the Mathematical Theory of Compressible Flow* (Oxford University Press, Oxford, 2004)
188. P.D. Panagiotopoulos, *Inequality Problems in Mechanics and Applications* (Birkhäuser, Basel, 1985)
189. V. Pata, A. Miranville, On the regularity of solutions to the Navier–Stokes equations. Commun. Pure Appl. Anal. **11**, 747–761 (2012)
190. G.J. Perrin, J.R. Rice, G. Zheng, Self-healing slip pulse on a frictional surface. J. Mech. Phys. Solids **43**, 1461–1495 (1995)
191. B.N.J. Persson, *Sliding Friction. Physical Principles and Applications* (Springer, Berlin, 2000)
192. K. Pileckas, Existence of solutions with the prescribed flux of the Navier-Stokes system in an infinite cylinder. J. Math. Fluid Mech. **8**, 542–563 (2006)
193. R. Pit, *Mesure locale de la vitesse à l'interface solide-liquide simple: Glissement et rle des interactions*. Thèse de doctorat, University of Paris XI, 1999
194. W. Pompe, Korn's First Inequality with variable coefficients and its generalization. Comment. Math. Univ. Carol. **44**, 57–70 (2003)
195. G. Prodi, On probability measures related to the Navier-Stokes equations in the 3-dimensional case. Air Force Resolution Division Contract AF61(052)-414, Technical Note, Trieste, 1961

196. J. Rencławowicz, W.M. Zajączkowski, Large time regular solutions to the Navier–Stokes equations in cylindrical domain. Topol. Methods Nonlinear Anal. **32**, 69–87 (2008)
197. J.C. Robinson, *Infinite-Dimensional Dynamical Systems* (Cambridge University Press, Cambridge, 2001)
198. J.C. Robinson, *Dimensions, Embeddings, and Attractors* (Cambridge University Press, Cambridge, 2011)
199. J.C. Robinson, Attractors and finite-dimensional behaviour in the 2D Navier-Stokes Equations. ISRN Math. Anal. **2013**, 29 pp. (2013). Article ID 291823
200. J.C. Robinson, J.L. Rodrigo, W. Sadowski, Classical theory of the three-dimensional Navier-Stokes equations (2016, to appear)
201. R. Rosa, The global attractor for the 2D Navier-Stokes flow on some unbounded domains. Nonlinear Anal. **32**, 71–85 (1998)
202. R. Rosa, Asymptotic regularity conditions for the strong convergence towards weak limit sets and weak attractors of the 3D Navier–Stokes equations. J. Differ. Equ. **229**, 257–269 (2006)
203. R. Rosa, Theory and application of statistical solutions of the Navier-Stokes equations, in *Partial Differential Equations and Fluid Mechanics*, ed. by J.C. Robinson, J.L. Rodrigo. London Mathematical Society Lecture Note Series, vol. 364 (Cambridge University Press, Cambridge 2009), pp. 228–257
204. T. Roubíček, *Nonlinear Partial Differential Equations with Applications* (Birkhäuser, Basel, 2005)
205. V. Scheffer, Turbulence and Hausdorff dimension, in *Turbulence and the Navier–Stokes Equations*, ed. by R. Temam. Lecture Notes in Mathematics, vol. 565 (1976), Springer, Berlin, New York pp. 94–112
206. B. Schmalfuss, *Attractors for Non-autonomous Dynamical Systems*, ed. by B. Fiedler, K. Gröger, J. Sprekels. Proceedings of Equadiff99 (World Scientific, Berlin, 2000), pp. 684–689
207. C.H. Scholz, *The Mechanics of Earthquakes and Faulting* (Cambridge University Press, Cambridge, 1990)
208. A. Segatti, S. Zelik, Finite-dimensional global and exponential attractors for the reaction-diffusion problem with an obstacle potential. Nonlinearity **22**, 2733–2760 (2009)
209. G.R. Sell, Non-autonomous differential equations and topological dynamics I, II. Trans. Am. Math. Soc. **127**, 241–262, 263–283 (1967)
210. G.R. Sell, Global attractors for the three dimensional Navier-Stokes equations. J. Dyn. Differ. Equ. **8**, 1–33 (1996)
211. G. Seregin, On a dynamical system generated by the two-dimensional equations of the motion of a Bingham fluid. J. Math. Sci. **70**(3), 1806–1816 (1994). (Translated from Zapiski Nauchnykh Seminarov LOMI, SSSR, vol. 188, 128–142, 1991)
212. J. Serrin, Mathematical principles of classical fluid mechanics, in *Fluid Mechanics I*, ed. by C. Truesdell. Encyclopedia of Physics, Springer, Berlin, Heidelberg, vol. 3/8/1 (1959)
213. I. Šestak, B.S. Jovanović, Approximation of thermoelasticity contact problem with nonmonotone friction. Appl. Math. Mech. **31**, 77–86 (2010)
214. J. Shieh, B.J. Hamrock, Film collapse in ehl and micro-ehl. J. Tribol. **113**, 372–377 (1991)
215. M. Shillor, M. Sofonea, J.J. Telega, *Models and Analysis of Quasistatic Contact: Variational Methods* (Springer, Berlin, Heidelberg, 2004)
216. J. Simon, Compact sets in the space $L^p(0,T;B)$. Ann. Mat. Pure Appl. (IV) **CXLVI**, 65–96 (1986)
217. M. Sofonea, A. Matei, *Mathematical Models in Contact Mechanics*. London Mathematical Society Lecture Note Series, vol. 398 (Cambridge University Press, Cambridge, 2012)
218. H. Sohr, *Navier-Stokes Equations* (Birkhauser, Basel, 2001)
219. R. Temam, *Navier–Stokes Equations. Theory and Numerical Analysis*, 3rd revised edn. (North-Holland, Amsterdam, New York, Oxford, 1984)
220. R. Temam, *Infinite Dimensional Dynamical Systems in Mechanics and Physics*, 2nd edn. (Springer, New York, 1997)

221. R. Temam, A. Miranville, *Mathematical Modeling in Continuum Mechanics*, 2nd edn. (Cambridge University Press, Cambridge, 2005)
222. R. Temam, M. Ziane, Navier–Stokes equations in three-dimensional thin domains with various boundary conditions. Adv. Differ. Equ. **1**, 499–546 (1996)
223. J.L. Tevaarwerk, The shear of hydrodynamic oil films. Ph.D. thesis, Cambridge, England, 1976
224. E.S. Titi, On approximate inertial manifolds to the Navier-Stokes equations. J. Math. Anal. Appl. **149**, 540–557 (1990)
225. M.I. Vishik, *Asymptotic Behaviour of Solutions of Evolutionary Equations* (Cambridge University Press, Cambridge, 1992)
226. M.I. Vishik, V.V. Chepyzhov, Trajectory and global attractors of three-dimensional Navier–Stokes system. Math. Notes **71**, 177–193 (2002)
227. M.I. Vishik, V.V. Chepyzhov, Trajectory attractors of equations of mathematical physics. Russ. Math. Surv. **66**, 637–731 (2011)
228. M.I. Vishik, A.V. Fursikov, *Mathematical Problems of Statistical Hydromechanics* (Kluwer Academic, Dordrecht, 1988)
229. X. Wang, Time averaged energy dissipation rate for shear driven flows in $\mathbb{R}^n$. Phys. D **99**, 555–563 (1997)
230. X. Wang, Effect of tangential derivative in the boundary layer on time averaged energy dissipation rate. Phys. D **144**, 142–153 (2000)
231. Y. Wang, S. Zhou, Kernel sections and uniform attractors of multivalued semiprocesses. J. Differ. Equ. **232**, 573–622 (2007)
232. Y. Wang, C. Zhong, S. Zhou, Pullback attractors of nonautonomous dynamical systems. Discrete Cont. Dyn. Syst. **16**(3), 587–614 (2006)
233. W.M. Zajączkowski, Long time existence of regular solutions to Navier-Stokes equations in cylindrical domains under boundary slip conditions. Stud. Math. **169**, 243–285 (2005)
234. E. Zeidler, *Nonlinear Functional Analysis and its Applications, Vol. II/A: Linear Monotone Operators* (Springer, New York, 1990)
235. E. Zeidler, *Nonlinear Functional Analysis and its Applications, Vol II/B: Nonlinear Monotone Operators* (Springer, New York, 1990)
236. S. Zelik, The attractor for a nonlinear reaction-diffusion system with a supercritical nonlinearity and it's dimension. Rend. Acad. Naz. Sci. XL Mem. Mat. Appl. **24**, 1–25 (2000)
237. S. Zelik, Spatially nondecaying solutions of the 2D Navier–Stokes equation in a strip. Glasg. Math. J. **49**, 525–588 (2007)
238. S. Zelik, Weak spatially nondecaying solutions of 3D Navier–Stokes equations in cylindrical domains, in *Instability in Models Connected with Fluid Flows II*, ed. by C. Bardos, A. Fursikov. International Mathematical Series, vol. 7 (Springer, Berlin, 2008), pp. 255–327
239. M.Z. Zgurovsky, P.O. Kasyanov, O.V. Kapustyan, J. Valero, N.V. Zadoianchuk, *Evolution Inclusions and Variation Inequalities for Earth Data Processing III, Long-Time Behavior of Evolution Inclusions Solutions in Earth Data Analysis* (Springer, Heidelberg, New York), 2012
240. C.K. Zhong, M.H. Yang, C.Y. Sun, The existence of global attractors for the norm-to-weak continuous semigroup and application to the nonlinear reaction-diffusion equations. J. Differ. Equ. **223**, 367–399 (2006)
241. M. Ziane, Optimal bounds on the dimension of the attractor of the Navier-Stokes equations. Phys. D **105**, 1–19 (1997)
242. M. Ziane, On the 2D-Navier-Stokes equations with the free boundary condition. Appl. Math. Opt. **38**, 1–19 (1998)

# Index




G. Łukaszewicz, P. Kalita, *Navier–Stokes Equations*, Advances in Mechanics and Mathematics 34, DOI 10.1007/978-3-319-27760-8

MIX
Papier aus verantwortungsvollen Quellen
Paper from responsible sources
FSC® C105338
FSC
www.fsc.org

If you have any concerns about our products, you can contact us on
**ProductSafety@springernature.com**

In case Publisher is established outside the EU, the EU authorized representative is:
**Springer Nature Customer Service Center GmbH**
**Europaplatz 3, 69115 Heidelberg, Germany**

Printed by Libri Plureos GmbH
in Hamburg, Germany